High-Performance Concrete

High-Performance Concrete

P.-C. Aïtcin
Université de Sherbrooke, Québec, Canada

E & FN SPON
An Imprint of Routledge
London and New York

First published 1998 by E & FN Spon, an imprint of Routledge
11 New Fetter Lane, London SE4P 4EE

Simultaneously published in the USA and Canada by Routledge
29 West 35th Street, New York, NY 10001

© 1998 Pierre-Claude Aïtcin

Typeset in 10/12 Palatino by Blackpool Typesetting Services Limited, UK
Printed in Great Britain by TJ International Ltd, Padstow, Cornwall

All rights reserved. No part of this book may be reprinted or
reproduced or utilized in any form or by any electronic, mechanical, or
other means, now known or hereafter invented, including photocopying
and recording, or in any information storage or retrieval system, without
permission in writing from the publisher.

British Library Cataloguing in Publication Data
A catalogue record for this book is available from the British Library

ISBN 0 419 19270 0

♾ Printed on acid-free paper, manufactured in accordance with ANSI/
NISO Z39.48-1992 and ANSI/NISO Z39.48-1984 (Permanence of Paper)

History would have it that the chiefs win the battles when, in actuality, the braves do the fighting.
                    Paraphrase of an American Indian saying

Accordingly, I dedicate this book to all my graduate students, past and present.

Contents

Foreword by Adam Neville, CBE, FEng		xx
Foreword by Yves Malier		xxii
Preface		xxv
Acknowledgements		xxx

1 Terminology: some personal choices ... 1
   1.1 About the title of this book ... 1
   1.2 Water/cement, water/cementitious materials or water/binder ratio ... 2
   1.3 Normal strength concrete/ordinary concrete/usual concrete ... 3
   1.4 High-strength or high-performance concrete ... 4
   References ... 5

2 Introduction ... 7
   Reference ... 21

3 A historical perspective ... 22
   3.1 Precursors and pioneers ... 22
   3.2 From water reducers to superplasticizers ... 26
   3.3 The arrival of silica fume ... 28
   3.4 Present status ... 29
   References ... 33

4 The high-performance concrete rationale ... 35
   4.1 Introduction ... 35
   4.2 For the owner ... 36
   4.3 For the designer ... 37
   4.4 For the contractor ... 38
   4.5 For the concrete producer ... 38
   4.6 For the environment ... 40
   4.7 Case studies ... 40
      4.7.1 Water Tower Place ... 40
      4.7.2 Gullfaks offshore platform ... 42
      4.7.3 Sylans and Glacières viaducts ... 46
      4.7.4 Scotia Plaza ... 50
      4.7.5 Île de Ré bridge ... 53
      4.7.6 Two Union Square ... 56
      4.7.7 Joigny bridge ... 59

viii Contents

		4.7.8	Montée St-Rémi bridge	62
		4.7.9	'Pont de Normandie' bridge	66
		4.7.10	Hibernia offshore platform	70
		4.7.11	Confederation bridge	76
	References			80
5	High-performance concrete principles			84
	5.1	Introduction		84
	5.2	Concrete failure under compressive load		85
	5.3	Improving the strength of hydrated cement paste		88
		5.3.1	Porosity	89
		5.3.2	Decreasing the grain size of hydration products	93
		5.3.3	Reducing inhomogeneities	93
	5.4	Improving the strength of the transition zone		93
	5.5	The search for strong aggregates		95
	5.6	Rheology of low water/binder ratio mixtures		96
		5.6.1	Optimization of grain size distribution of aggregates	96
		5.6.2	Optimization of grain size distribution of cementitious particles	97
		5.6.3	The use of supplementary cementitious materials	97
	5.7	The water/binder ratio law		97
	5.8	Concluding remarks		99
	References			99
6	Review of the relevant properties of some ingredients of high-performance concrete			101
	6.1	Introduction		101
	6.2	Portland cement		101
		6.2.1	Composition	101
		6.2.2	Clinker manufacture	103
		6.2.3	Clinker microstructure	106
		6.2.4	Portland cement manufacture	111
		6.2.5	Portland cement acceptance tests	113
		6.2.6	Portland cement hydration	115
			(a) Step 1 Mixing period	116
			(b) Step 2 The dormant period	117
			(c) Step 3 Initial setting	118
			(d) Step 4 Hardening	119
			(e) Step 5 Slowdown	120
		6.2.7	Concluding remarks on Portland cement hydration from a high-performance concrete point of view	120
	6.3	Portland cement and water		121
		6.3.1	From water reducers to superplasticizers	122
		6.3.2	Types of superplasticizer	126

	6.3.3	Manufacture of superplasticizers	126
		(a) First step: sulfonation	126
		(b) Second step: condensation	127
		(c) Third step: neutralization	128
		(d) Fourth step: filtration (in the case of calcium salt)	129
	6.3.4	Portland cement hydration in the presence of superplasticizers	130
	6.3.5	The crucial role of calcium sulfate	136
	6.3.6	Superplasticizer acceptance	137
	6.3.7	Concluding remarks	138
6.4	Supplementary cementitious materials		139
	6.4.1	Introduction	139
	6.4.2	Silica fume	140
	6.4.3	Slag	147
	6.4.4	Fly ash	152
	6.4.5	Concluding remarks	155
References			156

7 Materials selection — 162

7.1	Introduction		162
7.2	Different classes of high-performance concrete		162
7.3	Materials selection		163
	7.3.1	Selection of the cement	163
	7.3.2	Selection of the superplasticizer	170
		(a) Melamine superplasticizers	171
		(b) Naphthalene superplasticizers	171
		(c) Lignosulfonate-based superplasticizers	173
		(d) Quality control of superplasticizers	173
	7.3.3	Cement/superplasticizer compatibility	175
		(a) The minislump method	176
		(b) The Marsh cone method	178
		(c) Saturation point	180
		(d) Checking the consistency of the production of a particular cement or superplasticizer	182
		(e) Different rheological behaviours	183
		(f) Practical examples	185
	7.3.4	Superplasticizer dosage	189
		(a) Solid or liquid form?	190
		(b) Use of a set-retarding agent	190
		(c) Delayed addition	190
	7.3.5	Selection of the final cementitious system	190
	7.3.6	Selection of silica fume	192
		(a) Introduction	192
		(b) Variability	192
		(c) What form of silica fume to use	193

			(d) Quality control	193
			(e) Silica fume dosage	194
		7.3.7	Selection of fly ash	195
			(a) Quality control	196
		7.3.8	Selection of slag	196
			(a) Dosage rate	197
			(b) Quality control	197
		7.3.9	Possible limitations on the use of slag and fly ash in high-performance concrete	198
			(a) Need for high early strength	198
			(b) Cold weather concreting	198
			(c) Freeze–thaw durability	199
			(d) Decrease in maximum temperature	199
		7.3.10	Selection of aggregates	199
			(a) Fine aggregate	199
			(b) Crushed stone or gravel	200
			(c) Selection of the maximum size of coarse aggregate	202
	7.4	Factorial design for optimizing the mix design of high-performance concrete		203
		7.4.1	Introduction	203
		7.4.2	Selection of the factorial design plan	204
		7.4.3	Sample calculation	206
			(a) Iso cement dosage curves	207
			(b) Iso dosage curves for the superplasticizer	207
			(c) Iso cost curves	208
	7.5	Concluding remarks		210
	References			211
8	High-performance mix design methods			215
	8.1	Background		215
	8.2	ACI 211-1 Standard Practice for Selecting Proportions for Normal, Heavyweight and Mass Concrete		216
	8.3	Definitions and useful formulae		221
		8.3.1	Saturated surface dry state for aggregates	221
		8.3.2	Moisture content and water content	223
		8.3.3	Specific gravity	225
		8.3.4	Supplementary cementitious material content	225
			(a) Case 1	226
			(b) Case 2	226
		8.3.5	Superplasticizer dosage	226
			(a) Superplasticizer specific gravity	227
			(b) Solids content	227
			(c) Mass of water contained in a certain volume of superplasticizer	228
			(d) Other useful formulae	229

			(e) Mass of solid particles and volume needed	230
			(f) Volume of solid particles contained in V_{liq}	230
			(g) Sample calculation	230
		8.3.6	Water reducer and air-entraining agent dosages	231
		8.3.7	Required compressive strength	231
	8.4	Proposed method		233
		8.4.1	Mix design sheet	237
			(a) Mix design calculations	239
			(b) Sample calculation	241
			(c) Calculations	241
		8.4.2	From trial batch proportions to 1 m³ composition (SSD conditions)	246
			(a) Mix calculation	246
			(b) Sample calculation	248
			(c) Calculations	249
		8.4.3	Batch composition	252
			(a) Mix calculation	252
			(b) Sample calculation	253
			(c) Calculations	254
		8.4.4	Limitations of the proposed method	255
	8.5	Other mix design methods		257
		8.5.1	Method suggested in ACI 363 Committee on high-strength concrete	257
		8.5.2	de Larrard method (de Larrard, 1990)	259
		8.5.3	Mehta and Aïtcin simplified method	261
	References			263
9	Producing high-performance concrete			265
	9.1	Introduction		265
	9.2	Preparation before mixing		267
	9.3	Mixing		269
	9.4	Controlling the workability of high-performance concrete		271
	9.5	Segregation		272
	9.6	Controlling the temperature of fresh concrete		272
		9.6.1	Too cold a mix: increasing the temperature of fresh concrete	273
		9.6.2	Too hot a mix: cooling down the temperature of fresh concrete	274
	9.7	Producing air-entrained high-performance concretes		276
	9.8	Case studies		278
		9.8.1	Production of the concrete used to build the Jacques-Cartier bridge deck in Sherbrooke (Blais et al., 1996)	278
			(a) Concrete specifications	278

		(b) Composition of the high-performance concrete used	278
		(c) Mixing sequence	279
		(d) Economic considerations	280
	9.8.2	Production of a high-performance concrete in a dry-batch plant	281
		(a) Portneuf bridge (Lessard et al., 1993)	281
		(b) Scotia Plaza building (Ryell and Bickley, 1987)	282
	References		283
10	Preparation for concreting: what to do, how to do it and when to do it		285
	10.1	Introduction	285
	10.2	Preconstruction meeting	287
	10.3	Prequalification test programme	288
	10.3.1	Prequalification test programme for the construction of Bay-Adelaide Centre in Toronto, Canada	289
	10.3.2	Prequalification test programme for the 20 Mile Creek air-entrained high-performance concrete bridge on Highway 20 (Bickley, 1996)	290
	10.3.3	Pilot test	291
	10.4	Quality control at the plant	292
	10.5	Quality control at the jobsite	293
	10.6	Testing	294
	10.6.1	Sampling	295
	10.7	Evaluation of quality	295
	10.8	Concluding remarks	297
	References		297
11	Delivering, placing and controlling high-performance concrete		299
	11.1	High-performance concrete transportation	299
	11.2	Final adjustment of the slump prior to placing	300
	11.3	Placing high-performance concrete	301
	11.3.1	Pumping	301
	11.3.2	Vibrating	303
	11.3.3	Finishing concrete slabs	304
	11.4	Special construction methods	306
	11.4.1	Mushrooming in column construction	306
	11.4.2	Jumping forms	306
	11.4.3	Slipforming	307
	11.4.4	Roller-compacted high-performance concrete	309
	11.5	Conclusion	309
	References		309
12	Curing high-performance concrete to minimize shrinkage		311
	12.1	Introduction	311

Contents

xiii

12.2	The importance of appropriate curing	312
12.3	Different types of shrinkage	312
12.4	The hydration reaction and its consequences	313
	12.4.1 Strength	315
	12.4.2 Heat	316
	12.4.3 Volumetric contraction	317
	(a) Apparent volume and solid volume	317
	(b) Volumetric changes of concrete (apparent volume)	318
	(c) Chemical contraction (absolute volume)	319
	(d) The crucial role of the menisci in concrete capillaries in apparent volume changes	320
	(e) Essential difference between self-desiccation and drying	321
	(f) From the volumetric changes of the hydrated cement paste to the shrinkage of concrete	321
12.5	Concrete shrinkage	322
	12.5.1 Shrinkage of thermal origin	322
	12.5.2 How to reduce autogenous and drying shrinkage and its effects by appropriate curing of high-performance concrete	323
12.6	Why autogenous shrinkage is more important in high-performance concrete than in usual concrete	324
12.7	Is the application of a curing compound sufficient to minimize or attenuate concrete shrinkage?	327
12.8	The curing of high-performance concrete	327
	12.8.1 Large columns	329
	(a) Volumetric changes at A	329
	(b) Volumetric changes at B	330
	(c) Volumetric changes at C	331
	12.8.2 Large beams	331
	12.8.3 Small beams	332
	12.8.4 Thin slabs	332
	12.8.5 Thick slabs	333
	12.8.6 Other cases	334
12.9	How to be sure that concrete is properly cured in the field	334
12.10	Conclusion	335
References		336
13	Properties of fresh concrete	338
13.1	Introduction	338
13.2	Unit mass	340
13.3	Slump	340

		13.3.1	Measurement	340
		13.3.2	Factors influencing the slump	341
		13.3.3	Improving the rheology of fresh concrete	342
		13.3.4	Slump loss	343
	13.4	Air content		343
		13.4.1	Non-air-entrained high-performance concrete	343
		13.4.2	Air-entrained high-performance concrete	344
	13.5	Set retardation		345
	13.6	Concluding remarks		347
	References			347
14	Temperature increase in high-performance concrete			349
	14.1	Introduction		349
	14.2	Comparison of the temperature increases within a 35 MPa concrete and a high-performance concrete		350
	14.3	Some consequences of the temperature increase within a concrete		351
		14.3.1	Effect of the temperature increase on the compressive strength of high-performance concrete	352
		14.3.2	Inhomogeneity of the temperature increase within a high-performance concrete structural element	353
		14.3.3	Effect of thermal gradients developed during high-performance concrete cooling	353
		14.3.4	Effect of the temperature increase on concrete microstructure	354
	14.4	Influence of different parameters on the temperature increase		356
		14.4.1	Influence of the cement dosage	357
		14.4.2	Influence of the ambient temperature	360
		14.4.3	Influence of the geometry of the structural element	361
		14.4.4	Influence of the nature of the forms	363
		14.4.5	Simultaneous influence of fresh concrete and ambient temperature	364
		14.4.6	Concluding remarks	365
	14.5	How to control the maximum temperature reached within a high-performance concrete structural element		366
		14.5.1	Decrease of the temperature of the delivered concrete	366
			(a) Liquid nitrogen cooling	367
			(b) Use of crushed ice	367
		14.5.2	Use of a retarder	367
		14.5.3	Use of supplementary cementitious materials	368

	14.5.4 Use of a cement with a low heat of hydration	369
	14.5.5 Use of hot water and insulated forms or heated and insulated blankets under winter conditions	369
14.6	How to control thermal gradients	369
14.7	Concluding remarks	370
References		370

15 Testing high-performance concrete — 373

- 15.1 Introduction — 373
- 15.2 Compressive strength measurement — 374
  - 15.2.1 Influence of the testing machine — 375
    - (a) Testing limitations due to the capacity of the testing machine — 375
    - (b) Influence of the dimensions of the spherical head — 377
    - (c) Influence of the rigidity of the testing machine — 381
  - 15.2.2 Influence of testing procedures — 382
    - (a) How to prepare specimen ends — 383
    - (b) Influence of eccentricity — 388
  - 15.2.3 Influence of the specimen — 390
    - (a) Influence of the specimen shape — 390
    - (b) Influence of the specimen mould — 391
    - (c) Influence of the specimen diameter — 392
  - 15.2.4 Influence of curing — 395
    - (a) Testing age — 395
    - (b) How can high-performance concrete specimens be cured? — 396
  - 15.2.5 Core strength versus specimen strength — 397
- 15.3 Stress–strain curve — 398
- 15.4 Shrinkage measurement — 400
  - 15.4.1 Present procedure — 401
  - 15.4.2 Shrinkage development in a high water/binder concrete — 402
  - 15.4.3 Shrinkage development in a low water/binder concrete — 402
  - 15.4.4 Initial mass increase and self-desiccation — 404
  - 15.4.5 Initial compressive strength development and self-desiccation — 404
  - 15.4.6 New procedure for drying shrinkage measurement — 405
- 15.5 Creep measurement — 407
  - 15.5.1 Present sample curing (ASTM C 512) — 407
  - 15.5.2 Development of different concrete deformations during a 28 day creep measurement — 407

xvi   *Contents*

       15.5.3 Deformations occurring in a high water/binder ratio concrete subjected to early loading during a creep test   408
       15.5.4 Deformations occurring in a low water/binder ratio concrete subjected to early loading during a creep test   409
       15.5.5 Proposed curing procedure before loading a concrete specimen for creep measurement   410
  15.6 Concluding remarks on creep and shrinkage measurements   411
  15.7 Permeability measurement   412
  15.8 Elastic modulus measurement   415
  References   418

16 Mechanical properties of high-performance concrete   423
  16.1 Introduction   423
  16.2 Compressive strength   425
       16.2.1 Early-age compressive strength of high-performance concrete   426
       16.2.2 Effect of early temperature rise of high-performance concrete on compressive strength   427
       16.2.3 Influence of air content on compressive strength   428
       16.2.4 Long-term compressive strength   429
  16.3 Modulus of rupture and splitting tensile strength   431
  16.4 Modulus of elasticity   433
       16.4.1 Theoretical approach   433
       16.4.2 Empirical approach   437
       16.4.3 Concluding remarks on elastic modulus evaluation   440
  16.5 Poisson's ratio   442
  16.6 Stress–strain curves   442
  16.7 Creep and shrinkage   445
  16.8 Fatigue resistance of high-performance concrete   448
       16.8.1 Introduction   448
       16.8.2 Definitions   450
             (a) Wöhler diagrams   450
             (b) Goodman diagrams   451
             (c) Miner's rule   451
       16.8.3 Fatigue resistance of concrete structures   452
  16.9 Concluding remarks   453
  References   454

17 The durability of high-performance concrete   458
  17.1 Introduction   458
  17.2 The durability of usual concretes: a subject of major concern   460
       17.2.1 Durability: the key criterion to good design   461

	17.2.2	The critical importance of placing and curing in concrete durability	462
	17.2.3	The importance of the concrete 'skin'	463
	17.2.4	Why are some old concretes more durable than some modern ones?	466
17.3	Why high-performance concretes are more durable than usual concretes		467
17.4	Durability at a microscopic level		470
17.5	Durability at a macroscopic level		471
17.6	Abrasion resistance		472
	17.6.1	Introduction	472
	17.6.2	Factors affecting the abrasion resistance of high-performance concrete	473
	17.6.3	Pavement applications	476
	17.6.4	Abrasion–erosion in hydraulic structures	477
	17.6.5	Ice abrasion	477
17.7	Freezing and thawing resistance		477
	17.7.1	Freezing and thawing durability of usual concrete	478
	17.7.2	Freezing and thawing durability of high-performance concrete	479
	17.7.3	How many freeze–thaw cycles must a concrete sustain successfully before being said to be freeze–thaw resistant?	483
	17.7.4	Personal views	484
17.8	Scaling resistance		485
17.9	Resistance to chloride ion penetration		486
17.10	Corrosion of reinforcing steel		487
	17.10.1	Use of stainless steel rebars	489
	17.10.2	Use of galvanized rebars	489
	17.10.3	Use of epoxy-coated rebars	490
	17.10.4	Use of glass fibre-reinforced rebars	491
	17.10.5	Effectiveness of the improvement of 'covercrete' quality	492
		(a) Time to initiate cracking	492
		(b) Relationship between time to initiate cracking and initial current	494
	17.10.6	Concluding remarks	495
17.11	Resistance to various forms of chemical attack		496
17.12	Resistance to carbonation		497
17.13	Resistance to sea water		497
17.14	Alkali–aggregate reaction and high-performance concrete		497
	17.14.1	Introduction	497
	17.14.2	Essential conditions to see an AAR developing within a concrete	498

| | | (a) Moisture condition and AAR | 498 |
| | | (b) Cement content, water/binder ratio and AAR | 498 |

- 17.14.3 Superplasticizer and AAR — 499
- 17.14.4 AAR prevention — 499
- 17.14.5 Extrapolation of the results obtained on usual concrete to high-performance concrete — 500

17.15 Resistance to fire — 500
- 17.15.1 Is high-performance concrete a fire-resistant material? — 500
- 17.15.2 The fire in the Channel Tunnel — 502
  - (a) The circumstances — 502
  - (b) The damage — 503

17.16 Conclusions — 503
References — 504

18 Special high-performance concretes — 510

18.1 Introduction — 510

18.2 Air-entrained high-performance concrete — 511
- 18.2.1 Introduction — 511
- 18.2.2 Design of an air-entrained high-performance concrete mix — 512
  - (a) Sample calculation — 512
- 18.2.3 Improvement of the rheology of high-performance concretes with entrained air — 515
- 18.2.4 Concluding remarks — 516

18.3 Lightweight high-performance concrete — 516
- 18.3.1 Introduction — 516
- 18.3.2 Fine aggregate — 518
- 18.3.3 Cementitious systems — 518
- 18.3.4 Mechanical properties — 519
  - (a) Compressive strength — 519
  - (b) Modulus of rupture, splitting strength and direct tensile strength — 520
  - (c) Elastic modulus — 520
  - (d) Bond strength — 520
  - (e) Shrinkage and creep — 520
  - (f) Post-peak behaviour — 521
  - (g) Fatigue resistance — 521
  - (h) Thermal characteristics — 521
- 18.3.5 Uses of high-performance lightweight concrete — 522
- 18.3.6 About the unit mass of lightweight concrete — 522
- 18.3.7 About the absorption of lightweight aggregates — 524
- 18.3.8 About the water content of lightweight aggregates when making concrete — 525
- 18.3.9 Concluding remarks — 526

	18.4	Heavyweight high-performance concrete	526
	18.5	Fibre-reinforced high-performance concrete	527
	18.6	Confined high-performance concrete	530
	18.7	Roller-compacted high-performance concrete	534
	18.8	Concluding remarks	540
	References		540
19	Ultra high-strength cement-based materials		545
	19.1	Introduction	545
	19.2	Brunauer *et al.* technique	549
	19.3	DSP	549
	19.4	MDF	550
	19.5	RPC	551
	Selected references		553
20	A look ahead		556
	20.1	Concrete: the most widely used construction material	556
	20.2	Short-term trends for high-performance concrete	558
	20.3	The durability market rather than only the high-strength market	560
	20.4	Long-term trends for high-performance concrete	561
	20.5	High-performance concrete competition	562
	20.6	Research needed	562
	References		563

Afterword by Pierre Richard   564
Suggested reading   566
Index   569

Foreword

Adam Neville, CBE, FEng

In the past, many a book opened with a foreword written by a grand old man in the author's subject who took it upon himself to promote the work and to laud the author, not infrequently his former student or protegé. Nothing could be further from the truth in the present case.

First of all, I may be old but I am by no feat of imagination grand. It is, in fact, Pierre-Claude Aïtcin who is the grand man of high-performance concrete. It could therefore be strongly argued that his book needs no foreword, and yet it has not one but two. They were obviously written at the author's invitation and, while they may be no more than ornaments without commercial or sales value, they are a testimony to Aïtcin's innate modesty about his contribution.

When one begins to read the present book it becomes immediately obvious that it has been a labour of love, and a long love affair at that. The author tells us that he has been preparing this book for more than six years, but the seeds of high-performance concrete as a single, and yet wide-embracing, topic were first planted by Aïtcin three decades ago. There were many seeds: those of the technology of making high-performance concrete and those of the science of its behaviour. They were nurtured by Aïtcin and his team of students, undergraduate and postgraduate, by postdoctoral fellows and by academic colleagues at his own Université de Sherbrooke as well as in other institutions in Canada. In all this work it is easy to trace Aïtcin's influence, thoughts and inventiveness, as well as his dedication to the single goal of better and better high-performance concrete.

It has become better, not just because a higher strength or a higher durability has been achieved, but because the understanding of the behaviour of the material has been improved. It is his twin approach of scientific understanding and a knowledge of the technology of practical improvement that distinguishes Aïtcin from those who have investigated a single, narrow aspect of high-performance concrete and rest on the laurels of so many megapascals.

Very few people could write a book like this one. More than that: very few people even try to write a book which contains a single author's coherent understanding of a comprehensive topic such that a reader can readily benefit from it. Most so-called books have an editor, or even several editors, and contain a score or more of individual uncoordinated papers sharing half a dozen key words. It is left to the reader to wander in this jungle and to search for gems and pearls – but who is to set them into a single ornament?

Aïtcin, single-handedly writing this volume, has given us an enthusiastic but balanced picture of high-performance concrete, including all its facets from the constituents and their interaction to the concrete in service, and even a glimpse of it in the future. Thus the book will be welcomed by all those involved in high-performance concrete: the designer and the owner, the materials scientist, the concrete producer and the contractor, as well as by those civil engineering generalists who aspire to being *au fait* with the world around them.

Despite the author's closing words to Chapter 2, the present book *is* a treatise on high-performance concrete – there is none more exhaustive or more detailed, or better. So, despite what I said in my opening paragraph, I cannot but praise this book very warmly and recommend it wholeheartedly to all who seek a better concrete future.

A.M. Neville
London, England
5 February 1997

Foreword

Yves Malier

From prehistoric times, men's lives have been completely conditioned by their ability to master the use of materials. In fact, quite early on, historians differentiated the most active periods by giving them names such as 'Stone Age', 'Bronze Age' and 'Iron Age'.

Several tens of thousands of years later, it is striking to note that the advances in science (such as physics, chemistry and the life sciences) and technology (such as the engineering fields and computer science) that have marked the past 50 years and which will take us even further as the third millennium begins, are largely based on themes closely linked to materials.

Observing this evolution from prehistory to modern times, however, we note that, within this continuity of the role of materials, there has been a particularly marked difference in the approaches taken during the past decade, whatever the fields of application. At present, our approaches are more global – in certain cases, we speak of a 'systems approach' – whereby a scientific undertaking closely interrelates a number of different fields, such as mathematics, physics, chemistry, geology and mechanics. This kind of approach is interactive and free from the discrimination that characterized the past, opening the way to considering simultaneously construction design and implementation; components and their constituent materials; the environmental parameters that determine sustainability; and a global 'material–design–construction–maintenance' economic assessment.

Forty-five years ago, Albert Caquot, a member of the French Academy of Sciences, introduced Professor Courbon's structural calculus course at the École Nationale des Ponts et Chaussées in Paris with this insight: 'It is through improving the characteristics of materials, much more so than through advances being made in computing methods, no matter how significant these may be, that the most radical changes in the construction industry can be expected'.

Metals provide an example that, while somewhat beyond the scope of this remark, is illustrative. The progress experienced in heavy industry (automobiles, aeronautics, machine tools, household electrical appliances, etc.) have depended on advances in materials. Engineers and researchers have learned to master and diversify the composition and behaviour of materials, to adapt the raw form of iron, aluminium, copper and the like to the set of conditions specific to the operations, project durability, environment, workability and economic constraints for each element and aspect of technological developments.

Progress in the construction industry, often closely tied to large public-works projects, has remained quite dependent on two key characteristics, namely (1) that is based on prototypes and/or small-scale production (at least in comparison to the sectors cited above), and (2) that it relies on the use of major quantities of materials with a minimum of added value. To illustrate the latter point, each person on the planet 'generates' an annual average of nearly one cubic metre of concrete!

The resultant economic considerations, which incorporate the parameters impacting on construction costs, have traditionally led design engineers to view construction materials as 'black boxes' during the design process. The idea was that, since these materials should be as simple and consistent as possible, only two or three parameters (compressive strength, modulus of rupture and modulus of elasticity) were needed to characterize them.

The preceding decade has fuelled profound changes, just as the next will continue to do. The age of concrete, in the singular, has passed; we have definitely entered, just as we did for metals, the era of *concretes*.

Our best builders have already fully understood this lesson, in terms of optimizing facility sizing, improving processes, enhancing durability, refining design practices, and adapting material formulations and the behaviour of by-products from base materials and their combinations. An outstanding field of progress for the construction industry has thus been created and will go on to offer greater opportunities.

As Caquot foresaw, this progress can only be guided by materials-science specialists, provided they are able to develop, in addition to an understanding of purely physical, chemical and mechanical phenomena, at least a minimal knowledge of processes, design methodologies and environmental factors.

This was the outlook for us when, in 1980, Professor Roger Lacroix and I coined the expression 'high-performance concrete'. It was our view that, at a time when the future direction of French research was under discussion, only such a new, challenging concept would attract talented researchers, used to having a multidisciplinary and multi-scale approach, to study and develop a material that was not considered to be high-tech.

It was not at all easy to sell our global approach of 'material–design–construction–maintenance', sometimes called the 'French approach'. Even the English expression 'high-performance concrete' which we used was criticized, many preferring to stay with the more limiting but accepted 'high-strength concrete'.

It was at this juncture that Professor Pierre-Claude Aïtcin started playing what was to be a unique and determining role, due to his dual cultures and his belonging to both the North American and French scientific communities.

Between 1980 and 1990, when I directed the French National Project on high-performance concrete which brought together the range of French expertise (laboratories, contractors, engineering firms and owners) in a unique effort to promote the research, development and construction of high-performance concrete field applications, we carried out many projects. Some were modest, but others (bridges, dams, nuclear power plants, buildings, prefabricated elements, tunnels, jetties and so on) were quite impressive. Pierre-Claude Aïtcin was our North American liaison officer, and I always had a great deal of pleasure in having him take part in some of these achievements as well as in the many seminars and workshops during which the results of our field tests were presented. I also invited Professor Aïtcin to contribute to one of the first books on high-performance concrete that I was editing and which was published in English in 1992 by E and FN Spon.

Now, when high-performance concrete congresses, symposia and workshops are mushrooming all over the world, I am particularly pleased that E and FN Spon has decided to publish this synthesis work on high-performance concrete by Professor Aïtcin.

This book has been written in a language and style with which engineers, students, researchers, contractors, concrete producers and owners will be comfortable. It should rapidly become an indispensable tool.

I am totally convinced that this book will contribute to bringing about Caquot's prophecy, which I quoted at the beginning of this Foreword.

As a professor at the École Nationale des Ponts et Chaussées – the oldest civil engineering school in the world – and as a holder of the Concrete Chair inherited from Caquot, I can imagine no greater tribute for a French engineer such as Pierre-Claude Aïtcin than to have his exceptional work associated with the name of Albert Caquot, who was for all of us in France an outstanding scientist and an exceptional civil engineer in the area of concrete.

Professor Yves Malier
Paris

Preface

After years during which high-strength concrete was only perceived as a new concrete with a future limited to the construction of columns in high-rise buildings or offshore platforms, high-performance concrete is now viewed as an emerging type of concrete whose applications are growing both in volume and in diversity. High-performance concrete is not a cheap concrete that can be produced by anyone; on the contrary it is becoming an engineered, high-tech material whose properties, performance and possibilities continue to astonish us.

My first exposure to high-strength concrete dates back to 1970, when I first heard John Albinger, of Materials Service, make a presentation on the high-strength concretes he was delivering in the Chicago area at that time. He was so convincing and enthusiastic about high-strength concrete that I decided to end my concrete class at the University of Sherbrooke every year with a contest whose objective was for my students to make the strongest concrete possible with a maximum amount of cement and supplementary cementitious materials of 600 kg/m^3.

I remember that in the mid-1970s the winners were approaching 100 MPa and were rewarded with a cheque from the local office of the Canadian Portland Cement Association, on the basis of $1 CDN per MPa. This was more than enough to have a good beer party with my students at the end of the course.

A big jump in the compressive strength achieved was realized some years later, when one of my students asked my permission to use a dolomitic limestone aggregate that was produced by his uncle in the Montreal area. He won the contest with a compressive strength jump of about 25 MPa, in spite of the fact that he was using about the same water/binder ratio as the other students. This made me realize for the first time how important the selection of the coarse aggregate is when making high-performance concrete; 100 MPa was, and still is, the maximum strength that can be achieved with local Sherbrooke aggregates.

One advantage of being a professor is to learn so much from the errors as well as from the bright ideas of the students, when they do not listen to the lectures, or skip them, or do not read the literature. Then it is easy to explain the cause of their failure or of their fortuitous discoveries. One thing leading to the other, these first contacts with high-performance concrete induced me to write this book.

There were several difficulties in writing this book, a major one being the poor quality of my written English. I must admit that after the revisions of my initial drafts by S. Mindess and A. M. Neville, I was reading almost the same English words or a few new ones that they replaced or added, which I also knew well, but not exactly in the same order. I learned painfully that it is difficult to think in one language and to write in another. Hopefully, with their help my original thoughts have not been betrayed but have been transformed into a decent English.

I also faced some difficulties in preparing a scientific treatise on high-performance concrete. In spite of the apparent simplicity of usual concrete, it is in fact a very complex material, and this is even more true for high-performance concrete, in which several supplementary cementitious materials can be used at the same time in conjunction with several admixtures. Moreover, in some aspects high-performance concrete is both so similar to, and yet so different from, usual concrete, that it has sometimes been difficult to decide where to start and when to stop when describing and analysing the properties of high-performance concrete.

I also had difficulties with the great variety of definitions and terminology that are presently used world-wide. Consequently, I decided to start this book with a chapter devoted to terminology. As I say later in this first chapter, I am not claiming my terminology to be superior, but this is the one that I prefer and that must be accepted, momentarily at least, in order to understand what I have written.

During the past two or three years, I must admit, I have not been able to read with enough attention all of the literature published on high-performance concrete. This exercise would have been relatively easy in the late 1980s and early 1990s, when I started writing the first chapter of this book, but it is no longer true: as mentioned by Paul Zia, from 1974 to 1977, 97 papers were published on high-performance concrete, but from 1986 to 1989 the number went up to 471, and from 1990 to 1993 it shot up to 554! This situation forced me to accelerate somewhat the writing of this book, so as to prevent it from being outdated when printed.

I am aware that this book is somewhat incomplete, because I do not master at the same level of comprehension all aspects of high-performance concrete science and technology. I know that I have not cited all the papers that I should have, and that I have given a too much exposure to the work done at the Université de Sherbrooke by my graduate students, my research assistants and my postdoctoral fellows. This is so, not because I am convinced of the superiority of their work, but simply because it is their work that I know the best and because I have the easiest access to it. If some readers find that this book does not include the most up-to-date information available, I apologize.

In spite of the great research effort that has been carried out world-wide on high-performance concrete, it is clear that we do not yet have all the answers to the scientific and technological challenges that have to be faced when developing and using such a new material; high-performance concrete, in my opinion, is different enough from usual concrete to be considered a truly new type of concrete.

Considerable attention has been given in this book to the role of specifications and the use of appropriate test methods for determining concrete properties. For example, measuring high-performance concrete compressive strength is not as easy as many engineers believe. Also, many of the good old testing procedures that have been developed for some usual concrete properties should be questioned seriously in the case of high-performance concrete. In some cases, a part of these old standards should be reviewed and revised, so as to achieve the purpose of their existence: to give a fair numerical value of a property needed to make an appropriate and safe design, because structural design is only as good as the materials used and the knowledge of their properties.

Reference is made throughout this book to the American Society for Testing and Materials (ASTM) standards, and considerable use is made of the recommendations and reports of the American Concrete Institute (ACI), because these ASTM and ACI documents were the principal standards used in North America where I developed high-performance concrete technology with my students, and because they are increasingly being used world-wide. In the text, references are made to the up-to-date editions of the standards and recommendations available. However, these standards and recommendations are subject to frequent revisions, and the reader should therefore refer to the most recent editions, which may differ in some details from those referred to in this book.

High-performance concrete will not be 'Le concrete' of the future because in many applications engineers will continue to specify and use usual concrete, but it will be a type of concrete used more and more for its durability rather than for its strength, in future applications that even the most visionary concrete specialists can foresee or dream of. High-performance concrete is undoubtedly an engineered high-tech material whose potential has not yet been fully exploited.

Durability of concrete is a matter of timescale. Often I am asked how long a high-performance concrete will last; the only sensible answer I can give is that I shall not live long enough to see the end of my high-performance concretes. When you are approaching your 60s, the probability of being right is great enough to enable you to live the rest of your life totally unworried! Sometimes my answer is more adventurous, when I reply: do you believe that the builder of the Pantheon in Rome could have predicted that in 1998 his beautiful structure, built with a 10 MPa pozzolanic concrete, would still be in service and considered one of the most outstanding concrete structures in the world? The longevity of the Pantheon should remind us that concrete strength is not the key parameter of durability, but rather concrete compactness, good design and clever detailing.

When dealing with durability I know also that the timescale is very important; on a geological timescale, no human-made construction materials are durable; all concretes, as durable as we make them on a human timescale, will end their sevice life on Earth as a mixture of calcium carbonate, silica, calcium sulfate and clay, i.e. the initial materials that were transformed into clinker and Portland cement.

This book is not intented to be an exhaustive treatise on high-performance concrete, because high-performance concrete knowledge and technology are evolving very quickly. This book is written primarily to serve as a guide to engineers, graduate students and undergraduate students, to present the art, the science and the technology of high-performance concrete in a simple, clear and scientific manner (to paraphrase P. Kumar Mehta). It is not a theoretical book, but a practical one, though it refers as often as possible to the emerging new science of concrete. The production of high-performance concrete is still something of an art, but it is also the fruit of a complex science that we are starting to master. However, it must be admitted that, as has always been the case with concrete, concrete technology generally precedes concrete science. Most of our knowledge of the properties of high-performance concrete comes from the laboratory, field experience, errors and unexpected discoveries, which are later more or less easy to explain by theoretical considerations. It must not be forgotten that concrete also obeys the laws of physics, chemistry and thermodynamics. As for all other materials, the macroscopic properties of concrete are linked to its microstructural or nanostructural characteristics and properties. As we now start to explore these properties more scientifically, this must not be forgotten before we start to find on the world-wide web papers on 'virtual concrete'.

As with any material, the successful use of high-performance concrete depends upon an intelligent application of its properties in design. When making a high-performance concrete, if the mix is not properly designed for the appropriate service conditions, and if it is not properly handled and cured, the result will be a partial or total failure. It is thus vital that engineers acquire a thorough understanding of the material's properties and the procedures that are essential to make a material of the required quality and durability. There is a body of chemical, physical and thermodynamic principles underlying the behaviour of concrete. One aim of this book is to present a unified view of the behaviour of high-performance concrete in light of some of these principles, rather than to rely on a series of more or less unrelated empirical formulae coming from nowhere. For example, mechanical properties are discussed from the point of view of high-performance concrete as a composite multiphase material, which is more evident when observing high-performance concrete microstructure under an electron microscope.

With regard to the organization of the book, I have taken a somewhat different approach from the traditional. After general chapters on high-performance concrete, including a brief history of its development and selection to build structures around the world, the reader will find a chapter devoted to the main scientific principles that govern its properties. These principles are derived from the science of materials. Those readers more interested in matters practical could omit this chapter, as it is not essential to the proper production and use of high-performance concrete. The following chapters of the book correspond to the different steps followed by a high-performance concrete during its production, transportation, placing, curing, hardening and service. The last three chapters are devoted to special high-performance concretes, ultra high-strength cement-based materials, and to the future of concrete in general and high-performance concrete in particular.

Chapter 2 is an introduction to this book; it has been written as a kind of executive summary in order to make a selective reading of the different chapters easier for readers who might be short of time, or who do not want to follow the development of the chapters as presented.

The field of high-performance concrete is vast, and human effort is never perfect, as P.K. Mehta and P.J.-M Monteiro point out in the preface to their book *Concrete: Microstructure, Properties, and Materials*. Therefore, some readers may find shortcomings or discover omissions in this book. It is hoped that these deficiencies can be made up for by referring to the books, proceedings, reports and references that are listed at the end of each chapter. These references are provided as a guide to further reading and to provide a point of entry to the original scientific literature. Several important subjects have deliberately not been covered, including architectural high-performance concrete, repair and maintenance, and non-destructive testing. Regrettably, I have very little experience of these areas and prefer to leave it to others to write about them. Some of the latest work on shrinkage and creep has been presented even though it is not yet completed; however, I consider that the introduction of some of these new concepts is important owing to their relevance to curing of high-performance concrete.

When I was unable to distinguish between the significant and the insignificant in an area with which I was not too familiar, I decided simply to exclude this matter from this book. It will be the responsibility of the reader to make up on his or her own such deficiencies.

I would like to add a word of caution: high-performance concrete is evolving very fast, but our field experience is still limited. We must therefore open our eyes and our minds to take advantage of the laboratory and field experiences which have been the true drivers of the evolution of our knowledge of high-performance concrete. However, we will also have to take advantage of all of the sophisticated and powerful scientific equipment presently available to observe high-performance concrete at a micro- or even nanostructural scale, in order better to understand. As with any kind of material, macroscale properties are linked to the most minute microscale properties.

Pierre-Claude Aïtcin
Sherbrooke, February 1997

Acknowledgements

There are some individuals without whom this book could not have been written, and others without whom it might have been very different. First of all, there are my graduate students who year after year accumulated data on high-performance concrete, and it is the least I can do to dedicate this book to them. I am also in debt to John Albinger of Materials Services who somehow inspired me as early as 1970 when I heard him speak for the first time of high-strength concrete. Sidney Mindess has also played a key role, when he asked me, with enough insistence to clear away all of my objections, to write a book on high-performance concrete. I am not always sure that I did well to listen to him. However, I must admit that he supported me constantly over the years, not only by translating my 'Frenglish' into a decent English, but also by his criticisms, his remarks, his advice and his encouragement; I desperately needed all of them.

I was greatly helped by Adam M. Neville, whose perfect knowledge of both French and concrete enabled him to express in good English what he interpreted to be my thoughts in French, written in poor English. I appreciate his sharp and constructive criticisms, that improved significantly not only the wording but also the scientific and technological content of this book.

Another kind of indebtedness is to those who assisted me in the actual production of the book. They are too numerous to list by name, but Suzanne Navratil, Lise Morency, Sylvie Côté, Kathleen Lebeuf and Christine Couture, who typed and retyped chapter after chapter, are especially thanked for their work, as are my research assistants Mohamed Lachemi and Philippe Fonollosa, who assisted me in producing the final manuscript. Gilles Breton is responsible for the quality of the tables and figures, because I have to admit that I am one of the last members of a species on the way to extinction: a human being who doesn't know how to use a computer. I would like to thank Peter Gill for his careful proofreading which has greatly improved the final version.

My research assistants of the Industrial Chair, Pierre Laplante, Daniel Perraton, Michel Lessard, Moussa Baalbaki, Eric Dallaire, Serge Lepage and my post-doctoral fellows, Omar Chaallal, Buquan Miao, Pierre-Claver Nkinamubanzi, Guanshu Li and Shiping Jiang were also very helpful in monitoring, collecting and interpreting a great body of experimental results, with the help of the competent and dedicated technicians

Acknowledgements

of my research group: Claude Faucher, Claude Aubé, Claude Poulin, Ghislaine Luc, Irène Kelsey Lévesque, Maryse Darcy, Ghislain Roberge, Sylvain Roy, Jean-Yves Roy, Mario Rodrigue, Roland Fortin and Mylène Houle. The quality of the experimental results is a tribute to their skill.

I also do not want to forget to thank my French friends, Pierre Richard, Yves Malier, Micheline Moranville Regourd, François de Larrard, Paul Acker and Jacques Baron from whom I learned so much and so easily on high-performance concrete due to my Cartesian education. I must admit that being educated in France and having my teaching and research career in North America (in French) had a great influence on my way of thinking and my way of approaching concrete science and technology.

I must not forget to mention the fruitful input of all of my colleagues in Concrete Canada, the Canadian Network of Centres of Excellence on High-Performance Concrete which has been supported for seven years now by the Natural Sciences and Engineering Research Council of Canada. My involvement in this wonderful research programme definitely delayed the writing of this book, but at the same time much improved its content through the numerous interdisciplinary discussions and interchanges which I had with my colleagues. Rubbing shoulders with the structural and field engineers and scientists who are part of the Network has added so much to my perception of what high-performance concrete was, is and should be.

I also want to take this opportunity to thank the industrial partners of the Industrial Chair in Concrete Technology for their indefatigable support over nearly 10 years. Many field experiments reported in this book were carried out in collaboration with them: Claude Bédard, Marc Boudreau, Yvon Beaudoin, Yvan Bolduc, Marco Couture, André Bélanger, André Bouchard, Lino Pansieri, Michel Plante, Freddy Slim, Denys Allard, Philippe Pinsonneault, Bertin Castonguay, Antonio Accurso, Richard Parizeau, Marie-Christine Lanctôt, Gérard Laganière, Alain Dupuis, Jean Larivière, Daniel Vézina, Guy Roberge, Nélu Spiratos and Marco Dumont.

I am also in debt to all of my colleagues at the Université de Sherbrooke – Richard Gagné, Gérard Ballivy, Brahim Benmokrane, Carmel Jolicoeur, Kamal Khayat and Arezki Tagnit-Hamou – and the Université Laval of Québec, who are members of the Centre de Recherche Interuniversitaire sur le Béton (CRIB) for the numerous discussions we have shared on high-performance concrete and their help in the final revision of the manuscript.

I would like to thank the Natural Sciences and Engineering Research Council of Canada for its continuous support of my research activities over the last thirty years at Université de Sherbrooke. I also received firm support from the Lavalin Corporation, the Department of Transportation of the Province of Québec, Hydro-Québec, and the City of Sherbrooke, for experiments on the use of high-performance concrete in real-life conditions. These organizations were not afraid to undertake within the

Industrial Chair in Concrete Technology or with Concrete Canada demonstration projects which have been so valuable in obtaining practical experience. It is not often that public agencies are willing to take the risks involved in co-operating with the development of a new technology. It is my pleasure to thank personally Bernard Lamarre, Yvan Demers, Yves Filion, Claude Cinq-Mars and Jean Perrault for their help.

My students have been able to achieve quite easily very low water/binder ratio concretes with very high strength with the help of the different superplasticizers developed by Handy Chemicals from La Prairie, Québec, Canada, and I would like to thank Nelu Spiratos, its CEO, for his continuous technical support and the constant improvement of the quality of the superplasticizers needed to produce all these high-performance concretes.

Special mention should be made of the co-operation of the American Concrete Institute, ASTM, the American Ceramic Society and the Laboratoire Central des Ponts et Chaussées, which allowed me to draw freely on the material in their publications.

Many other organizations and individuals have given me permission to adapt or reproduce material from their publications for this book and I thank them for this courtesy. These are: the American Society of Civil Engineers; Canadian Government Publishing Center (Ministry of Supply and Services, Canada); Concrete Construction Publications Inc. (329 Interstate Road, Addison, Il 60101); Elsevier Scientific Publishers; Federal Highway Administration (US Department of Transportation); Institut Technique du Bâtiment et des Travaux Publics (France); National Institute of Standards and Testing (NIST, USA); the Transportation Research Board and Dr Iroshi Uchikawa from Onoda Cement.

I would like also to thank my wife, Gisèle, for her patience; she has endured so much in the cause of high-performance concrete!

CHAPTER 1

Terminology: some personal choices

It may seem strange to begin a book on high-performance concrete with a chapter devoted to terminology. Discussions on terminology are tricky and can be endless, but it must be admitted that often the quality of the information in a technical book is diminished by the lack of consensus on the exact meaning of the terms used.

The author makes no claim for the superiority of the terminology he uses; he wants only to make clear the exact meaning of the terms he employs. The reader is free to disagree with the pertinence and the validity of the proposed terminology but, by accepting it momentarily, he will better understand the concepts and ideas expressed in this book.

There is no clear consensus about the meaning of expressions such as high-strength and high-performance concrete. Even the simple concept of water/cement ratio, that has been the pillar of concrete technology for years, is losing most of its meaning since several supplementary cementitious materials and more or less reactive fillers are replacing some of the cement in modern concretes. It is therefore essential to be more precise about these concepts and expressions in order to eliminate the most frequent source of misunderstanding in science and technology: the existence of several meanings to a simple word, expression or concept.

The acceptance of these definitions is essential to make the most of reading this book. As said by A.M. Neville: 'The choice of one term over another is purely a personal preference, and does not imply a greater accuracy of definition' (Neville, personal communication, 1996).

1.1 ABOUT THE TITLE OF THIS BOOK

It was without any hesitation that the author selected *High-Performance Concrete* as the title of this book, rather than *High-Strength Concrete*. There is no doubt that this new type of concrete offers more than just a high compressive strength, as discussed in section 1.4. In spite of the fact that up to now high-performance concrete has been used primarily in high-strength applications, it is inevitable that in the very near future high-performance concrete will be mostly specified and used for its durability rather than specifically for its high compressive strength. When the civil engineering community comes to understand this, and changes its perception of high-performance concrete, the construction industry will definitely take a big step forward.

In the present state of the art, it is not possible to make a durable high-performance concrete that does not also have a high compressive strength, but this situation can evolve. Moreover, the author is well aware that the expression 'high-performance' concrete is not yet universally accepted by the scientific and engineering communities and that in some countries it has a totally different meaning, but the term is his personal choice and he will use it accordingly.

1.2 WATER/CEMENT, WATER/CEMENTITIOUS MATERIALS OR WATER/BINDER RATIO

It took much longer for the author to select the expression to be used to describe the fundamental and universal concept that is hidden behind such an expression. The water/cement ratio concept has been the pillar of concrete technology for almost a century; it is a simple and convenient concept as long as concrete does not contain any cementitious materials other than Portland cement.

When the very simple but very important concept of water/cement ratio was initially developed, at a time when concrete technology was in its infancy, it was an unquestionable expression; concrete was made using solely Portland cement. This is no longer completely true because, when concrete is made with a modern Portland cement, it can contain a small amount of limestone or silica filler that the cement producer is allowed by many national standards to blend within the Portland cement and still call his product Portland cement. Thus the expression 'water/cement ratio' has already lost some of its basic significance even when concrete is made solely with a modern Portland cement (Barton, 1989; Kasmatka, 1991; Shilstone, 1991). Moreover, with the passage of the years, though to different extents in different parts of the world, the use of so-called 'supplementary cementitious materials' or 'fillers' has become a more common practice so that many modern concretes now incorporate fly ashes, slags, natural pozzolans, silica fumes, limestone fillers, silica fillers or rice husk ash, etc. These finely divided materials can be part of a so-called 'blended' cement or simply be added to the concrete at the batching plant. The expression 'supplementary cementitious material' does not unquestionably include all of these materials because, *stricto sensu*, it is not clear that a limestone or a silica filler qualifies as a cementitious material. As discussed by A.M. Neville, the expression water/cementitious ratio that has been used is absolutely inappropriate because cementitious is an adjective (Neville, 1994). What expression should be used, then, to replace the obsolete expression of water/cement ratio? The author thinks that it is time to propose a new expression that will embrace the expressions 'supplementary cementitious material', 'filler' and 'Portland cement'. Of course, such a term covering materials with such a wide variety of properties will have a meaning that is more vague, but it would nevertheless be convenient in order to translate the fundamental concept that was hidden in the original water/cement ratio expression. This fundamental concept was related to the ratio between water and the fine particles that give the concrete its strength, and it is still valid when all of these fine materials are now added to modern concretes.

As the use of blended cement was developed much earlier in Europe than in North America, European authors faced this dilemma much earlier, so that a number of different expressions have been proposed to translate the water/cement ratio expression into a more correct one. Influenced by his French culture, the author has decided to use the English equivalent of the French expression *'rapport eau/liant'*, which may be translated as water/binder ratio (W/B) instead of the lengthy expression 'water/cementitious materials ratio'. However, the author recognizes here, too, that the use of one expression over another is purely a matter of personal preference, not accuracy of definition.

When this expression is used the following definition is implicit: the term 'binder' as used in this book represents any finely divided material that is used in a concrete mixture having about the same fineness as or finer than Portland cement. However, the use of 'water/binder ratio' does not mean that the author does not recommend the simultaneous use or calculation of the water/cement (W/C) ratio, because, as will be seen, both ratios are very important from a technological point of view. As most of the supplementary cementitious materials and fillers used in conjunction with Portland cement in high-performance concrete are much less reactive than Portland cement, during setting and early hardening the actual value of the water/cement ratio is, of course, very important, because the early strength and impermeability of the hardening concrete are almost entirely a function of the bonds created by the early hydration of the Portland cement part of the binder. In the case of blended cements, it is not always easy to calculate the exact water/cement ratio because the exact amount of cement contained in the blended cement is not always known, some national standards specifying only a range of potential compositions and not a precise composition.

1.3 NORMAL STRENGTH CONCRETE/ORDINARY CONCRETE/USUAL CONCRETE

The author does not like the expression 'normal strength concrete' used to describe the usual concretes that are used presently by the construction industry. This expression implies that concretes other than these are abnormal. Very often it is these usual concretes that are abnormal because they are used in environmental conditions under which they are unable to fulfil, on a long-term basis, their structural function. As this type of concrete will continue to be used in the future since there are many applications in which a high water/binder ratio concrete, having a not-so-high compressive strength, is perfectly adequate and economical, the author has had to select another expression to describe them. In spite of the fact that, with present technological standards, the expression 'low-strength concrete' would have been more appropriate, the author has finally decided to use the expression 'usual concrete' rather than ordinary concrete or normal strength concrete. The expression 'ordinary concrete' was rejected because it might imply that other concretes could be extraordinary, while the expression 'unusual concrete' still applies well in most parts of the world to designate high-performance concrete. Ten years from now, when high-performance concrete will not be as unusual, the expression 'usual concrete' will not be appropriate to designate low-strength concrete, but this is not very important, because this book will be obsolete long before the concrete industry uses high-performance concrete on a routine basis.

1.4 HIGH-STRENGTH OR HIGH-PERFORMANCE CONCRETE

In the 1970s, when the compressive strength of the concrete used in the columns of some high-rise buildings was higher than that of the usual concretes used in construction, there is no doubt that it was legitimate to call these new concretes 'high-strength' concretes. They were used only because their strength was higher than that of the usual concretes generally specified at that time. In fact, by present standards, they were only improved usual concretes. They were made using the same technology as that used to make usual concrete except that the materials used to make them were carefully selected and controlled (Freedman, 1971; Perenchio, 1973; Blick, Petersen and Winter, 1974).

However, when superplasticizers began to be used to decrease the water/cement or water/binder ratios rather than being exclusively used as fluidifiers for usual concretes, it was found that concretes with a very low water/cement or water/binder ratio also had other improved characteristics, such as higher flowability, higher elastic modulus, higher flexural strength, lower permeability, improved abrasion resistance and better durability. Thus the expression 'high-strength concrete' no longer adequately described the overall improvement in the properties of this new family of concretes (Malier, 1992). Therefore the expression 'high-performance concrete' became more and more widely used. However, the acceptance of this expression is not yet general; for example, the name of ACI Committee 363 is still the High-Strength Concrete committee and not High-Performance Concrete committee.

Most of the detractors of the expression 'high-performance concrete' criticize this expression because it is too vague. What is the 'performance' of a concrete? How can it be measured? When using the expression 'high-strength concrete' there is no misunderstanding possible, except that the limit at which a concrete is no longer a usual concrete and becomes a high-strength one is not the same for everybody. This kind of endless discussion can be overwhelming (or can generate a new one!) if concrete is considered in terms of its water/binder ratio. For the author, a high-performance concrete is essentially a concrete having a low water/binder ratio. But how low? A value of about 0.40 is suggested as the boundary between usual concretes and high-performance concretes.

This 0.40 value, which might be perceived as being totally arbitrary, is based on the fact that it is very difficult, if not impossible, to make a workable and placeable concrete with most ordinary Portland cements that are presently found on the market, without the use of a superplasticizer if the water/binder ratio is lower than 0.40. Moreover, this value is close to the theoretical value suggested by Powers to ensure full hydration of Portland cement (Powers, 1968) and, as will be seen later in this book, it seems that this value denotes concretes that are starting to present autogenous shrinkage.

If this definition is accepted, it is evident that a 0.38 water/binder ratio concrete is not very much stronger and will not exhibit much better performance than a 0.42 one. But as soon as the water/binder ratio deviates significantly from the 0.40 value, usual concretes and high-performance concretes have not only quite different compressive strengths but also quite different microstructures (quite different shrinkage behaviour) and quite different overall performances.

Therefore, if it is the water/binder ratio that is to be used to differentiate high-performance concrete from usual concrete, why have I not selected as a title for this book *Low Water/Binder Ratio Concrete* instead of *High-Performance Concrete*? In spite of the fact that this title would be more correct from a scientific point of view, it would be much less appealing and it would probably be rejected by the publishers from marketing considerations.

REFERENCES

Barton, R.B. (1989) Water–cement ratio is passé. *Concrete International*, **11**(11), November, 75–8.

Blick, R.L., Petersen, C.F. and Winter, M.E. (1974) *Proportioning and Controlling High-Strength Concrete*, ACI SP-46, pp. 141–63.

Freedman, S. (1971) *High-Strength Concrete*, IS1 76–OIT, Portland Cement Association, Skokie, IL, 19 pp.

Kasmatka, S.H. (1991) In defense of the water–cement ratio. *Concrete International*, **13**(9), September, 65–9.

Malier, Y. (1992) Introduction, in *High-Performance Concrete – From Material to Structure* (ed. Y. Malier), E and FN Spon, London, pp. xiii–xxiv.

Neville, A. (1994) Cementitious materials – a different viewpoint. *Concrete International*, **16**(7), July, 32–3.

Perenchio, W.F. (1973) *An Evaluation of Some Factors Involved in Producing Very High-Strength Concrete, Portland Cement Association Research and Development Bulletin* RD 104, Skokie, IL, 7 pp.

Powers, T.C. (1968) *The Properties of Fresh Concrete,* John Wiley and Sons, Inc., New York, 664 pp.

Shilstone, J.M. (1991) The water–cement ratio – which one and where do we go? *Concrete International,* **13**(9), September, 64–9.

CHAPTER 2

Introduction

The purpose of this introduction is to provide an overview of the different chapters that make up this book. It can be considered as a kind of executive summary for readers who want to have a general idea of the content of the book, for readers who are not interested in all the chapters, or for readers who do not like the order of the chapters. It follows a chapter on terminology, in which the exact meaning of certain terms and expressions has been clearly defined, because it is the author's experience that very often a lack of understanding is simply the result of the use of terms or expressions that have different meanings for different individuals in different countries. These choices do not pretend to be the perfect ones, because in matters of terminology the pros and cons of a definition can be discussed endlessly, but the precise sense of the terms used in this book must be accepted in order to have a comprehensive reading.

In Chapter 3, a historical perspective of the development of high-performance concrete over the past 25 years is presented. It shows how, slowly, high-strength concrete was introduced to the high-rise building market in the Chicago area in the late 1960s and early 1970s, and how it then spread throughout the world and came increasingly to be called high-performance concrete. In 1997, high-performance concrete is used in most parts of the world. Many, in fact too many, workshops, symposia and conferences are held almost every year on the subject; similarly, many papers, too many papers, are published in technical journals, so that readers are submerged by a flood of information. What is needed right now is more papers synthesizing what is really known, what is not well enough known and what should be known, rather than detailed academic papers having very little impact on the science and technology of high-performance concrete. It is one of the ambitious objectives of this book to present a comprehensive synthesis of what I presently know, or at least what I think I know, about high-performance concrete.

In Chapter 4, the rationale for high-performance concrete is developed, followed by a presentation, in some detail, of 11 case studies. Why should an owner, why should a designer, why should a concrete producer become interested in high-performance concrete? What are the technical and economic advantages of specifying a high-performance concrete? There is no unique answer. The 11 case studies selected show that high-performance concrete was selected to build both outstanding structures and some modest ones because it was found, for various reasons, that the use of high-performance concrete could achieve the most economical and durable structure under the circumstances. The cases selected do not pretend to cover all past, present and possible future uses of high-performance concrete. However, they suffice to explain the main advantages and characteristics of high-performance concrete that have led to its increasing use as a construction material. The discussion is given from the point of view of the owner, the designer, the contractor and the concrete producer. The conclusion of this chapter discusses some of the environmental advantages of using high-performance concrete. It demonstrates how future development of the use of high-performance concrete could result in a better use and conservation of our shrinking natural resources and decrease the environmental damage associated with any production of concrete.

Chapter 5 develops the principles of high-performance concrete. First, there is a review of some fundamental principles that govern the compressive strength and the tensile strength of materials. Then it is noted how the strength of high-performance concrete can be increased at each of the following levels: the hydrated cement paste, the transition zone and the aggregates. There is a discussion of the conflicting demands concerning the amount of mixing water required to make a workable concrete. This conflict can be solved through the use of recently developed organic polymers, known as superplasticizers, fluidifiers or high-range water reducers. It must be realized that in high-performance concrete the reduction of the water/binder ratio is achieved by reducing the quantity of mixing water, and not by adding more cement. The fluidity loss of the concrete caused by this reduction of the water dosage is compensated for by increasing the superplasticizer content. While concrete produced in this way does not contain enough water to hydrate all the cement particles completely, the use of a low water/binder ratio tends to reduce significantly the inherent porosity of the hardened cement paste, thereby confirming Féret's work published in 1892. Different means of improving the rheology of very low water/binder pastes are discussed. The trend towards using more and more supplementary cementitious materials raises a question about the significance of the water/cement ratio. Should the water/cement ratio or the water/binder ratio be used to characterize a high-performance concrete?

Chapter 6 is a review of the relevant properties of some of the ingredients used to make high-performance concrete. Although readers of this book probably already possess a general knowledge of the properties of materials used in making concrete, the fabrication of high-performance concrete requires the use of materials with certain specific properties. It is therefore useful to review the relevant properties of Portland cement, superplasticizers and supplementary cementitious materials.

In Chapter 7, it is shown that making high-performance concrete is not a hit-or-miss operation, but rather a process that involves rigorous selection of the best materials available in a given region and establishing optimal quality assurance programmes for selecting materials. It is emphasized that Portland cements meeting the various national standards, which are broadly similar, can behave quite differently when used to make low water/binder ratio concretes. These differences can be accounted for by cement fineness, the overall reactivity of the interstitial phase (particularly that of C_3A), the reactivity of the C_3S, the solubility of the different forms of calcium sulfate found in the cement, and the solubility of alkalis in concrete falling in the 0.20 to 0.35 water/binder range and containing high levels of superplasticizers.

Often, making a practical and economical high-performance concrete is linked to what is known in the industry as cement–superplasticizer compatibility. In that respect, it is shown that present Portland cement characteristics, and particularly its 'gypsum' content, are optimized through a standard procedure which can be misleading when this cement is used in conjunction with a superplasticizer at a very low water/binder ratio. From experience, it is found that some commercial superplasticizers and normal Portland cements are not satisfactory at this level in terms of rheological properties and compressive strength. For example, when used for high-performance applications, some superplasticizer and cement combinations cannot maintain a sufficient slump for even an hour to enable the placing of high-performance concrete to be as easy as that of usual concrete. Some cement–superplasticizer combinations cannot reach 75 MPa even when the cement content is increased or the water/binder ratio is reduced by adding more superplasticizer. Producing such concretes is too expensive and, to make matters worse, the set is retarded to the point that the formwork often cannot be stripped even after 24 hours.

Finding aggregates that meet the minimum standard requirements for usual concrete in the 20 to 40 MPa range is fairly easy; however, when targeting 75 MPa, a number of problems arise. Performance can be limited by certain aggregates, such as gravels that are too smooth and not clean enough, those containing too many soft and crumbly particles, soft limestones and hard aggregates with a poor shape characterized by flat or elongated particles. Aiming for a design compressive strength of 100 MPa imposes even greater restrictions on selecting aggregates, cements and superplasticizers. Successfully producing a 100 MPa concrete requires:

- a very strong, clean, cubical coarse aggregate (with some exceptions for some glacial gravels);
- a cement that performs outstandingly well, both rheologically and in terms of strength;
- a superplasticizer that is totally compatible with the selected cement.

At present, taking the next step to produce concretes in the 125 to 150 MPa range is too often beyond the capability of most ready-mix operations, except in some rare instances in which the required exceptional materials can be found in the same location. When this exceptional situation presents itself, the coarse aggregate usually constitutes the weak link of a 0.20 to 0.25 water/binder ratio high-performance concrete, in which failure will be initiated under compressive loading.

Nevertheless, we already have a number of 'rules of thumb' (for which we have at least some scientific explanations) that concrete producers can use in the selection of their raw materials. Some of these rules are stated below:

- The higher the targeted compressive strength, the smaller the maximum size of the coarse aggregate should be. While 75 MPa compressive strength concretes can easily be produced with a good coarse aggregate of a maximum size ranging from 20 to 28 mm, aggregate with a maximum size of 10 to 20 mm should instead be used to produce 100 MPa compressive strength concretes. Concretes with compressive strengths of over 125 MPa have been produced to date, with coarse aggregate having a maximum size of 10 to 14 mm.
- Sand coarseness must increase proportionally to compressive strength and cement dosage; a fineness modulus in the 2.70 to 3.00 range is preferred if available.
- Using supplementary cementitious materials, such as blast-furnace slag, fly ash and natural pozzolans, not only reduces the production cost of concrete, but also provides answers to the slump loss problem. The optimal substitution level is often determined by the loss in 12 or 24 hour strength that is considered acceptable, given climatic conditions or the minimum strength required.
- While silica fume is usually not really necessary for compressive strengths under 75 MPa, most cements require it in order to achieve 100 MPa. Given the materials available to date, it is almost impossible to exceed the 100 MPa threshold without using silica fume.

The cost of usual concrete depends basically on the cost of the cement, which accounts for the economic attractiveness of substituting industrial by-products (slag, fly ash), inert materials (limestone fillers) and powdered natural products (natural pozzolans and calcined clay) for a portion of the cement. The same holds true for high-performance concrete, but in addition there is a strong rheological advantage due to the slower reactivity of these supplementary cementitious materials. Such by-products react more slowly than Portland cement; they scarcely react at all during the first hours after mixing, and therefore their use generally reduces the amount and the cost of the superplasticizer. By-products can also be used to decrease further the water/binder ratio to yield a more compact and stronger concrete.

When selecting the raw materials to make high-performance concrete, it must not be forgotten that the economic difference between high-performance and usual concretes relates directly to the additional cost of cement, superplasticizer and silica fume (when applicable). Since high-performance concrete requires a significant superplasticizer dosage to achieve enhanced levels of performance, its cost can substantially affect concrete production costs. Moreover, a more or less good compatibility between cements and superplasticizers can result in the doubling of the superplasticizer content for a given water/binder ratio, which translates into production-cost overruns (or savings) of US$10 to $20/m^3$, which is significant.

While the cost of silica fume varies according to regional and local market conditions, in some parts of the world specifying a compressive strength of 100 MPa, instead of 90 MPa, can double the production cost of a high-performance concrete, since silica fume can be 10 times more expensive than cement. When making a 90 MPa concrete, the producer often does not need to use the 10% silica fume required to achieve a 100 MPa compressive strength.

In Chapter 8 it is shown how little has been done in the area of mixture proportioning methods for high-performance concrete. Until recently, researchers and concrete producers tended to adapt or modify time-tested formulations developed using traditional methods. Recently, however, an original theoretical approach involving grout fluidity and Farris's theory has been proposed, along with some more pragmatic methods. In any event, there are still no hard-and-fast guidelines for formulating high-performance concrete, and the topic remains open. Through the years, a semi-empirical method that is simple to understand and that can easily be put into application has been developed and used successfully at the Université de Sherbrooke; it is presented in this chapter with some sample calculations.

Chapters 9 and 11 deal with the production, placing and control of high-performance concrete. High-performance concrete has been successfully produced with all types of equipment encountered at ready-mix plants. For example, the non-air-entrained concrete used to build Nova Scotia Plaza, in Toronto, that had an average compressive strength of 94 MPa at 91 days, was batched in a dry-batch plant, i.e. completely mixed in a truck mixer. This mixing method obviously required optimizing the sequence for introducing materials, and the mixing time could take as long as 15 minutes per truck load. In the case of the Hibernia offshore platform, which was built with an 86 MPa air-entrained high-performance concrete, the concrete was batched in only 90 seconds in very efficient twin mixers of 2 and 2.5 m^3 capacity.

To date, high-performance concrete has been placed using the same field methods and equipment as its usual counterparts. As an example, the Scotia Plaza high-performance concrete was pumped to the top of the 68th storey throughout the year in ambient temperatures ranging from

−20 °C to +30 °C with no major problems. Both the 130 MPa concrete for Seattle's Two Union Square and the 96 MPa concrete for 225 W. Wacker Drive in Chicago were also pumped into place. The Hibernia air-entrained high-performance concrete was pumped through lines up to 800 m long and 60 m high with an air content and a spacing factor making it freeze–thaw-resistant.

Chapter 10 is dedicated to the preparation that must be done before concreting, because those who do not have any experience with high-performance concrete, and who believe that using a high-performance concrete does not require more attention than usual concrete, will learn rapidly that they are very wrong. As most of the selected ingredients are working at a very high level, close to their maximum capacity, using a high-performance concrete requires some special attention. It is strongly recommended, whatever the experience of the concrete producer and of the contractor, that there is a preconstruction meeting during which all the specifications and details are reviewed and checked one by one.

For important projects, where an appreciable amount of high-performance concrete will be used, or when it is the first time that the concrete producer and contractor have used high-performance concrete, it is recommended that a prequalification test be set up. During this exercise it will be possible to perform a rehearsal and to be sure that on D-day everybody will know exactly what do to, how to do it and when to do it.

During the preconstruction meeting and the prequalification test programme, the testing methods should be reviewed carefully, as well as the methods specified to control the quality not only of the concrete, but also of the placing and curing. When such steps have been carefully followed there are very few chances of facing a mess or, if unforeseen conditions are creating some trouble, it will be easy to react and to correct the situation. As in all human operations, and more specifically in the critical ones, the essence of success resides in the establishment of a good communication system and a clear sharing of responsibility. In that sense a preconstruction meeting and prequalification test reinforce the quality of the communications between the principal actors on the project.

Chapter 12 is one of the most important in this book because it explains autogenous shrinkage. If the conditions leading to autogenous shrinkage and also its danger are not well understood, this type of shrinkage, the effects of which are almost totally negligible for usual concretes, can jeopardize the quality of the concrete structure. More so than in the case of usual concretes, it is important to start moist curing as early as possible to provide an external supply of water to the hydrating cement paste, so that its coarser capillaries do not start to dry out due to the transfer of the water they contain into the very small pores created by the volume contraction developed within the hydrating cement paste.

In usual concrete a great part of the mixing water is contained in large capillaries, so the drying due to self-desiccation creates menisci that are so large that they develop only weak tensile stresses and therefore induce negligible autogenous shrinkage. When the drying phenomenon associated with self-desiccation develops in a high-performance concrete that contains very few large capillaries, the menisci rapidly develop in small capillaries, where they create large stresses and therefore induce a rapid and significant autogenous shrinkage in the absence of any external supply of water. **Therefore all high-performance concrete must be water-cured as soon as possible before hydration starts**. When can it be known that hydration has started? Usually, when the high-performance concrete temperature starts to increase. If external water is not provided to the concrete by that time, autogenous shrinkage will develop very quickly and will be responsible for cracks that will constitute a preferential penetration path for any corrosion agent. This will decrease the life cycle of the structure significantly or necessitate an early rehabilitation programme, in spite of the fact that high-performance concrete was specified and used specifically to avoid this situation.

Those who have been used to relying on the application of a curing membrane on the surface of usual concrete, and who believe that this will work well with high-performance concrete, are making a big mistake because they are not providing any external water to the concrete when the autogenous shrinkage is developing. Moreover, waiting 24 hours to start wet curing is a big mistake because high-performance concrete autogenous shrinkage starts as soon as the hydration begins, and this occurs well before 24 hours.

The lower the water/binder ratio and the lower the water/cement ratio, the more attention must be paid to implement the wet curing as soon as possible. The retarding effect of superplasticizer or of retarders introduced during the mixing of high-performance concrete results in a delay of the hydration reaction, but autogenous shrinkage is developed by 24 hours, which is usually about the time that people begin to think about curing. Of course, the longer the water curing the better, but after 7 days high-performance concrete can withstand a lack of further water curing without any major problems. In any case, wet curing of less than 3 days should never be authorized.

Some specific field situations are briefly analysed at the end of this chapter to serve as a guide for field applications, and a proposal to introduce curing as a separate operation from the placing and checking of concrete properties is presented. It is the author's opinion that the curing of any concrete is as important as its properties when a durable structure has to be built. Therefore it requires at least the same level of attention.

The properties of fresh concrete are reviewed in Chapter 13. The techniques normally used to measure these properties are the same as those used with usual concretes, so there are very few problems with their implementation. However, it must be emphasized that the slump test is not the best way to check the workability of a specific high-performance concrete, though in the absence of any simple method to replace it we have to live with it. In spite of their merits, the author does not think that in the near future the so-called field rheometers developed by various researchers and laboratories will fill this gap between the present slump test and what is really needed to appreciate properly the workability of a high-performance concrete. In that area we still have to live with a qualitative description of workability.

In Chapter 14 the hardening process of high-performance concrete is discussed. Given the high cement content in such concretes, one would expect significantly higher temperature increases in large elements, such as columns in high-rise buildings, than in similar structures made with usual concrete. Field work has demonstrated that, while the internal temperature of a concrete element can rapidly reach 65 °C, the thermal gradient never exceeds 20 °C/m, beyond which network cracking due to unequal cooling can occur. Nevertheless, very recent experiments carried out by the concrete research groups at the Université de Sherbrooke and McGill University have shown that the heat generated by a 30 MPa concrete in a massive concrete column rivalled that of 70 and 100 MPa concretes in similar columns.

This rise in temperature at the outset of hydration raises a number of questions that have not, as yet, attracted much research:

1. What is the effect of this temperature rise on hydration kinetics in low water/binder ratio concretes? Does a concrete hardening at 65 °C, or even 75 °C, have the same microstructure as a concrete of the same composition cured under standard conditions at 20 °C?
2. What are the reactivities of supplementary cementitious materials that usually hydrate more slowly than Portland cement at room temperature?
3. Can the rapid initial temperature rise in concrete initiate alkali–aggregate reactivity in aggregates that would remain dormant at normal temperatures, or even under accelerated testing conditions in which aggregate reactivity with alkalis in cement is examined after exposure at 38 °C?
4. How does this homogeneous initial temperature rise, followed by a relatively long non-homogeneous cooling period to ambient conditions, affect the mechanical properties of concrete in structural elements?
5. Can test results (compressive strength, elastic modulus and others) for standard specimens cured under standard conditions be used to characterize the mechanical properties of concrete in actual structural elements?

As things stand, the answers we have are incomplete and only partially documented. Nevertheless, the results of a number of research programmes in progress around the world look towards rapidly finding the answers needed to these fundamental questions, on which the future of high-performance concrete depends.

Chapter 15 focuses on characterization testing before discussing high-performance concrete properties *per se*, because testing of such a strong material has to be carried out very carefully.

The main concern in this respect is determining whether or not high-performance concrete can be assessed using tests developed over the years for usual concrete. The first unsolved problem presently is how to cure concrete specimens before testing to avoid or control the effects of autogenous shrinkage.

It has been shown that concrete cast in place can experience an unusual temperature rise. Unfortunately, this issue has not yet received any attention. Moreover 'conventional' compressive strength and permeability tests can give rise to significant problems when applied to high-performance concrete:

- High-performance concrete is so dense that it is impossible for water to percolate through a well-prepared specimen under any pressure. The so-called 'rapid chloride-ion permeability' or 'gas permeability' testing is therefore more appropriate for evaluating transport properties of concrete permeability.
- Most testing laboratories today do not have testing machines of sufficient capacity to test standard 150×300 mm cylinders, and therefore 100 200 mm specimens must, of necessity, be used for high-performance concrete.
- The most commonly used capping compounds do not have the compressive strength required for testing high-performance concrete. Even reducing the capping thickness to a minimum in order to maximize the platen effect as much as possible often results in the compound exploding by the end of the test. Whenever concretes with strengths of 100 MPa or greater are to be tested, both ends of the cylinders should be ground.
- Modulus of rupture and tensile splitting tests have posed no particular problems.
- While measuring the elastic modulus causes no difficulties, it is surprising that the stress–strain curve of some concretes is convex with respect to the strain axis. In fact, the stress–strain curve of the coarse aggregate strongly influences that of the concrete, since high-performance concrete exhibits excellent stress transfer at the paste/aggregate interface.
- Most testing machines, even with specimens of reduced size, do not have adequate rigidity for uniaxial compressive testing: the specimens tend to fail by exploding. This lack of rigidity also causes serious problems in determining the post-peak part of the stress–strain curve. Some experiments have revealed behaviour such that, in terms of rock mechanics, the concretes can be categorized as Class II rocks: specimen

failure can be slightly prolonged, requiring reprogramming of all of the testing machine's servo-controls, in order to get the complete stress–strain curve.

A recent study has shown that there is a simple linear relationship between the static and dynamic elastic moduli (the latter being easier to measure by non-destructive ultrasonic methods). This leads one to hope that using the dynamic elastic modulus may remove the problems in measuring the static modulus.

Chapter 16 is devoted to a discussion of the principal mechanical and physical characteristics of high-performance concrete, starting with compressive strength. The main factors affecting the compressive strength of high-performance concrete are reviewed, especially the long-term compressive strength.

When some researchers noted some long-term strength loss in small air-cured specimens made of usual concrete containing silica fume, there was concern as to what this meant for high-performance concrete. It should be recalled that usual concrete without silica fume exhibits similar losses, although to a slightly lesser degree. Since high-performance concrete usually incorporates silica fume, detractors attempted to discredit it by drawing attention to the so-called strength regression. However, core testing of actual structures made of high-performance concrete has not revealed any significant strength reduction at all. The few data published to date on core strength results versus specimen strength values will be discussed in order to cast some light on this controversial subject.

Modulus of rupture and splitting tests are also discussed, as well as the validity of the $f_r = \psi(f'_c)$ and $E_c = \psi(f'_c)$ relationship commonly used by designers for usual concrete.

It is shown that aggregates, particularly the coarse ones, influence not only the compressive strength and modulus of rupture, but also the elastic properties of high-performance concrete. In fact, owing to the excellent bond that is developed between the very dense hydrated cement paste and the aggregates, it can be said for the first time that concrete behaves like a real composite material. The stress transfer is so improved that aggregates strongly influence the usual stress–strain curve and hysteresis loop developed when testing high-performance concrete for elastic modulus.

Since measuring the elastic modulus of concrete requires expensive and complex equipment, attempts have been made to establish a relationship between the elastic modulus and the compressive strength, since the latter is fairly easily obtained. All national codes suggest the use of relationships such as $E_c = \psi(f'^{1/n}_c)$, where n varies from 2 to 3. These empirical relationships have been established over the years for usual concretes and allow a reasonable ($\pm 30\%$) prediction of elastic modulus based on compressive strength. This is possible because both elastic modulus and compressive strength in usual concrete depend mostly on the strength of the hydrated cement paste, which in turn depends on the water/binder ratio. The water/binder ratio controls the strength and rigidity of the hydrated cement paste throughout the concrete, but more particularly in the weak transition zone between the paste and aggregates. In usual concrete, this zone is porous and weak to the point that there is practically little transfer of stress from the paste to the aggregate, but the reverse is the case as strength increases. Experimental results demonstrate that aggregates play as important a role as the water/binder ratio in terms of modulus of elasticity, rendering the universal formulae for prediction of the elastic modulus from compressive strength inapplicable, unless a correction factor taking into account the nature of the aggregate is introduced into these formulae.

A number of researchers have proposed a variety of models of relative complexity to represent concrete elastic behaviour and to predict concrete elastic modulus from that of the cement paste and/or the mortars and the coarse aggregate. Unfortunately, none of the models described in the literature is able to predict accurately the elastic modulus of high-performance concrete which contains aggregates different from those commonly used in producing usual concrete. Therefore specific relationships have to be developed for almost every high-performance concrete due to local variations in the aggregates and cementitious materials used. However, two models recently proposed by Baalbaki seem to be very promising in predicting the elastic modulus of a high-performance concrete when knowing either the elastic modulus of the paste, of the mortar and of the coarse aggregate, or the compressive strength of a particular high-performance concrete.

The durability of high-performance concrete is discussed in Chapter 17. This is of course an issue of concern to designers and owners who certainly are interested not only in high compressive strength but also in long life cycle. Most types of distress from which usual concretes suffer are reviewed. It is shown how high-performance concrete behaves when facing the same kind of aggressive environments. For example, high-performance concrete is so impervious that it is expected not to undergo surface carbonation. This topic, however, has received little attention so far in terms of research.

Controversy reigns at the moment concerning the factors governing the freeze–thaw resistance of high-performance concrete. Some researchers claim that high-performance concrete does not have good freeze–thaw resistance unless air-entrained, others maintain the opposite. We now know that even high compressive strength ($>80\,\mathrm{MPa}$) is not a reliable criterion for judging internal microcracking due to repeated cycles of freezing and thawing when Procedure A (freezing and thawing in water) of the ASTM C666 rapid freeze–thaw testing method is used. But the question is: is this testing method applicable and valid for high-performance concrete?

It has also been found that the 230 μm maximum specification for the spacing factor to guarantee the freeze–thaw durability and scaling resistance of usual concrete found in the Canadian code does not always apply to all high-performance concretes and that there is instead a threshold value higher than that for different high-performance concretes.

The problem is relatively complex because high-performance concrete can be made from a great variety of materials in different proportions which, as would be expected, can result in very different physico-chemical characteristics. While the freeze–thaw resistance of non-air-entrained high-performance concrete is not always excellent, it is fairly easy to produce air-entrained high-performance concrete in the laboratory and in the field with outstanding freeze–thaw resistance.

Resistance to frost action in high-performance concrete depends both on the water/binder ratio and on the type of binder (cement and mineral addition) used. These factors directly influence the permeability and porosity of the paste, the two main parameters controlling the concrete's ability to resist the internal cracking that can be initiated by freeze/thaw cycles. The results from a number of laboratory experiments indicate that simply reducing the water/binder ratio substantially, below 0.30, is not enough: it is also necessary to achieve the finest, lowest capillary porosity possible so that no freezable water, or at least very little of it, is present in the high-performance concrete. Should the capillary porosity remain too great, the paste will contain too much freezable water, which will rapidly result in cracking.

Some non-air-entrained high-performance concretes made with high supplementary cementitious material contents (fly ash, slag or silica fume) or with relatively high water/binder ratios (greater than 0.30), display a poorer resistance to freezing cycles despite their often high compressive strengths of over 80 MPa. On the other hand, some laboratory tests have shown that non-air-entrained, high-performance concrete (80 to 100 MPa) made with a high early strength cement, 6% silica fume and a high-quality limestone aggregate can easily withstand more than 1000 freezing cycles even after its demolding 24 hours after being cast. The autogenous shrinkage in such concretes is so high during the first 24 hours that almost all the big capillary pores that could contain freezable water are dried out.

The scaling resistance of high-performance concrete has received much less attention than its resistance to internal cracking (freezing/thawing in water). This is surprising since, in most countries, attack by deicing salts causes the greatest amount of damage to concrete structures exposed to freezing. In the few documented studies of this aspect of durability it has been shown that the high-performance concrete tested performed well and without any need for air entrainment.

High-performance concrete is so compact and hard that it exhibits a high resistance to abrasion. It should be mentioned, however, that the aggregate abrasion resistance has the greatest influence on the concrete abrasion resistance. Laboratory and field results have already shown that high-performance concrete could be as resistant to abrasion and wear as the best granites.

Chapter 17 also discusses another crucial issue for concrete durability, i.e. alkali–aggregate reaction. Very little work has been published in this field for high-performance concrete; it is usually assumed that, as there is little or no free water within high-performance concrete, the alkali–aggregate reaction cannot develop, as one of the three conditions necessary for the development of such a reaction is missing. However, very recently it has been found that, owing to the substantial rise in the early temperature of concrete cast in certain massive elements, some aggregates can become reactive and develop reaction rims. Therefore, the present rules that apply to selecting unreactive aggregates for usual concrete should also be used when selecting aggregates for high-performance concrete.

As far as fire resistance is concerned, owing to a lack of serious and exhaustive studies, we have to live with some contradictory experimental results. Unless such studies are carried out, we may have to wait for a large conflagration to occur in a concrete structure built with a high-performance concrete in order to obtain a reliable answer. As I was writing this section, such a fire occurred in the 'Chunnel' (connecting England and France under the English Channel), where high-performance concrete was used to build the precast lining elements of the Chunnel in the French section. Therefore we should learn some very interesting things. It seems, however, that the introduction of some polypropylene fibres in high-performance concrete drastically improves the fire resistance of high-performance concrete. When melting or burning, the polypropylene fibres create small channels in the high-performance concrete through which water vapour can be released, avoiding the creation of internal pressure which results in high-performance concrete spalling.

In Chapter 18 some special high-performance concretes are presented, but not all of them. In spite of the fact that high-performance concrete is a new type of concrete, it has already developed into different categories which are different enough from the most commonly used high-performance concretes at the present time that they have to be presented separately. All of these special high-performance concretes exhibit one or several characteristics and/or properties that justify the term 'special' used to characterize them. This is very hopeful, because it exemplifies one of the best qualities of concrete in general: its adaptability and flexibility in particular situations. It is absolutely certain that this trend will expand in the future and that new types of special high-performance concrete will be developed to fulfil the special needs of a designer, of a contractor or of an owner.

In Chapter 18 a paragraph is specifically devoted to air-entrained high-performance concrete, in spite of the fact that the creation of air-entrained high-performance concrete and its properties have been described in several other chapters. I want the reader to be able to find most of our present knowledge of this important matter in a single location rather than have to discover this information throughout the book.

Air entrainment does not only make high-performance concrete more durable to freezing and thawing, it also greatly improves the rheology of very low water/binder ratio concrete, especially when silica fume is used. It is not a mistake to introduce a limited volume of small air bubbles into a high-performance concrete, because the strength decrease (1% air increase decreases compressive strength by 5%), can easily be overcome by reducing slightly the water/binder ratio, while the improvements to the workability, placeability and finishability are difficult and costly to achieve. It is my opinion that in the future more and more very low water/binder ratio concretes, or high-performance concretes in which, for economic reasons, it is not possible to use any supplementary materials, will have a total air content of 3.0 to 4.5% to improve their rheology.

In spite of their limited use, heavyweight high-performance concretes are briefly mentioned. This is not the case for lightweight high-performance concretes, which are finding more and more economical uses. In some cases the use of saturated lightweight aggregate is promoted to fight self-desiccation. The saturated porous lightweight aggregates constitute not only a source of external water for the hydrated cement paste, but also a source that has the great advantage of being well distributed throughout the total volume of the concrete. In spite of the fact that the use of a weak aggregate decreases both compressive strength and elastic modulus (to a greater extent), having a lighter concrete and a well-distributed external source of water for the hydrating cement paste are largely advantageous.

Some fibre-reinforced high-performance concretes have already found some niche applications in repairs, or in the construction of overlays. As already mentioned in Chapter 17, some researchers think that the addition of polypropylene fibres to high-performance concretes could make these more fire resistant.

A brief mention is made of confined high-performance concrete, but only to remind designers that this is a very simple solution to increase further the compressive strength of high-performance concrete, though not its elastic modulus. Several high-rise buildings have been built taking advantage of this following the construction of the Two Union Square high-rise building in Seattle, which is presented in Chapter 4.

Finally, a very recent development in the placing of high-performance concrete that is developing in Québec is presented, concerning roller-compacted high-performance concrete. This technique has been successfully and economically used in several major industrial projects in metallurgical and pulp and paper plants. It permits the placing of large quantities of high-performance concrete in a very short period of time using asphalt paving technology and machinery. In one particular project, 78 000 m² of a 300 mm slab, the equivalent of 16 adjacent football or soccer fields, was completed by a limited crew within one and a half months with no reinforcement and no joints. This represents a tremendous field of application for high-performance concrete.

Chapter 19 shows that in spite of its infancy, high-performance concrete is already challenged by a new type of concrete, even more resistant and more ductile: the reactive powder concrete developed by Pierre Richard. In this new type of concrete the coarser aggregate has a maximum grain size of 300 to 600 μm, and is composed of several well-graded powders playing different roles during the strengthening of the hydrated cement paste. The mechanical properties of reactive powder concrete can be improved by some very simple heat treatment, by the introduction of steel fibres or by its confinement in a thin steel tube. In spite of the fact that it is too early to see all the potential of this new type of concrete, it is now realistic to dream of a 1000 MPa concrete.

Chapter 20 has been entitled 'A Look Ahead'. I do not pretend to be a good fortune teller or particularly skilled in drawing cards or reading tea leaves, but I could not resist expressing my own views and to make predictions on the future of high-performance concrete. Making a wrong prediction for the future is not, after all, as bad as making a wrong prediction in an already well-known and well-developed field. In any event, I am ready to take credit for my correct predictions, just as I am willing to bear the shame of my wrong ones.

Finally, the book ends with a list of monographs, state-of-the-art papers and proceedings of international conferences, whose reading should complement the reading of this book. In spite of the fact that many of the papers that can be found in these pieces of literature are cited in the different chapters of the book, I do not pretend to have given all of them a fair treatment or treated them deeply enough to satisfy the special needs of readers interested in very specific points. This book on high-performance concrete science and technology is intended to present a synthesis view of the principal knowledge that I presently have on high-performance concrete. It is not intended to be a full and detailed treatise on high-performance concrete.

REFERENCE

Féret, R. (1892) Sur le compacité des mortiers hydrauliques. *Annales des Ponts et Chaussées*, Vol. 4, 2nd semester, pp. 5–161.

CHAPTER 3

A historical perspective

3.1 PRECURSORS AND PIONEERS

It must have been due to a surfeit of the 'pioneering spirit' for a small group of designers and concrete producers to have the will to launch high-strength concrete in the mid-1960s. Why innovate in concrete technology by increasing compressive strength? At that time, most designers were satisfied to design structures based on the 15 to 20 MPa concretes that were well understood, economical and safe. Similarly, concrete producers were making enough money by selling their concrete in 'horizontal-type' structures. It was not obvious to most engineers that concrete would one day displace steel in the construction of high-rise buildings. The conventional wisdom at that time was that concrete was only good for use in high-rise buildings for foundations and the construction of floors or to protect structural steel elements against fire.

However, as in all fields of human endeavour, there are always a few individuals who are not frightened by traditional taboos and are willing to innovate. This is what happened in the early 1960s in the Chicago area, where high-strength concrete first started to be used in significant quantities in major structures (Freedman, 1971).

This development was made possible in Chicago because, at that time, a courageous designer and an innovative concrete producer had just started to work together (Moreno, 1987). Even though the strength of the first high-strength concretes they used seems quite modest by current standards, it should be remembered that, at that time, the usual concrete used by the building industry had a compressive strength of only 15 to 30 MPa. Proposing to double this compressive strength overnight was a real challenge. It should also be remembered that the cements and admixtures that were available at that time were not necessarily as suitable for the manufacture of high-strength concrete as some of those that are available today (Perenchio, 1973).

Most commercial cements were ground much coarser than at present, and the commercial water reducers used at that time were mostly lignosulfonate-based. These lignosulfonates varied considerably in their composition and 'purity', which also led to significant variability in performance characteristics. Moreover, lignosulfonate-based water reducers had a strong tendency to entrap or entrain air, as well as to retard set, when used in high dosages. Moreover, it should also be remembered that in the 1960s, the ready-mix concrete industry in North America was only just starting to use fly ash in significant amounts. Fly ash producers had just begun to ensure the quality and consistency of their product, and concrete producers had not yet realized the full economic potential of good quality fly ash.

It is in such a context that the first high-strength concrete was developed. In order to give this high-strength concrete a chance to demonstrate its value, designers and concrete producers had to figure out a way to convince owners to let them use such a new material, of which they had no prior experience and very little data. The approach they decided to follow was a clever one (Albinger and Moreno, 1991). Whenever a high-rise building was being planned, the concrete producer asked the permission of the owner to include (at no extra cost, of course) one or two columns made from experimental concretes having compressive strengths 10 to 15 MPa above those the designer had already selected for the main columns. What was the risk in accepting one or two stronger columns among so many others? Once it had been proven that it was possible to make, deliver and place a concrete having a compressive strength of 10 to 15 MPa over what was usually delivered, without any complaint from the contractor and his placing crew and without any structural catastrophe, it was simple to see the use of such a concrete in the construction of the next building that was already on the designer's desk. It was easy to take this step forward, because increasing the compressive strength of the concrete resulted in cost savings, in terms of the increased space on office and parking floors made available for rent due to the reduction in column size, especially in the lower floors which are usually the most profitable.

Of course, during the construction of this new building the same trick was repeated: the concrete producer again proposed to include a new experimental column, at no extra cost, with a concrete having 10 to 15 MPa higher compressive strength than the design strength required for this high-rise building. This time, the new experimental concrete had in fact a compressive strength 20 to 30 MPa higher than the regular concrete that was delivered by other ready-mix companies. In order not to discourage the owner by pointing out the innovative character of this concrete, the concrete producer referred to the relatively small strength increase between this new concrete and the one that was selected by the designer.

By using such a stratagem, the maximum compressive strength of the concrete used in high-rise buildings in the Chicago area was multiplied by three, step by step, slowly and progressively, over a period of about 10 years, raising compressive strength from 15–20 MPa to 45–60 MPa (Figure 3.1).

However, concrete compressive strengths stopped increasing at about 60 MPa because a technological barrier was reached that could not be surmounted with the available raw materials. In the early 1970s it was

Lake Point Tower 1965
(f'_c = 53 MPa)

River Plaza 1976
(f'_c = 77 MPa)

Fig. 3.1 Examples of high-rise buildings built in the 1960s and 1970s in the Chicago area.

impossible to make concrete with a compressive strength in excess of about 60 MPa because the water reducers that were normally available at that time were not capable of reducing the water/binder (W/B) ratio any further (Blick, Petersen and Winter, 1974).

In fact, the main factor in bringing about the large compressive strength increase that had already been achieved was a decrease in the W/B ratio, made possible by the appropriate use of efficient water reducers. Moreover, in the search to decrease the W/B ratio to the range of 0.35 to 0.40, it appeared that the selection of the cement was critical. Such a cement had to perform well not only from a mechanical point of view but also from a rheological point of view. In other words, it had to give rise to a relatively small slump loss during the first hour after mixing.

The first cements used to make high-strength concrete were Type I, Type II or preferably the so-called 'modified Type II' cements. The details of how to make a 'modified Type II' cement, which had a low rheological reactivity but a good strength gain (like a Type I) were a commercial secret for many cement and concrete producers. Moreover, in order to minimize the slump loss problem, concrete producers were substituting a certain quantity of high-quality fly ash for Portland cement. This reduced water demand and slump loss, in many cases allowing the W/B ratio to be decreased more than enough to compensate for the lower early strengths caused by the cement substitution.

In order to decrease the W/B ratio, the water reducer dosage was increased, although it was only slightly higher than that usually recommended for concrete having a compressive strength of 20 to 30 MPa. The dosage could not be increased much further because it caused set retardation and entrapped an excessive quantity of air bubbles, which resulted in lower strengths. It was not advantageous from the contractor's point of view to retard the setting of the concrete too much, especially since 20 to 25% of the cement was already replaced by fly ash, which always resulted in a lower-than-normal rate of strength gain even in the absence of water reducers (Blick, Petersen and Winter, 1974).

In fact, a reasonable rate of gain of compressive strength at early ages is required by the contractor, so that construction can progress rapidly. If the use of long-term high-performance concrete results in low short-term compressive strengths, then it is likely to be rejected by the contractor owing to his understandable preference for a concrete that may be weaker at 28 days but which can have the formwork removed earlier. In the same way, the entrainment of too much air should be avoided because the presence of air bubbles in concrete reduces its ultimate compressive strength. Air entrainment was not necessary in these first applications of high-strength concrete, because the freezing and thawing resistance of these concretes was of no concern. Such concretes were only used for interior applications, in which the risk of freezing in a saturated state was negligible. At that time the only air-entrained high-performance concretes were those used in the splash zone in offshore structures (Ronneberg and Sandvik, 1990).

Another secret known to the producers of high-performance concrete by that time was that it was necessary to find, among the available water reducers, the one that showed the least of these undesirable secondary effects. This allowed them to increase its dosage as much as possible before too much set retardation occurred or too much air was entrained. Moreover, in order to lower the W/B ratio as much as possible, concretes were delivered with a fairly low slump, typically 75 to 100 mm (Blick, Petersen and Winter, 1974).

Finally, the sampling of high-performance concrete was done with particular care (for example, using steel moulds), in order to prevent the loss of even a single MPa of potential compressive strength as a result of the sampling technique (Freedman, 1971; Blick, Petersen and Winter, 1974).

This was the state of development of high-performance concrete technology in the early 1970s, when superplasticizers were first introduced to the concrete market (Chicago Committee on High-Rise Buildings, 1977; Ronneberg and Sandvik, 1990).

3.2 FROM WATER REDUCERS TO SUPERPLASTICIZERS

It was in the late 1960s that superplasticizers were first used in concrete, their introduction occurring almost simultaneously in both Japan and Germany (Hattori, 1981; Meyer, 1981). It is surprising that the concrete industry started to use superplasticizers so late, since the first US patent covering the fabrication and utilization of water reducers based on polycondensates of naphthalene sulfonate was obtained in 1938 (Tucker, 1932). However, at that time lignosulfonates were so cheap and their level of performance was considered so satisfactory for 15 to 25 MPa concrete that it was not necessary to look for water reducers with higher efficiency and much higher cost.

It should be mentioned that the first applications of superplasticizers were as fluidifiers, rather than as water reducers. They were used on the construction site to fluidify concretes that very often already contained a lignosulfonate-based water reducer introduced during the initial mixing at the ready-mix plant. The original reason for using superplasticizers in this way was to facilitate the placing of the concrete without the risk of segregation and strength loss which occurs when concrete is retempered with water. However, because the period during which these first-generation superplasticizers effectively fluidified concrete was limited, superplasticizers were usually added in the field just before placing of the concrete. From a practical point of view, it should also be mentioned that it can be dangerous to transport a very fluid concrete in a ready-mix truck, not only because of the possibility of segregation but also for safety considerations. It can be difficult to control a truck when driving with a liquid mass of 15 to 20 tons in the mixer, and there is a high risk of concrete spilling out of the truck when braking. In order to reduce these dangers, the trucks have to carry a reduced volume if the concrete is very fluid.

During the 1980s, by increasing little by little the superplasticizer dosages over the range usually recommended by manufacturers, it began to be realized that superplasticizers could be used as high-range water reducers. In that respect, they were much more powerful than lignosulfonates and could be used at much higher dosages before any significant retardation could occur and before too much air would be entrained in the concrete (Ronneberg and Sandvik, 1990).

As increased dosages of superplasticizers began to be used for concretes having lower and lower W/B ratios, the slump loss problem became more and more important. By using large enough dosages of superplasticizer it was found possible to lower the W/B ratio of concrete down to 0.30 and still get an initial slump of 200 mm. This was a great advance when compared with what was possible with lignosulfonate-based water reducers, with which concrete had to be delivered at only 75 to 100 mm slump (Albinger and Moreno, 1991; Aïtcin, 1992). However, slump loss problems were exacerbated by the use of such low W/B ratios. In order to try to overcome this problem, superplasticizer manufacturers developed, with more or less success, superplasticizer formulations incorporating retarders.

It is interesting to note that, when superplasticizers first started to be used as high-range water reducers, they were nevertheless used in such a way that the W/B ratio was never reduced below the 'psychological barrier' of about 0.30. This value was thought to be the minimum W/B ratio for adequate hydration of the Portland cement used in the mix. Reducing the W/B ratio below this was taboo until H.H. Bache reported that by lowering the W/B ratio of a particular micro-concrete down to 0.16, using a very high dosage of superplasticizer and a new ultrafine cement substitute (silica fume), he had been able to reach a compressive strength of 280 MPa (Bache, 1981). Of course, Bache's results were obtained in a laboratory and required special curing procedures, but nevertheless these results shocked the concrete community, which still had trouble in regularly achieving 25 to 30 MPa in the field.

Bache's micro-concrete had essentially no potential for use by the ready-mix industry because it involved the use of calcined bauxite as aggregate (calcined bauxite can cost more than $1000/t), and it also required very heavy external vibration for compaction and a very special curing technique. However, the successful development of such a material proved that the psychological W/B ratio barrier could be overcome. In fact, Bache's work showed that the ultimate compressive strength of concrete depends not only on the quality, quantity and efficiency of the cementitious materials used, but also on the degree of compaction and ultimate porosity of the solid matrix formed after the hardening process is completed. About 100 years after first being proposed (Féret, 1892), the W/B law concerning compressive strength can be said to have been extended!

After this, researchers and concrete producers started to lower W/B ratios below 0.30 into the range where there was not enough water introduced during the mixing to hydrate all of the cement in the mix fully (Aïtcin, Laplante and Bédard, 1985; Moreno, 1987, 1990). This meant that the cement in such concretes could never hydrate completely. Only the finer cement grains became fully hydrated, the interior of the coarser ones and the grains richest in C_2S, mainly played the role of a filler (Aïtcin, Regourd and Bédard, 1983).

It has been found from experience that, at such low W/B ratios, some commercial Portland cements start to fall short in terms of compressive strength, and even among the ones that still exhibit good compressive strengths, some nevertheless present serious slump loss problems. Often, these problems were already noticeable (but to a lesser degree) in the 0.30 to 0.35 W/B ratio range.

However, by careful selection of the cement and the superplasticizers, it was possible to decrease the W/B ratio to 0.30, then to 0.27, then to 0.25 and finally, recently, down to 0.23 to obtain a compressive strength of

130 MPa (Godfrey, 1987). In the very near future, high-performance concretes having W/B ratios of around 0.20 could be delivered in the field if they prove to be useful in terms of performance and economics. By using a very carefully chosen combination of Portland cement and superplasticizer, it has been possible at the Université de Sherbrooke to make a 0.17 W/B ratio concrete with a 230 mm slump 1 hour after its mixing. Such a concrete gave a compressive strength of 73.1 MPa at 24 hours but failed to increase its strength to more than 125 MPa after long-term wet curing (Aïtcin et al., 1991). This relatively low ultimate strength (considering the low W/B ratio) may be ascribed to an inadequate water content in terms of hydration, but the true reason is probably more complex than this.

3.3 THE ARRIVAL OF SILICA FUME

It was mentioned above that Bache used silica fume in order to be able to make very low W/B ratio concretes. The ready-mix industry soon took advantage of the arrival of this new cementitious material, once it began to be supplied in a usable form and at a price that was acceptable.

Although the first concrete utilization of this by-product of the fabrication of silicon or ferrosilicon was reported in 1952 by a Norwegian researcher (Bernhardt, 1952), it was only in the late 1970s that silica fume started to be used as a supplementary cementitious material in concrete in Scandinavia. It was not until the early 1980s that it began to be used in this way in North America (Aïtcin, 1983; Malhotra et al., 1987).

For a long time, silicon and ferrosilicon producers had allowed the by-product silica fume to escape into the atmosphere. They only started to collect it after being subjected to strong environmental regulations from their governments. They then had to find ways to minimize or eliminate the emission of these very fine dust particles. So, in a number of industrial countries, silicon and ferrosilicon producers were forced to invest a great deal of money in sophisticated flue-gas dust collection systems in order to recover this by-product for which there was apparently no market. Moreover, the problem of handling such a fine powder was so difficult that even the most optimistic silicon producer could not at that time have conceived of the idea that one day he could take advantage of such a nuisance dust!

The first interesting results obtained by the Scandinavians in usual concrete, the very impressive discoveries of Bache and co-workers in Denmark and a significant research effort in the early 1980s in several other countries, resulted in the rapid acceptance of silica fume as a supplementary cementitious material for concrete almost everywhere in the world in less than 5 years.

The particular advantage of using silica fume as a very fine and reactive pozzolan for use in high-performance concrete was rapidly recognized (Aïtcin, 1986). In fact, this recognition occurred to such an extent that many people now believe that the use of silica fume is the *sine qua non* of high-performance concrete. This is only partially true, but it will be seen in the following chapters that silica fume is really a very advantageous supplementary cementitious material from which beneficial action is developed at several different periods in the life of concrete. By using silica fume, it has been shown that it is possible to make workable concretes with compressive strengths in the 100 to 150 MPa range (Detwiler, 1991).

3.4 PRESENT STATUS

High-performance concrete acceptance and use is growing slowly in many countries; indeed, its use still represents a very low percentage of the concrete market. However, several countries launched major specific research programmes on high-performance concrete in the late 1980s, among which are the USA (Hoff, 1993), Norway (Holand, 1993), Canada (Aïtcin and Baalbaki, 1996), France (Malier, 1992; de Larrard, 1993) Switzerland (Alou, Charif and Jaccoud, 1988), Australia (Burnett, 1989; Potter and Guirguis, 1993), Germany (König, 1993), Japan (Aoyama *et al.*, 1990), Korea (Sung-Woo Shin, 1990), China (Zhu Jinquam and Hu Qingchang, 1993) and Taiwan (Chern, Hwang and Tsai, 1995). With the present information available from seminars, symposia, short courses and articles published in various journals, it is no longer a real challenge to make and use high-performance concrete. What is difficult to change is attitudes.

It is the author's hope that this book will contribute to making high-performance concrete a less mysterious material, which is easy to make, safe to use and can result in more durable and appealing structures. High-performance concrete is not a 'super material', without any weaknesses and drawbacks. But as we learn more through its growing use, research, discoveries, successful experiments and, of course, failures, it will be more effectively utilized, to the advantage of us all. High-performance concrete makes concrete a better performing material, allowing designers to use it efficiently in increasingly slender structures. Architects prefer using high-performance concrete in high-rise construction in order to design thinner floor slabs and slimmer columns, in addition to it being aesthetically more appealing. Small-diameter concrete columns in high-rise construction translate into more rental space, and hence into more income for the owner. Some contractors favour the use of high-performance concrete because they can remove the formwork earlier. In addition to reducing creep and shrinkage, increasing the use of high-performance concrete in high-rise construction increases the stiffness of the structure. As a result, deflections of concrete members are reduced. Whenever high-performance concrete is used for columns in high-rise buildings, the lateral stiffness of the building is increased, thus reducing lateral sway caused by wind loading and increasing the occupant comfort level. In addition to producing more slender structures, the use of high-performance concrete in high-rise construction can also decrease the amount of steel, and thus the dead load.

Therefore structures may be built on soils with marginal load-carrying capacities. Bridge piers and decks can be designed with high-performance concrete to construct structural elements that are slender and elegant and that can be integrated harmoniously into the landscape.

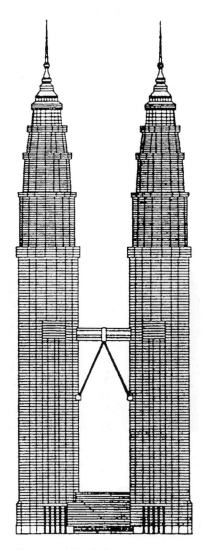

Fig. 3.2 The Petronas Towers of Kuala Lumpur – the highest concrete building in the world (451 m).

Present status

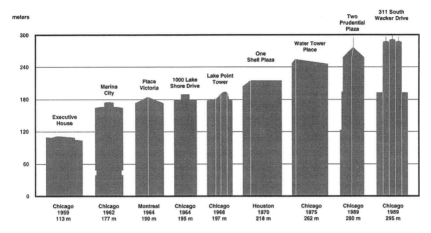

Fig. 3.3 Concrete in high-rise construction (from Concrete Reinforcing Steel Institute, *Bulletin* No. 4, 1990).

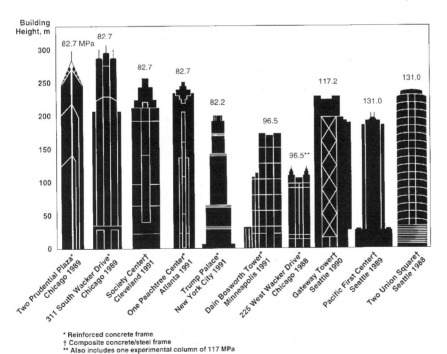

Fig. 3.4 High-strength concrete shapes, new US skylines (from *High-Strength Concrete*, EB114.01T, Portland Cement Association, 1994).

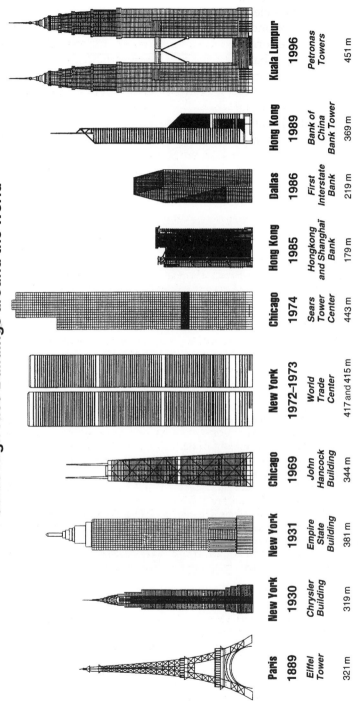

Fig. 3.5 Some high-rise buildings around the world, after *Le Monde* (Edelmann, 1996).

The benefits of using high-performance concrete in the construction of high-rise buildings is clearly illustrated in Figures 3.2 to 3.5, which show the increasing height and strength of concrete high-rise buildings around the world. High-performance concrete retains the versatility of usual concrete. However, it has reached the strength and durability of natural rock, but a rock that can easily be shaped because it remains fluid long enough to form elaborate and complicated shapes, a rock that can be easily reinforced with steel bars, prestressed or post-tensioned with cables, or mixed with all kinds of fibres.

This material exists, and it is the responsibility of engineers and the construction industry to use it as efficiently as possible.

REFERENCES

Aïtcin, P.-C. (1983) *Condensed Silica Fume*, Éditions de L'Université de Sherbrooke, Sherbrooke, Québec, Canada, ISBN 2-7622-0016-4, 52 pp.

Aïtcin, P.-C. (1986) New high-tech, ready-mix concretes. *Engineering Digest*, **32**(10), November, 32–3.

Aïtcin, P.-C. (1992) The use of superplasticizers in high-performance concrete, in *High-Performance Concrete: from Material to Structure* (ed. Y. Malier), E and FN Spon, London, ISBN 0-419-17600-4, pp. 14–33.

Aïtcin, P-C. and Baalbaki, M. (1996) *Canadian Experience in Producing and Testing HPC*, ACI SP-159, pp. 295–308.

Aïtcin, P.-C., Laplante, P. and Bédard, C. (1985) *Development and Experimental Use of 90 MPa (13,000 psi) Field Concrete*, ACI SP-87, pp. 51–70.

Aïtcin, P.C., Regourd, M. and Bédard, C. (1983) *Microstructural Study of a 135 MPa Ready Mix Concrete*, 5th International Conference on Cement and Concrete Microscopy, Nashville, pp. 164–79.

Aïtcin, P.-C., Sarkar, S.L., Ranc, R. and Lévy, C. (1991) A high silica modulus cement for high performance concrete, in *Ceramic Transactions – Advances in Cementitious Materials* Vol. 16, (ed. by S. Mindess), American Ceramic Society, Westerville, OH, USA, ISBN 0-944904-33-5, pp. 103–20.

Albinger, J. and Moreno, J. (1991) High-strength concrete: Chicago style. *Concrete Construction*, **26**(3), March, 241–5.

Alou, F., Charif, H. and Jaccoud, J.-P. (1988) Bétons à hautes performances. *Chantiers/Suisse*, **19**, Septembre, 725–30.

Aoyama, H., Murato, T., Hiraishi, H. and Bessho, S. (1990) *Outline of the Japanese National Project on Advanced Reinforced Concrete Buildings with High-Strength and High-Quality Materials*, ACI SP-121, pp. 21–31.

Bache, H.H. (1981) *Densified Cement/Ultra Fine Particle-Based Materials*, Second International Conference on Superplasticizer in Concrete, Ottawa, June, pp. 1–35.

Bernhardt, C.J. (1952) SiO_2 – Dust as admixture to cement. *Betongen Idag*, April, pp. 29–53.

Blick, R.L., Petersen, C.F. and Winter, M.E. (1974) *Proportioning and Controlling High-Strength Concrete*, ACI SP-46, pp. 141–63.

Burnett, I. (1989) High-strength concrete in Melbourne, Australia. *Concrete International*, **11**(4), April, 17–25.

Chern, J.C., Hwang, C.L. and Tsai, T.H. (1995) Research and development of HPC in Taiwan. *Concrete International*, **17**(10), October, 71–7.

Chicago Committee on High-Rise Buildings (1977) *High-Strength Concrete in Chicago High-Rise Buildings*, Report No. 5, 45 pp.

Concrete Reinforcing Steel Institute (1990) Case History Report, *Bulletin* No. 4, 933 North Plum Grove Road, Schaumburg, IL 60173–4758.

de Larrard (1993), A survey of recent research performed in French, LPC Network on High-Performance Concrete, *Utilization of High-Strength Concrete* (eds I. Holland and E. Sellevold), Norwegian Concrete Association, Oslo, Norway, ISBN 82-91341-00-1, pp. 57–67.

Detwiler, G. (1991) High-strength silica fume concrete – Chicago style. *Concrete International*, **14**(10), October, 32–6.

Edelmann, F. (1996) Les plus hauts gratte-ciel du monde s'achèvent à Kuala Lumpur. *Le Monde*, 6 August, p. 22.

Féret, R. (1892) Sur la compacité des mortiers hydrauliques. *Annales des Ponts et Chaussées*, Vol. 4, 2nd semester, pp. 5–161.

Freedman, S. (1971) *High-Strength Concrete*, IS1 76–OIT, Portland Cement Association, Skokie, IL, 19 pp.

Godfrey, K.A. Jr (1987) Concrete strength record jumps 36%. *Civil Engineering*, **57**(10), October, 84–6.

Hattori, K. (1981) *Experiences with Mighty Superplasticizer in Japan*, ACI SP-62, pp. 37–66.

High-strength Concrete (1994), E894.OIT, Portland Cement Association, Skokie, IL.

Hoff, G. (1993) Utilization of high-strength concrete in North America, *Utilization of High-Strength Concrete* (eds I. Holland and E. Sellevold), Norwegian Concrete Association, Oslo, Norway, ISBN 82-91341-00-1, pp. 28–36.

Holand, I. (1993) High-strength concrete in Norway – Utilization and Research, *Utilization of High-Strength Concrete* (eds I. Holland and E. Sellevold), Norwegian Concrete Association, Oslo, Norway, ISBN 82-91341-00-1, pp. 68–79.

König, G. (1993) Utilization of High-strength concrete in Germany, *Utilization of High-Strength Concrete* (eds I. Holland and E. Sellevold), Norwegian Concrete Association, Oslo, Norway, ISBN 82-91341-00-1, pp. 45–56.

Malhotra, M., Ramachandran, R., Feldman, R. and Aïtcin, P.-C. (1987) *Condensed Silica Fume in Concrete*, CRC Press, Boca Raton, Florida, ISBN 0-8493-5657-1, 221 pp.

Malier, Y. (ed.) (1992) *High-Performance Concrete – From Material to Structure*, E and FN Spon, London, ISBN 0-419-17600-4, 542 pp.

Meyer, A. (1981) *Experiences with the Use of Superplasticizers in Germany*, ACI SP-62, pp. 21–36.

Moreno, J. (1987) High-strength concrete in Chicago high-rise buildings, in *Concrete Structures for the Future*, Proceedings of IABSE Symposium, Versailles, Paris, IABSE-AIPC-IVBH, Zurich, ISBN 3-85748-053-1, pp. 407–12.

Moreno, J. (1990) 225 W. Wacker Drive. *Concrete International*, **12**(1), January, 35–9.

Perenchio, W.F. (1973) *An Evaluation of Some of the Factors Involved in Producing Very High Strength Concrete*, Portland Cement Association *Research and Development Bulletin* RD 014, 7 pp.

Potter, R.J. and Guirguis, S. (1993) 'High-strength concrete in Australia', *Utilization of High-Strength Concrete* (eds I. Holland and E. Sellevold), Norwegian Concrete Association, Oslo, Norway, ISBN 82-91341-00-1, pp. 581–9.

Ronneberg, H. and Sandvik, M. (1990) High-strength concrete for North Sea platforms. *Concrete International*, **12**(1), January, 29–34.

Sung-Woo Shin (1990) High-strength concrete in Korea. *Engineered Concrete Structures*, **3**(2), August, 3–4, Portland Cement Association, Skokie, IL.

Tucker, G.R (1932) *Concrete and Hydraulic Cement*, US patent 2, 141, 569. Application Wednesday 21 November, Serial No. 643,740. Patent granted in 1938.

Zhu Jinquam and Hu Qingchang (1993) High strength concrete in China. *Engineered Concrete Structures*, **6**(2), August, 1–3, Portland Cement Association, Skokie, IL.

CHAPTER 4

The high-performance concrete rationale

4.1 INTRODUCTION

Instead of a comparison of high-performance concrete with usual concrete, I have chosen broadly to describe a number of selected international projects involving the successful use of high-performance concrete. Each case study discusses the rationale for selecting high-performance concrete over other structural materials.

After building code restrictions have been taken into account, the final selection of a particular structural material usually hinges on economic factors. In some rare instances, a single technological characteristic may govern the decision. More often, several specific technical advantages are involved. While one can probably say that weighing the pros and cons ultimately decides the issue according to structural type, function, location, local economic considerations, and so on, the decision is not always based on the same criteria.

Eleven world-wide projects involving high-performance concrete, which are familiar to the author, were selected. These projects are used to illustrate different factors, including compressive strength, which have influenced the use of high-performance concrete for economic reasons. While it is evident that a sample of this size cannot claim adequately to cover this subject (CEB, 1994), these selected cases provide an insight as to why owners, designers, contractors and concrete producers have found their own particular reasons to use high-performance concrete.

In order to use a given structural material to its best economic advantage, it is important to understand its properties fully. According to Professor Jorg Schlaich (1987): 'One cannot design with and work with a material which one does not know and understand thoroughly. Therefore, design quality starts with education'.

The aim of this book is to familiarize concrete producers, contractors, designers and owners with the properties of high-performance concrete. Such knowledge can enable them to benefit from the advantages of using this new material, as well as make them aware of its limitations.

Although various characteristics of high-performance concrete can offer several advantages over those of usual concrete, there is still room for improvement. High-performance concrete, like all other materials, has shortcomings. Therefore, there will continue to be plenty of uses for usual concrete with compressive strengths of 20 to 30 MPa for a variety of applications. The development of high-performance concrete will not make the use of usual concrete obsolete.

4.2 FOR THE OWNER

The owner's ultimate objective is to get the greatest possible return on investment during the lifetime of the construction. While a dollar figure for the return can be set in the private sector, less well-defined social benefits apply to governmental structures or some combinations of dollars and social benefits for parapublic agencies, crown corporations, and the like.

In most cases, structural materials are of little interest to the owner as long as they satisfy the functional requirements at an acceptable price. While some owners insist on specific cladding for reasons of appearance, the structural material, generally hidden from view, is a secondary concern at best.

Nevertheless, owner requirements can strongly influence the final choice of structural material. Increasing the compressive strength from 60 to 75 MPa on the TROLL offshore platform, which is designed for a water depth of 300 m, will result in a reduction of some 50 000 m^3 of concrete and a saving of US$77 million.

By stipulating that all tenants should enjoy the same comfort regardless of wind velocity, the owners of Seattle's Two Union Square indirectly tilted the scales towards a stiff material such as high-performance concrete, rather than steel. Steel-framed high-rise buildings sway too much in strong winds and the engineering solutions proposed to counteract sway tend to be costly and only partially effective.

By imposing a tight construction schedule for the Île de Ré bridge and Northumberland Strait Crossing Fixed Link, the owners were indirectly stipulating high-performance concrete since it was the only possible choice for speeding up casting of the 40 to 80 tonne prefabricated box girders in the case of Île de Ré bridge and the 7500 tonne one of the Northumberland Strait Crossing Fixed Link.

Other owner preferences or requirements that directly or indirectly bias the final selection of the structural material could be cited. In fact, while certain of them may indicate, at the outset, a variety of theoretically possible materials, a single economically practical solution often emerges.

Given the competition raging in the construction industry between steel, usual concrete and high-performance concrete, high-performance concrete solutions have prevailed when they have offered the owner the best return on investment under the circumstances.

4.3 FOR THE DESIGNER

At first glance, the designer would appear to have the final say in the selection of the construction materials. When you think about it, however, this decision must accommodate a variety of inputs. For example, the designer has to satisfy the owner's functional requirements and the architect's aesthetic requirements, taking into account the technological constraints imposed by the building codes. Sometimes there may be no choice at all. It is hard to imagine a designer in Pittsburgh, the steel capital of the USA, selecting anything but a steel structure for the industrial corporate office of a steel company. On the other hand, what cement company would house its corporate offices in a steel-frame building? Earthquake-prone areas may favour choices of ductile material such steel instead of concrete.

Such special cases aside, the designer's final determination is usually based on the technical and economical perception of the construction market in which the structure is to be built (Smith and Rad, 1989). In a certain sense, however, personal preference for one particular structural material can make a difference, because the design may be more efficient if the designer uses a material that he or she likes and knows.

In the field cases that follow, high-performance concrete was not always selected on the grounds of compressive strength. Exceptional elastic modulus, more rapid approach to ultimate creep, high durability, enhanced impermeability, or a combination of these factors came into play; even the high abrasion resistance of high-performance concrete results in its use in pavements in Norway, where studded tyres are used in winter.

In 1960, opting for high-strength concrete for Water Tower Place in Chicago resulted in a decrease in column cross-section for the lower storeys. This in turn significantly decreased the dead load of the building and increased the rental space. By incrementally decreasing concrete compressive strength in the upper storeys, the same prefabricated column forms could be used throughout the building, thereby reducing construction costs (Johnson, 1984). Concrete compressive strength decreased incrementally from 60 MPa at the ground floor to 30 MPa at the top of the building.

Choosing high-performance concrete for Two Union Square in Seattle was based more on high elastic modulus than high strength, even though these two properties are somewhat related. The high elastic modulus was required to increase the building rigidity to dampen swaying in high winds. According to Gordon (1991), 'The top of the Empire State Building sways about two feet (600 mm) during a storm'.

The French company Bouygues developed high-performance concrete-based innovative and economical designs for the Île de Ré bridge and the Sylans/Glacières viaduct. Construction costs were cut drastically by the on-site prefabrication and fast form stripping allowed by initial high early strength: the heavy prefabricated bridge segments (weighing from 40 to 80 tonnes in the case of Île de Ré) were stripped and handled within 12 hours.

High-performance concrete not only reduces the dead load of offshore platforms but also ensures outstanding durability, especially in the critical splash zone, where it is subjected to very severe exposure (this was the case for the Hibernia offshore platform in Canada).

Another interesting feature of high-performance concrete is its rapid attainment of the final level of creep in comparison with usual concrete, a key factor for the Île de Ré bridge and Sylans/Glacières viaduct. In fact, time-dependent deformation in a structural element can give rise to non-negligible differential stresses, requiring special countermeasures in the initial design. These provisions complicate the design, increase reinforcement (added material costs), boost labour costs (placing, steel inspection) and slow down placing of the concrete, and so on.

4.4 FOR THE CONTRACTOR

While contractors rarely play a significant role in selecting the structural materials for a particular structure, they are sometimes allowed to propose more economical alternative designs based on their own experience. For example, a contractor might convince an owner to use a higher strength concrete or could determine how much its use would save on the final total cost of the structure.

When Bouygues of France decided to cast the prefabricated box girders for the Île de Ré bridge with 70 MPa concrete instead of the design 40 MPa, the unit-cost increase per cubic metre of concrete was small compared with the savings resulting from faster casting, the same situation as the design of the Northumberland Strait Crossing Fixed Link.

4.5 FOR THE CONCRETE PRODUCER

Concrete producers have long been used to selling concrete for horizontal-type structures, leaving the vertical market to steel. Since producing and delivering 30 MPa concrete requires no special skills or high-technological quality-control measures, the ready-mix concrete industry is rife with producers in major urban areas where concrete is used extensively. Moreover, usual concrete is such a well-established traditional structural material that it requires no aggressive, innovative marketing techniques. Its qualities are well known and its properties well established. Its use is strictly codified in various building codes, complemented by abundant literature on how and when to use it. Consequently, competition in the usual concrete market is almost exclusively based on unit price, rather than on quality. This remains true in spite of the relatively successful efforts of certain ACI committees and public agencies towards implementing quality control programmes. Of course, in such a competitive market, it is difficult to make money with a material that is so ubiquitous.

Faced with this situation, concrete producers might turn to promoting high-performance concrete in their areas, if they can control the more difficult production, delivery and marketing. High-performance concrete is high-tech, unforgiving to 'hit-or-miss' operations. It demands research to determine the most suitable constituents available in a given area. Quality control, in terms of both raw materials and the finished products, is imperative and must be backed by sound promotion aimed at owners, architects and designers to educate them in its effective use in order to create more elegant and economical structures.

From a practical standpoint, it is generally assumed that an average of 10 to 20% of the total volume of concrete used in, for example, a high-rise building, should be high-performance concrete. Although this represents a mere fraction of the concrete volume used in the construction of the building, being able to supply high-performance concrete puts a concrete producer in an enviable position for getting the contract for the remaining 80 to 90% usual concrete needed to complete the building. Moreover, the competition between serious concrete producers in the high-performance concrete market should be much fairer, since all abide by the same quality rules.

Some concrete producers involved in high-performance concrete fabrication and promotion have indicated that the know-how and quality control programmes developed in seeking higher strength usually positively influence the productivity and profitability of the usual concrete industry. Increased pressure on the suppliers and closer scrutiny of usual concrete production and delivery can make usual strength concrete production more profitable.

Interestingly enough, by increasing compressive strength by a factor of four, the ready-mix industry succeeded in knocking steel out of the top slot in the high-rise market in just a few years. That just was not possible with 25 MPa concrete.

Since high-performance concrete fabrication necessitates a high-tech approach, concrete producers interested in this market have to bear in mind that their ultimate success begins with developing a good quality-control team, a strong technical department, and a well-defined and focused marketing strategy. This, of course, is in addition to investments in materials, equipment and personnel. This kind of bet on the company's future has already paid off for a number of concrete producers (Malier, 1991).

4.6 FOR THE ENVIRONMENT

Whenever high-performance concrete is used instead of usual concrete, this demonstrates that Portland cement's binding power has been used more effectively. The higher water content (much more than strictly required to hydrate the cement fully) of usual concrete results in a porous, weak microstructure. Since Portland cement production is energy-intensive, making usual concrete can be considered partially wasteful. To demonstrate this, one needs merely to compare the cost of all the materials necessary to support a certain load in a particular structural element with high-performance concrete and usual concrete: high-performance concrete uses less cement and less aggregate.

4.7 CASE STUDIES

The 11 selected case studies represent a small fraction of projects that make direct use of high-performance concrete. The CEB–FIP working group on high-strength/high-performance concrete has published a bulletin where a list of the main high-performance concrete structures built around the world can be found (CEB, 1994). Each case begins with a general project description that touches on, among other things, the suitability of using high-performance concrete for that particular application.

These 11 civil engineering projects are:

1. Water Tower Place, built in 1970 in Chicago, Illinois, USA.
2. Norway's Gullfaks offshore platform, built in 1981.
3. Sylans and Glacières viaduct, built in France in 1986.
4. Scotia Plaza, built in 1988 in Toronto, Canada.
5. Île de Ré bridge, built in 1988 in France.
6. Two Union Square, built in 1988 in Seattle, Washington, USA.
7. Joigny bridge, built in 1989 in France.
8. Montée St-Rémi bridge built near Montreal, Canada in 1993.
9. The 'Pont de Normandie' bridge, completed in 1993.
10. Hibernia offshore platform, completed in Newfoundland, Canada in 1996.
11. Confederation bridge completed between Prince Edward Island and New Brunswick in Canada in 1997.

4.7.1 Water Tower Place

The Water Tower Place building was built in 1970 (Anon, 1976). It is an 86–storey building located in downtown Chicago (Figure 4.1). Although by present standards the 60 MPa compressive strength concrete used in the columns of the lowest floors does not represent a large achievement, one must remember that this compressive strength was obtained at a

Case studies

time when superplasticizers were not yet used in the production of high-performance concrete. At the time this building was constructed, only lignosulfonate-based water reducers were being used by the concrete industry, and it was not easily possible to obtain very high compressive strengths. In order to reach high strength levels using a regular water reducer, the concrete composition had to be very carefully optimized and a stringent quality control programme had to be implemented. The idea was conceived to produce the highest possible compressive strength through lowering the water/binder ratio as much as possible.

The cement was carefully selected from those available in the Chicago area at the time. About ten different cements were tested to determine their rheological and mechanical characteristics. Several commercially available water reducers were tested with the selected cement in order to choose the most suitable additive. The objective of this testing programme was to produce a slump value of 100 mm at the job site without causing excessive entrapment of air or excessive set retardation. In order to lower the amount of mixing water necessary to obtain that 100 mm slump on the job site, approximately 15% of a high-quality Class F fly

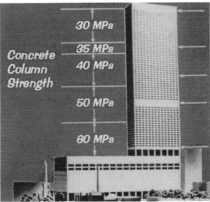

Fig. 4.1 Water Tower Place, Chicago 1970. This 262 m high-rise building was built with a 60 MPa concrete before superplasticizers began to be used in high-performance concrete. Some innovative features were developed during the construction of the building. All the columns were cast using the same prefabricated steel forms from bottom to top, but the concrete compressive strength varied according to variation of loads in the structure, resulting in cost savings and speeding up of construction. Courtesy of J. Albinger.

ash was substituted for an equal weight of cement. This ASTM Class F fly ash had a pale grey colour and had a low loss on ignition and a low alkali content. The fly ash proved effective in allowing a sufficient reduction of mixing water, which was needed to produce the desired slump. The sand used was a relatively coarse natural siliceous sand. The coarse aggregate was crushed dolomitic limestone with a nominal size of 10 mm. It was clean, reasonably cubic and quite strong.

This high-performance concrete was only used for the columns of the first 13 floors. For the next floors the cross-sectional area of the columns was kept constant but the 28 day compressive strength was incrementally reduced to 50, 40 and 35 MPa, as shown in Figure 4.1. In the top storeys, the compressive strength was reduced to 30 MPa, which was the usual compressive strength used for the construction of columns in high-rise buildings at that time. Keeping the cross-sectional area of the columns constant enabled the contractor to use only one set of metal forms for the entire height of the building and thereby to save on construction costs. Moreover, keeping the cross-sectional area of the columns constant resulted in savings in the finishing time and cost of the floors, since each floor had exactly the same geometry and same surface (Anon, 1976).

Some of the columns were instrumented to monitor the long-term behaviour of this first high-strength concrete (Russell and Corley, 1977; Russell and Larson, 1989).

4.7.2 Gullfaks offshore platform

This text was essentially written from technical brochures provided by Norwegian contractors (1986, 1988, 1990) and from papers by Haug and Sandvik (1988), Moksnes (1990) and Ronneberg and Sandvik (1990).

High-performance concrete has been extensively used for the construction of offshore platforms in Norway since the early 1970s. More than 20 such concrete structures have been built involving 2 million cubic metres of high-performance concrete, part of which was prestressed. All of these platforms have, up to the present, good service records despite the harsh environment of the North Sea, where the designers had to take into account a 100 year wave of 30 m amplitude. All of these massive structures are installed in deep water ranging from 70 m for the Ekofisk platform to 216 m in the case of the Gullfaks platform. Most of them were built near Stavanger in southern Norway.

Construction usually starts in a dry dock where the lower domes, the skirt (Figure 4.2) and the cell walls are completed before the structure is floated out to a 400 m deep fjord where the rest of the concrete structure is slipformed (Figure 4.3 (a)). Once the concrete platform is constructed, its base is ballasted and sunk to install the operating mechanical unit, which can weigh up to 50 000 tonnes (Figure 4.3 (b)). The completed platform is then refloated and towed to its final location, where it is sunk

(Figure 4.3 (c)). The concrete skirt is usually anchored deep into the sea-bed. For example, the Gullfaks platform has a concrete skirt that penetrates 22 m into the sea bed. Once in place, such platforms are expected to operate for 30 years.

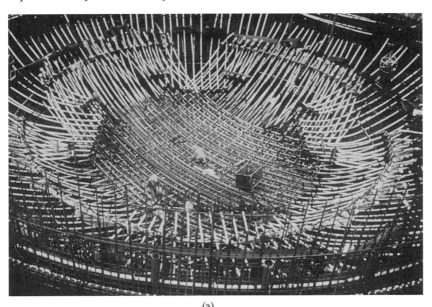

(a)

(b)

Fig. 4.2 Construction of the platform in a dry dock: (a) a congested area in the lower dome; (b) slipforming of the walls of the different cells. Courtesy of Norwegian contractors.

Fig. 4.3 Offshore platforms: (a) construction of the legs by slipforming; (b) deck mating; (c) towing to the oil field. Courtesy of Norwegian contractors.

Over the past 20 years, the specified compressive strength of concrete used to build such offshore platforms has steadily increased from 45 MPa in the early 1970s to up to 70 MPa for the Gullfaks platform. Presently, offshore concrete platforms involving the use of 100 MPa concrete are in the developmental stage.

This achievement in the placing of more than 2 million cubic metres of high-performance concrete had been made possible through the extensive research and quality control programmes implemented by Norwegian contractors, which worked in close collaboration with research institutions at the Norwegian Institute of Technology (NTH) and the Foundation for Scientific and Industrial Research at the University of Trondheim (SINTEF), which, along with other research institutions and the cement and oil companies, were involved in an extensive development project on the properties and characterization of high-performance concrete.

Not only did the specified compressive strength increase through the years, but the specified slump of the concrete increased as well. This greatly facilitated the placing of concrete in highly congested areas where the amount of reinforcement can reach 1000 kg/m^3 (Figure 4.2). In order to increase both the compressive strength and the slump, while reducing the cement content, it was necessary to improve the cement quality in terms of the rheological and mechanical characteristics of the concrete. This called for using more efficient admixtures that increased the total material cost by 8 to 10%. Furthermore, the sand was hydraulically processed to separate it into eight different fractions which were then carefully recombined to obtain a favourable continuous gradation.

For example, in the case of the Beryl A platform which was cast between 1973 and 1975, each cubic metre of concrete contained 430 kg of cement, 175 l of water, and 4 l of admixture. The slump was 120 mm, and the compressive strength measured on 100 mm cubes was 55 MPa. In the late 1980s, for the Gullfaks platform, the cement content was reduced to 400 kg/m^3, the water content was reduced to 165 l/m^3, and 6 l of admixtures and 10 kg of silica fume were added to improve the pumpability and cohesiveness of the high-workability concrete. The average slump was 240 mm, and the average compressive strength 79 MPa. In this latter case, uniform production quality has been achieved with a standard deviation of 3.4 MPa, which corresponds to a coefficient of variation of 4.3%.

In order to ensure a constant quality for the thousands of cubic metres of high-performance concrete, its composition and properties had to be adjusted so that they could accommodate the unavoidable variations in the materials as well as human error. Therefore, instead of opting for a well-determined optimized concrete in terms of materials and composition, the Norwegian contractors developed instead what is called a 'tolerant' concrete, i.e. a concrete that does not demand too much of the user. The properties of this concrete are not greatly affected by the usual range of variations of materials and/or by human error. However, in order to produce such a 'flexible' concrete, an extensive quality control programme had to be implemented, including 'acceptance testing at all stages of production, supervision of workmanship, early warning of quality deviations and documentation of results obtained for future applications'.

Not only was the compressive strength of standard 100 mm cube samples high, but an extensive testing programme later showed that concrete cores cut from platforms also exhibited high strength levels. During the construction of the Oseberg A platform, the average core compressive strength was 82.7 MPa compared with the 90 day cube strength of 83.9 MPa for a 60 MPa specified concrete.

As an illustration of this particular use of high-performance concrete in offshore platforms, some of the main characteristics of the world's largest offshore concrete structure, the Gullfaks platform, are presented.

The Gullfaks oil field was developed by means of three concrete platforms: Gullfaks A installed in 135 m of water, Gullfaks B in 141 m of water and, in 1989, Gullfaks C installed in 216 m of water. The construction of Gullfaks C started in a dry dock in January 1986. The skirt, which covers a surface of 16 000 m^2, and lower domes were completed up to a height of 40 m, before being floated and towed out to the Jars fjord in

June 1989 (where the water depth exceeded 400 m) to proceed with the construction of the rest of the platform in quiet water. The slipforming of the 24 storage cell walls (each cell has a diameter of 28 m) involved placing 110 000 m^3 of concrete and 30 000 tonnes of reinforcing steel. After the completion of the upper domes, the four conical shafts with a diameter ranging from 12.8 m to 28 m, were slipformed for a total length of 170 m. The total volume of the 65 to 70 MPa specified strength concrete necessary to build this platform was 244 000 m^3. It was reinforced by 70 000 tonnes of reinforcing steel and 3500 tonnes of prestressing steel.

The platform was ready for deck mating in December 1988. It was towed out (1.5 million tonnes displacement) and installed in the field in May 1989. The total height of the concrete structure is 262 m, and its storage capacity is 2 million barrels of oil, which corresponds to an 8 day production for the 52 wells of that field.

The construction and installation of this platform led to the development of some innovative engineering concepts, such as:

- the slipforming of 114 000 m^3 of 65 MPa concrete within 42 days during the construction of the storage cells;
- the slipforming of three inclined shafts over a period of 50 days at an average rate of 3.3 m/day. To be able to sustain such a slipping rate, the concrete had to be pumped from the floating batching plant first horizontally for more than 150 m, then vertically 180 m to the top of the shafts.
- the 22 m tall concrete skirts built underneath the main caisson penetrate the sea floor, which has a low bearing capacity. This skirt also increased the stability of the platform. Its penetration into the sea floor was achieved by water ballast and suction. The penetration was followed by the grouting of the underbase, which necessitated the pumping of 33 000 m^3 of grout within 5 days.
- the airlift out of dock for the 220 000 tonne, 40 m high base structure. Natural buoyancy was not enough to lift and float the structure. Some 800 000 m^3 of air was trapped under the lower domes and the skirt.

4.7.3 Sylans and Glacières viaducts

The Sylans and Glacières viaducts in France are part of Highway A40 linking Macon to Geneva and the Mont Blanc tunnel (Figure 4.4). The viaducts are 1268 m long and 215 m long, respectively (Richard, 1988). The walls of the gorge in these areas are so sheer that the separate decks had to be constructed at different levels (Figure 4.4).

In order to obtain a lightweight structure which would allow a 60 m span, which is about 20% longer than could have been obtained with a conventional design, it was decided to opt for a triangulated truss arrangement which had already been used in two previous projects, the

Case studies 47

Teheran Olympic Stadium and the Bubiyan bridge in Kuwait (Baudot, Thao and Radiquet, 1987; Richard, 1988).

The Sylans and Glacières viaducts were built with triangulated deck webs designed as a series of precast X elements (Figure 4.5a) (Cadoret, 1987). In this particular case, the saving on foundations was significant to offset the higher cost of fabrication of the deck itself. Owing to the

Fig. 4.4 Overview of Sylans viaduct. Courtesy of Pierre Richard from Bouygues.

48 The high-performance concrete rationale

presence of a thick and variable layer of loose rock, all piles were 4 m in diameter and were anchored in the bedrock at depths varying from 6 to 35 m. On top of each pile there was a pier, the base of which was

(a)

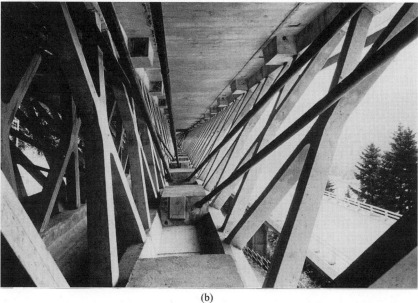

(b)

Fig. 4.5 Construction of the Sylans viaduct: (a) detail of a prefabricated segmental box girder; (b) inside the viaduct. Courtesy of Pierre Richard from Bouygues.

protected from falling loose rock by a strong semi-circular collar. The reduction of the number of necessary piles and piers had a significant influence on the final cost of the entire viaduct, so that the final design cost was estimated to be 7% less than it would have been for a conventional precast concrete segmental structure.

In the case of a triangulated truss, the web weight is only a small portion of the total weight of the structure. Therefore, for the same dead load, the depth of the Sylans and Glacières viaduct deck structure was increased, thus allowing the construction of longer spans. The selected depth of the structure was 4.17 m. Had it been a precast box girder structure, the depth would have been 3.45 m. However, with the solutions selected the post-tensioning could not be included in the web and had to be placed completely outside of the concrete (Figure 4.5b) (Thao, 1990). This presented numerous advantages both during the construction and later for maintenance of the structure. Each external post-tensioning cable could be changed at any time if some changes were required during the construction of the prefabricated elements.

Owing to the considerable length of the structure (the project totalled nearly 3 km of two-lane deck, each 10.75 m wide), it was decided to use precasting methods for the construction of the 640 deck segments, each 4.66 m long and weighing 58 tonnes. The relatively narrow deck bed required the use of only four triangulated deck webs made of post-tensioned concrete X elements.

Using prestressing, it was possible to produce the different structural elements separately, then to post-tension them together to form the bridge deck. In addition to materials savings, a lower dead load was reported to save money in handling and erection operations.

The construction principle adopted was related more to steel construction practice than to the usual precast concrete construction of present-day methods (Roussel, 1989). Owing to variations in the longitudinal and transverse profiles of the highway and the different positions of the deck segments within each span, each segment was unique in terms of its exact dimensions. Moreover, the X elements were not interchangeable. Sixteen different kinds of X element were manufactured; they differed slightly from each other in terms of their geometry, their prestressing and the amount of reinforcing steel. It was necessary to ensure that the proper reinforcement was placed in the correct X and that each X was accurately placed in its corresponding location and segment.

The most difficult concreting operation was the construction of the X elements because of their small cross-sectional area (200 × 200 mm) in which a high amount of reinforcing steel (eight reinforcing bars 8 mm in diameter) had to be placed as well as the two 40 mm diameter post-tensioning ducts that crossed each other at the centre of the X. The fresh concrete had to be fluid enough to flow through the reinforcing steel, and also had to achieve an early strength to permit a fast reuse of the moulds. It was found that, in order to reach 17 MPa at 12 hours (measured on

150 × 300 mm specimens), it was necessary to adjust the 28 day compressive strength to 60 MPa for the X and to 50 MPa for the deck, although the required design strength for the whole project was only 40 MPa. No low-pressure steam curing was used. These high levels of strength were achieved using 400 kg/m^3 of cement and no silica fume. Enough melamine-based superplasticizer was used to obtain a slump of 200 to 250 mm. The concrete had a water/cement (W/C) ratio of 0.37. In order to put in place such a rapid hardening concrete in so small and congested a cross-section, a very stringent quality control programme had to be developed, covering all aspects of the operation from manufacturing to final erection.

In the Sylans and Glacières project, the high cost of a temporary bridge was not justified. Therefore, the more conventional method of erecting the deck structure by cantilevering from the piers was used. If the joints had been cast in place, this would have slowed the erection process by the period necessary for the cast concrete to reach an adequate strength. Therefore, instead of opting for mid-span 50 to 200 mm-wide cast-in-place joints to complete the spans, it was decided to erect an entire span consisting of two cantilevered half-spans on either side of a pier at about 200 mm from their correct ultimate position. The assembly was then slid into its final position where the two mating ends were permanently joined by post-tensioning. Eliminating the closure using the cast-in-place joint, saved 1 or 2 days per span.

As there were no cast-in-place closure joints at the centre of each span to allow for geometric adjustments for imperfect vertical and horizontal positions of the superstructure, a sophisticated software package had to be developed for analysis of the cumulative effects of geometrical imperfections. Using this software, it was possible to design the Sylans and Glacières viaduct with only eight continuous deck-structure sections, 475 m long between expansion joints, with extremely good precision. For example, after the erection of 160 segments totalling 692 m, the observed error between design and execution was 15 mm transversely and 40 mm longitudinally. This is an average of 0.25 mm per segment.

4.7.4 Scotia Plaza

Scotia Plaza is a 68-storey building 275 m high, constructed in 1986 and 1987 in downtown Toronto (Figure 4.6). This entirely concrete high-rise building was designed using concrete with a specified strength of 70 MPa. It was the first Canadian high-rise building for which such a high compressive strength was specified. Recently, several other high-rise buildings have been built with 70 MPa or even higher compressive strength concrete, but in many respects the construction of Scotia Plaza represents a significant achievement in the domain of high-performance concrete technology in Canada, and one of the first uses of a slag in a high performance concrete mixture (Ryell and Bickley, 1987; Quinn, 1988).

(a)   (b)

Fig. 4.6 Scotia Plaza in Toronto: (a) during construction using the jumping form method; (b) completed building on the left. Courtesy of John Bickley.

The concrete developed for this specific project was in fact a mixture of Portland cement, finely ground blast-furnace slag and silica fume, the composition of which is given in Table 4.1.

During the two years of construction, concrete was placed at temperatures varying between $-20\,°C$ and $35\,°C$. During the summer, in order to keep the temperature of the delivered concrete below $25\,°C$, the concrete had to be cooled with liquid nitrogen.

This concrete was prepared in a dry-batch plant, meaning that the concrete was truck-mixed. A special loading sequence and procedure had to be developed in order to obtain a reproducible mixture. The procedure developed was successful because the compressive strength of the 142 loads of concrete that were tested averaged 93.6 MPa at 91 days with a coefficient of variation of 7.3%. This means that, if one test out of

Table 4.1 Composition of the Scotia Plaza concrete (Ryell and Bickley, 1987)

	Cementitious materials			Aggregates		Admixtures	
Water	Cement	Silica fume	Slag	Coarse	Fine	WR*	Super**
	kg/m^3					l/m^3	
145	315	36	135	1 130	745	0.835	6.0

*Water reducer. **Superplasticizer.

Table 4.2 Some characteristics of the hardened concrete of Scotia Plaza (Ryell and Bickley, 1987)

Age, days	2	7	28	56	91
(n) of loads tested	124	149	149	146	142
f'_c (MPa)	61.8	67.1	83.7	89.5	93.6
σ (MPa)	5.5	4.7	6.1	6.1	6.8
V (%)	8.9	7.0	7.3	6.8	7.3

ten lower than the specified strength is considered the minimum acceptance criterion, then the actual design strength of the concrete was 85 MPa, which is well above the 70 MPa specified compressive strength.

An extensive quality control programme was developed during construction. Results showed that the concrete producer was able to maintain exceptional control of the mixture, given that concreting was spread over 20 months with outside temperatures varying between $-20\,°C$ and $35\,°C$. An overview of the test results is presented in Table 4.2.

The owner stated that the fact that the concrete strengths were higher than they needed to be gave him great comfort. Moreover, the builder stated that the cost of high-performance concrete per square metre was moderate, while the benefits were significant.

This extra strength was provided free of charge by the concrete producer because he was aware that the construction of a second high-rise building was planned in downtown Toronto, the Two Place Adelaide Building. In order to resist loads caused by high-velocity wind, a 70 MPa concrete would result in a non-competitive concrete alternative to a steel-framed building, whereas, with an 85 MPa compressive strength, the concrete alternative appeared to be more economical and could provide some additional parking places in the underground storeys.

Before the construction of this second high-rise building, all the concrete producers operating in the Toronto area were asked to take a prequalification test in order to avoid an unpleasant situation similar to that encountered in Chicago where the lowest bidder was once unable to provide a concrete strong enough to fulfil the design requirements.

The rationale for prequalification trials of potential concrete suppliers is to:

1. Assure the engineer that the specified strength, higher than any previously specified in the area, can be met by the existing supply industry.
2. Qualify enough suppliers to provide competitive bidding for the owner.
3. Confirm that all the specified properties, such as high strength, high modulus of elasticity and acceptably low temperature rises and gradients, can be met with the materials used by all the suppliers.

Prequalification included tests on fresh and hardened concrete, the casting of a monolith of similar dimensions to major column cross-sections, and in-place tests to determine strength and temperature history. Three concrete producers applied for that prequalification test and passed it successfully.

4.7.5 Île de Ré bridge

Île de Ré, a small island lying 2 km off the coast of France near La Rochelle, is a popular vacation resort. In the past, cars had to wait up to 10 hours to cross the channel during the summer months when there was a large influx of tourists. In spite of the high passage fare, the ferry operation had to be subsidized (US$3 million per year).

The construction of the 3 km-long bridge started in September 1986 and was completed less than 20 months later (Figure 4.7(a)) (Cadoret and Richard, 1992). It was opened to traffic on 19 May 1988, just before the 1988 summer tourist season (Figure 4.7(b)). The toll for the bridge was fixed at the same rate as the fare charged by the ferries. Almost immediately, traffic increased by 80%.

The bridge was designed to have a 15.5 m-wide platform and provisions had to be made to pass the Île de Ré water supply pipe (0.60 m in diameter), telephone lines and electric lines (90 000 volts) inside the box girders.

In order to limit the effect of collision by large ships, the bridge was separated in six independent viaducts. The span length was designed to allow four, 100 m-wide ship-channels for navigation.

From an economic point of view it was decided to employ an extremely repetitive construction method (Figure 4.8), using precast box-girders placed by the cantilever method and external prestressing (Figure 4.7(a)) (Causse, 1990).

The separation between the two adjacent halves of the arch was made by a single joint and a neoprene support in the span placed one segment before mid-span. The timescale prompted the contractor to prefabricate the box girders. A typical segment length was 3.80 m, weighing 40 to 80 tonnes. Each on-pier segment was made of two half segments 2.60 m long and weighing 12 tonnes each (Figure 4.8(c)). Eight short-line steel moulds were used, six of them for typical segments, one for the on-pier segment, and one for special segments such as the articulation segments and the segments at the two abutments of the bridge. The prefabrication site turned out seven box girders per day with eight steel forms in use. On average, the girders were launched 1 or 2 months after casting.

While the design strength used in calculating the girders was 40 MPa (Virlogeux, 1990), they were actually built of a concrete with a characteristic compressive strength of 59.5 MPa, well above the specification. The mean 28 day compressive strength for the 798 cylinders tested was 67.7 MPa with a standard deviation of 6.3 MPa. This level of compressive

54 *The high-performance concrete rationale*

(a)

(b)

Fig. 4.7 Île de Ré bridge: (a) under construction; (b) after completion. Courtesy of Pierre Richard from Bouygues.

strength was selected because a 20 MPa compressive strength was needed at 10 to 12 hours of age to strip the box girders. The average slump of the concrete was 150 mm (Cadoret, 1987). Concrete compressive strength was uniform during the whole precasting because a tight quality control programme was established, as shown in Table 4.3.

The time of formwork removal was based on maturity measurements. Experimental curves were developed from on-site laboratory tests carried out by the owner. It was decided to apply a safety factor of 1.10 before implementing the maturity method. Owing to the rigorous application of this method, it was possible, without taking any risk, to determine the time at which the precast box-girders could be demoulded without risking damage. Over a 4 month period the initial 15 hour delay for mould stripping was lowered to 12 hours and even to 10 hours.

Another major advantage of using high-performance concrete was the reduced long-term creep of the prestressed concrete box-girders. It was possible to adjust the dimensions of each box girder to take into account the largely instantaneous strain of the concrete when the post-tensioning was applied, so that the final dimensional precision obtained was of the

(a)

(b)

(c)

Fig. 4.8 Île de Ré bridge: (a) aerial view of the prefabrication yard; (b) forms for the casting of a segmental box girder; (c) a segmental box girder. Courtesy of Pierre Richard from Bouygues.

Table 4.3 Compressive strength of the Île de Ré concrete (Cadoret and Richard, 1992)

Year 1987	n*	Compressive strength in MPa	
		7 d	28 d
March	3	42.6	53.5
April	33	50.8	59.5
May	61	52.2	63.4
June	98	58.4	68.9
July	129	58.8	66.5
August	140	55.9	66.4
September	154	56.1	67.3
October	114	60.2	72.8
November	66	61.6	72.5
Total/Averages	798	56.9	67.7

*Number of samples tested.

order of 1 mm. Instead of being obliged to cast in place the neutral joint to take into account the off-alignment and level difference, the last element was cast like a key stone, accelerating the completion of each span by 2 to 3 days.

A very high-strength, superplasticized silica fume grout was injected into the post-tensioning ducts. No segregation or bleeding was observed.

4.7.6 Two Union Square

The Two Union Square Building located in Seattle, Washington, offers a variety of interesting features. Its 58-storey structure is braced by an innovative composite core of four 3 m-diameter steel pipes filled with 130 MPa concrete (Figure 4.9). More than 100 pipes manufactured in Korea were transported by barge across the Pacific Ocean and welded together on site. Each pipe has a 600 mm-wide steel plate projecting from either side to attach the core to a brace moment-resisting frame that also includes composite columns along the perimeter, as shown in Figure 4.9. In addition, between the 35th and 38th floors diagonal braces run between the core and the perimeter (Godfrey, 1987).

To stiffen the 216 m-high building to limit swaying in strong winds and earthquake vibration, it was necessary to fill these steel pipes with a high-performance concrete with an elastic modulus of 50 GPa, which is twice that of a usual concrete. However, for liability concerns, the contractor was asked only to comply with the more usual compressive strength specification. It was found from laboratory data that in order to obtain such a high elastic modulus it was necessary to raise the compressive strength of the concrete to 130 MPa, in spite of the fact that a 90 MPa concrete would have been sufficient from the strength point of view.

Case studies

In fact, high-rise sway is a major concern, given that some buildings oscillate to the point of causing motion sickness to occupants in upper storeys. The designed-in rigidity was claimed to offer occupants at the top almost the same comfort as their ground floor colleagues, but with a better view. The composite piles contain no reinforcing steel, since the concrete was confined by the steel pipe. Nelson studs, 300 mm in length, were welded inside the steel pipes at 300 mm centres to help in transferring shear forces to the steel shell from the concrete core (Figure 4.10). The concrete pumping and steel erection were carried out simultaneously, with concrete being pumped two storeys below the steel erection which rose up to the 58th storey. Each 7.2 m-high pipe section was filled with concrete pumped up from its base to achieve a good compaction without any internal or external vibration (Ralston and Korman, 1989).

The concrete placement was done within 9 months, with concrete delivered at night in order to obtain a smooth and on-time delivery schedule avoiding the traffic jams of rush hours. Up to 750 m³ of normal strength concrete and high-performance concrete could be placed each night without any difficulties.

(a)   (b)

Fig. 4.9 Two Union Square, Seattle: (a) under completion; (b) completed. The concrete cast in the columns of the Two Union Square building is the strongest that has ever been used in any construction project. In spite of the fact that the building was designed for a compressive strength (f_c) of 90 MPa, it was necessary to achieve a 130 MPa compressive strength in order to obtain an elastic modulus of 50 GPa. Courtesy of Weston Hester.

Fig. 4.10 Two Union Square, Seattle. View of the inside of a section of the steel pipe used to confine the high-performance concrete. Courtesy of Weston Hester.

According to the structural engineer, the savings in steel cut costs by 30% even though the high-performance concrete price came to about $160 per cubic metre.

Making a 130 MPa field concrete was not an easy task. This level of compressive strength was achieved in Seattle because of the availability of high-quality raw materials. A Type I/II Portland cement, low in alkalis and exhibiting very low rheological reactivity in the presence of the large dosage rates of superplasticizer, was used. In this particular case, the compatibility of the naphthalene superplasticizer with the selected cement was excellent so that a water/cementitious ratio of 0.22 could be used (Howard and Leatham, 1989).

Such a strength level was achievable also because outstanding aggregates were available in Seattle. The selected coarse aggregate had a nominal size of 10 mm. It was a fluvioglacial pea gravel, very strong, very clean and rough enough to develop a good mechanical bond with the hydrated cement paste (Aïtcin, 1989). The sand came from the same pit and had a rather coarse gradation corresponding to an average fineness modulus of 2.80. It consisted of sharp angular particles.

The moisture contents of both the sand and the coarse aggregate were determined for every 40 m^3 of concrete. Prior to each night's placement, the aggregate stock piles were sprinkled with water to keep the surface of the sand and coarse aggregate cool and moist. The slump of the concrete was checked at the plant because of the small margin for error in water dosage when using slump in excess of 250 mm.

As soon as the truck arrived at the job site, the slump was measured and the superplasticizer was added as needed. Most of the time, the high-strength mix required a second dosage of superplasticizer at the job site to achieve the final desired slump of 200 to 250 mm. Periodic checks for slump were made on waiting trucks, and subsequent additions of superplasticizer were made as required. It was found that the timing of the addition was critical to achieve and maintain high slump without segregation or premature slump loss.

Testing such high-performance concrete raises some critical issues (Simons, 1989). The testing was done on 100×200 mm cylinders after a preliminary study showed that, although they were more sensitive to abuse, 100×200 mm cylinders cast in steel moulds and rodded yielded the highest and most consistent results. It was found that these smaller specimens gave a compressive strength that was on average 8% higher than that of 150×300 mm cylinders. It was also decided that, for optimal curing, concrete cylinders had to be placed immediately after casting into a temperature-controlled water-saturated lime bath at the job site. Prior to testing, both ends of the concrete cylinders were ground smooth and faced to eliminate the necessity of capping.

A quality assurance manual was developed by the contractor, the testing laboratory and the concrete producer, and was submitted for review to the owner, the engineer and the city engineer, after which it became part of the contract documents.

One of the most important aspects of this programme was the section outlining the training and qualifications of the personnel involved in the production of the high-performance concrete. The drivers, batchmen and testing personnel received personalized training and instruction so that they would know what problems to look out for, not only in their specific areas, but in other areas as well. By keeping the job personnel informed and involved, a genuine team spirit was established, and this was carried through to other aspects of the job.

4.7.7 Joigny bridge

The Joigny bridge, located on the Yonne river 150 km southeast of Paris, appears to be just an ordinary, unremarkable, modern bridge (Figure 4.11) (Malier, Brazilier and Roi, 1991). It is a modest 114 m long and is composed of three balanced spans 34, 46 and 34 m long, with an overall width of 15.80 m. However, for the engineer who looks at the bridge in more detail, it is quite remarkable as it is one of the first bridges built with high-performance concrete. When it was built, the bridge was fully instrumented and has been monitored since the placing of the concrete and the post-tensioning (Malier, Brazilier and Roi, 1992).

60   *The high-performance concrete rationale*

The Joigny bridge is located in the countryside far from any large town or industrial area where the concrete market is large and advanced. For this reason, the bridge was built entirely from locally

Fig. 4.11 Joigny bridge. Note the two standby holes that could receive two additional post-tensioning cables to strengthen the bridge if exceptional loads (French Army and French power utility) had to cross the bridge or to replace a defective post-tensioning cable. Courtesy of Y. Malier.

available materials by two small ready-mix plants which had not produced high-performance concrete before. The Joigny site was purposely selected for this full-scale experiment to demonstrate the feasibility of constructing a typical bridge using high-performance concrete with unsophisticated means and materials that could be found almost anywhere in France. From an economic point of view, this is a very interesting case study because an alternative design using usual concrete was also considered in order to make a fair economic comparison between the usual concrete and high-performance concrete alternatives. In this connection, it must be mentioned that, for reasons that are too long to detail here, the two piers were constructed with usual concrete so that high-performance concrete was used only for the construction of the bridge deck. The bridge deck was designed for a design strength of 60 MPa instead of the usual 35 to 40 MPa design strength required by the French code.

The bridge has a double-tee cross-section made of two beams with trapezoidal cross-sections and an upper slab. The distance between the axes of the two beams is 8 m and their minimum width at the bottom is 0.50 m. The structure is longitudinally post-tensioned by 13 external tendons (Figure 4.11). The use of external post-tensioning offers three main advantages. The first is that, in the case of an experimental structure, this layout allows a simpler and more accurate post-tensioning force measurement in the tendons. The second advantage is that the width of the webs can be reduced to the minimum, chiefly determined by stress considerations, so that the dead weight of the structure can be reduced. The final advantage of external post-tensioning is that the tendons can be easily replaced if necessary. Two stand-by holes can be seen in Figure 4.11 where, within a working day, a new tendon can be placed.

As previously mentioned, a comparison was carried out between the usual concrete and high-performance concrete designs. It was found that the concrete volume was reduced from 1395 m^3 for a 35 MPa usual concrete down to 985 m^3 for a 60 MPa high-performance concrete. This 30% reduction in concrete volume led to a load reduction in the piers, abutments and foundations.

The decrease in the dead weight also included savings in the quantities of post-tensioning strands but, since the height to span ratio was not the same in the two alternative designs, the steel savings were not as large as they could have been.

Laboratory trial mixes were made in order to define the optimum composition of a typical mix that had a high workability, the possibility to pump over 120 m-long pipes, an average compressive strength of 70 MPa, and which could withstand a 30 km transportation distance from the two concrete plants that were to provide the concrete (neither of which was able to produce enough concrete per day to cast the whole bridge within 24 hours).

The average slump was 220 mm, with a minimum of 190 mm and a maximum of 250 mm. Compressive strength measured on 160 x 320 mm cylindrical specimens was:

3 days = 26.2 MPa
7 days = 53.6 MPa
28 days = 78.0 MPa
1 year = 102.0 MPa

At 57 days, 150 mm cores gave an average strength of 86.1 MPa. The standard deviation at 28 days was 6.8 MPa. The average splitting strength measured on the same size of specimen was 5.1 MPa at 28 days.

The temperature rise of the concrete varied greatly. Depending on the geometry of the different parts of the bridge deck, a maximum temperature of 73 °C was recorded in the centre of the massive end blocks, while it was only 32 °C in the upper slab, which has a large surface for heat loss; an intermediate value of 57 °C was found in the webs.

The strains of a cross-section located near the middle of the central span are continuously recorded, and up to now they are 15% less than the calculated ones. This difference could be attributed to an underestimate of the actual modulus of elasticity of the hardened concrete in the structure. The tensile forces of 5.1 and 5.2 MN measured in the longitudinal tendons are also in good agreement with the theoretical forces of 5.1 and 4.9 MN near the anchorage and at mid-span, respectively.

Creep and shrinkage measurements are also being monitored (Schaller et al., 1992). Results will be published once they are available.

4.7.8 Montée St-Rémi bridge

The Montée St-Rémi overpass located near Montreal has two continuous 41 m spans made of high-performance concrete having a specified 28 day compressive strength of 60 MPa (Figure 4.12). The high-performance concrete was poured entirely *in situ* into forms on shoring and prestressed by means of external longitudinal prestressing tendons. At the end of the deck, internal transverse prestressing tendons counter the stress of the anchorage of the external tendons. The concrete was a W/C = 0.29, air-entrained high-performance concrete containing 450 kg of blended silica fume cement. The concrete was made in a mobile dry-batch plant and cast in place on 23 June 1993, when the maximum temperature reached 28 °C. In order to keep the temperature of the fresh concrete below the specified 20 °C maximum temperature, an average of 40 kg of crushed ice was substituted for the water. The superplasticizer used was of the naphthalene type; its dosage was 7.5 l/m^3. An air-entraining agent and a retarder were also used. The composition of the concrete is given in Table 4.4.

In order to avoid plastic shrinkage, concreting began at 6.00 p.m. and finished at 6.00 a.m. The concrete surface was covered with a curing compound as soon as it was finished. The quality of the delivered concrete was very good (Table 4.5): the average compressive strength was 80.7 MPa with a standard deviation of 6.1 MPa, and the average air content was 6.3% with a standard deviation of 1%. The average temperature and slump of the delivered concrete were 17.7 °C and 180 mm.

Fig. 4.12 Montée St-Remi bridge.

Table 4.4 High-performance concrete for Montée St-Rémi bridge (Aïtcin and Lessard, 1994). Reproduced by permission of ACI

Water/cement + silica fume	0.29
Water* (l/m^3)	90
Ice (kg/m^3)	40
Blended silica-fume cement (kg/m^3)	450
Coarse aggregate (kg/m^3)	1100
Fine aggregate (kg/m^3)	700
Superplasticizer (l/m^3)	7.5
Air-entraining agent (ml/m^3)	325
Retarding agent (ml/m^3)	450

*Includes the water in the superplasticizer.

Finally, the average spacing factor was 180 µm, which is well below the 230 μm value required by the CSA A23.1 standard for freeze–thaw durability.

This bridge was instrumented with 32 thermocouples to monitor the rise in temperature in the different structural parts of the bridge (Lachemi, Lessard and Aïtcin, 1996). From the data collected, simulations at different ambient temperatures and different temperatures of the fresh concrete were made in order to determine the best concreting conditions from a temperature elevation and thermal gradient point of view (Lachemi, Bouzoubaâ and Aïtcin, 1996; Lachemi and Aïtcin, 1997). From this thermal study it was found that on the one hand, the ambient temperature has a much greater effect on tensile stresses at the surface of massive concrete elements than does the fresh concrete temperature. On the other hand the temperature of the fresh concrete is more important for the peak temperature reached at the centre of the element. The results obtained clearly demonstrate the effectiveness of decreasing the initial temperature to decrease the peak temperature. The smallest thermal gradient was obtained for the temperature conditions (fresh concrete/ambient) 10 °C/28 °C, and the greatest for the conditions 25 °C/10 °C.

Another very interesting feature of this overpass is that a detailed analysis of the construction cost was carried out, as well as that of two

Table 4.5 Results of statistical analysis of high-performance concrete (Montée St-Rémi bridge; Aïtcin and Lessard, 1994). Reproduced by permission of ACI

Concrete properties	*Average*	*Standard deviation*
Temperature (°C)	17.7	1.9
Slump (mm)	180	34
Air content (%)	6.3	1.0
Spacing factor (μm)	184	*
Compressive strength (MPa)	80.7	6.1

*Insufficient data.

Case studies

Table 4.6 High-performance concrete versus conventional prestressed concrete decks (Montée St-Rémi bridge). Courtesy of L.-G. Coulombe

	60 MPa high-performance concrete	Prestressed concrete (35 MPa)	
		Poured in situ post-tensioned	AASHTO V precast beams
Thickness of deck (mm)	1600	1490	1800
Thickness of deck/span	1/25.2	1/27.5	1/22.4
Concrete (m^3)	462*	557*	513
Prestressing steel (tonnes)	22.8	23	24.5
Replacement of tendons	Yes	No	No

*Ratio of 557/462 = 1.21.

other 35 MPa overpasses built by the same contractor on the same project. The characteristics of the three overpasses are given in Table 4.6 (Coulombe and Ouellet, 1994).

The usual way of assessing bridge construction costs consists of expressing the total cost in terms of the unit cost per square metre of deck. After standardizing the total costs of the three overpasses, the unit costs were expressed as ratios using the overpass with the smallest span as a reference with a value of 1.000. Figure 4.13 gives the standardized unit cost as a function of the span length in the case of the three overpasses that were built. If a line is drawn between the two points representing the two 35 MPa overpasses, for a 41 m span built with a 35 MPa concrete, it should have given a 1.027 standardized unit cost, when in fact it was found to be 0.977 for the high-performance concrete Montée St-Rémi overpass. The savings in the initial cost brought about

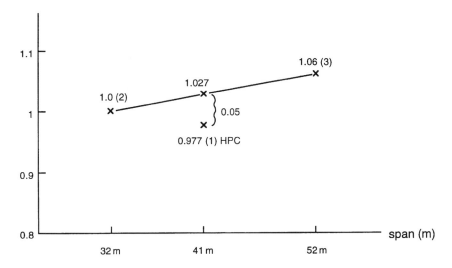

Fig. 4.13 Standardized unit costs.

by the use of high-performance concrete in this particular project were therefore 5%.

4.7.9 'Pont de Normandie' bridge

The 'Pont de Normandie' cable stay bridge was the longest in the world when it was built in 1993. The bridge has an overall length of 2141 m and a centre span of 856 m (Figure 4.14) (Virlogeux, 1993).

The 21.20 m-wide deck accommodates four traffic lanes for cars and two pedestrian/bicycle traffic lanes. The main span is composed of three parts: two concrete (116 m long) sections erected by the cantilever method from both pylons, and a 624 m-long central section built of segmented steel box-girders, also built using the cantilever method. The same cross section was used for the concrete and steel segments between the pylons.

The two pylons have an inverted 'Y' shape, their total height is 215 m above the low-tide level and their weight is 20 000 tonnes (Figure 4.15). They are made of 7500 m^3 of a 60 MPa high-performance concrete up to a height of 152 m with the cables attached to the vertical leg of the Y which is made of steel segments.

Around 70 000 m^3 of concrete were used to construct the whole structure, 35 000 m^3 of which were a so-called grade 60 high-performance concrete, which means that its design strength was 60 MPa at 28 days. In fact the average strength of the concrete based on the result of the testing of 938 160 × 320 mm cylinders was equal to 79.0 MPa, with a standard deviation of 5.4 MPa and a coefficient of variation of 6.9% (Monachon and Gaumy, 1996). Interestingly, it should mentioned that all the testing was done using the sand-box method (Boulay and de Larrard, 1993). An excellent correlation was found between the results obtained with this testing technique and the end-grinding technique.

The splitting strength that this high-performance concrete uses was also extensively checked. From the testing of 311 160 × 320 mm cylinders the average splitting strength was found to be equal to 5.6 MPa, with a standard deviation of 0.5 MPa and a coefficient of variation of 8.8% (Monachon and Gaumy, 1996).

The mix design was developed in order to match not only the 60 MPa 28 day design strength, but also the quite high early strength requirements presented in Table 4.7. Early strength values were read on a maturimeter that was calibrated before the concreting operations started. Every month the calibration curve was checked in order to take into account any variation in the characteristics of the cement. Maturimeters are based on Arrhenius' law, which links strength, time and concrete temperature.

The composition of the high-performance concrete is presented in Table 4.8. The fine aggregate was a 0.4 mm natural sand and the coarse aggregate, which was a semi-crushed gravel, had a maximum size of

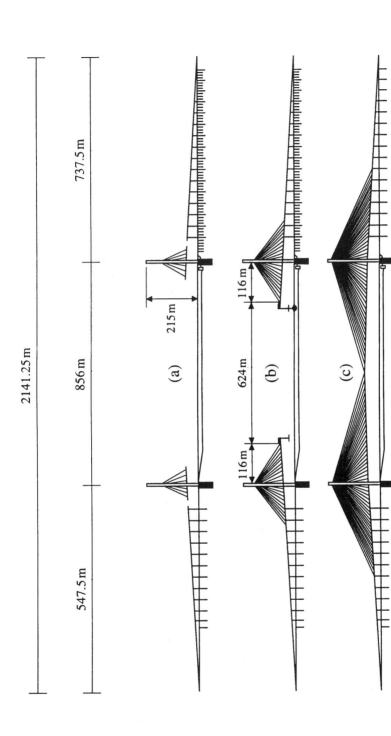

Fig. 4.14 Construction of 'Pont de Normandie' bridge: (a) the completed approach and the construction of the high-performance concrete bridge deck by the cantilever method; (b) placement of the steel box girders by the cantilever method; (c) completed bridge.

Fig. 4.15 'Pont de Normandie' under construction. One of the two pylons in construction during a visit of the author (pale hat) in December 1992 with Jean-François de Champs (darker hat) of Campenon Bernard Company.

Case studies

Table 4.7 Early strength requirements for the construction of the Pont de Normandie bridge (Monachon and Gaumy, 1996)

Hour equivalent*	11	12	15	20	28
MPa	10	15	20	25	35
Operation done	1	2	3	4	5

1. Removal of formwork for pylon lifts and jumping forms
2. Tension of the deck single strands
3. Removal of formworks for the deck slab
4. Moving the cantilever mobile equipment
5. Tensioning of temporary cable-stays

*Concrete strength at time 't' expressed in hour equivalents was read on a maturimeter.

20 mm. The superplasticizer used was a modified melamine type. The cement was a blended silica fume cement containing 8% silica fume.

The effective mixing time was 2 minutes and the consistency was checked by means of the wattmeter-recorder method in the two batching plants (one on each side of the Seine river). The workability of the concrete was checked not by measuring the slump but rather by means of the flow table test according to EN-ISO 9812. Standard consistency was adjusted to vary between 450–530 mm and 510–560 mm according to the segments to be built and the equipment to be used.

The concrete was either pumped for the construction of the incrementally launched bridge segments (Figure 4.16) or placed with a bucket for the construction of the pylons and the cantilever segments. The concrete was usually vibrated internally and externally for the raker strut.

In order to lower heat gradients and thermal cracking, preventive measures were taken to homogenize the temperature of the concrete. All external forms were insulated, with inner deck cells closed and heated with hot air so that the average temperature within the whole was maintained at 30 °C. Finally the deck slab was protected by means of hot air curing units covered with isothermal sheets. Hot curing was stopped after 18 hours, but it was only 6 hours later that the concrete was exposed to the ambient temperature, when it had cooled down (Monachon and Gaumy, 1996).

Table 4.8 Composition of the grade 60 high-performance concrete used to build the Pont de Normandie bridge (Monachon and Gaumy, 1996)

W/B	kg/m³				l/m³
	Water	Cement	Aggregate		Superplasticizer
			Fine	Coarse	
0.36	150 to 155	425	770	1065	10.6 to 11.7

Fig. 4.16 The 605 m-long north approach built as an incrementally launched structure. The final weight of the approach was 26 000 tonnes. Courtesy of Paul Acker.

4.7.10 Hibernia offshore platform

The Hibernia offshore platform is a gravity-based structure (GBS) located in the Grand Banks area, 315 km east of St John's, Newfoundland, Canada. The climate in the Grand Banks area is very harsh and icebergs pushed southwards by the cold Labrador current can enter the Hibernia area and hit the platform. The Hibernia platform has been designed to resist such an impact. After being towed to its final location, it was ballasted with 600 000 tonnes of magnetite and has a final mass of 1 400 000 tonnes, including the 50 000 tonnes of the Topside (Woodhead, 1993a,b).

To build the GBS, 450 000 tonnes of high-performance concrete, 93 000 tonnes of reinforcing steel and 7000 tonnes of prestressing cables were used. The GBS is in fact a hollow cylinder having a diameter of 106 m and a height of 85 m; it is topped by four shafts 26 m high, on which rests the Topside, as shown in Figure 4.17. Eighty-three development wells have been drilled in two different productive reservoirs in the oil field from the platform.

The vital part of the platform, comprising the oil storage reservoirs and the four shafts, is protected by an icebelt comprising 16 teeth, as shown in Figure 4.18. These teeth are designed not only to absorb the impact of potential icebergs, but also to crush the icebergs into pieces.

Case studies

Most of the GBS's vertical parts have been built by slipforming: rebars and concrete were placed continuously at an average rate of one metre per day, so that there is not a single cold joint in these vertical parts.

Different types of concrete were used to build the whole GBS, but the one that was most widely used was a so-called 'modified normal weight' high-performance concrete having a water/binder ratio of 0.31 for the splash zone and 0.33 for the submerged zone for a design compressive strength of 69 MPa (Hoff and Elimov, 1995).

The design of the 'modified normal density' mix had to be carefully adjusted to satisfy design and placing constraints:

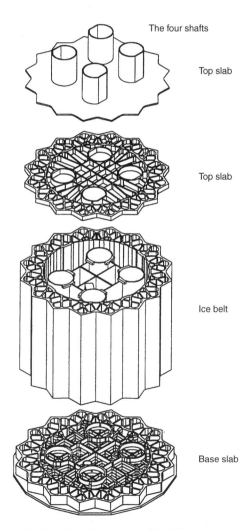

The four shafts

Top slab

Top slab

Ice belt

Base slab

Fig. 4.17 GBS – exploded view. Courtesy of R. Elimov.

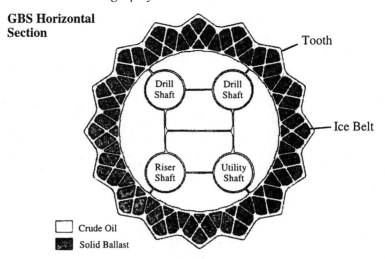

Fig. 4.18 The oil storage facilities and four shafts protected by the ice belt. Courtesy of HMDC.

- design constraints: its unit mass had to be between 2200 and 2250 kg/m^3 for buoyancy, but at the same time its elastic modulus had to be greater than 32 GPa;
- placing constraints: in many areas the concrete had to be placed around more than 1000 kg of steel reinforcement per cubic metre (Figure 4.19(a)), and even much more than that in some very congested areas. Moreover, the setting of the concrete had to be carefully adapted to the slipforming rate (Figure 4.19(b)). Finally, the concrete had to be fluid enough to be placed under vibration in congested areas without any segregation.

In order to achieve such a low water/binder ratio, a silica fume blended cement, containing 8.5% of silica fume and a naphthalene superplasticizer were used. In order to obtain such a low unit mass, necessary to achieve the right buoyancy for the platform during its towing phase, half the volume of the coarse aggregate was replaced by a lightweight aggregate and some air was entrained. All the aggregates were non-reactive to alkalis. Entrained air was also useful to meet the freeze–thaw resistance criteria in the splashing zone; it was also found to be very helpful for the pumping, placing and finishing of the concretes. The mixing of the concrete was done in two very efficient high-speed mixers, producing 2.5 m^3 of air-entrained high-performance concrete every 1 minute and 25 seconds (40 to 45 seconds for loading; 35 to 40 seconds for mixing). As discussed in Chapter 11, this concrete was pumped without any problems over horizontal distances longer than 200 m and vertical distances sometimes totalling more than 100 m. This pumping was done with the ambient temperatures as low as −20 °C and as high as 25 °C. The slump of the concrete at the end of the pump line had to be higher than

Case studies

220 mm in order to be put into place without too much difficulty by the placing crew around the most congested areas.

Owing to the application of a tight quality control programme, not only on the materials used to produce this sophisticated high-performance

(a)

(b)

Fig. 4.19 Slipforming Hibernia GBS: (a) a congested area; (b) slipformed surfaces. Courtesy of R. Elimov.

concrete, but also on the sampling and testing methods, an 86.7 MPa average strength air-entrained high-performance concrete was produced with a standard deviation of 3.2 MPa, which corresponds to 3.7% coefficient of variation (Hoff and Elimov, 1995).

As shown in Figure 4.20, the lower part of the GBS was built, as is usually done, in a dry dock until an elevation of 18.8 m was reached. Then the more than 100 000 tonne base raft was towed to the deep water site where the construction was completed (Figure 4.21 (a)), the concrete being produced in a floating batching plant moored to the side of the platform (Figure 4.21(b)).

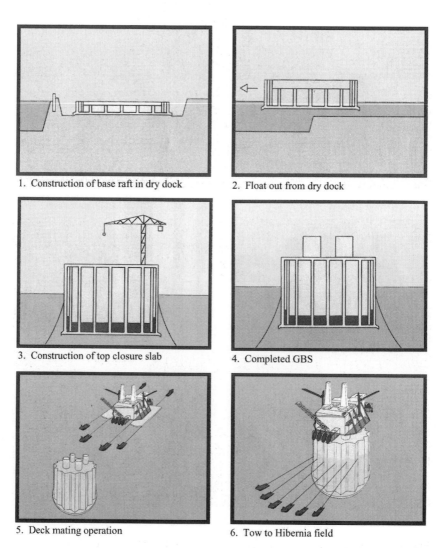

1. Construction of base raft in dry dock
2. Float out from dry dock
3. Construction of top closure slab
4. Completed GBS
5. Deck mating operation
6. Tow to Hibernia field

Fig. 4.20 GBS construction (courtesy of HMDC).

Case studies

For the deck mating operation, the platform was ballasted with sea water and solid ballast. After deballasting, it was towed to the Hibernia field, where its final ballasting with the iron ore took place. The storage

(a)

(b)

Fig. 4.21 Hibernia GBS: (a) almost completed GBS, May 1995; (b) the floating batch plant. Courtesy of R. Elimov.

capacity of the platform is 1.3 million barrels of oil. Its design life is 25 years.

4.7.11 Confederation bridge

This bridge is also known as the PEI bridge because it links Prince Edward Island to New Brunswick in continental Canada (Tadros et al., 1996). The structure is approximately 13 km long. It consists of seven approach spans on Prince Edward Island with a total length of 555 m, 45 marine spans with a total length of 11 080 m and 14 approach spans on the New Brunswick side with a total length of 1275 m. The vertical clearance at the centre of the navigation span is 49 m.

A typical marine span has a length of 250 m. Each span consists of a variable depth concrete box-girder measuring 14 m in depth at the pier locations and tapering parabolically to 4.5 m at middle span. The box width ranges from 5.0 m at the bottom of the section to 7.0 m at the top. The marine span is constructed as a 192.5 m-long double cantilever (Figure 4.22), each unit made of 18 line-cast segments built in 10 fixed locations in the assembly yard (Figure 4.23). The pier segment is always cast at the same location and then transported to the first casting line and fixed to the first two forms. These two first segments are cast one on each side, cured and post-tensioned. This part is then moved to the next casting line, and so on all the way to the sixth one, where the double cantilever unit is completed. Its final weight is about 7500 tonnes: 200 tonnes more than the Eiffel Tower or as much as the weight of 125 000 people. Drop-in spans either 52 m or 60 m long connect two cantilevers to create a frame or to close the gap between two frames.

The support piers for the marine spans consist of an 8 m diameter hollow octagonal concrete shaft ranging in height up to 45.8 m. The top of the shaft is joined rigidly to the superstructure. To protect the bridge against ice forces, the pier shafts have an ice shield at water level to break and deflect ice located between 4 m below and 3.5 m above mean sea level. This ice shield is protected against ice abrasion using either a 10 mm-thick steel plate or a 100 MPa high-performance concrete.

The footing on which the pier shaft sits is 22 to 26 m in diameter; it is situated on a tremie concrete base placed on the founding surface dredged down to bedrock.

All these marine elements are placed by a heavy-lift floating catamaran called the Svanen (Figure 4.24).

The design criteria for the project are intended for a 100 year service life. To achieve such a service life different concretes were developed to build the different structural elements. The mix design of these concretes was tailored to satisfy the design and durability requirements. For example, the marine spans have been built using an air-entrained high-performance concrete (for durability), having a design strength of 55 MPa. In fact the average strength of this 4 to 6% air-entrained

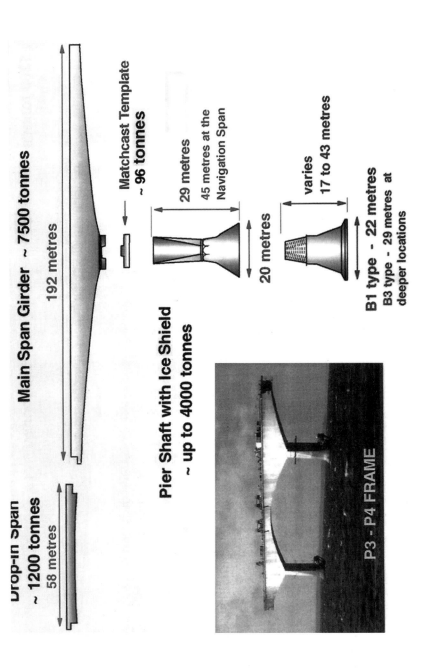

Fig. 4.22 Schematic representation of the main bridge component. Adapted from a document of Public Works and Government Services, Canada.

78 *The high-performance concrete rationale*

high-performance concrete was 72 MPa. The composition of the different concretes used for the completion of the project are given in Table 4.9. After pumping, this concrete did not meet the Canadian A23.1

Fig. 4.23 Assembly yard (Prince Edward Island side) for the Northumberland Strait Crossing Fixed Link. Courtesy of G. Tadros.

Case studies

Fig. 4.24 The Svanen in action. Courtesy of G. Tadros.

Table 4.9 Average composition of three of the concretes used during the construction of the Northumberland Strait Crossing Fixed Link precast elements (courtesy of W. Langley)

	Marine girders	Ice shield	Thick sections
W/B	0.30	0.25	0.31
Water (kg/m^3)	145	142	142
Cement (blended silica-fume cement) (kg/m^3)	430	520	330
Fly ash (class F) (kg/m^3)	45	60	130
Aggregate (kg/m^3): Coarse, $\varnothing$ max = 20 mm	1030	1100	1050
Fine	0.18	0.35	0.19
Air-entraining agent (l/m^3)	1.80	1.60	1.38
Water reducer (l/m^3)	–	0.58	0.30
Set retarder (l/m^3)	–	0.58	0.30
Superplasticizer (l/m^3)	3.20	6.00	1.80

specification for freeze–thaw durability (i.e. average spacing factor lower than 230 mm) but successfully passed 500 cycles of the ASTM C666 Procedure A (usually this standard requires that the concrete only passes 300 cycles).

The cement used was a low alkali blended silica fume cement (7.5%) and the superplasticizer was naphthalene-based. All the aggregates used in the project were non-reactive to alkalis.

The total volume of high-performance concrete used in this project was about 280 000 m^3 out of a total volume of concrete of 315 000m^3 that had been used upon completion of the project.

REFERENCES

CEB *Bulletin d'information* No. 222 (1994) *Application of High-Performance Concrete*, Report of the CEB-FIP Working Group on High-Strength/High-Performance Concrete, CEB, Lausanne, Switzerland, 75 pp.

Gordon, J.E. (1991) *Structures or Why Things Don't Fall Down*, Penguin Books, London, ISBN 014 01.3628 2, 174 pp.

Johnson, R.B. (1984) Concrete column cost optimization. *Concrete Technology Today*, **5**(3), September, 3.

Malier, Y. (1991) The French approach to using HPC. *Concrete International*, **13**(7), July, 28–32.

Schlaich, J. (1987) Quality and economy, in *Concrete Structures for the Future*, Proceedings of IABSE Symposium, Versailles, Paris, IABSE-AIPC-IVBH, Zurich, ISBN 3-85748-053-1, pp. 31–40.

Smith, G.J. and Rad, F.N. (1989) Economic advantages of high-strength concrete in columns. *Concrete International*, **11**(4), April, 37–43.

References

Water Tower Place

Anon (1976) Water Tower Place – high strength concrete *Concrete Construction*, **21**(3), March, 100–104.

Russell, H.G. and Corley, W.G. (1977) *Time Dependent Behaviour of Columns in Water Tower Place*, ACI SP-55, pp. 347–73.

Russell, H.G. and Larson, S.C. (1989) Thirteen years of deformations in Water Tower Place. *ACI Structural Journal*, **86**(2), March–April, 182–91.

Gullfaks offshore platform

Concrete Platform a Modern Fairytale, Norwegian Concrete Engineering, Concrete for the World, published in May 1988 by the Norwegian Concrete Association, Kronprinsensgate 17, N-0251 Oslo, Norway, pp 8–9.

Haug, A.K. and Sandvik, M. (1988) *Mix Design and Strength Data for Concrete Platforms in the North Sea*, ACI SP-109, pp. 495–524.

Moksnes J. (1990) *Norwegian Concrete Engineering – Concrete for the World*, Norwegian Concrete Association, Kronprinsensgate 17, N-0251 Oslo, Norway, pp. 19–21.

NC News (1986), No. 2, August, published by Norwegian Contractors Information and Public Relations Department, Holtet 45, N-1320 Stabbek, Norway, pp. 14–15.

Ronneberg, H. and Sandvik, M. (1990) High strength concrete for North Sea platforms. *Concrete International*, **12**(1), January, 29–34.

Sylans and Glacières viaducts

Baudot, J., Thao, P.X. and Radiquet, B. (1987) *Les viaducs de Sylans et des Glacières*, in *Concrete Structures for the Future*, Proceedings of IABSE Symposium, Versailles, Paris, IABSE-AIPC-IVBH, Zurich, ISBN 3-85748-053-1, pp. 493–9.

Cadoret, G. (1987) Béton à Haute Performance, in *Concrete Structures for the Future*, Proceedings of IABSE Symposium, Versailles, Paris, IABSE-AIPC-IVBH, Zurich, ISBN 3-85748-053-1, pp. 401–6.

Richard, P. (1988) Swiss cheese box girders. *Civil Engineering*, **58**(3), March, 40–43.

Roussel, J.P. (1989) Un premier bilan de le construction du viaduc de Sylans ou les difficultés de l'innovation. *Revue Générale des Routes et des Aérodromes*, no. 660, Février, pp. 31–9.

Thao, P.X. (1990) External prestressing in bridges – the example of the Sylans and Glacières viaducts (France), in ACI SP-120, *External Prestressing in Bridges*, pp. 145–57.

Scotia Plaza

Quinn, P.J. (1988) Silica fume concrete in super high rise buildings. *Construction Canada*, **30**(2), March–April, 7–8.

Ryell, J. and Bickley, J.A. (1987) Scotia-Plaza: high strength concrete for tall buildings, in *Utilization of High Strength Concrete*, Tapir, Stavanger (ed. I. Holland *et al.*), N-7034 Trondheim NTH, Norway, ISBN 82-519-0797-7, pp. 641–53.

Île de Ré bridge

Cadoret, G. (1987) Béton à Haute Performance, in *Concrete Structures for the Future*, Proceedings of IABSE Symposium, Versailles, Paris, IABSE-AIPC-IVBH, Zurich, ISBN 3-85748-053-1, pp. 401–6.

Cadoret, G. and Richard, P. (1992) Full scale use of high performance concrete in building and public works, in *High Performance Concrete – From Material to*

Structure (ed. Y. Malier), E & FN Spon, London and New York, ISBN 0-419-17600-4, pp. 379–411.

Causse, G. (1990) Ré Island bridge external prestressing, in ACI SP-120, *External Prestressing in Bridges*, pp. 175–84.

Virlogeux, M. (1990) The Ré Island bridge, in *Bridges and Tunnels*, Proceedings of the XI International Congress on Prestressed Concrete, F18, Hamburg, pp. 186–92.

Two Union Square

Aïtcin, P.C. (1989) From gigapascals to nanometers, in *Engineering Science Foundation Conference on Advances in Cement Manufacture and Use*, (ed. E. Gartner), American Society of Civil Engineers, Potosi, Mo., pp. 105–30.

Godfrey, K.A., Jr (1987) Concrete strength record jumps 36%. *Civil Engineering*, **57**(10), October, 84–8.

Howard, N.L. and Leatham, D.M. (1989) The production and delivery of high-strength concrete. *Concrete International*, **11**(4), April, 26–30.

Ralston, M. and Korman, R. (1989) Put that in your pipe and cure it. *Engineering News Record*, **22**(7), February, 44–53.

Simons, B.P. (1989) Getting what was asked for with high-strength concrete. *Concrete International*, **11**(10), October, 64–6.

Joigny bridge

Malier, Y., Brazilier, D. and Roi, S. (1991) The bridge of Joigny, a high performance concrete experimental bridge. *Concrete International*, **13**(5), May, 40–42.

Malier, Y., Brazilier, D. and Roi, S. (1992) The Joigny bridge: an experimental high performance concrete bridge, in *High Performance Concrete – From Material to Structure* (ed. Y. Malier), E & FN Spon, London and New York, ISBN 0-419-17600-4, pp. 424–31.

Schaller, I., de Larrard, F., Sudret, J.P., Acker, P. and Le Roy, R. (1992) Experimental monitoring of the Joigny bridge, in *High Peformance Concrete – From Material to Structure* (ed. Y. Malier), E & FN Spon, London and New York, ISBN 0-419-17600-4, pp. 432–57.

Montée St-Rémi bridge

Aïtcin, P.-C. and Lessard, M. (1994) Canadian experience with air-entrained high-performance concrete. *Concrete International*, **16**(10), October, 35–8.

Coulombe, L.G. and Ouellet, C. (1994) The Montée St-Rémi overpass crossing autoroute 50 in Mirabel. The savings achieved by using HPC. *Concrete Canada Newsletter*, **2**(1), December, 2–5 (available from the Network Administrative Centre at Sherbrooke University).

Lachemi, M. and Aïtcin, P.-C. (1997) Influence of ambient and fresh concrete temperatures on the maximum temperature and thermal gradient in a high performance concrete. *ACI Materials Journal*, **94**(2), March–April, 102–10.

Lachemi, M., Lessard, M. and Aïtcin, P.-C. (1996) *Early Age Temperature Developments in a High Performance Concrete Viaduct*, ACI SP-167, pp. 149–74.

Lachemi, M., Bouzoubaâ, N. and Aïtcin, P.-C. (1996) *Thermally Induced Stresses During Curing in a High Performance Concrete Bridge: Field and Numerical Studies*, Proceedings of the 2nd International Conference in Civil Engineering on Computer Applications: Research and Practice, Bahrain, April, Vol. 2, University of Bahrain Press, pp. 451–7.

References

'Pont de Normandie' bridge

Boulay, C. and de Larrard, F. (1993) The sand-box: a new capping method for testing high-performance concrete cylinders. *Concrete International*, **15**(4), April, 63–6.

Monachon, P. and Gaumy, A. (1996) *The Normandie Bridge and the Société Générale Tower*, HSC Grade 60, 4th International Symposium on the Utilization of High-Strength/High-Performance Concrete, Paris, May, paper no. 125, 13 pp.

Virlogeux, M. (1993) *Le Point sur le Projet du Pont de Normandie, Annale de l'ITBTP*, no. 517, série génie civil 216, October, pp. 2–78.

Hibernia offshore platform

Hoff, G.C. and Elimov, R. (1995) *Concrete Production for the Hibernia Platform*, Supplementary Papers, 2nd CANMET/ACI International Symposium on Advances in Concrete Technology, Las Vegas, Nevada, 11–14 June, pp. 717–39.

Woodhead, H.R. (1993a) Hibernia development project – development of the construction site. *Canadian Civil Engineering Journal*, **20**(3), June, 528–35.

Woodhead, H.R. (1993b) Hibernia offshore oil platform. *Concrete International*, **15**(12), December, 23–30.

Confederation bridge

Tadros, G., Combault, J., Bilderbeek, D.W. and Fotinos, G. (1996) *The Design and Construction of the Northumberland Strait Crossing Fixed Link in Canada*, 15th Congress of IABSE, Copenhagen, Denmark, 16–20 June, 24 pp.

Other reading

Cook, J.E. (1989) 10,000 psi concrete. *Concrete International*, **11**(10), October, 67–75.

de Larrard, F. (1993) Application des bétons à hautes performances aux ouvrages d'art. Condition pour une mise en oeuvre de qualité. *Bulletin de Liaison des Laboratoires des Ponts et Chaussées*, No. 187, September–October, pp. 37–44.

Gerwick, B.C., Jr (1985) Lessons from an exciting decade of concrete sea structures. *Concrete International*, **7**(8), August, 34–7.

Hoff., G.C. (1985) The challenge of offshore structures. *Concrete International*, **7**(8), August, 12–22.

Keck, R. and Casey, K. (1991) A tower of strength. *Concrete International*, **13**(3), March, 23–5.

Laning, A. (1992) One Peachtree Centre dominates Atlanta skyline. *Concrete Construction*, **37**(2), February, 71–3.

Moreno, J. (1990) 225 W. Wacker Drive. *Concrete International*, **12**(1), January, 35–9.

Pistilli, M., Cygan, A. and Burkart, L. (1992) Concrete suppliers' fills. *Concrete International*, **14**(10), October, 44–7.

Randall, V. and Foot, K. (1986) High-strength concrete for Pacific First Center. *Concrete International*, **11**(4), April, 14–16.

Shydlowski, M. (1992) High-performance concrete plays a key role in Atlanta Tower construction. *Concrete Construction*, **37**(2), February, 81–4.

CHAPTER 5

High-performance concrete principles

5.1 INTRODUCTION

It must be admitted that up to the present time, progress in the field of high-performance concrete has been the fruit of an empirical approach rather than a fundamental and scientific one. As has often been the case in concrete technology, advances in practice have proceeded thorough scientific investigations. Moreover, at the present time, it is possible to explain the better performance of high-performance concrete on the basis of principles that can be scientifically supported, although it is not yet possible to explain every aspect of high-performance concrete in minute detail. As a consequence, it will be seen in Chapter 7 that the selection of concrete-making materials, and in Chapter 8 that mix proportions, are no longer governed by pure empiricism, but that it is possible to follow practical guidelines in order to avoid starting over.

However, in spite of this progress in the state of the art, we cannot expect that in the near future it will be possible to select 'on paper' the materials and their proportions to make economical and high-performance concrete in a given place. In fact, as long as high-performance concrete is made of about the same simple and low-cost materials that are used to make usual concrete, their actual composition will not necessarily be the best one for making high-performance concrete. As high-performance concrete still represents a small volume of the concrete market, cement producers are not interested in investing too much in modifying their production processes. Moreover, in a given place, the selection of the materials used to make high-performance concrete will always be limited by economic considerations because, in order to stay technically competitive with usual concrete, the production cost of high-performance concrete will have to be as low as possible.

It will be the art of concrete engineers and scientists to find the best (or the least limiting) system of materials to achieve the lowest possible water/binder (W/B) ratio from a concrete mixture with a workability that can be controlled long enough to make the placing, consolidation and finishing of high-performance concrete as easy as that of usual concrete.

As will be seen, making high-performance concrete is more complicated than producing usual concrete. The reason for this is that, as the compressive strength increases, the concrete properties are no longer related only to the water/binder ratio, the fundamental parameter governing the properties of usual concrete by virtue of the porosity of the hydrated cement paste. In usual concrete, so much water is put into the mixture that both the bulk hydrated cement paste and the transition zone represent the weakest links in concrete microstructure, where mechanical collapse starts to develop when concrete is subjected to a compressive load.

In this chapter, we will discuss some theoretical principles explaining how to increase concrete compressive strength while controlling rheological properties in full (or at least mostly). In particular, we will show that most strength properties of high-performance concrete are related to the hydration of silicates, while most of the time the rheological behaviour is controlled by the hydration of the interstitial phase in the presence of calcium and sulfate ions. In both cases, the amount of water available at the beginning of the hydration process is critical but in conflicting terms. Solving this conflict economically is the key to success in making high-performance concrete.

5.2 CONCRETE FAILURE UNDER COMPRESSIVE LOAD

In spite of the fact that compressive strength is not the only concrete property making the use of high-performance concrete advantageous, it is nevertheless important because, as will be seen, it is closely related to the same concrete microstructural features that also govern other properties, such as elastic modulus and durability. Moreover, from an historical point of view this is the way high-performance concrete was developed.

When the failure surface of a usual concrete specimen tested in uniaxial compression is examined with the naked eye, it can be seen that the rupture developed either within the mortar (Figure 5.1) or along the interface between the mortar and coarse aggregate particles, since these two zones constitute the weakest link in usual concrete. However, in some cases, if the concrete contains weak aggregates, some of the failure planes propagate through the aggregate particles (Figure 5.2).

If the failure surface is more closely observed using an optical or a scanning electron microscope, it can be seen that mortar failure developed either within the hydrated cement paste or along the interface between the hydrated cement paste and the aggregate. This special zone, called the transition zone, is known to be a weak area in usual concrete. Under close examination, the transition zone appears to be composed of

86   *High-performance concrete principles*

Fig. 5.1 Failure surface of a usual concrete.

Fig. 5.2 Failure surface of concrete containing a weak coarse aggregate. The failure planes have propagated through the coarse aggregate particles.

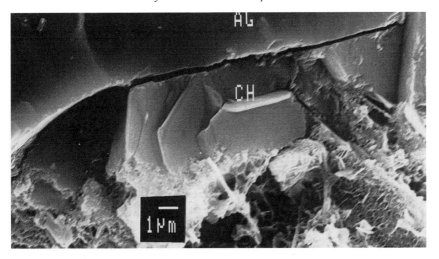

Fig. 5.3 Transition zone in a low-strength concrete (17.5 MPa). AG: aggregate; CH: hydrated lime.

a porous hydrated cement paste containing numerous and well-developed lime crystals (Figure 5.3).

Concrete failure will always develop in the weakest part of one of these three phases. Therefore, in order to increase the compressive strength of concrete, great care must be taken to strengthen these three phases (Aïtcin and Mehta, 1990). When testing concrete in compression it must be remembered that, as stated by Mehta and Aïtcin (1990):

> While most other properties of concrete are more closely related to averages rather than extremes of the microstructural components, strength and fracture depend critically on microstructural extremes rather than averages. In other words, fracture, especially in compression, is a weak-link type process. Therefore, in addition to the number, size, and shape of pores, their spatial distribution or local concentration is a major factor in failure.

The Griffith and Weibull theories on the strength of solids, although not directly applicable to concrete, are nevertheless useful in understanding the control of microstructure and strength of the material. According to Illston, Dinwoodie and Smith (1987):

> The statistical approach views the testpiece as a chain of elements, each at the same stress level, with the same probability of failure. This is equivalent to the probability of the presence of a Griffith crack of critical length, and if the crack propagates the element fails instantaneously. Further, since the testpiece is made up of a chain of elements, the failure of one element signals the failure of the whole

testpiece. Thus, the strength of the testpiece is the strength of its weakest link.

From a fracture mechanics approach, concrete can be considered as a non-homogeneous material composed of three separate phases:

- the hydrated cement paste;
- the transition zone between aggregate and hydrated cement paste;
- the aggregates (which can themselves be polycrystalline, as in the case of granite).

In the following, we will see how the failure of the hydrated cement paste in the transition zone and of the aggregates can be delayed as much as possible for a given set of materials.

5.3 IMPROVING THE STRENGTH OF HYDRATED CEMENT PASTE

Hydrated cement paste (C–S–H) can, as a first approximation, be considered to be a single-phase crystalline material to which the fundamental principles governing the behaviour of brittle solids, such as ceramics, can be usefully applied to control the microstructure and properties of concrete. This is especially true for high-performance concrete, which shows more microstructural similarities with ceramics than usual concrete.

The porosity dependence of the tensile strength of a single-phase crystalline material is generally expressed by an exponential relationship, such as $S = S_o\, e^{-bp}$, where S is the tensile strength of the material which contains a volume fraction porosity p, S_o is the intrinsic tensile strength at zero porosity, and b is a factor depending on the size and shape of pores (Aïtcin and Mehta, 1990).

The compressive strength of brittle materials is much greater than the tensile strength because in tension the material can fail by rapid propagation of a single flaw or microcrack. Since a number of tensile cracks must coalesce to cause a compressive failure, in compression much more energy is needed for the formation and propagation of the microcrack system. Assuming the compressive failure to be a multiple of several tensile failures, the Griffith tensile fracture theory and concepts of continuum mechanics can be used to predict that the compressive strength of a homogeneous ceramic is eight times its tensile strength. This ratio between compressive and tensile strengths applies quite well in concrete.

Besides this, there has been no attempt specifically to derive a theory of the microstructural dependence of the compressive strength of porous materials. However, from empirical compressive strength studies many researchers have reported that the experimental data can be fitted to the equation: $f'_c = f_o'\, (1 - p)^m$, where f'_c is the compressive strength of the material containing the volume fraction porosity p, f_o' is the intrinsic compressive strength at zero porosity, and m is a factor depending on intercrystalline bonding in the solid, the shape and size of pores or flaws, grain size and the presence of impurities. In general, compressive strength decreases with increasing pore size and increases with decreasing grain size. Studies involving microstructure–strength relations in ceramics have also shown that, besides porosity, grain size and the presence of inhomogeneities, there are the other important factors that control tensile strength.

In conclusion, the strength of hydrated cement paste can be improved by paying close attention to the following parameters:

- porosity: a large number of big pores or voids (diameter > 50 nm), especially when concentrated in one location, are detrimental to strength;
- grain size: in general, the strength of a crystalline phase increases with decreasing grain size;
- inhomogeneities: with multiphase materials, microstructural inhomogeneities are a source of strength loss.

In order to improve the strength of the hydrated cement paste it is necessary to work on the microstructure of the hydrated paste at these three levels (Nielsen, 1993).

5.3.1 Porosity

When the anhydrous silicate compounds come into contact with water their hydration starts to take place through solution. In other words, the liquid phase becomes saturated with various ions which combine to form different hydration products that start filling the space originally occupied by water. At any point during the hydration process, the unfilled space between cement particles will consist of voids or capillary pores. With increasing hydration and decreasing capillary porosity, the moisture movement in the system becomes sluggish. The subsequent hydration of the unhydrated portions of large cement particles is believed to be a slow process through solid-state reactions.

Scanning electron microscope studies have shown that the early hydration products formed when there is plenty of water and empty space in the cement–binder system consist of flocs of large crystals which generate a considerable volume of voids (Figure 5.4). Since the early hydration products crystallize out in the water-filled space surrounding the cement particles (i.e. outside the original boundary of a hydrating cement particle), researchers refer to them as 'outer products'. On the other hand, the hydration products from solid-state reactions, formed within the original boundaries of hydrating cement particles, are called the 'inner products' and are more compact and poorly crystalline (Figure 5.5) (Mehta and Monteiro, 1993). As the strength of hydrated Portland

Fig. 5.4 Outer type products of hydration.

cement paste is derived mainly from van der Waals's forces of attraction, it follows that the more compact (i.e. free from large pores) and poorly crystalline the hydration products the higher the strength. This explains why in a hydrated cement paste the path of fracture is through the outer products rather than the inner products.

It should be obvious now that, from the standpoint of the strength of hydrated cement paste, it is highly desirable to obtain a microstructure resembling the inner products by somehow eliminating the formation of outer products (Mehta and Monteiro, 1993). The concept of outer- and

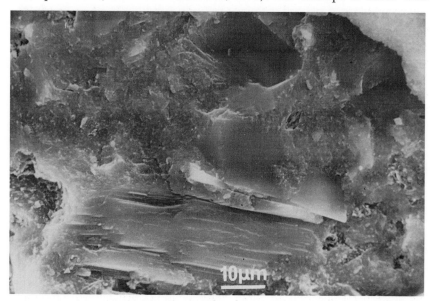

Fig. 5.5 Inner type products of hydration.

inner-type products in cement hydration is useful in appreciating the role of a low water/binder ratio, superplasticizers and supplementary cementitious materials for making high-performance concrete, as discussed later.

However, there is one type of porosity that cannot be avoided, namely the chemical contraction accompanying the hydration of the silicate phases. The final volume of hydrated cement paste is 8 to 10% smaller than the combined volume of anhydrous silicates and water. This volume contraction causes the hydrated cement paste to shrink during the hydration process when it is not properly cured. As part of this shrinkage will be restrained by the aggregate skeleton, a network of ultrafine microcracks develops within the hydrated cement paste. The extent of this network of very fine cracks depends on the amount of the anhydrous silicate phase that has become hydrated and on the amount of restraint by the aggregate. These points will be covered in more detail in Chapter 12, dealing with the curing of high-performance concrete.

The main factors affecting the porosity of the hydrated cement phase are the ratio of the volume of water available, the volume of the silicate phase to be hydrated and the amount of air entrapped during mixing. This was recognized as early as 1892 by Féret, who formulated his famous expression:

$$f'_c = k\left(\frac{c}{c+w+a}\right)^2$$

where f'_c is the compressive strength of the hydrated cement paste, c, w, and a are the volumes of cement, water and air, respectively, and k is a constant depending on the type of cement (Féret, 1892). As Féret would have been dealing, without any doubt, with pure Portland cement, we will, like him, use the letter c to represent the volume of Portland cement.

Féret's expression can be rewritten if both the numerator and denominator are divided by c as:

$$f'_c = k\frac{1}{(1+(w/c)+(a/c))^2}$$

As in hydrated cement paste or concrete, the volume of entrapped air is usually less than 1 or 2% of the total volume of concrete, so the term a/c can be neglected. Therefore Féret's expression can be rewritten as:

$$f'_c = k\frac{1}{(1+(w/c))^2}$$

It is clear that in order to increase the compressive strength, the water/cement ratio must be reduced.

When the water/cement ratio of the hydrated cement paste is reduced, the cement particles come closer together in the freshly mixed cement paste, as shown in Figure 5.6. As a result, there is less capillary porosity

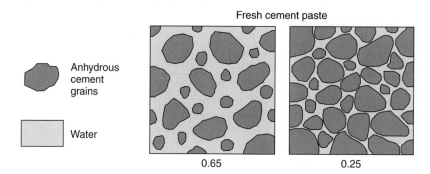

Fig. 5.6 Schematic representation of two fresh cement pastes having a water/cement ratio of 0.65 and 0.25. In this representation the ratio of the surface of the water to that of the cement is equal to the massic water/cement ratio.

and less free space for the outer products to develop. Moreover, as there is less water available, water becomes more rapidly saturated with the ions responsible for the development of the outer hydration products. As the cement particles are now closer to each other, the outer product has less of a gap to bridge in order to develop early bonding between cement particles. This explains why low water/cement ratio cement pastes develop strength more rapidly. Finally, as the cement particles are close to each other and rapidly bonded, moisture movement becomes sluggish, favouring the formation of inner products during hydration.

From this discussion we can see that, in order to decrease the porosity of the hydrated cement paste, it is necessary to lower, as much as possible, the amount of entrapped air and the water/cement ratio in the freshly mixed cement paste. This has to be achieved while the freshly mixed paste remains fluid enough to give concrete the workability necessary for transport and placement.

Taking into account the pre-eminent influence of the water/cement ratio on hydrated cement paste, some researchers have set up records of strength, which unfortunately have not yet been reported in the *Guinness Book of Records*! Using a high-pressure compaction technique (350 MPa pressure) and high-temperature curing (250 °C for 2 hours) Roy *et al.* made cement pastes with a water/binder ratio as low as 0.093, and a compressive strength of up to 470 MPa at 28 days (Roy, Gouda and Bobrowsky, 1972). Regourd *et al.* reported an attempt to achieve ultra high-strength cement paste: 204 MPa at 8 days (Regourd, Hornain and Mortureux, 1978). This paste contained an ultrafine cement with a Blaine specific surface area of 750 m^2/kg, 0.5% diethylcarbonate used as a grinding aid, 1% lignosulfonate and 0.5% potassium carbonate used as a set-controlling admixture. The porosity was reported to be as low as 5% by volume, corresponding to a hydration degree of the anhydrous cement lower than 70%.

Bache (1981), using more conventional concrete techniques, made concrete with a compressive strength of 270 MPa using a very strong aggregate, a high dosage of superplasticizer and silica fume in a 0.16 water/binder ratio mixture. More recently, following the pioneering work of Birchall and Kelly (1983), Young (1994) and his research team have developed a 'Macro Defect Free' concrete with compressive and tensile strengths that could not have been foreseen a few years ago. Tensile stresses at failure of 70 MPa are currently achieved when properly mixing cement, cellulose hydroxypropylmethyl cellulose or hydrolysed polyvinylacetate and water.

These sophisticated compositions show that, when previous theoretical concepts are applied to hydrated cement paste, ultra high-strength Portland cement-based materials can be made. A special chapter at the end of this book (Chapter 19) is devoted to these Portland cement-based materials.

Thus from a theoretical point of view, when making high-performance concrete, one of the key factors is the maximum possible reduction of the water/cement or water/binder ratio.

5.3.2 Decreasing the grain size of hydration products

Decreasing the water/binder ratio favours the formation of inner products characterized by a fine texture. The C–S–H of such inner products looks like a highly compact glassy phase when observed under a scanning electron microscope, as shown in Figure 5.5. At very low water/binder ratios, hydrated cement paste does not contain any large stacks of hexagonal platelets of lime, long needles of ettringite or long hairy filaments of C–S–H that are characteristic of high water/binder ratio pastes.

5.3.3 Reducing inhomogeneities

Entrapped air bubbles can be considered as microstructural inhomogeneities that should be minimized in high-performance concrete when high strength is the final objective. This can be achieved through proper consolidation. The natural tendency of cement particles to flocculate in loose form should also be eliminated. In that respect, it will be shown that the superplasticizers necessary to lower the water/binder ratio play a key role in improving the dispersion of cement particles in freshly mixed paste. However, it will be seen later on that the entrainment of a small quantity of air (2 to 3%) is very helpful in the placing and finishing of high-performance concrete.

5.4 IMPROVING THE STRENGTH OF THE TRANSITION ZONE

In the 0.40 to 0.70 water/binder ratio range, it has been found that the experimental values of concrete compressive strength are considerably lower than those predicted by the strength–porosity relationship for hydrated cement paste or mortar with the same water/binder ratio. The combined effects of the aggregate grain size distribution and water/binder ratio on concrete strength must be investigated further in order to explain this anomalous behaviour. During consolidation coarse aggregate particles, depending on their size, shape and surface texture, prevent a homogeneous distribution of water in fresh concrete. Because of this localized 'wall effect', some bleed-water tends to accumulate at the surface of coarse aggregate particles (Neville, 1995). As a result, the local water/binder ratio in the cement paste next to coarse aggregates, called the transition zone, becomes substantially higher than the water/binder ratio in the cement paste some distance away. Compared with the bulk cement paste, the microstructure of the transition zone is characterized by the presence of large pores and large crystalline hydration products (Figure 5.3). This phenomenon is responsible for the microstructural inhomogeneities, which, as discussed below, have serious implications for concrete strength.

Microstructural differences between the hydrated cement paste in the transition zone and the bulk cement paste play an important role in determining the strength characteristics of concrete via the Weibull weakest link theory. In usual concrete, the transition zone is typically 0.05 to 0.1 mm wide and contains relatively large pores and large crystals of hydration products. Therefore, according to this theory, when the concrete is subjected to a given stress, the cracks first begin to develop in the transition zone.

With usual concrete (0.50 to 0.70 water/binder ratio), tensile stresses induced by drying shrinkage or thermal shrinkage strains are large enough to cause microcracking in the transition zone even before subjecting the concrete to any service loads. The fracture strength of the element under load is thus controlled by the propagation and joining together of a part of the microcrack system in the hydrated cement paste (Illston, Dinwoodie and Smith, 1987).

In general, the strength of concrete increases with age as long as the unhydrated cement particles continue to form hydration products, which tends to reduce the size and the total volume of voids, especially in the transition zone. Moreover, with aggregates derived from limestone, certain siliceous rocks and calcinated clays or shales, it seems that, in addition to van der Waals's bonds, some chemical bonding at later ages contributes to the transition zone strength. Generally speaking, in the 0.50 to 0.70 water/binder ratio range, it can be said that the inherently weak microstructure of the transition zone prevents concrete from behaving like a true composite material. As long as large pores and microcracks continue to be present in the transition zone, the strength of aggregate particles plays no part in determining the strength of the composite material (concrete), since there is little effective stress transfer between the bulk cement paste and the aggregate.

Of course, the situation would change if the weakest link in usual concrete, namely the transition zone, were somehow strengthened so that under increasing stress it would not be the first component to fail. Under these conditions, the strength and elastic properties of aggregate become important in determining the behaviour of concrete when subjected to increasing levels of stress (Aïtcin, 1988; Baalbaki et al., 1991; Ezeldin and Aïtcin, 1991; Baalbaki, Aïtcin and Ballivy, 1992). Reducing the water/binder ratio and using silica fume tend to reduce the thickness and weakness of the transition zone.

5.5 THE SEARCH FOR STRONG AGGREGATES

The selection of particularly strong aggregates is not necessary when producing usual concrete. Generally, it is only necessary to check that standard performance requirements for aggregates are met. On the other hand, in high-performance concrete, the hydrated cement paste and the transition zone can be made so strong that, if the aggregates, particularly the coarse ones, are not strong enough, they can become the weakest link within the concrete, a situation similar to that encountered in lightweight aggregate concrete (Figure 5.2).

The aggregates used to make high-performance concrete are natural sand and gravel or crushed aggregate (Aïtcin, 1988). The strength of natural aggregates depends on the nature of the parent rock, which was reduced to its present size through various natural weathering processes. As a result, nothing can be done to improve the strength of natural aggregates: they must be used as they are. Petrographic studies can be useful when using natural aggregates because they give an indication of the strength of the different particles constituting the aggregate. Another approach is to include these aggregates in a high-performance concrete and to look at the fracture surface after compression failure.

If crushed aggregates are used to make high-performance concrete, their processing leads to individual particles containing the minimum possible concentration of weak elements. As blasting and crushing are not particularly gentle treatments, in order to obtain an ideal defect-free aggregate particle, a fine-textured strong rock that can be fractured in particles containing a minimum amount of microcracks should be selected. This rock can be a single-phase material, such as limestone, dolomitic limestone and syenite, or a polyphasic material such as granite. Rocks containing weak cleavage planes or severely weathered particles must be avoided. Detailed geological and petrographic studies can help in the search for a strong aggregate (Aïtcin and Mehta, 1990).

Experience shows that in some places it is the strength of the aggregate that constitutes the weakest link in high-performance concrete. This will be discussed in more detail in the next chapter. For higher compressive strengths in the future, more attention will need to be paid to the selection of aggregates, which used to be considered to be almost inert fillers in concrete.

So far we have viewed high-performance concrete merely as concrete with a very low water/binder ratio, often containing such ultrafine materials as silica fume to improve the bulk paste and transition strength (Goldman and Bentur, 1989) and made with strong aggregates. Binder particles, being closer to each other, develop very strong bonds during their hydration, but this closeness may also create rheological problems in fresh concrete.

5.6 RHEOLOGY OF LOW WATER/BINDER RATIO MIXTURES

While from a strength point of view it is essential to use the lowest possible water/binder ratio, we must remember that high-performance concrete has to be transported and placed with relative ease using conventional construction procedures (reviewed in Chapter 11).

In field applications, concrete usually must retain an adequate workability for about an hour and a half. In precast plants, where placing is more rapid, it is generally sufficient to ensure a high workability for up to half an hour.

Concrete rheology is essentially governed by physical and chemical factors. Among the physical factors are the grain size distribution and shape of the aggregates. In very low water/binder ratio systems, the grain size distribution and shape of the cement particles can also play an important role in determining the rheology of fresh concrete. Among the chemical factors affecting the rheology of fresh concrete are the initial reactivity of the cement and supplementary cementitious materials when in contact with water, and the duration of the so-called 'dormant period'.

These two fundamental aspects of concrete rheology will be examined separately in sections 5.6.2 and 5.6.3. Among other factors affecting the rheology of concrete are the general condition of the mixer and its mixing efficiency, particularly its shearing action, the temperature of high-performance concrete after it has been mixed and the ambient temperature.

5.6.1 Optimization of grain size distribution of aggregates

A considerable amount of work has been undertaken on the optimization of the grain size distribution of powders or aggregate skeletons. However, the objective in these cases was usually the reduction of the porosity of the final product rather than improvement in concrete workability. Most of the present state of the art is the result of early work by Fuller and Thompson (1907) in North America and by Bolomey (1935), Faury (1953) and Caquot (1937) in France. Most of these authors arrived at different recommendations which are still in use. It is well established, for example, that flat and elongated aggregate particles are detrimental to workability, and that cubic or spherical aggregates produce better workability. However, little research has been carried out on the influence of the grain size distribution of aggregates on high-performance concrete workability (de Larrard, 1987; de Larrard and Buil, 1987). Therefore, the rules that were developed and successfully used in usual concrete have often been extended to high-performance concrete.

5.6.2 Optimization of grain size distribution of cementitious particles

Fundamental research has been carried out to highlight the need for optimizing the grain size distribution of cementitious materials for improving workability and strength. Bache was one of the first to point out the advantages of adding silica fume to very low water/binder ratio concrete mixtures to improve workability (Bache, 1981). He explained the beneficial effect of adding such a fine powder in such a concrete mixture by the fact that when minute spherical silica fume particles are well dispersed in the cement–water system, they can displace water molecules from the vicinity of cement grains, so that entrapped water molecules between flocculated cement particles can be freed and thus contribute to fluidizing the mixture. Detwiler and Mehta (1989) later showed by using an inert carbon black having similar grain size to silica fume that the chemical nature of the very fine particles was not critical.

5.6.3 The use of supplementary cementitious materials

By replacing some of the cement by a supplementary cementitious material containing no C_3S, C_3A or C_4AF, the rheology of any given concrete mixture becomes easier to control, provided that the grain size distribution and shape of the particles of the supplementary cementitious material are almost the same as those of the replaced cement. Such replacement is attractive from an economic viewpoint since the price of a supplementary cementitious material is usually lower than that of cement (except for silica fume, in most cases). Moreover, it will be seen later that savings can also be obtained from the reduction in the dosage of superplasticizer necessary to achieve the desired workability. However, the optimum dosage of cementitious material has to be found, taking into account strength requirements at early age.

5.7 THE WATER/BINDER RATIO LAW

As already mentioned, strength increase is obtained mainly through a drastic reduction of the porosity of hydrated cement paste. This porosity reduction is obtained by adding more cement while reducing the amount of mixing water through the use of superplasticizers and by replacing, whenever it is feasible and economical, some cement with an equal volume of a supplementary cementitious material, and, perhaps in the near future, with any filler having the right particle size distribution and shape.

Generally, there is no argument when dealing with the amount of water. Water on the surface of the aggregate or in the admixture is added to the actual mixing water, and any water absorbed by the aggregate is subtracted from it. But are we really sure that in mixtures having very low water/binder ratios, all this water combines with the cement and supplementary cementitious materials? At the present time, there is increasing evidence that this is not the case.

Should the water/binder ratio be calculated taking into account only the cement used and neglecting the amount of supplementary cementitious materials, or should the ratio of water/(cement + supplementary cementitious materials) be used? It is known that not all supplementary cementitious materials have the same binding properties as that of Portland cement and that they do not all react at the same rate as cement. Should the amount of different supplementary cementitious materials be multiplied by a correction factor to take into account the difference in their binding properties, as has been done by de Larrard (1990)?

There is also increasing evidence that at many years of age, there are still many unhydrated cement particles and unreacted supplementary cementitious particles in high-performance concrete. Therefore, if we are sure that: (1) not all of the water present in concrete, W, will finally combine with any part of the cementitious system; (2) not all of the cement particles, C, will hydrate; and (3) not all of the supplementary cementitious materials, S, will react, is it still worthwhile calculating the following different ratios: W/C or $W/(C + S)$ or W/B with $(B = C + S)$ or $W(C + kS + k'S')$, k and k' being factors depending on the efficiency of the supplementary cementitious material (Hassaballah and Wenzel, 1995; Baron and Ollivier, 1996)?

It is the author's opinion that the first two ratios should be systematically calculated and given for any high-performance mixture, but that they are not as fundamentally characteristic of a particular high-performance concrete as they are of usual concrete.

It is interesting to calculate the exact water/cement ratio, W/C, of any particular concrete since it gives a fairly good idea of the conditions under which the early setting and hardening of the concrete mixture will develop, because at an early age the supplementary cementitious material is not yet very active. This last statement is not always true because the supplementary cementitious material (e.g. fly ash) can bring with it soluble alkalis that will increase the 12 hour concrete compressive strength when compared with a reference mixture without any fly ash. This has also been found to be true for limestone filler. However, in such a case, this beneficial effect of the fly ash or limestone filler does not last very long because the 24 hour strength of fly ash or limestone filler concrete is lower than that of a reference concrete made with Portland cement.

It is also interesting to calculate the W/B ratio because it gives an idea of the mixture's solid content in terms of the total amount of fine particles and water available to hydrate them; this can be used to quantify the closeness of the fine particles within the liquid phase.

It is, however, the author's opinion that in the future the most important values to characterize high-performance concrete will be: (1) the content of mixing water, (2) the content of cement and (3) the content of total fine particles used to make a high-performance, flowing concrete with slump greater than 200 mm.

5.8 CONCLUDING REMARKS

From a materials point of view, high-performance concrete is nothing more than a concrete with a very low porosity. This very low porosity is achieved mainly by using much less mixing water than in usual concrete so that cement and supplementary cementitious particles are much closer to each other than in usual concrete mixtures.

As the porosity of the paste decreases, the strength of the concrete increases as long as the aggregates, particularly the coarse ones, are strong enough. Thus the selection and proportioning of high-performance concrete ingredients is a much more critical issue than in the case of usual concrete. This point is developed in Chapters 7 and 8.

REFERENCES

Aïtcin, P.-C. (1989), From gigapascals to nanometres, in *Engineering Foundation Conference on Advances in Cement Manufacture and Use* (ed. E. Gartner), American Society of Civil Engineers, Potosi, Mo., pp. 105–30.

Aïtcin, P.-C. and Mehta, P.K. (1990) Effect of coarse-aggregate characteristics on mechanical properties of high-strength concrete. *ACI Materials Journal*, **87**(2), March–April, 103–7.

Baalbaki, W., Aïtcin, P.-C. and Ballivy, G. (1992) On predicting elastic modulus of high-strength concrete. *ACI Materials Journal*, **89**(5), September–October, 517–20.

Baalbaki, W., Benmokrane, B., Chaallal, O. and Aïtcin, P.-C. (1991) Influence of coarse aggregate on elastic properties of high-performance concrete. *ACI Materials Journal*, **88**(5), 499–503.

Bache, H.H. (1981) Densified cement/ultra-fine particle-based materials. Presented at the Second International Conference on Superplasticizers in Concrete, 10–12 June, Ottawa, Canada, published by Aalborg Cement, Aalborg, PO Box 163, DK-9100 Aalborg, Denmark, 12 pp.

Baron, J. and Ollivier, J.-P. (1996) *Les Bétons – Bases et Données pour leur Formulation*, Eyrolles, ISBN 2-212-01316-7, pp. 288–305.

Birchall, J.D. and Kelly, A. (1983) New inorganic materials. *Scientific American*, **248**(5), May, 104–15.

Bolomey, J. (1935) Granulation et prévision de la résistance probable des bétons. *Travaux*, **19**(30), 228–32.

Caquot, A. (1937) Le rôle des matériaux inertes dans le béton. *Mémoire de la Société des Ingénieurs Civils de France*, Fasc. No. 4, July–August, pp. 562–82.

de Larrard, F. (1987) Modèle linéaire de compacité des mélanges granulaires, in *Structure and Materials Properties, Proceedings of the First RILEM Congress, Versailles* (ed. J. C. Mass), Vol. 1, Chapman & Hall, London, pp. 325–32.

de Larrard, F. (1990) A method for proportioning high-strength concrete mixtures. *Cement, Concrete, and Aggregates*, **12**(1), 47–52.

de Larrard, F. and Buil, M. (1987) Granularité et compacité dans les matériaux de génie civil. *Matériaux Construction*, RILEM, **20**(116), 117–26.

Detwiler, R.J. and Mehta, P.K. (1989) Chemical and physical effects of condensed silica fume in concrete. Supplementary paper, in *Third International Conference on Fly Ash, Silica Fume, Slag and Natural Pozzolans in Concrete*, Trondheim, (ed. M. Alasali), CANMET, Ottawa, pp. 295–306.

Ezeldin, A. and Aïtcin, P.-C. (1991) Effect of coarse aggregate on the behavior of normal and high-strength concretes. Technical Note in ASTM Journal *Cement, Concrete, and Aggregates*, **13**(2), 121–4.

Faury, J. (1953) *Le Béton, Influence de ses Constituants inertes. Règles à Adopter pour sa Meilleure Composition. Sa Confection et son Transport sur les Chantiers*, 3rd edn, Dunod, Paris, pp. 66–7.

Féret, R. (1892) Sur la compacité des mortiers hydrauliques. Mémoires et documents relatifs à l'art des constructions et au service de l'ingénieur. *Annales des Ponts et Chaussées*, Vol. 4, 2nd semester, pp. 5–161.

Fuller, W.B. and Thompson, S.E. (1907) The laws of proportioning concrete. *Transactions of the American Society of Civil Engineers*, **59**, 67–143.

Goldman, A. and Bentur, A. (1989) Bond effects in high-strength silica fume concretes. *ACI Materials Journal*, **86**(5), September–October, 440–47.

Hassaballah, A. and Wenzel, T.H. (1995) *A Strength Definition for the Water to Cementitious Materials Ratio*, ACI SP-153, pp. 417–37.

Illston, J.M., Dinwoodie, J.M. and Smith, A.A. (1987) *Concrete, Timber and Metals: The Nature and Behaviour of Structural Materials*, Van Nostrand Reinhold, New York, ISBN 0-442-30145-6, 663 pp.

Mehta, P.K. and Aïtcin, P.-C. (1990) *Microstructural Basis of Selection of Materials and Mix Proportions for High-Strength Concrete*, ACI SP-121, pp. 265–86.

Mehta, P.K. and Monteiro, P.J.M. (1993) *Concrete – Structure, Properties and Materials*, Prentice-Hall, 2nd edn, ISBN 0-07-041344-4, pp. 190–97.

Neville, A.M. (1995) *Properties of Concrete*, 4th edn, Pitman, London, ISBN 0-582-23070-5, 844 pp.

Nielsen, L.F. (1993) Strength development in hardened cement paste: examination of some empirical equations. *Materials and Structure*, **26**(159), June, 255–60.

Regourd, M., Hornain, H. and Mortureux, B. (1978) Influence de la Granularité des Ciments sur leur Cinétique d'hydratation. *Ciments, Bétons, Plâtres et Chaux*, No. 712, pp. 137–43.

Roy, D.M., Gouda, G.R. and Bobrowsky, A. (1972) Very high strength cement pastes prepared by hot pressing and other high pressure techniques. *Cement and Concrete Research*, **2**(3), May, 349–65.

Young, J.F. (1994) *Engineering Microstructures for Advanced Cement-Based Materials*, tribute to Micheline Moranville Regourd, Sherbrooke, published by the Network of Centres of Excellence on High-Performance Concrete. Available from Concrete Canada, Faculty of Applied Sciences, University of Sherbrooke, Sherbrooke, J1K 2R1, Canada.

CHAPTER 6

Review of the relevant properties of some ingredients of high-performance concrete

6.1 INTRODUCTION

Readers of this book probably already possess a general knowledge of the properties of the materials used in making concrete; detailed knowledge can be found in specialist books. Although high-performance concrete does not contain any special or unusual ingredients, it does require the use of materials with certain specific properties, at least in some respects. It is therefore useful to review the relevant properties, and this will be done in this chapter, starting with Portland cement and then continuing with superplasticizers and finally with supplementary cementitious materials. Aggregates will be considered in Chapter 7.

In the following, we will use the usual simplified cement chemistry notation for the Portland cement phases: C for CaO, S for SiO_2, A for Al_2O_3, F for Fe_2O_3 and H for H_2O. Therefore, $C_3S = 3CaO \cdot SiO_2$, $C_2S = 2CaO \cdot SiO_2$, $C_3A = 3CaO \cdot Al_2O_3$ and $C_4AF = 4CaO \cdot Al_2O_3 \cdot Fe_2O_3$. The usual expressions 'alite' and 'belite' will also be used to refer to the impure forms of C_3S and C_2S that are found in Portland cement clinker.

6.2 PORTLAND CEMENT

6.2.1 Composition

It is the burning of a well-proportioned mixture of raw materials containing the four main oxides – CaO, SiO_2, Al_2O_3 and Fe_2O_3 – that produces Portland cement clinker, one of the two basic ingredients required to make Portland cement. The other is calcium sulfate in the form of gypsum ($CaSO_4 \cdot 2H_2O$) or hemihydrate ($CaSO_4 \cdot \frac{1}{2}H_2O$) or anhydrite or calcium sulfate ($CaSO_4$), or a mixture of two or three of them (Bye, 1983).

The chemical composition of the raw materials in terms of CaO, SiO_2 and Al_2O_3 is adjusted so that the final mixture falls in the composition triangle limited by C_3S, C_2S and C_3A in the SiO_2—CaO—Al_2O_3 ternary phase diagram (Figure 6.1), because it is only in this area that these three phases can exist simultaneously (Philips and Muan, 1959). For simplification from a phase point of view, it can be assumed that Fe_2O_3 and Al_2O_3 play similar roles, but Figure 6.2 shows that if Portland cement were composed only of these three oxides it would have to be burned at a very high temperature, so iron oxide is added into the raw mix in order to lower the firing temperature.

Limestone provides the necessary amount of CaO in the mix, and clay or shale supplies SiO_2, Al_2O_3 and Fe_2O_3. Some natural siliceous limestones contain SiO_2, Al_2O_3 and Fe_2O_3 in such proportions that Portland cement clinker can be almost exclusively produced from them. In such cases sand, bauxite and iron oxide as a source of SiO_2, Al_2O_3 and Fe_2O_3,

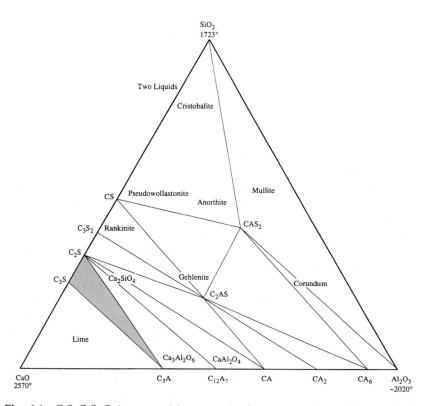

Fig. 6.1 C_3S–C_2S–C_3A composition zone in the ternary phase diagram SiO_2–CaO–Al_2O_3 (after Osborn and Muan, 1960).

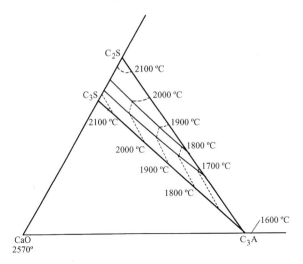

Fig. 6.2 Detail of the SiO_2–CaO–Al_2O_3 phase diagram in the C_3S–C_2S–C_3A area showing melting temperature (after Osborn and Muan, 1960).

respectively, are added as corrective materials to adjust the chemical composition of the raw mix.

When the optimization of the proportions of these four main oxides has been accomplished, Portland cement manufacturers have to deal with some impurities present in the raw materials which during the burning process will lead to the formation of more or less desirable phases that may or may not interfere with the regular hydration process of a mixture of pure C_3S, C_2S, C_3A and C_4AF phases.

6.2.2 Clinker manufacture

Portland cement clinker is the end product of a highly complex pyro-processing technology that transforms the raw materials into calcium silicates and calcium aluminate/iron phases. The nature of the fuel used in the cement kiln, or to be more precise the nature of the impurities present in it, is also relevant. In particular, sulfur and ash content can play a critical role during the formation of clinker: ash as a source of impurities and sulfur in the formation of volatile alkali sulfates.

From a composition point of view, Portland cement clinker comes out of the rotary kiln as a mixture of two well-crystallized silicate phases, C_3S and C_2S, and an interstitial phase composed of C_3A and C_4AF, more or less crystallized (Figure 6.3), and a few 'impurities' such as periclase, (MgO), hard burnt lime (CaO) and alkali sulfates.

The SiO_2–CaO binary phase diagram (Figure 6.4) clearly shows that, if Portland cement clinker were made of a mixture, M, consisting of CaO and SiO_2 lying in the C_2S–C_3S range, slowly cooled to room temperature, it will be composed of γ-C_2S and lime (CaO) having no hydraulic properties, because C_3S is not a stable compound at room temperature. Figure 6.4 shows that above 1450 °C mixture M is composed of C_3S and α-C_2S, but during cooling to room temperature the following phase transformations are likely to occur within the solid phase:

- 1450 °C: transformation of the mixture into C_3S and α'-C_2S;
- 1250 °C: transformation of the mixture into α'-C_2S and CaO;
- 725 °C: transformation of the mixture into γ-C_2S and CaO.

Therefore, in order to stabilize the silicates in their reactive form, the clinker must be cooled rapidly (quenched) after passing through the clinkering zone in the kiln (Figure 6.3). The quenching freezes the different phases in their high-temperature form, not allowing them to follow the natural phase transformation which would have occurred under very slow cooling. The microstructure of the clinker after this

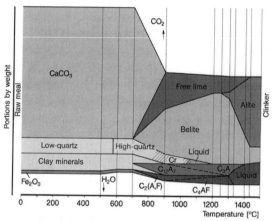

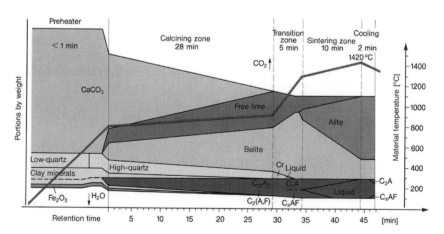

Fig. 6.3 Schematic representation of the phase transformations occurring in a cement kiln (courtesy of Humboldt Wedag).

quenching influences significantly its hydraulic properties: two Portland cements can have exactly the same chemical composition but quite different microstructures and hydraulic behaviour.

It has been also mentioned that the effective role of Al_2O_3 and Fe_2O_3 is to lower the temperature at which the raw materials have to be burned so as to be transformed into clinker. The aluminate and ferrite phases melt during the clinkering process and the presence of this liquid phase results in a significant decrease in the temperature of C_3S formation; it allows Ca^{++} ions to diffuse more rapidly and more easily into the previously formed C_2S to facilitate its transformation into C_3S (Figure 6.3).

From a thermodynamic point of view, the higher the amount of interstitial phase the lower the burning temperature, but from a manufacturing point of view, the optimum amount of interstitial phase has to represent 12 to 20% of the total mass of the clinker phases. Usually it is rather close to 15 to 16% in North American clinkers.

If there is too much liquid phase during the clinkerization process it can exude out of the silicate phases, resulting in the build up of a ring within the kiln (Palmer, 1990), that can block it and result in an attack of the refractories. On the other hand, if there is an insufficient amount of

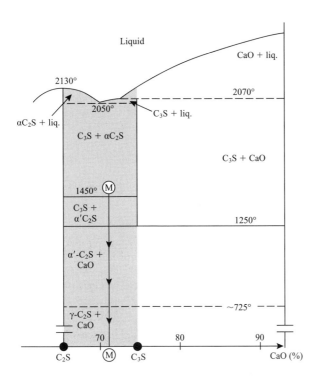

Fig. 6.4 SiO_2–CaO binary diagram in the C_2S–C_3S area (after Philips and Muan, 1959).

liquid phase, the diffusion of Ca^{++} ions in C_2S is more difficult, so the clinker contains less C_3S than it should and too much uncombined lime. Moreover, the resulting clinker is very abrasive, resulting in a rapid wear of the refractory lining in the clinkering zone of the kiln.

Thus the cement manufacturer has a certain latitude in the amount of interstitial phase that can be contained in the clinker that is produced. Moreover, changing the ratio of Al_2O_3 to Fe_2O_3 can modify the hydraulic properties of the clinker, allowing the production of ASTM Type I, II, III or V from almost the same raw materials.

In this brief presentation of the main phases of clinker, the influence of the firing conditions (Maki et al., 1990), whether oxidizing or reducing, and of several other factors that affect clinker quality have purposely not been addressed in order to stay within reasonable limits of space.

6.2.3 Clinker microstructure

Thus Portland cement clinker is a complex multiphase material, the characteristics and properties of which depend not only on the chemical composition of the raw meal but also on the pyrotechnology that transforms it into clinker. Two clinkers can have exactly the same chemical composition but quite different microstructural features (Figure 6.5). They can have exactly the same phase composition and still have different hydraulic properties because, for example, after quenching, the average size of alite crystals can be quite different from one clinker to the other, a characteristic that strongly influences its reactivity and strength after grinding (Figure 6.6).

Moreover, in all these previous discussions it was assumed that all the phase transformations that were supposed to occur were in fact occurring. In reality, perfect equilibrium conditions are never reached within a rotary kiln, particularly if the raw meal contains too many coarse siliceous particles of quartz, when numerous belite 'nests' (i.e. regions particularly rich in belite) will form in the clinker (Figure 6.7). During final sintering, these particular areas are too rich in silica, so the C_2S cannot be transformed into C_3S due to the lack of nearly unavailable CaO. Similarly, if the meal contains too coarse limestone particles, hard burnt lime-rich areas develop in the clinker (Figure 6.8). Finally, depending on the firing conditions, C_2S can contain different amounts of impurities that will influence its characteristics.

Depending on the viscosity of the interstitial phase in the sintering zone, it can either be well dispersed between C_2S and C_3S, or be more concentrated in some specific area. Depending on the temperature in the clinkering zone and the severity of quenching, the solidified interstitial phase can be in an almost entirely vitreous state in the case of a high temperature and rapid quenching, because C_3A and C_4AF crystals did not have adequate time to crystallize from the liquid phase, or it can be crystalline if the quenching phase were slow enough to allow C_3A and

C_4AF to crystallize (Figure 6.9). Usually the interstitial phase solidifies as a mixture of C_3A and C_4AF crystals within a vitreous matrix.

Moreover, the C_3A present in the interstitial phase can be found in different polymorphic forms depending on the amount of trapped Na^+ and K^+ ions. If the amount of Na^+ ions in the C_3A is less than 2.4%,

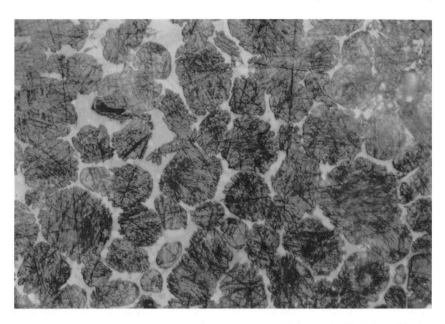

Fig. 6.5 Micrographs of two clinkers having the same composition but different microstructural features.

108   *Review of relevant properties*

the C_3A will retain the cubic form of pure C_3A (Regourd, 1978), whereas if this amount is higher than 5.3%, it will form monoclinic C_3A. In between these two values, C_3A has an orthorhombic lattice. However, as the transformation from cubic to orthorhombic form is gradual between 2.4 and 3.8%, generally speaking the C_3A found in Portland cement is a mixture of cubic and orthorhombic forms. Whether the C_3A is cubic or

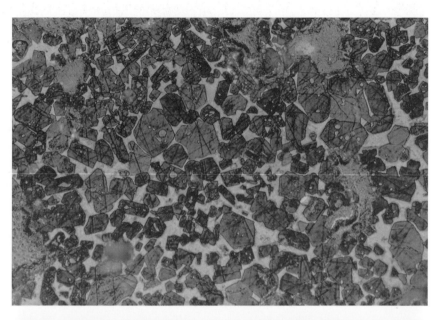

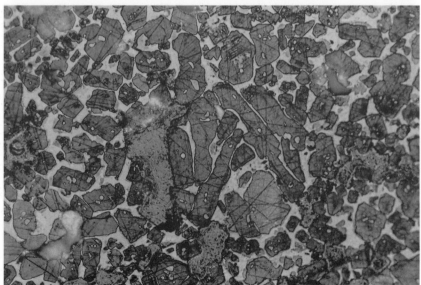

Fig. 6.6 Alite crystals (C_3S) of different sizes within a clinker.

Portland cement

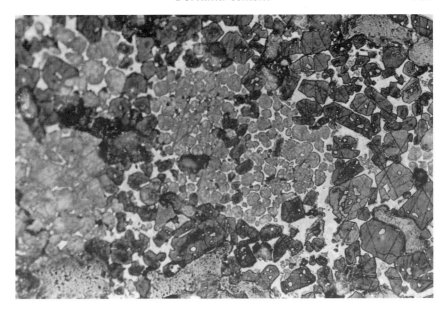

Fig. 6.7 Belite nests in a clinker.

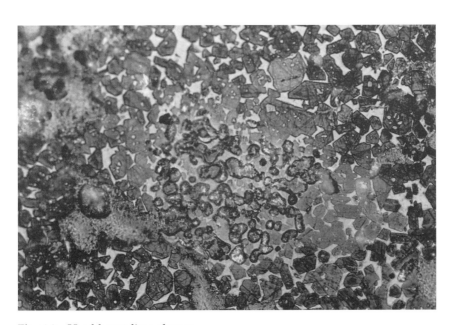

Fig. 6.8 Hard burnt lime cluster.

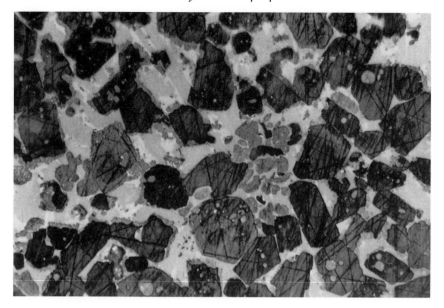

Fig. 6.9 Well-crystallized interstitial phase.

orthorhombic, or a mixture of both, is a very important feature from a rheological point of view during cement hydration, especially in very low water/binder ratio mixtures in the presence of superplasticizers.

The amount of Na^+ ions trapped in C_3A (that mainly influence the crystalline form of C_3A) depends on the Na^+/SO_3^- balance within the kiln. If there is an excess of SO_3, because the fuel used is rich in sulfur, the alkalis will immediately combine with this SO_3 to form first alkali sulfates (Grzeszczyk, 1994; Miller and Tang, 1996). Thus only a few Na^+ ions will enter the C_3A and it will crystallize in the cubic form. In the contrary case, if the amount of alkalis exceeds the amount of SO_3 needed to form sodium sulfate, Na^+ ions in excess will enter the C_3A lattice and lead to the formation of orthorhombic C_3A. The alkali sulfates that are formed in such conditions are deposited in the clinker, either as droplets near C_3S and C_2S, or are intimately associated with the interstitial matrix, as shown in Figure 6.10.

Finally, it is important to remember that C_4AF is in fact a solid solution of C_2A and C_2F where the molar ratio of A/F is generally almost equal to 1. However, in some cases C_4AF can be closer to C_2A or C_2F, or even contain C_6A_2F crystals and be more or less reactive during Portland cement hydration.

Several other microstructural features can vary from one clinker to another; this is why in some sense each clinker is unique (Hornain, 1971; Gebauer and Kristmann, 1979). The phase composition of clinker between one kiln and another (or even between one batch and another) can be different since, on the one hand, the raw materials are never

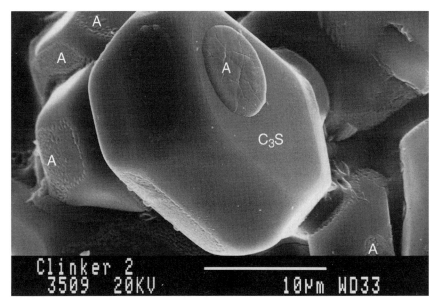

Fig. 6.10 Alkali sulfates (A) in a clinker (courtesy of I. Kelsey-Lévesque).

identical and, on the other, because it is very difficult to reproduce exactly the same firing conditions which so much influence clinker microstructure (Odler, 1991).

The complexity of Portland cement clinker phasic composition presented in this simplified version of reality explains why, most unfortunately, the knowledge of only the chemical composition of a particular clinker is not enough to predict its hydraulic behaviour, especially when making high-performance concrete with a very low water/binder ratio.

6.2.4 Portland cement manufacture

In order to produce Portland cement, clinker has to be interground with an optimum amount of calcium sulfate. The role of this calcium sulfate is to control the initial hydration of Portland cement. In the absence of calcium sulfate, ground clinker undergoes a flash set which is an irrecoverable stiffening of the mixture. Because C_3A reacts very rapidly with water to form hydrogarnet (Regourd, 1978; Mindess and Young, 1981), if some calcium sulfate is interground with clinker, then C_3A reacts with it and water to form a more or less impervious shell of ettringite that inhibits the reaction of C_3A with water (Collepardi et al., 1979).

During Portland cement manufacture, the amount of calcium sulfate-bearing minerals, their solubility and the cement fineness are adjusted so that the Portland cement produced satisfies all the requirements of relevant national standards.

Usually, the calcium sulfate-bearing material is gypsum, $CaSO_4 \cdot 2H_2O$, but the use of anhydrite ($CaSO_4$) or calcium sulfate ($CaSO_4$) or a mixture of both gypsum and anhydrite is becoming popular for economic and production reasons. These forms of calcium sulfate present about the same final solubility when they dissolve in water, but their dissolution **rate** is quite different: gypsum releases its Ca^{++} and SO_4^{--} much faster than does anhydrite.

Moreover, during the final grinding, it is possible that the temperature within the ball mill becomes higher than 110 °C, so some gypsum dehydrates partially and is transformed into hemihydrate (or plaster of Paris), $CaSO_4 \cdot \frac{1}{2}H_2O$, which has a solubility higher than that of gypsum. In order to have an easier control of the rheological properties of grout, fresh mortar or concrete, in some cases conversion of some gypsum into hemihydrate is favourable, because the rate of dissolution of SO_4^{--} ions is higher for hemihydrate than for gypsum and so it promotes the formation of ettringite instead of hydrogarnet during early hydration. However, if too much gypsum is transformed into hemihydrate during grinding, **false set** can be experienced: concrete stiffens rapidly during mixing with water. This false set is due to rehydration and precipitation of some hemihydrate into gypsum. In this case, if mixing is lengthened, most of the plasticity of the mix can be recovered because some of the gypsum formed during the rehydration of the hemihydrate redissolves in the remaining mixing water.

This explains why during the grinding of clinker and calcium sulfate the temperature within the grinding mill has to be closely controlled.

In some cases, in order to limit the temperature of Portland cement during its final grinding, water is sprayed inside the grinding mill so that the excess of hemihydrate that could be responsible for a false set rehydrates to gypsum. Usually, it is found that a 40 to 50% transformation rate of gypsum into hemihydrate is optimum. Currently, however, with the use of more high-efficiency air separators in modern cement plant, sometimes clinker does not experience a sufficiently high temperature during its grinding, and in some cases not enough gypsum is dehydrated into hemihydrate. Thus if the sulfate content is optimized, taking into account only the SO_3 content in the cement, the early dissolution of SO_4^{--} ions (especially in very low water/binder mixtures in the presence of a superplasticizer) can be too low to control C_3A hydration efficiently and avoid a more or less severe flash set situation.

As will be seen later, the rheology of average or high water/binder ratio mixtures (W/B $\geq$ 0.50) is not affected too much by variations in the nature of the calcium sulfate present in Portland cement at the end of its grinding, but this is not the case as the water/binder ratio decreases.

Moreover, Portland cement clinker always contains alkali sulfates generated during the clinkering process owing to the reaction of the sulfur contained in the fuel and the alkalis present in the raw materials. Therefore, arcanite (K_2SO_4) and aphthitalite ($Na_2SO_4 \cdot 3K_2SO_4$) are found in Portland cement. These sulfates could have precipitated at the surface of clinker C_2S or C_3S crystals or have been trapped within the interstitial phase.

The solubility of alkali sulfate is generally greater than that of calcium sulfate, so the interstitial water can contain a high amount of SO_4^{--} ions in solution but very few Ca^{++} ions. In such a case ettringite cannot be formed and setting problems can be experienced. Usually this situation is overcome by adjusting, at the cement plant, the calcium sulfate dosage and the calcium sulfate phase composition. However, this fragile equilibrium can be destroyed when using a lignosulfonate-based water reducer (Paillère et al., 1984; Dodson and Hayden, 1989) or polysulfonate-based superplasticizer (Ranc, 1990), as will be seen later in dealing with the so-called cement/superplasticizer compatibility.

Finally, during grinding, cement manufacturers often introduce minute amounts of chemicals known as grinding aids to increase the output of the grinding mill operation in terms of tonnes/hour, and at the same time reduce the pack-set factor of the cement, permitting a smooth filling and discharge of the Portland cement within the silos. These chemicals also influence Portland cement hydration characteristics, even though this subject is poorly documented in the literature.

Therefore, even in the case of uniform clinker production, it is not certain that a cement uniform enough for low water/binder application will be produced from a rheological point of view if the nature and proportions of the sulfate-bearing materials change or if the temperature within the grinding mill is not well controlled.

6.2.5 Portland cement acceptance tests

As Portland cement is a multiphasic material, its hydraulic properties depend on:

- the chemical composition of the raw meal and the fuel used in terms not only of the four major oxides, but also of other minor components;
- the pyroprocessing technology that transforms this raw meal into silicate and alumino-ferrite phases and the so-called impurities;
- its final grinding (Regourd, Hornain and Mortureux, 1978).

Therefore it has been necessary to develop a set of acceptance tests that guarantee users a certain level of uniformity in relation to the performance and binding properties of Portland cement.

While these acceptance tests vary slightly from one country to another, they are similar in broad terms. Tests consist of making a standardized paste and/or a standardized mortar, using a standard sand and a fixed water/binder ratio of a given flowability, and verifying that the rheology in the fresh state, the hardening rate and the strength stay within prescribed limits. Some other chemical soundness features also form part of the acceptance tests to control the level of some potentially detrimental impurities, such as periclase (MgO).

As an example, the ASTM standards related to the acceptance of Portland cement will be very briefly discussed in order to show how these sets of rules have been developed for conditions found in usual concrete, but not necessarily for conditions that are found in high-performance concrete.

For a cement to be marketed as an ASTM Type I to V Portland cement, tests according to the following ASTM standard test methods are used:

ASTM C 109/ C 109M-95	Standard Test for Compressive Strength of Hydraulic Cement Mortars (Using 2 inch or 50 mm Cube Specimens)
ASTM C 114-94	Standard Test Method for Chemical Analysis of Hydraulic Cement
ASTM C 115-94	Standard Test Method for Fineness of Portland Cement by the Turbidimeter
ASTM C 151-93a	Standard Test Method for Autoclave Expansion of Portland Cement
ASTM C 186-94	Standard Test Method for Heat of Hydration of Hydraulic Cement
ASTM C 204-94a	Standard Test Method for Fineness of Hydraulic Cement by Air Permeability Apparatus
ASTM C 266-89	Standard Test Method for Time of Setting of Hydraulic Cement by Gillmore Needles
ASTM C 348-93	Standard Test Method for Flexural Strength of Hydraulic Cement Mortars
ASTM C 349-94	Standard Test Method for Compressive Strength of Hydraulic Cement Mortars (Using Portions of Prisms Broken in Flexure)
ASTM C 451-89	Standard Test Method for Early Stiffening of Portland Cement (Paste Method)
ASTM C 150-95	Standard Specification for Portland Cement

The ASTM standard mortar in the ASTM C 109/C 109M-95 acceptance test is composed of 1 part cement, 2.75 parts Ottawa sand (by weight) and a certain amount of water. The W/B ratio is 0.485 for Type I Portland cement. For other types of cement, the amount of water must be adjusted to produce a flow of 110 ± 5 using a flow-table.

The rheological behaviour is verified using the so-called Gillmore ASTM C266-89 or Vicat ASTM C191-92 needles that give the initial and final setting time, while the strength is checked by measuring the compressive strength of 50 mm cubes (2 inch cubes) at different ages according to the type of cement and at the sacrosanct age of 28 days.

Whatever the acceptance values, it must be emphasized that all these tests are performed on a mixture of Portland cement and water with a water/binder ratio of around 0.50. For many years, these conditions were representative of normal usage of Portland cement in concrete, or, even better, represented the lower end of the usual water/binder ratio range used in most commercial concretes. Therefore it was a very safe way of accepting cement.

In order to satisfy these acceptance tests, cement manufacturers know how to modify the chemical composition of the raw meal, adjust the burning process, fine-tune the specific surface area of the Portland cement and optimize the calcium sulfate content during the final grinding of clinker.

However, who now mixes water and Portland cement without any chemical admixtures when making concrete?

Because the acceptance tests deliberately ignore the possible interactions during hydration between certain admixtures and Portland cement, concrete producers have, from time to time, experienced strange behaviour in some freshly mixed concrete, generally referred to as Portland cement/superplasticizer 'incompatibility'. Also, the 0.50 water/binder ratio value no longer represents the lowest value used in the industry; many concretes are made with a water/binder ratio lower than this.

6.2.6 Portland cement hydration

The literature on this subject has inspired many researchers, and it appears that the present outflow of research papers related to this subject will continue, (Mindess and Young, 1981; Vernet and Cadoret, 1992; Nonat, 1994; Eckart, Ludwig and Stark, 1995; Persson, 1996). This interest in understanding a very simple technological behaviour, namely the delay of setting time from the great variety of Portland cement characteristics, even when they comply with the same acceptance standard, is of great interest from a technological point of view. The details of the chemical reactions that take place during hydration are very complex, and it has been said that concrete is the fruit of a very simple technology but a very complex science.

It must be admitted that at present the details of the chemical processes that transform Portland cement paste into a hard solid mass are not fully understood in their most detailed features (Van Damme, 1994). However, progress has been made in this field and the main steps of cement hydration are well understood. It is also well known how to alter, to a certain extent at least, the kinetics of cement hydration using specific chemical admixtures such as accelerators, retarders, water reducers and superplasticizers. It is not my intention to write a state-of-the-art report on this subject, but rather to focus attention on the specific steps that are of special interest for high-performance concrete.

In order to describe the hydration process in a schematic way, the presentation of this subject by C. Vernet (1995) will be used. Figures 6.11

116   *Review of relevant properties*

to 6.14 are reproduced from this document to describe the five initial steps of Portland cement hydration.

(a) STEP 1   Mixing period

During this step, the different ions liberated by the various phases enter into solution. This dissolution is quite rapid and exothermic in nature, and two fast-reacting hydrates germinate. The surface of the cement particles becomes partially covered with hydrated calcium silicate (C—S—H) formed from Ca^{++}, $H_2SiO_4^{--}$ and OH^- ions originating from the silicate phases of the clinker, and with ettringite (a hydrated trisulfo-aluminate salt of calcium) formed by the combination of Ca^{++}, AlO_2^-, SO_4^{--} and OH^- ions originating from the interstitial phase and from the different forms of the calcium sulfate which are present in the cement.

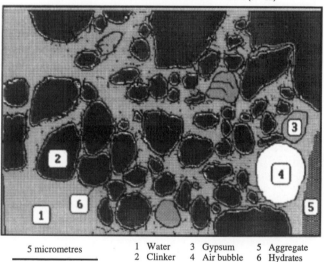

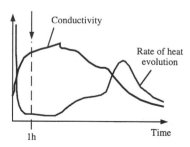

Fig. 6.11   Microstructure of the cement paste at 1 hour (after Vernet, 1995).

Portland cement

(b) STEP 2 The dormant period (Figures 6.11 and 6.12)

The rapid increase in both the pH and Ca^{++} ion content of the mixing water further slows down the dissolution of the clinker phase. The thermal flux slows down considerably, but it never stops. A small amount of $C-S-H$ is formed during this period, and if there is the right balance between aluminium and sulfate ions, reduced amounts of ettringite or hydrogarnet are also formed. During this period, the aqueous phase becomes saturated in Ca^{++}, but there is no lime $(Ca(OH)_2)$ precipitation, most probably because of its slow germination rate in comparison with that of the competing $C-S-H$. Some flocculation of the cement grains also occurs during this period.

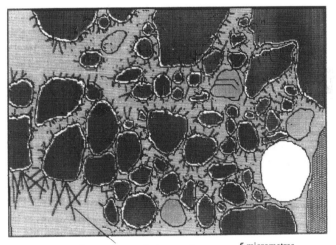

The long needles represent ettringite crystals   5 micrometres

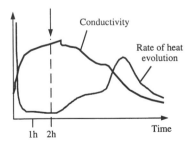

Fig. 6.12 Microstructure of the cement paste at 2 hours (after Vernet, 1995).

(c) STEP 3 Initial setting (Figure 6.13)

The hydration reaction is suddenly activated when lime starts to precipitate. This occurs when there is practically no more silicate in the aqueous phase. This sudden consumption of Ca^{++} and OH^- ions speeds up the dissolution of all the Portland cement components. The thermal flux increases slowly at the beginning (because CH precipitation is endothermic and consumes some heat) and becomes faster at a later stage.

Usually, initial set falls within this period, except when some stiffening of the mortar occurs due to the development of ettringite needles and some C—S—H. The hydrated silicate and aluminate phases start to create some interparticle bonding, resulting in a progressive stiffening of the paste.

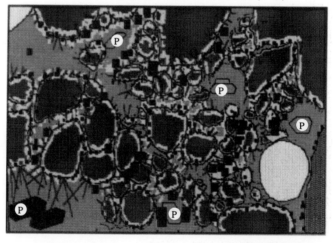

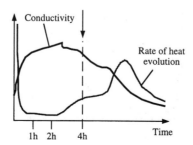

Fig. 6.13 Microstructure of the cement paste at 4 hours (after Vernet, 1995).

Portland cement

(d) STEP 4 Hardening (Figure 6.14)

In most Portland cements, there is less calcium sulfate than the amount necessary to react with the aluminate phase, so that during the setting SO_4^{--} ions are initially totally consumed by the formation of ettringite. This usually occurs between 9 and 15 hours after the initial mixing. At that time, ettringite becomes the source of sulfate to form monosulfoaluminate with the remaining aluminate phase. This reaction generates heat and results in the acceleration of the hydration of silicate phases.

Note: The hydration products formed during these first steps are often referred to as external products because they grow out of the cement grains into the interstitial aqueous phase. They appear as a porous and

HARDENING

MICROSTRUCTURE OF THE CEMENT PASTE (t = 9h)

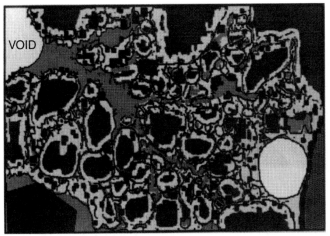

When there is no more calcium sulphate, ettringite crystals dissolve and calcium monosulphoaluminate is precipitated

5 micrometres

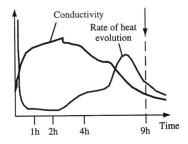

Fig. 6.14 Microstructure of the cement paste at 9 hours (after Vernet, 1995).

loose network of fibrous C—S—H, ettringite needles, monosulfoaluminate platelets and hexagonal stacked portlandite crystals.

(e) STEP 5 Slowdown

At this stage of hydration, the cement grains are covered by a layer of hydrates which becomes thicker and thicker, and it is progressively more difficult for water molecules to reach the still unhydrated part of the cement particles through this thick layer. Hydration slows down because it is mostly controlled by the rate of diffusion of water molecules through the hydrate layers, and the hydrated cement paste appears as a very compact 'amorphous' massive paste known as internal product.

Portland cement hydration stops either when there is no more anhydrous phase (well-cured high water/binder ratio concrete), or when water can no longer reach the unhydrated phases (very dense and highly deflocculated systems) or when there is no more water available, if that happens (very low water/binder ratio).

6.2.7 Concluding remarks on Portland cement hydration from a high-performance concrete point of view

From a strength point of view it is important that Portland cement develops as much dense C—S—H as it can, because calcium silicates (which represent about 80% of the total mass of Portland cement) are responsible for strength development in concrete. This can be achieved by decreasing the water/binder ratio, but, at the same time, this decrease in the water/binder ratio, which results in the decrease of the amount of water available to hydrate Portland cement, should not impair the rheology of the fresh concrete. Also, an appropriate balance between sulfate, calcium and aluminium ions must exist within the fresh cement paste in order to avoid the formation of hydrogarnet which results in flash set.

Therefore, the polymorphic nature of the C_3A becomes very important as far as the rheology of very low water/binder ratio mixtures is concerned. If the C_3A is cubic, it reacts rapidly with sulfate ions to form an almost impervious ettringite shell around C_3A which prevents its further hydration. In such a case SO_4^{--} ions must be available very rapidly. If the C_3A is orthorhombic, it reacts less rapidly but forms a loose network of long ettringite needles which keep growing without creating the same kind of tight barrier as when the C_3A is cubic. In such a case it is important that a steady release of SO_4^{--} ions occurs in order to control the rheology.

From a practical point of view, when making high-performance concrete with present Portland cements, it can be said that it is often easier to get the targeted strength than to gain an easy control of the rheology.

6.3 PORTLAND CEMENT AND WATER

It has been known for more than a century that for a given cement content, the lower the water content of concrete, the stronger it is.

Water is clearly an essential ingredient of concrete which fulfils two basic functions: a physical function, to give concrete the required rheological properties, and a chemical function, to produce the reactions of hydration. The ideal concrete should contain only enough water to develop the maximum possible strength of cement while providing the rheological properties needed for placement (Grzeszczyk and Kucharska, 1990).

Unfortunately, present Portland cements preclude attaining this ideal concrete. On the one hand, cement particles, characterized by many unsaturated surface charges, have a strong tendency to flocculate when in contact with a liquid as polar as water (Kreijger, 1980; Chatterji, 1988). Adopting such a flocculated structure implies that cement particles trap a certain amount of water within the flocks and this water is then no longer available to lubricate the mix, as can be seen in Figure 6.15. On the other hand, the hydration reactions do not wait until concrete is in the forms before starting. Hydration begins as soon as Portland cement comes into contact with water because, on the one hand, some of the cement compounds are very reactive and, on the other, cement contains very fine particles, so a large surface area of reactive phases is in contact with water.

In order to give a certain level of workability to concrete when only cement and water are used, it is necessary to use more water than is necessary to hydrate all the cement particles fully. This additional mixing water, which will never be linked to any cement particle, generates

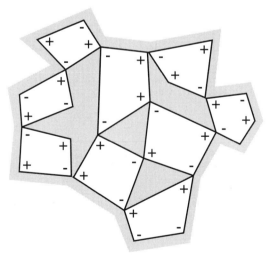

Fig. 6.15 Cement paste particles in a flocculated structure (after Kreijger, 1980).

Review of relevant properties

porosity within the hydrated cement paste, and results in a weakening of the mechanical properties of concrete and a decrease in its durability.

As it is impossible to manufacture a Portland cement that does not flocculate, in order to enhance hydration it is necessary to find chemical admixtures able to reduce this natural tendency to flocculation, and thus to reduce the amount of mixing water required.

6.3.1 From water reducers to superplasticizers

Some 60 years ago it was found that certain organic molecules known for their dispersing properties could also be used to neutralize the electrical charges present at the surface of cement particles and thus reduce their tendency to flocculate. These molecules are still used and marketed as water reducers, superplasticizers and dispersing agents. To a chemist, water reducers can be anionic, cationic or even non-ionic (Kreijger, 1980). Anionic and cationic reducers are composed of molecules having a highly charged end that neutralizes an opposite electrical charge on a cement particle. In the case of non-ionic water reducers, these molecules act like dipoles glued to the cement grains. Figure 6.16 illustrates the mode of action of these three types of water reducer molecule. In an effort to reduce the amount of mixing water in concrete, it was found that lignosulfonate molecules, which are obtained from a pulp and paper mill waste, act as excellent dispersing agents (Figure 6.17).

This by-product is inexpensive and requires only simple processing to be used successfully in concrete. Using lignosulfonates it is possible to obtain a 5 to 10% water reduction before some drawbacks occur; these include strong retardation due to the presence of wood sugars and entrapment of large air bubbles caused by surfactants present in the wood.

Of course, this reduction in the mixing water is not high enough to eliminate all the mixing water unnecessary to hydrate all cement particles fully. It is possible to refine lignosulfonates further, by removing more sugars and more surfactants (Mollah *et al.*, 1995), but the final product can sometimes be too expensive for the concrete industry.

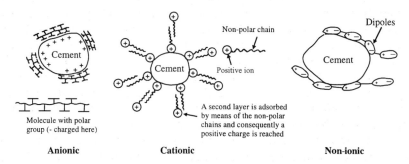

Fig. 6.16 Different types of dispersing agent (after Kreijger, 1980).

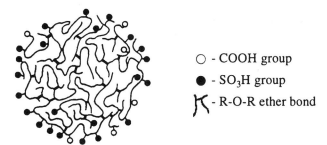

Fig. 6.17 Schematic representation of a lignosulfonate polyelectrolyte microgel unit (after Rixom and Mailvaganam, 1986).

Using lignosulfonates, it is possible to produce concretes with a compressive strength of about 50 to 60 MPa (Blick, Petersen and Winter, 1974).

In the search for a higher water reduction, it was found as early as 1932 (when Tucker applied for a patent which was delivered only in 1938 (Tucker, 1938)) that polycondensate naphthalene sulfonate molecules possessed the requisite properties. Unfortunately, at that time these synthetic products were not competitive on a price basis with the usual water reducers and lignosulfonates. Moreover, at that time, the concrete industry was fully convinced that there was no advantage in increasing concrete compressive strength above the 20 to 30 MPa range (Black, Rossington and Weinland, 1963).

For almost 40 years, the concrete industry world-wide was satisfied with the performance of these first-generation water reducers and lignosulfonates, until, almost simultaneously, the dispersing virtues of naphthalene sulfonate formaldehyde condensate were rediscovered in Japan and a new family of powerful water reducers based on melamine sulfonate formaldehyde condensates was introduced in Germany (Hewlett and Rixom, 1977; Hattori, 1978; Meyer, 1978). These products are known either as superplasticizers, high-range water reducers or fluidifiers.

Initially, these molecules were almost always used to fluidify ordinary concrete in the field just prior to placing, lignosulfonates still being used during the mixing at the plant. One of the main advantages of these new molecules, apart from their efficiency in fluidifying concrete mixtures without segregation, was that they could be used in much higher dosages than the previous water reducers because they were synthetic products made of raw materials that did not contain sugar or surfactant impurities. However, one of the major drawbacks of these first-generation superplasticizers was the very brief duration of their fluidifying action: at best, they were efficient for 15 to 30 minutes only (Young, 1983; Ramachandran, Beaudoin and Shilua, 1989), so they had to be incorporated into the mix just prior to its placing in the forms (Bonzel and Siebel, 1978; Malhotra, 1978).

As superplasticizer usage became more common, it was realized that these molecules could be used to reduce the amount of mixing water to a level never experienced before, and still produce a mix with a high workability. These synthetic molecules are so powerful in dispersing cement particles that, for the first time in concrete technology, it is possible to make a fluid concrete having a water/binder ratio below 0.30. Owing to the action of these molecules, it is possible indirectly to create the conditions of the ideal cement, as stated earlier. Consequently, in recent years the compressive strength of concrete has increased to a level previously unthought of.

Over time, superplasticizer manufacturing technology has improved so that at present some superplasticizers can maintain a high slump for 45 to 90 minutes. However, there are still some Portland cements that do not behave well with superplasticizers and low-quality superplasticizers that do not perform well with any cement, as will be shown later in this chapter.

In order to visualize the tendency of cement grains to flocculate in the presence of water and compare the efficiency of water reducers and superplasticizers to deflocculate cement particles, the following simple experiment can be performed: 50 g of Portland cement is put into each of three 1 l beakers. The first one, identified by the letter W in Figure 6.18, contains only cement and water; the second one, identified by the letter L, contains 10 ml of a lignosulfonate-based water reducer mixed with water; and the third one, identified by the letters SUP contains 10 ml of superplasticizer mixed with water. The lignosulfonate and superplasticizer dosages are purposely 10 times higher than their usual dosage in concrete. It has been found that with such a lignosulfonate or superplasticizer dosage it is possible to annihilate any tendency for the cement particles to flocculate in such dispersed solutions.

The three beakers are shaken for 1 minute so as to obtain a homogeneous suspension and left to rest for 48 hours. After this time all the cement particles have sedimented, as seen in Figure 6.18(a), but if we look more closely at Figure 6.18(b), which represents a close-up of the bottom of the three beakers, it is seen that the height of the volume occupied by the same amount of cement is different in each case:

$h_W > h_L > h_{SUP}$;
$h_L \cong 0.7\, h_W$ and $h_{SUP} = 0.5\, h_W$.

This simple experiment illustrates on the one hand the tendency of Portland cement to flocculate, and on the other hand the difference in the efficiency of a lignosulfonate-based water reducer and of a naphthalene superplasticizer to deflocculate cement particles.

Moreover, if we observe the top of graduate L, containing the lignosulfonate, it is seen that during the shaking some 'foam' developed along a certain height h_f, but none in the case of the superplasticizer (Figure 6.18(a)). Lignosulfonate water reducers always contain some tensioactive components that entrain air; on the contrary superplasticizers that are made from the synthesis of pure chemicals do not contain such tensioactive agents.

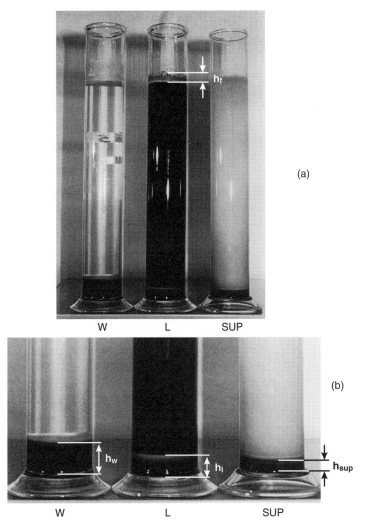

Fig. 6.18 Setting of Portland cement: (a) view of the beakers after 48 hours; (b) detailed view of the volume occupied by the settled particles. W: in water; L: in water + water reducer; SUP: in water + superplasticizer.

6.3.2 Types of superplasticizer

There are four main families of commercial superplasticizer (Bradley and Howarth, 1986; Rixom and Mailvaganam, 1986):

1. Sulfonated salts of polycondensate of naphthalene and formaldehyde, usually referred to as polynaphthalene sulfonate or more simply as naphthalene superplasticizers.
2. Sulfonated salts of polycondensate of melamine and formaldehyde, usually referred to as polymelamine sulfonate or more simply as melamine superplasticizers.
3. Lignosulfonates with very low sugar and low surfactant contents.
4. Polyacrylates.

At present, the most widely used bases when making superplasticizers are of the first two types, but in their formulations commercial superplasticizers can contain a certain amount of normal water reducers, such as lignosulfonates and gluconates. Commercial superplasticizers can be used in conjunction with water reducers, with set retarders or even with accelerators.

6.3.3 Manufacture of superplasticizers

In order to understand why within the same family a particular superplasticizer can be more efficient than another and how superplasticizers work in concrete, it is useful to know, at least in broad terms, how they are manufactured (Lahalih, Absi-Halabi and Ali, 1988). In the following, the manufacture of naphthalene superplasticizer will be briefly described because it is most familiar to the author.

In the case of naphthalene sulfonate, the manufacturing steps are sulfonation, condensation, neutralization and filtration.

(a) First step: sulfonation (Figure 6.19)

In this step, naphthalene and sulfuric acid are mixed in appropriate proportions in the reactor, which is heated. The acidic sulfonate group HSO_3^- is fixed in one of the two possible positions on the two carbon rings of the naphthalene molecule. The 12 o'clock position is called α and the 2 o'clock position is called β (Figure 6.19). The 4, 6, 8 and 10 o'clock positions can be deduced from the α and β positions by rotation around the axis of symmetry of the naphthalene group. From a molecular model of the sulfonated naphthalene molecule, it can be seen that when the sulfonate group is in the α position it has little freedom to rotate because it is almost blocked by the two hydrogens that are in the 2 and 10 o'clock positions. However, when the sulfonate group is in the β position it can rotate freely on itself over 360°.

It is now well established that the β position of the sulfonate group makes the superplasticizer molecule more efficient; therefore, the superplasticizer manufacturer has to adjust thermodynamic parameters

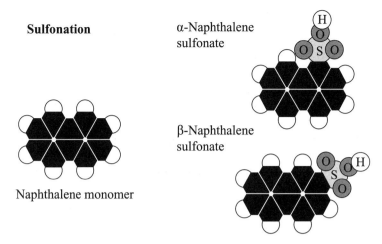

Fig. 6.19 Sulfonation is the first step in the fabrication of a polynaphthalene superplasticizer (courtesy of Martin Piotte).

during sulfonation so as to place the maximum number of sulfonate groups in the β position, though 100% success is never achieved. In some cases, when there is a poor control of the sulfonation, the β positions can represent as little as 50% of the total. In cases where an excellent control of the sulfonation reaction is achieved, 85 to 90% of the β positions can be filled. (The percentage of α and β positions in a particular superplasticizer can be evaluated by magnetic resonance.)

Also important in sulfonation is the number of available sites that are sulfonated. Although there are two sites that could theoretically be sulfonated in each naphthalene molecule (one in each carbon ring), in practice there is almost always only one site (or slightly less) per naphthalene molecule that is sulfonated. If the saturation ratio is equal to the ratio of the number of sulfonated groups divided by the number of available positions in the polymer, a saturation ratio of 0.90 to 0.95 is considered excellent, which means that a little less than one of the two positions has been sulfonated.

(b) *Second step: condensation* (Figure 6.20)

The polymerization of the naphthalene groups is realized through a condensation reaction between two naphthalene groups and a formaldehyde molecule, as shown in Figure 6.20. The reactive group involved in the condensation is always taken on the non-sulfonated ring in any of the available positions. As two condensation reactions can occur on the non-sulfonated ring, some branching can take place during condensation. In order to produce molecular chains as long as possible, the superplasticizer manufacturer has to adjust the condensation conditions. In general, an average degree of polymerization of 9 to 10 is achieved when the polymerization process is under control, i.e. on average a

Condensation

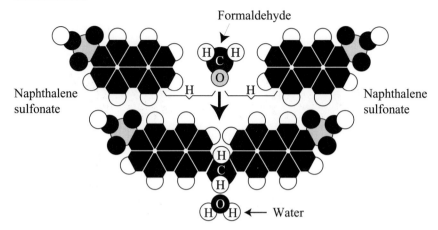

Fig. 6.20 Condensation is the second step in the fabrication in a polynaphthalene superplasticizer (courtesy of Martin Piotte).

superplasticizer molecule contains 9 to 10 naphthalene molecules, but monomers or di- and trimers almost always exist at the end of the condensation reaction. There are also longer chains having a degree of polymerization higher than 10.

Lengthening the polymer chain, generally speaking, increases the viscosity of the superplasticizer; however, an increase in the viscosity can also be the result of an increase in the branching and cross-linking of the polymeric chain. Thus when manufacturing a superplasticizer, an increase in the viscosity of the superplasticizer does not preclude a real lengthening of the polymer chain and an increase in the coverage efficiency of the superplasticizer. From a practical point of view, it seems that beyond an average polymerization number of 9 to 10, naphthalene superplasticizer molecules start to lose efficiency because this molecular weight increase is obtained through cross-linking rather than by linear lengthening of the polymer. (Measuring the length of the polymer is not an easy task; it is quite long and involves the use of a sophisticated process of ultrafiltration, chromatography in a liquid phase and light-scattering measurement) (Piotte *et al.*, 1995).

(c) *Third step: neutralization* (Figure 6.21)

The pH of the naphthalene sulfonic acid is between 2 and 3, which is too low for a material that has to be introduced into a high pH medium, such as a mixture of Portland cement and water. The polymerized sulfonic acid must be neutralized using a base. The most commonly used base is NaOH, but $Ca(OH)_2$ is also used. This neutralization process results in the formation of a sodium or calcium salt. In the literature, it is found

Neutralization

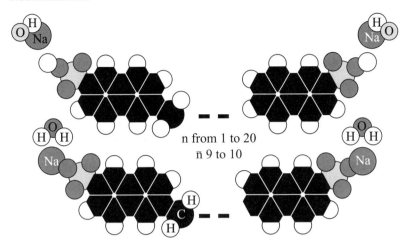

Fig. 6.21 Neutralization is the third step in the fabrication of a polynaphthalene superplasticizer (courtesy of Martin Piotte).

that other cations have been used to neutralize the sulfonic acid, such as Li, K, Mg, NH_3, and, mono-, di- and triethanolamine (Piotte, 1983).

(d) Fourth step: filtration (in the case of calcium salt)

This step is carried out to remove, if necessary, the remaining sulfates so as to produce a superplasticizer in which the amount of active solids (the polymerized chains) is the highest possible. Usually, superplasticizer producers quote only the total amount of solids, and rarely the amount of active solids.

At the end of the manufacturing process, the final product, the naphthalene superplasticizer, appears as a dark brown liquid (and as a translucent liquid in the case of the melamine sulfonate). The total solid content of naphthalene sulfonate is generally about 40 to 42%, while for melamine sulfonate it is in the range of 22 to 42%. However, in order to be shipped over long distances the two types of superplasticizer can be spray-dried so as to produce a powder: brown in the case of naphthalene superplasticizer, white in the case of melamine.

Making an efficient superplasticizer necessitates good raw materials and a good control of the most important parameters that influence the efficiency of a naphthalene superplasticizer:

- the relative ratio of the α and β positions occupied by the sulfonate group (the more β position the better);
- the number of sulfonated naphthalene groups per naphthalene ring (the nearer to 1 the better);

- the degree of polymerization (in the case of naphthalene superplasticizers an average polymerization number of 9 to 10 seems to be optimal in order to avoid too much cross-linking and branching);
- the amount of active solids (and not necessarily the total amount of solids).

Unfortunately, very few data sheets of commercial superplasticizers contain such information. Very often, a commercial data sheet indicates that the particular brand of superplasticizer is a brown liquid containing 40 to 42% total solid content, having a pH of between 7.5 and 8.5 and a viscosity of 60 to 80 centipoises. Such a description can hide great differences in the most important parameters that really influence the superplasticizer efficiency (Garvey and Tadros, 1972; Chibnowski, 1990).

As the physico-chemical tests necessary to evaluate the true quality of a commercial superplasticizer are sophisticated, involving testing apparatus not usually found in commercial laboratories and often not even in universities, the most economical and rapid way to evaluate the efficiency of a particular superplasticizer is to test its overall efficiency through a series of rheological tests conducted on the cement that will be used.

6.3.4 Portland cement hydration in the presence of superplasticizers

At the present time, there is no clear and accepted theory explaining in full detail the action of superplasticizers on cement particles during the mixing of concrete and the initial hydration of Portland cement (Petrie, 1976). When superplasticizers began to be used, some researchers believed that the superplasticizer–cement interaction was only physical in nature; they consequently used non-reactive powders with different surface states and studied the dispersion of these non-hydraulic systems in order to explain superplasticizer action in dispersing solid particles in water (Foissy and Pierre, 1990).

Other researchers took a chemical approach, studying the effect of superplasticizers on the rate of dissolution of the different ionic species present in the mixture (Andersen, 1986; Diamond and Struble, 1987; Odler and Abdul-Maula, 1987; McCarter and Garvin, 1989; Paulini, 1990). Yet others took a more cement-oriented approach, studying separately the effect of superplasticizers on the different phases of Portland cement (C_3S, C_2S and C_3A), hoping that the hydration of Portland cement is due to the addition or a combination of the effects of a superplasticizer on these individual phases (Massazza and Costa, 1980). In this particular approach, some problems occur because Portland cement does not always contain the same phases in the same proportions. Moreover, it is also well known that these phases interact with each other during hydration and that the interaction between the cement and the superplasticizer is complicated by the simultaneous interaction between the cement and the sulfates and the sulfates and the superplasticizer, as is schematically represented in Figure 6.22.

It has also been found that superplasticizers interfere not only with Portland cement hydration but also with sulfate dissolution and the value of the SO_4^{--}/AlO_2^- ratio, as shown in Figure 6.23 from Vernet (1995). In this figure it is shown how the value of this ratio can lead to a system having a 'normal slump loss' to a false set when it is very high, or rather, a flash set when it is very low. For intermediate values, weak false set or weak flash set can be experienced. In a recent paper on the dissolution rates of different forms of calcium sulfate in the presence or absence of superplasticizers, Vernet (1995) showed that a naphthalene superplasticizer strongly interacted with hemihydrate, delaying in some cases the precipitation of gypsum. This is a reason why superplasticizers are used in making gypsum board.

In spite of the merits of these different approaches, it must be admitted that much has to be learned about the interaction between superplasticizers and cement phases in order to understand better why some

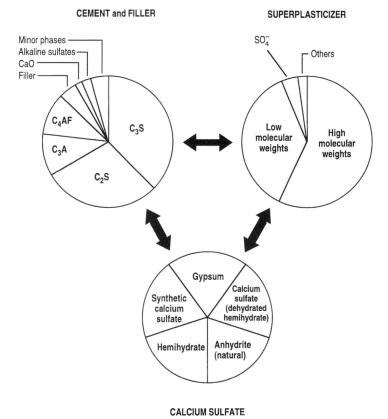

Fig. 6.22 The complexity of Portland cement, calcium sulfate and admixture interaction (after C. Jolicoeur and P.-C. Aïtcin).

superplasticizers work better with some cements, and vice versa (Buil, et al., 1986; Cunningham, Dury and Gregory, 1989; Uchikawa, Uchida and Okamura, 1990; Fernon, 1994; Vernet, 1995).

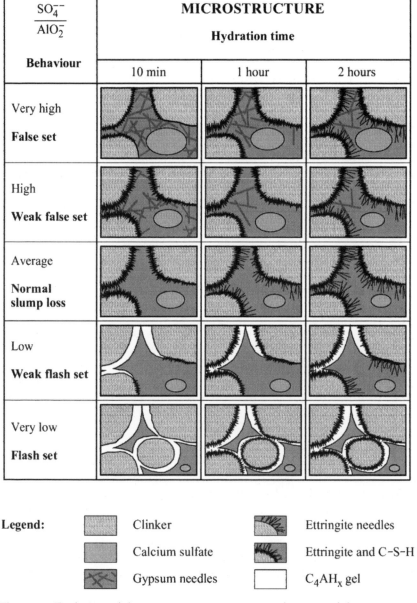

Fig. 6.23 Evolution of the paste microstructure as a function of the concentration gradients of sulfate and aluminate at the solid–liquid interface (from C. Vernet).

However, some facts have already been clearly established, providing some understanding of the interaction between cement and superplasticizers (Hewlett and Young, 1987; Andersen, Roy and Gaidis, 1988; Jolicoeur et al., 1994).

It is well established that, for a given Portland cement, the dosage rate of superplasticizer necessary to obtain a paste of definite fluidity increases with the specific surface area of the cement. The finer the cement, the higher the amount of superplasticizer needed to obtain a given fluidity or workability.

Superplasticizer molecules can be adsorbed on the C_3S. This has been clearly shown by direct observation of a radioactive sulfur-marked superplasticizer (Onofrei and Gray, 1989) and by indirect observation: as the superplasticizer dosage is increased, a delay in the development of the heat of hydration is observed (Aïtcin et al., 1987). This delay in the setting of concrete has also been observed in the field when, deliberately or otherwise, high dosages of superplasticizer were used, resulting in a heat development delay of as much as 1 day in some cases (Aïtcin, Laplante and Bédard, 1985).

In their experiments Onofrei and Gray (1989) have also clearly established that some superplasticizer was fixed by the hydrated interstitial phase. Luke and Aïtcin have also shown evidence of the modification of the rate and process of ettringite formation from C_3A in the presence of superplasticizer molecules (Luke and Aïtcin, 1991).

This strong interaction between the interstitial phase and superplasticizer molecules has also been reported by Hanna et al. (1989) and Khalifé (1991) in their studies on the rheology of cement grouts made with cements containing different amounts of C_3A and C_4AF and different types of superplasticizer (see also Carin and Halle, 1991).

In an attempt to study the influence of composition on the different phases of Portland cement, Aïtcin et al. (1991) reported that the use of a special Portland cement containing less than 10% of interstitial phase (3.6% of C_3A and 6.9% of C_4AF) was very economical in terms of superplasticizer dosage to make almost fluid high-performance concrete at a very low water/binder ratio. Using such a cement, it was possible to produce a high-performance concrete with a water to cementitious ratio of 0.17 and a 230 mm slump, 1 hour after mixing using slightly less than 100 l of water and 28.6 l of naphthalene superplasticizer. The setting of this concrete was not overdelayed because its compressive strength was 72 MPa at 24 hours.

Based on these observations and observations from others, Jolicoeur et al. (1994) have proposed the following schematic mechanisms (Figures 6.24 to 6.27) to explain the action of superplasticizers during cement hydration.

Figure 6.24 shows the adsorption mechanism of a polymerized molecule of superplasticizer on the surface of a cement grain which is negatively charged. This adsorption is made possible by the presence of

SURFACE ADSORPTION

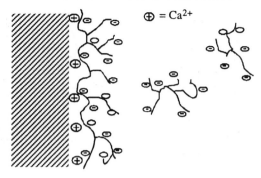

Cement particle

Fig. 6.24 Adsorption of the superplasticizer molecules due to Van der Waals' and electrostatic forces (after Jolicoeur et al., 1994).

Ca^{++} ions that have been liberated by a partial dissolution of some calcium ions from the cement.

Figure 6.25 represents the electrostatic repulsion of two cement particles, one positively and the other negatively charged, which otherwise would have been attracted in the absence of the adsorbed superplasticizer molecules on their respective surfaces.

It is well accepted that naphthalene and melamine superplasticizers work essentially in this manner (Uchikawa, 1994; Uchikawa, Hanehara and Sawaki, 1997).

Figure 6.26 represents a case of steric repulsion between two molecules of superplasticizer adsorbed on two adjacent cement grains. Superplasticizer molecules form a kind of coating on cement grains which annihilates electrostatic attraction. Polyacrylate superplasticizers act by

ELECTROSTATIC REPULSION

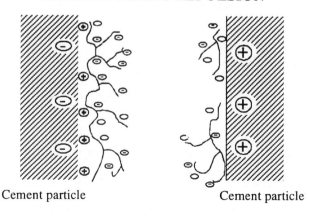

Cement particle   Cement particle

Fig. 6.25 The charged surface induces interparticular repulsive forces over a long distance (after Jolicoeur et al., 1994).

STERIC REPULSION

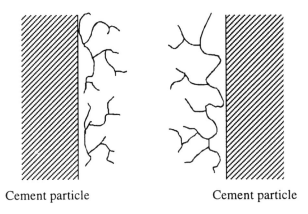

Cement particle   Cement particle

Fig. 6.26 Steric repulsion between molecules of superplasticizer adsorbed on two cement particles (after Jolicoeur *et al.*, 1994).

steric and electrostatic repulsion (Uchikawa, 1994; Uchikawa, Hanehara and Sawaki, 1997).

Figure 6.27 represents the interaction between superplasticizer molecules and some active sites on the cement grains, essentially sites having an affinity for RSO_4^{--} and SO_4^{--} ions. In such a case the superplasticizer is competing with calcium sulfate to neutralize these sites. This explains

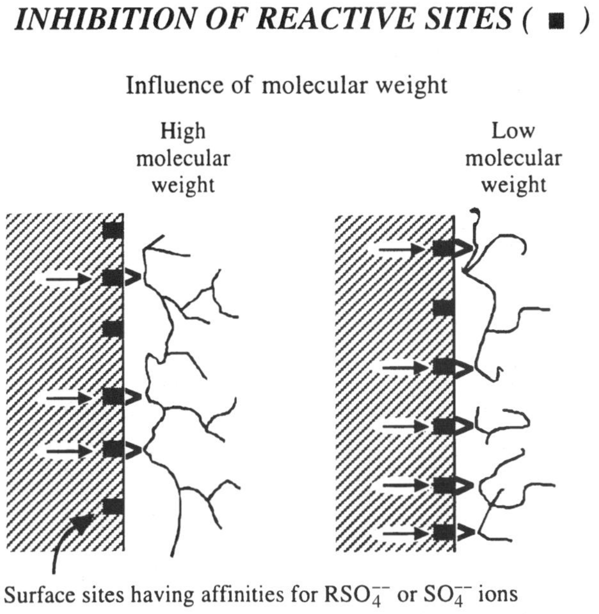

INHIBITION OF REACTIVE SITES (■)

Influence of molecular weight

High molecular weight   Low molecular weight

Surface sites having affinities for RSO_4^{--} or SO_4^{--} ions

Fig. 6.27 Inhibition of some reactive sites (after Jolicoeur *et al.*, 1994).

why, in some cases, the very strong initial dispersive action of a superplasticizer is rapidly lost because a non-negligible part of it reacts with these active sites and is no longer available to disperse the cement particles (Baussant, 1990; Vichot, 1990).

6.3.5 The crucial role of calcium sulfate

Calcium sulfate is added to Portland cement clinker to control the setting time of Portland cement; therefore, calcium sulfate can be considered as a powerful retarding agent. In the presence of calcium sulfate C_3A is transformed into ettringite ($C_3A \cdot 3CaSO_4 \cdot 32H_2O$). This first layer of ettringite seems to be impervious and blocks any further hydration of C_3A, so during the dormant period concrete can be transported and placed without too much loss of slump.

However, this mechanism of action of calcium sulfate is not always this simple, because Portland cement contains different forms of sulfate that can interfere with the calcium sulfate and control C_3A hydration (Nawa and Eguchi, 1987). Modern clinkers made with fuels rich in sulfur (Miller and Tang, 1996) can contain non negligible amounts of arcanite (K_2SO_4) and aphthitalite ($Na_2SO_4 \cdot 3K_2SO_4$) and calcium langbeinite ($2CaSO_4 \cdot K_2SO_4$) can be found in some Portland cement. These sulfates crystallize within the interstitial phase or are found on C_3S and C_2S.

Years ago, gypsum was almost the only calcium sulfate phase added during the manufacture of Portland cements, but for many reasons modern Portland cements can also contain other forms of calcium sulfate, such as:

- crystallized natural anhydrite $CaSO_4$
- gypsum $CaSO_4 \cdot 2H_2O$
- hemihydrate $CaSO_4 \cdot \tfrac{1}{2}H_2O$
- dehydrated gypsum (also called soluble anhydrite) $CaSO_4$
- synthetic calcium sulfate $CaSO_4$

The solubility rate of all these sulfates is not the same and can be strongly modified in the presence of a superplasticizer, so the equilibrium between the solubility rate of the C_3A phase of the cement and that of the calcium sulfate can be strongly modified. This can result in either a flash set or a false set situation in spite of the fact that the cement producer adjusted the calcium sulfate content according to the standard testing procedure, though this adjustment was done in the absence of any water reducer or superplasticizer. Therefore, in some cases a cement and a superplasticizer that fulfil the specifications of acceptance standards are not compatible at all (Dodson and Hayden, 1989). Thus in the near future it will be necessary to review the acceptance standards for both cements and superplasticizers because incompatibility problems will become more frequent as the use of high-performance concrete grows (Tagnit-Hamou, Baalbaki and Aïtcin, 1992).

6.3.6 Superplasticizer acceptance

In North America, superplasticizers must comply with the requirements of ASTM C494-92 Standard Specification for Chemical Admixtures for Concrete.

According to this standard, high-range water reducers (the ASTM name for superplasticizers) are classified into two categories: type F (water-reducing high-range admixture) and type G (water-reducing high-range and retarding admixture). In order to determine to which family it belongs, a superplasticizer has to be tested in the following manner:

> The cement used in any series of tests shall be either the cement proposed for specific work in accordance with 11.4, a Type I or Type II cement conforming to Specification C 150, or a blend of two or more cements, in equal parts. Each cement of the blend shall conform to the requirements of either Type I or Type II, Specification C 150. If when using a cement other than that proposed for specific work, the air content of the concrete made without admixture, tested as prescribed in 14.3, is more than 3.0%, select a different cement, or blend, so that the air content of the concrete will be 3.0% or less ...
>
> The cement content shall be 307 ± 3 kg/m^3.
>
> Adjust the water content to obtain a slump of 88 ± 12 mm.
>
> Add the admixture in the manner recommended by the manufacturer and in the amount necessary to comply with the applicable requirements of the specifications for water reduction or time of setting or both.

According to Table 6.1 (Table 1 of ASTM C494-92), the superplasticizer must be tested with a maximum water content equal to 88% of the control.

Moreover, the superplasticizer is tested in a usual concrete mixture made with a coarse aggregate meeting the grading requirements size for No. 57 of the ASTM C33 Specification (25 to 4.5 mm). This means that in order to obtain a slump of 88 ± 12 mm an average water content of 175 l/m^3 shall be used for air-entrained concrete or 195 l/m^3 for non-air-entrained concrete. These water contents correspond, in concrete mixtures containing 307 kg of cement per cubic metre, to W/B ratios of around 0.63 for non-air-entrained concrete, and around 0.57 for air-entrained concrete. As at least an 88% water reduction must be achieved with a usual Type I ASTM cement when a superplasticizer is used; therefore, the final W/B ratio at which the test is performed is finally around 0.55 and 0.50 for the non-air-entrained and air-entrained concretes.

These testing conditions are very far from the conditions in which superplasticizers are used in high-performance concrete, where the cement content generally varies between 450 and 550 kg/m^3 rather than being 307 kg/m^3. This explains why compliance with the ASTM C150 Standard Specification for a particular cement and C 494-92 for a

Table 6.1 Physical requirements[a] (ASTM C 494). ©ASTM. Reprinted with permission

	Type F, water reducing, high range	Type G, water reducing, high range and retarding
Water content, max, % of control	88	88
Time of setting, allowable deviation from control, h: min		
Initial: at least	...	1:00 later
not more than	1:00 earlier	3:30 later
Final: at least	...	...
not more than	1:00 earlier nor 1:30 later	3:30 later
Compressive strength, min, % of control:[b]		
1 day	140	125
3 days	125	125
7 days	115	115
28 days	110	110
6 months	100	100
1 year	100	100
Flexural strength, min, % of control:[b]		
3 days	110	110
7 days	100	100
28 days	100	100
Length change, max shrinkage (alternative requirements):[c]		
Percentage of control	135	135
Increase over control	0.010	0.010
Relative durability factor, min[d]	80	80

[a] The values in the table include allowance for normal variation in test results.
[b] The compressive and flexural strength of the concrete containing the admixture under test at any test age shall not be less than 90% of that attained at any previous test age. The objective of this limit is to require that the compressive or flexural strength of the concrete containing the admixture under test shall not decrease with age.
[c] Alternative requirements, see 17.1.4, % of control limit applies when length change of control is 0.030% or greater; increase over control limit applies when length change of control is less than 0.030%.
[d] This requirement is applicable only when the admixture is to be used in air-entrained concrete which may be exposed to freezing and thawing while wet.

particular superplasticizer does not imply that the cement and the superplasticizer will be compatible at a very low water/binder ratio and will not result in a premature slump loss or in an excessive retardation.

6.3.7 Concluding remarks

Superplasticizer efficiency is controlled by physico-chemical parameters that are not simple to measure. The manufacturing process of superplasticizers must be under tight control in order to produce a reproducible and efficient superplasticizer. The mode of action of superplasticizers is very complex and more or less well understood, making it difficult to evaluate the theoretical potential of a particular superplasticizer to disperse a particular cement. Therefore, as will be seen in the next chapter, the best way to evaluate the compatibility between a particular brand of superplasticizer and a particular brand of cement is to study directly the rheological characteristics of a particular grout, mortar or concrete in relation to the cement and superplasticizer.

Not all commercial superplasticizers perform as efficiently with all Portland cements in spite of the fact that their data sheets can be very similar, because the information given in these data sheets (which are the easiest to characterize) is not necessarily the most important to evaluate the potential efficiency of a superplasticizer (Uchikawa, Sone and Sawaki, 1977). The more relevant properties – number of β sites sulfonated, degree of sulfonation, average molecular length and molecular length range – are difficult to measure.

Moreover, the standard methods used at present to rate superplasticizers are not adequate to evaluate the performance of a particular superplasticizer when making a high-performance concrete.

6.4 SUPPLEMENTARY CEMENTITIOUS MATERIALS

6.4.1 Introduction

High-performance concrete can be made using Portland cement alone as a cementitious material. However, a partial substitution of Portland cement by one or a combination of two or three supplementary cementitious materials, when available at competitive prices, can be advantageous, not only from an economic point of view but also from a rheological, and sometimes strength, point of view (Uchikawa, 1986; Uchikawa, Uchida and Okamura, 1987; Uchikawa et al., 1992).

Most supplementary cementitious materials have one feature in common: they contain some form of vitreous reactive silica which, in the presence of water, can combine with lime, at room temperature, to form calcium silicate hydrate of the same type as that formed during the hydration of Portland cement. However, some fillers that are presently used in some modern Portland cement do not do this.

Basically, a pozzolanic reaction can be written in the following manner:

Pozzolan + lime + water → calcium silicate hydrate

It must be noted that at room temperature this reaction is generally slow and can take several months for completion. However, the finer and the more vitreous the pozzolan, the faster its reaction with lime.

As has been previously seen, Portland cement hydration liberates a large amount of lime as a result of the hydration of C_3S and C_2S (30% of the anhydrous cement mass). Such lime contributes very little to the strength of the hydrated cement paste and can be responsible for durability problems since it can be leached out easily by water. This leaching action results in an increase in the porosity of the cement paste matrix, and thus in a higher leachability, and so on. The only positive feature of this lime in concrete is that it maintains a high pH, which enhances the stability of the oxide layer that protects and passivates the reinforcing steel.

When making concrete, if pozzolan is mixed with Portland cement in an adequate proportion (from 20 to 30%), theoretically all the lime produced by the hydration of Portland cement can be transformed into C—S—H. As a result, the hydration reaction of an adequate mixture of Portland cement and pozzolan may be written as follows:

$$\text{Portland cement} + \text{pozzolan} + \text{water} \rightarrow \text{C—S—H}$$

However, the actual conditions in concrete are usually very far from this ideal situation, and the pozzolanic reaction is never completed.

Although some natural pozzolans are still in use in some countries such as Italy, Greece, Chile and Mexico (Mehta, 1987), they do not seem to have been used in high-performance concrete. Most of the pozzolans used in high-performance concrete are industrial by-products (Malhotra, 1987; Malhotra and Mehta, 1996). Among those used in large volumes are fly ash and silica fume (Malhotra, Carette and Sivasundaram, 1984; Berry and Malhotra, 1987; Sellevold and Nilsen, 1987; Khayat and Aïtcin, 1992; Malhotra and Ramezanianpour, 1994). Slag, which is not *per se* a pozzolanic material (Hooton, 1987; Ryell and Bickley, 1987), has also been used to make high-performance concrete. All these materials will be referred to as supplementary cementitious materials, as stated in Chapter 1.

These supplementary cementitious materials can be placed in the same ternary diagram as used to present the chemical and phase composition of Portland cement, shown in Figure 6.28. Since these materials are industrial by-products, their chemical composition is much less well defined than that of Portland cement, which explains the large area which they cover in the SiO_2—CaO—Al_2O_3 ternary diagram. Moreover, their phase compositions do not necessarily correspond to the phase compositions given by the ternary phase diagram SiO_2—CaO—Al_2O_3, because only a part of these oxides was fused during the burning process in which they were formed, whereas the compounds indicated in the phase diagram correspond to crystallized species obtained by the slow cooling of a melted mixture of the three oxides.

6.4.2 Silica fume

Silica fume is a by-product of the fabrication of silicon metal, ferrosilicon alloys and other silicon alloys. Silicon and silicon alloys are produced in submerged electric arc furnaces where quartz is reduced in the presence of coal (and iron during the production of ferrosilicon alloy) (Figures 6.29 and 6.30). During the reduction of silica, within the electric arc, a gaseous

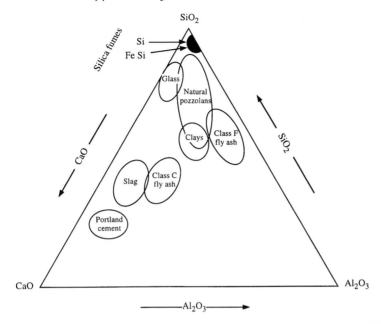

Fig. 6.28 Representation of the principal cementitious materials in a SiO_2–CaO–Al_2O_3 ternary diagram.

silicon suboxide, SiO, is produced (Aïtcin, 1983). As this gas escapes towards the upper part of the burden, it cools down, condenses and oxidizes in the form of ultrafine silica particles. These particles are collected by a dedusting system (Figure 6.31).

A few other particles from the burden are also entrained with the silica fume particles (very fine quartz particles, coal particles, and graphite particles from the electrodes) and some half-burnt wood chips from the burden. Some of these wood chips which are not collected by the so-called 'spark arrestors' can be found in the silica fume. However, all these impurities represent a very small percentage of the solids collected

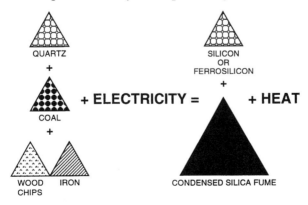

Fig. 6.29 Principle of the manufacture of silicon or ferrosilicon.

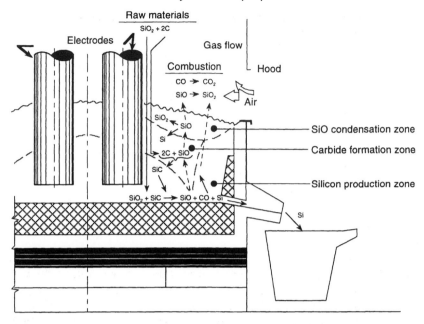

Fig. 6.30 Chemical reactions taking place in the reaction zone of a furnace.

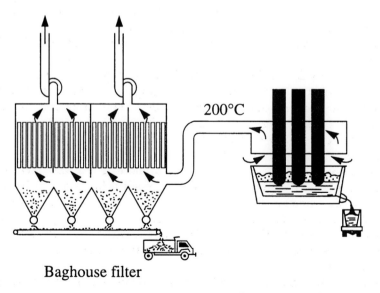

Baghouse filter

Without a heat recovery system

Fig. 6.31 Recovery of silica fume in the baghouse.

Table 6.2 Typical chemical composition of some silica fumes (after Aïtcin, 1983)

	Grey silicon	Grey ferrosilicon	White ferrosilicon
SiO_2	93.7	87.3	90.0
Al_2O_3	0.6	1.0	1.0
CaO	0.2	0.4	0.1
Fe_2O_3	0.3	4.4	2.9
MgO	0.2	0.3	0.2
Na_2O	0.2	0.2	0.9
K_2O	0.5	0.6	1.3
Loss on ignition	2.9	0.6	1.2

in the dedusting system, which is also called the 'baghouse' (Figure 6.31).

From a chemical point of view, silica fume is mostly composed of silica (Table 6.2). The SiO_2 content of silica fume varies, depending on the type of alloy produced. The higher the silicon content of the alloy, the higher the SiO_2 content of the silica fume. Silica fume produced during the manufacture of silicon metal generally contains more than 90% SiO_2. Silica fume produced during the manufacture of 75% Fe—Si alloy has an SiO_2 content greater than 85%.

From a structural point of view, silica fume is composed mostly of vitreous silica, as shown in the X-ray diffractogram in Figure 6.32. Cooling of the SiO vapour and its oxidation to SiO_2 happen so quickly that SiO_2 tetrahedra have no time to organize themselves in an orderly fashion so as to yield a form of crystalline silica. The flatter the hump observed on the X-ray diffractogram the more amorphous the silica fume.

From a physical point of view, silica fume particles appear to be perfectly spherical, with diameters ranging from less than 0.1 μm to about 1 or 2 μm, so that the average silica fume sphere is 100 times smaller than an average cement particle. Therefore, a transmission electronic microscope must be used in order to obtain photographs of silica fume particles, as shown in Figure 6.33.

The specific gravity of silica fume is about 2.2, a usual value for vitreous silica. The specific surface area of silica fume cannot be measured in the same way as for Portland cement owing to its extreme fineness, and is determined by nitrogen adsorption. Typical values reported range from 15 000 to 25 000 m^2/kg. Using the same technique, the specific surface area of an ordinary Portland cement is approximately 1500 m^2/kg.

Since the raw materials used to make silicon or ferrosilicon are quite pure, the silica fume collected from a given furnace usually has a consistent composition. This is the case as long as the type of alloy produced in the furnace does not change.

Moreover, when most of the dedusting systems were designed the silicon and ferrosilicon industry considered silica fume to be a waste and troublesome material of no commercial value, so that often the silica fume produced by different furnaces producing different alloys is collected in a single 'baghouse'. In such cases, the silica fume collected can be a mixture of different types with different chemical composition and pozzolanic properties. Therefore, it is very important that the chemical composition of any commercial silica fume is checked on a regular basis in order to ensure consistent use of the material. It is therefore important to establish the true origin of the silica fume that is to be used in high-performance concrete or to buy it from a supplier who is aware of the intended use or to test its reactivity (Pistilli, Rau and Cechner, 1984; Pistilli, Winterstein and Cechner, 1984; Vernet and Noworyta, 1992).

Silica fume is at present available in four different forms: in bulk as produced, in slurry form, in densified form and blended with Portland

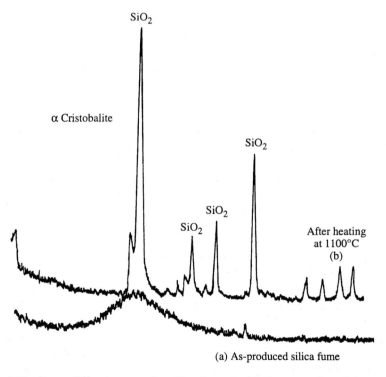

Fig. 6.32 X-ray diffractogram of a silica fume in (a) the as-produced form and (b) after heating at 1100 °C. After heating, silica fume crystallizes as cristobalite. The hump in the diffractogram of the as-produced silica fume is located at the greatest peak of the diffractogram of α cristobalite, indicating that the silica tetrahedra in the vitreous particles are organized over a short distance range, as in α cristobalite.

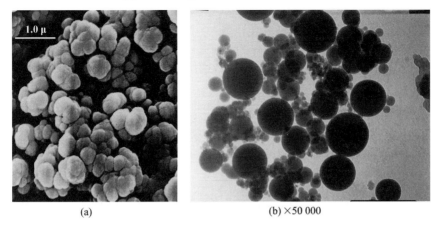

Fig. 6.33 Electron microscope pictures of silica fume: (a) scanning electron microscope – naturally agglomerated silica fume particles in an as-produced silica fume; (b) transmission electron microscope – dispersed individual particles.

cement. The advantages and disadvantages of using silica fume in one form rather than in another will be discussed in Chapter 7.

Compared with other supplementary cementitious materials, the peculiar characteristics that make silica fume a very reactive pozzolanic material are its very high SiO_2 content, its amorphous state and its extreme fineness.

The beneficial effects of silica fume on the microstructure and mechanical properties of concrete are due not only to a rapid pozzolanic reaction, but also to the physical effect of the silica fume particles, which is known as the 'filler effect' (Sellevold, 1987; Rosenberg and Gaidis, 1989; Khayat, 1996). Moreover, silica fume has a chemical effect related to the germination of crystals of portlandite, $Ca(OH)_2$ (Groves and Richardson, 1994).

Durekovic and Tkalcic-Ciboci (1991) advocate that the presence of a superplasticizer influences the size dispersity of silica anions by increasing the proportion of higher polymers.

Because of their fineness, silica fume particles can fill the voids between the larger cement particles, when they are well deflocculated in the presence of an adequate dosage of superplasticizer, as shown in Figure 6.34. The filler effect is also said to be responsible for the increase in fluidity of concretes with a very low water/binder ratio. Therefore, owing to its unique physical characteristics, the resulting solid matrix including silica fume is dense even before any chemical bonds between the cement particles have developed.

Silica fume particles, as other ultrafine particles, can act as a natural site of germination of crystals of $Ca(OH)_2$, which are developed as a multitude of minute portlandite crystals which are not detectable with a

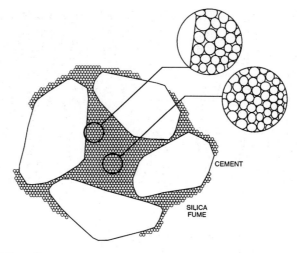

Fig. 6.34 Filler effect of silica fume according to H. Bache.

scanning electron microscope or even when making an X-ray diffractogram. However, when the same paste is submitted to a thermogravimetric and differential thermal analysis, the characteristic loss of mass of portlandite at 450 °C is well identified (Figure 6.35).

Owing to the extremely small size of silica fume particles, the addition of silica fume sharply reduces both the internal and the superficial bleeding in the mixture. This reduced bleeding is very important from a microstructural point of view because it drastically transforms the microstructural characteristics of the transition zone between cement paste and aggregates, and cement paste and reinforcing steel (Goldman and Bentur, 1989). These transition zones are more compact than the relatively porous one usually obtained when concrete does not contain any silica fume.

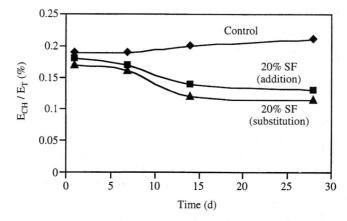

Fig. 6.35 Mass loss of portlandite with and without silica fume.

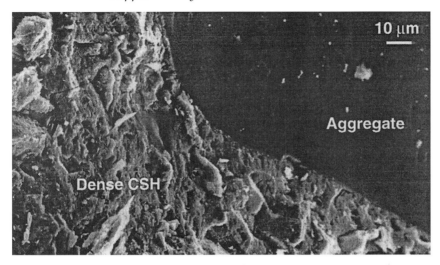

Fig. 6.36 Dense C–S–H in a silica fume concrete around an aggregate. The absence of a transition zone between the aggregate and the paste can be noticed.

Moreover, silica fume particles have a fluidifying effect on very low water/binder ratio mixtures, which otherwise are quite sticky. This rheological behaviour has not yet been fully explained but some researchers relate it to the action of tiny spheres of silica fume acting as very small ball-bearings. Silica fume particles also displace some of the water present between flocculated cement grains, thus increasing the amount of water available to fluidify concrete.

The combination of these different modes of action of silica fume in concrete results in a very dense microstructure (Regourd, 1983; Durekovic, 1995) with a very tight bond between aggregates and hydrated cement paste, as shown in Figure 6.36. With such an enhanced microstructure, silica fume increases the compressive strength of concrete, especially between 7 and 28 days. Moreover, as silica fume reduces the porosity of the cement paste at its interface with aggregate, the concrete permeability is greatly reduced.

6.4.3 Slag

Slag, or ground granulated blast-furnace slag as it should properly be called, is the by-product of the manufacture of pig iron in a blast furnace (Figure 6.37). All the impurities contained in the iron ore and in the coke pass into the blast-furnace slag (Figure 6.38). As these impurities could result in a mixture with a very high melting point, which could be uneconomical, fluxing agents are added to the blast-furnace burden so that the resulting chemical composition of the impurities stays within a very definite area of the $SiO_2-CaO-Al_2O_3$ phase diagram, corresponding to one of the two lower melting temperature areas within this

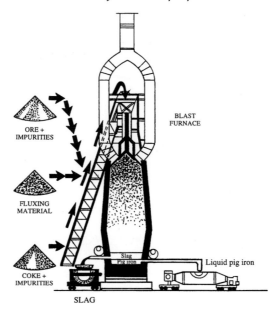

Fig. 6.37 Schematic representation of a blast furnace.

diagram. Thus, from a chemical point of view, slag has a quite constant composition to which metallurgists pay attention, because any deviation from it translates into a significant energy requirement and additional cost (Table 6.3).

Melted slag has a much lower specific gravity, about 2.8, than that of pig iron, which is above 7.0, so molten slag floats on top of molten pig iron and can be tapped separately (Figure 6.37).

Slag can be cooled in two different ways. First, it can be left to cool slowly so that it crystallizes mainly in the form of melilite, a solid solution of ackermanite and gehlenite (Figure 6.39). When cooled in such a way, blast-furnace slag is crystallized and can be used as aggregate in concrete, asphalt and roofing stone or to build roads and embankments,

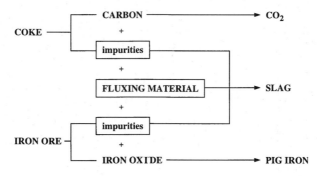

Fig. 6.38 Schematic representation of the production of blast-furnace slag.

Supplementary cementitious materials

Table 6.3 Typical chemical composition of some blast-furnace slags (Aïtcin, 1968)

	French slag	North American slag
SiO_2	29 to 36	33 to 42
Al_2O_3	13 to 19	10 to 16
CaO	40 to 43	36 to 45
Fe_2O_3	< 4%	0.3 to 20
MgO	< 6%	3 to 12
S^-	< 1.5%	–

but it has practically no hydraulic value and cannot be used as a supplementary cementitious material, even when finely ground (Figure 6.40) (Aïtcin, 1968).

However, if slag is quenched when it comes out of the blast furnace, it solidifies in a vitreous form and can then develop cementitious properties if properly ground and activated (Figure 6.41). The quenching of the slag can be carried out in three different ways. The molten slag can be:

1. poured into a huge water basin where it disintegrates into small particles in the form of coarse sand, also referred to as 'granulated slag';
2. quenched directly by powerful water jets as it runs out of the furnace into metallic gutters. Here, it is also transformed into a kind of sand, also referred to as 'granulated slag';

Fig. 6.39 Air-cooled or crystallized slag under polarized light in an optical microscope. The white prismatic crystals are melilite.

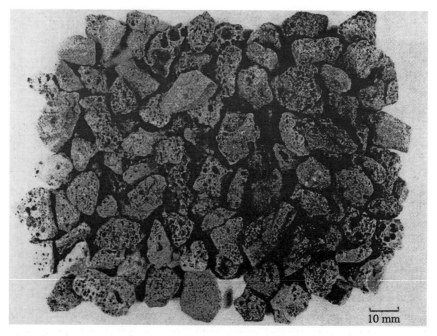

Fig. 6.40 Porous aggregates made of air-cooled slag.

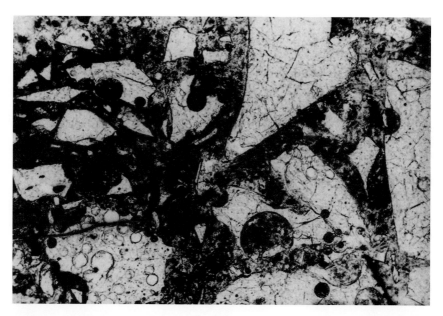

Fig. 6.41 Vitreous slag particles (in white). Note the shape of the slag particles and their porous aspect due to quenching in water. The totally vitreous particles of slag denote that the temperature of this slag was high when it as quenched (such slag is sometimes referred to as 'hot slag').

3. projected through the air by a special wheel so that the quenching is done by a combination of the action of water and air. In this case, the cooled slag has the shape of more or less porous spherical pellets, and is usually called 'pelletized slag'. These pellets can be used as lightweight aggregate to manufacture concrete blocks, or they can be ground to make a cementitious powder.

Thus as a supplementary cementitious material, slag possesses some useful features (Hinrichs and Odler, 1989): it has a chemical composition which does not vary too much because it must be within a well-defined composition area in the $SiO_2-CaO-Al_2O_3$ phase diagram. There can be some differences in the chemical content of MgO and Al_2O_3 of slags depending on the use of olivine as a fluxing agent rather than limestone, but this does not dramatically change the hydraulic properties of slag when used as a supplementary cementitious material.

The critical feature to be checked carefully when using slag is its vitreous state because its hydraulic properties are closely related to this feature (Figure 6.41). If the temperature of the slag was somewhat low, meaning that some crystals could be present in the molten phase (Figure 6.42), upon quenching, the slag can be less reactive than a hotter one which is more vitreous. Well-quenched slags have a pale yellow, beige or grey colour while cold slags have a much darker colour varying from dark grey to dark brown (Aïtcin, 1968).

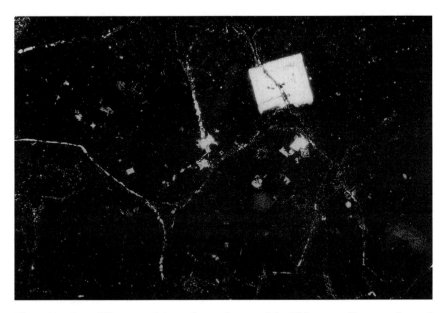

Fig. 6.42 A melilite crystal in a glassy slag particle. This crystallite was formed in the molten slag before its quenching, showing that the temperature of the slag was not too high when it was quenched (such slag is sometimes referred to as a 'cold slag').

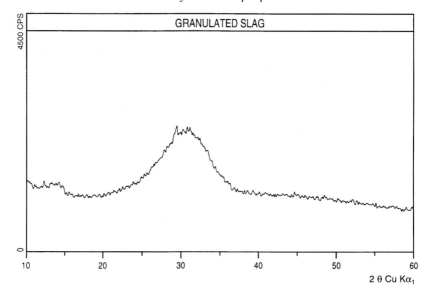

Fig. 6.43 X-ray diffractogram of a granulated blast-furnace slag.

One easy way to check if the slag has been well quenched is to obtain an X-ray diffractogram. In the absence of any crystallites, the X-ray diagram presents a hump centred on the main peak of melilite (Figure 6.43).

Slag can be blended with cement after separate grinding or by intergrinding with clinker, or it can be sold separately to concrete producers as a supplementary cementitious material. Blended material is more common in Europe while the use of slag as a separate ingredient is prevalent in North America.

6.4.4 Fly ash

Fly ashes are the small particles collected by dedusting systems of coal-burning power plants, as shown schematically in Figure 6.44. Fly ashes can have different chemical and phase compositions because they are exclusively related to the type and amount of impurities contained in the coal burnt in the power plant (Figure 6.45). Coal from the same source used in the same plant will produce almost the same fly ash. However, as shown in Table 6.4, the chemical composition of fly ashes from different plants can vary.

From a physical point of view, fly ashes can also be very different from one another. They can appear as plain spherical particles (Figure 6.46), with a grain size distribution similar to that of Portland cement, or they can contain some cenospheres, i.e. hollow spheres (Figure 6.47). In some cases, they can also contain angular-shaped particles (Figure 6.48).

From a chemical point of view, the different fly ashes available can be classified into broad families; for example, the ASTM recognizes two

Supplementary cementitious materials

types of fly ash in its C618-94a Standard Specification for Coal Fly Ash and Raw or Calcined Natural Pozzolan for Use as Mineral Admixtures in Portland Cement Concrete: Class F and Class C fly ashes. Class F fly ash is usually produced in power plants burning anthracite or subbituminous coal extracted, for example, in the eastern part of the USA. On the other hand, Class C fly ash is produced by burning lignite or bituminous coal from, for example, the southern or western part of the USA. Such fly ashes are characterized by a high calcium content.

In France, fly ashes are classified into three groups: the silico-aluminous group, which corresponds mainly to ASTM Class F, the

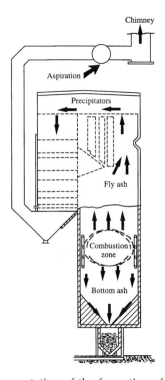

Fig. 6.44 Schematic representation of the formation of fly ashes.

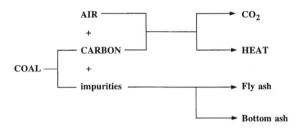

Fig. 6.45 Schematic representation of the formation of fly ashes.

Table 6.4 Typical chemical composition of some fly ashes (Aïtcin et al., 1986). Reproduced by permission of ACI

	Class F	Class F	Class C	Sulfocalcic	Sulfocalcic
SiO_2	59.4	47.4	36.2	24.0	13.5
Al_2O_3	22.4	21.3	17.4	18.5	5.5
Fe_2O_3	8.9	6.2	6.4	17.0	3.5
CaO	2.6	16.6	26.5	24.0	59.0
MgO	1.3	4.7	6.6	1.0	1.8
Na_2O equivalent	2.2	0.4	2.2	0.8	–
SO_3	2.4	1.5	2.8	8.0	15.1
Loss on ignition	2.0	1.5	0.6	–	–
$SiO_2 + Al_2O_3 + Fe_2O_3$	90.7	74.9	60.0	59.5	22.5
Free lime	–	–	–	–	28.0

silicocalcic group, which corresponds mainly to Class C, and the sulfocalcic group, which has at the same time a high calcium and a high sulfur content.

In spite of the merits of these different classifications, it is not always easy to fit any given fly ash into a particular category and predict its pozzolanic behaviour. It has been found that most fly ashes are pozzolanic materials, but some can be non-pozzolanic, while others are self-cementitious (Aïtcin et al., 1986).

In any case, in order to participate in any pozzolanic reaction, a particular fly ash must contain a significant amount of vitreous material and the best way to check this is to make an X-ray diffractogram, as shown in Figure 6.49.

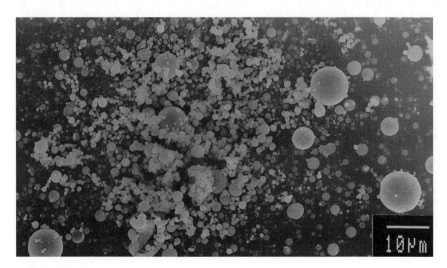

Fig. 6.46 Spherical fly ash particles.

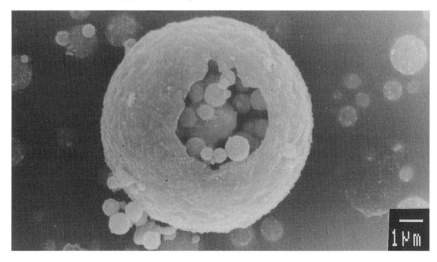

Fig. 6.47 Plerosphere containing cenospherical particles of fly ash.

6.4.5 Concluding remarks

The use of supplementary cementitious materials, when available at a competitive price, is beneficial in making high-performance concrete because they can result in a cost saving. Their dosage in the final mix depends on the desired early strength of the high-performance concrete, taking into account ambient temperature.

Fig. 6.48 Angular-shaped fly ash particles (courtesy of I. Kelsey-Lévesque).

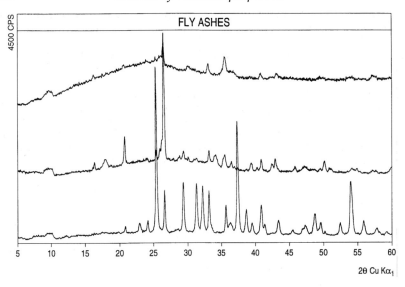

Fig. 6.49 X-ray diffractogram of different fly ashes.

The use of a combination of two cementitious materials, slag and silica fume, or fly ash and silica fume, is beneficial because silica fume reactivity can compensate for the slower reactivity of slag of fly ash.

Of the three cementitious materials that have been briefly reviewed, fly ashes are the most variable and least reactive. This does not mean that they are not of use in making high-performance concrete, but that they should be used with care and not on the basis of any generalization.

REFERENCES

Aïtcin, P.-C. (1968) Sur les propriétés minéralogiques des sables de laitier de haut-fourneau de fonte Thomas et leur utilisation dans les mortiers en bétons, *Revue des Matériaux de Construction*, Mai, pp. 185–94.

Aïtcin, P.-C. (1983) *Condensed Silica Fume*, Les Éditions de l'Université de Sherbrooke, Université de Sherbrooke, Québec, Canada, ISBN 2-7622-0016-4, 52 pp.

Aïtcin, P.-C., Laplante, P. and Bédard, C. (1985) *Development and Experimental Use of a 90 MPa (13 000 psi) Field Concrete*, ACI SP-87, October, pp. 51–70.

Aïtcin, P.-C., Autefage, F., Carles-Gibergues, A. and Vaquier, A. (1986) *Comparative Study of the Cementitious Properties of Different Fly Ashes*, ACI SP-91, pp. 91–114.

Aïtcin, P.-C., Sarkar, S.L., Regourd, M. and Volant, M. (1987) Retardation effect of superplasticizer on different cement fractions. *Cement and Concrete Research*, **17**(6), December, 995–9.

Aïtcin, P.-C., Sarkar, S.L., Ranc, R. and Levy, C. (1991) A high silica modulus cement for high-performance concrete, in *Ceramic Transactions: Advances in Cementitious Materials* (ed. S. Mindess), Vol. 16, pp. 103–20.

Andersen, P.J. (1986) The effect of superplasticizers and air-entraining agents on the zeta potential of cement particles. *Cement and Concrete Research*, **16**, 931–40.

Andersen, P.J., Roy, D.M. and Gaidis, J.M. (1988) The effect of superplasticizer molecular weight on its adsorption and dispersion of cement. *Cement and Concrete Research*, **18**, 980–86.

Baussant, J.-B. (1990) Nouvelle méthode d'étude de la formation d'hydrates des cunebts. Applications à l'analyse de l'effet adjuvants organiques. PhD Thesis, University of Franche-Condé, No. 156, 194 pp.

Berry, E.E. and Malhotra, V.M. (1987) *Fly Ash in Concrete*, Supplementary Cementing Materials for Concrete, Canadian Government Publishing Centre, Supply and Services Canada Ltd, Ottawa, Canada, K1A 0S9, ISBN 0-660-12550-1, pp. 37–163.

Black, B., Rossington, D.R. and Weinland, L.A. (1963) Adsorption of admixtures on Portland cement. *Journal of the American Ceramic Society*, **46**(8), August, 395–9.

Blick, R.L., Petersen, C.F. and Winter, M.E. (1974) *Proportioning and Controlling High-Strength Concrete*, ACI SP-46, pp. 141–63.

Bonzel, J. and Siebel, E. (1978) *Flowing Concrete and its Application Possibilities*, First International Symposium on Superplasticizers in Concrete, Ottawa, Canada, May, 31 pp.

Bradley, G. and Howarth, J.M. (1986) Water soluble polymers: the relationships between structure, dispersing action and rate of cement hydration. *Cement, Concrete, and Aggregates*, **8**(2), Winter, 68–75.

Buil, M., Witier, P., de Larrard, F., Detrez, M. and Paillère, A.M. (1986) *Physicochemical Mechanism of the Action of the Naphthalene Sulfonate Based Superplasticizers on Silica Fume Concretes*, ACI SP-91, pp. 959–71.

Bye, G.C. (1983) *Portland Cement Composition, Production and Properties*, Pergamon Press, New York, ISBN 0-08-039965-2, 149 pp.

Carin, V. and Halle, R. (1991) Effect of matrix form on setting time of belite cement which contains tricalcium aluminate. *Ceramic Bulletin*, **70**(2), 25l–3.

Chatterji, V.S. (1988) On the properties of freshly made Portland cement paste: part 2. Sedimentation and strength of flocculation. *Cement and Concrete Research*, **18**, 615–20.

Chibnowski, S. (1990) Effect of molecular weight of a polymer on the structure of a layer adsorbed on the surface of titania. *Powder Technology*, **63**, 75–9.

Collepardi, M., Baldini, G., Pauri, M. and Conradi, M. (1979) Retardation of tricalcium aluminate hydration by calcium sulfate. *Journal of the American Ceramic Society*, **62**(1–2), January–February, 33–5.

Cunningham, J.C., Dury, B.L. and Gregory, T. (1989) Adsorption characteristics of sulfonated melamine formaldehyde condensates by high performance size exclusion chromatography. *Cement and Concrete Research*, **19**, 919–26.

Diamond, S. and Struble, L.J. (1987) *Interaction Between Naphthalene Sulfonate and Silica Fume in Portland Cement Pastes*, Materials Research Society, Fall Meeting, Boston, 19 pp.

Dodson, V.H. and Hayden, T. (1989) Another look at the Portland cement/chemical admixture incompatibility problem. *Cement, Concrete, and Aggregates*, **11**(1), Summer, 52–6.

Durekovic, A. (1995) Cement pastes of low water to solid ratio: an investigation of the porosity characteristics under the influence of a superplasticizer and silica fume. *Cement and Concrete Research*, **25**(2), 365–75.

Durekovic, A. and Tkalcic-Ciboci, B. (1991) Cement pastes of low water to solid ratio: an investigation of the polymerization of silicate anions in the presence of a superplasticizer and silica fume. *Cement and Concrete Research*, **21**(6), 1015–22.

Eckart, V.A., Ludwig, H.-M. and Stark, J. (1995) Hydration of the four main Portland cement clinker phases. *Z.K.G. International*, No. 8, 443–52.

Fernon, V. (1994) Caractérisation de produits d'interaction adjuvants/hydrates du ciment. *Journée Technique Adjuvants, Les Technodes Guerville*, September, 14 pp.

Foissy, A. and Pierre, A. (1990) Les Mécanismes d'Action des Fluidifiants. *Ciments, Bétons, Plâtres, Chaux*, No. 782, pp. 18–19.

Garvey, M.J. and Tadros, T.F. (1972) Fractionation of the Condensates of Radium Naphthalene 2-Sulfonate and Formaldehyde by Gel Permeation Chromatography. *Kolloid-Z.u.Z. Polymere*, **250**(10), pp. 967–72.

Gebauer, J. and Kristmann, M. (1979) The influence of the composition of industrial clinker on cement and concrete properties. *World Cement Technology*, **10**(2), March, 46–51.

Goldman, A. and Bentur, A. (1989) Bond effects in high-strength silica fume concrete. *ACI Materials Journal*, **86**(5), September–October, 440–47.

Groves, G.W. and Richardson, I.G. (1994) Microcrystalline calcium hydroxide in pozzolanic cement pastes. *Cement and Concrete Research*, **24**(6), 1191–6.

Grzeszczyk, S. (1994) Distribution of potassium in clinker phases. *Silicate Industries*, **61**(7–8), 241–6.

Grzeszczyk, S. and Kucharska, L. (1990) Hydrative reactivity of cement and rheological properties of fresh cement pastes. *Cement and Concrete Research*, **20**, 165–74.

Hanna, É., Luke, K., Perraton, D. and Aïtcin, P.-C. (1989) *Rheological Behavior of Portland Cement in the Presence of a Superplasticizer*, Third CANMET/ACI International Conference on Superplasticizer and Other Chemical Admixtures in Concrete, Ottawa, ACI SP-119, pp. 171–88.

Hattori, K. (1978) *Experience with Mighty Superplasticizer in Japan*, ACI SP-62, pp. 37–66.

Hewlett, P.C. and Rixom, R. (1977) Superplasticized concrete. *ACI Journal*, **74**(5), May, N6–12.

Hewlett, P.C. and Young, J.F. (1987) Physico-chemical interactions between chemical admixtures and Portland cement. *Journal of Materials Education*, **9**(4), 389–436.

Hinrichs, N. and Odler, I. (1989) Investigation of the hydration of Portland blastfurnace slag cement: hydration kinetics. *Advances in Cement Research*, **2**(5), January, 9–13.

Hooton, R.D. (1987) *The Reactivity and Hydration Products of Blast-Furnace Slag*, Supplementary Cementing Materials for Concrete, Canadian Government Publishing Centre, Supply and Services Canada Ltd, Ottawa, Canada, K1A 0S9, ISBN 0-660-12250-1, pp. 247–88.

Hornain, H. (1971) *Sur la Répartition des Eléments de Transition et leur Influence sur quelques Propriétés du Clinker et du Ciment*, CERILH Publication, No. 212, August–September, 15 pp.

Jolicoeur, C., Nkinamubanzi, P.-C., Simard, M-A. and Piotte, M. (1994*) Progress in Understanding the Functional Properties of Superplasticizer in Fresh Concrete*, ACI SP-148, pp. 63–88.

Khalifé, M. (1991) Contribution à l'étude de la compatibilité ciment/superplastifiant. Université de Sherbrooke, Master's Degree Thesis No. 645, 125 pp.

Khayat, K. (1996) *Effect of Silica fume on Fresh and Mechanical Properties of Concrete*, CANMET-ACI Intensive Course on Fly Ash, Slag, Silica Fume, Other Pozzolanic Materials and Superplasticizers in Concrete, Ottawa, Canada, April, 34 pp.

Khayat, K.H. and Aïtcin, P.-C. (1992) *Silica Fume in Concrete – An Overview*, ACI SP-132, Vol. 2, pp. 835–72.

Kreijger, P.C. (1980) *Plasticizers and Dispersing Admixtures*, Admixtures Concrete International, The Construction Press, London, UK, pp. 1–16.

Lahalih, S.H., Absi-Halabi, H. and Ali, M.A. (1988) Effect of polymerization conditions of sulfonated – melamine formaldehyde superplasticizers on concrete. *Cement and Concrete Research*, **18**, 513–31.

Luke, K. and Aïtcin, P.-C. (1991) Effect of superplasticizer on ettringite formation, in *Ceramic Transactions: Advances in Cementitious Materials* (ed. S. Mindess), Vol. 16, pp. 147–66.

Maki, I., Taniska, T., Imoto, Y. and Ohsato, H. (1990) Influence of firing modes on the fine textures of alite in Portland clinker. *Il Cemento*, No. 2, 71–8.

Malhotra, V.M. (1978) *Effect of Repeated Dosages of Superplasticizers on Workability, Strength and Durability*, First International Symposium on Superplasticizers on Concrete, Ottawa, Canada, May, 34 pp.

Malhotra, V.M. (ed.) (1987) *Supplementary Cementing Materials for Concrete*. Canadian Government Publishing Centre, Supply and Services Canada, Ottawa, Canada, K1A 0S9, ISBN 0-660-12550-1, 428 pp.

Malhotra, V.M. and Ramezanianpour, A.A. (1994) *Fly Ash in Concrete*. Published by CANMET Natural Resources Canada, 562 Booth Street, Ottawa, Ontario, Canada, K1A 0G1, ISBN 0-660-15764-0, 307 pp.

Malhotra, V.M. and Mehta, P.K. (1996) *Pozzolanic and Cementitious Materials*, Gordon and Breach, ISBN 2-88-449-211-9, 191 pp.

Malhotra, V.M., Carette, C.G. and Sivasundaram, V. (1984) Role of silica fume in concrete: a review, in *Advances in Concrete Technology* (ed. V.M. Malhotra), CANMET, Natural Resources Canada, 562 Booth Street, Ottawa, Canada, K1A 0G1, ISBN 0-660-5393-9, pp. 915–90.

Massazza, F. and Costa, V.B. (1980) *Effect of Superplasticizers on the C_3A Hydration*, 7th International Conference on the Chemistry of Cement, Vol. 4, Paris, pp. 529–34.

McCarter, W.J. and Garvin, S. (1989) Admixtures in cement: a study of dosage rates on early hydration. *Materials and Structures*, Vol. 22, pp. 112–20.

Mehta, P.K. (1987) *Natural Pozzolans*, Supplementary Cementing Materials for Concrete, Canadian Government Publishing Centre, Supply and Services Canada Ltd, Ottawa, Canada, K1A 0S9, ISBN 0-660-12550-1, pp. 3–33.

Meyer, A. (1978) *Experience in the Use of Superplasticizers in Germany*, ACI SP-62, pp. 21–36.

Miller, F.M. and Tang, F.J. (1996) The distribution of sulfur in present-dry clinkers of variable sulfur content. *Cement and Concrete Research*, **26**(12), 1821–9.

Mindess, S. and Young, J.F. (1981) *Concrete*, Prentice Hall, Englewood Cliffs, NJ, 671 pp.

Mollah, M.Y., Palta, P., Hers, T.R., Vempati, R.K. and Coche, D.L. (1995) Chemical and physical effects of sodium lignosulfonate superplasticizer on the hydration of Portland cement and solidification/stabilization consequences. *Cement and Concrete Research*, **25**(3), 671–82.

Nawa, T. and Eguchi, H. (1987) *Effects of Types of Calcium Sulfate on Fluidifying of Cement Paste*, Review of the 41st General Meeting, The Cement Association of Japan, ISBN 4-88175-002-X, pp. 40–43.

Nonat, A. (1994) Interactions between chemical evolution (hydration) and physical evolution setting in the case of tricalcium silicate. *Materials and Structures*, **27**(186), 187–95.

Odler, I. (1991) Strength of cement (final report). *Materials and Structures*, **24**, 143–57.

Odler, I. and Abdul-Maula, S. (1987) Effect of chemical admixtures on Portland cement hydration. *Cement, Concrete, and Aggregates*, **9**(1), Summer, 38–43.

Onofrei, M. and Gray, M. (1989) *Adsorption Studies of 35s-labelled Superplasticizer in Cement-based Grout*, ACI SP-119, pp. 645–60.

Osborn, E.F. and Muan, A. (1960) *Phase Equilibrium Diagrams of Oxide Systems, Plate 1*, published by the American Society and the Edward Orton, Jr Ceramic Foundation.

Paillère, A.M., Alègre, R., Ranc, R. and Buil, M. (1984) *Interaction Entre les Réducteurs d'Eau-Plastifiants et les Ciments*, Rapport Lafarge/Laboratoire Central des Ponts et Chaussées, Paris, France, pp. 105–8.

Palmer, G. (1990) Ring formation in cement kilns. *World Cement*, **21**(12), December, 533–43.

Paulini, P. (1990) Reaction mechanisms of concrete admixtures. *Cement and Concrete Research*, **20**, 910–18.

Persson, B. (1996) Hydration and strength of high-performance concrete. *Advanced Cement Based Materials*, **3**(3), April–May, 107–23.

Petrie, E.M. (1976) Effect of surfactant on the viscosity of Portland cement – water dispersions. *Ind. Eng. Chem., Prod. Res. Dev.*, **15**(4), 242–9.

Philips, B. and Muan, A. (1959) System $CaO–SiO_2$. *Journal of the American Ceramic Society*, **42**(9), 414.

Piotte, M. (1983) *Caractérisation d'un Poly Naphthaline Sulfonate. Influence de son Contre-son et de sa Masse Molaire sur l'Interaction avec le Ciment*. PhD Thesis, Université de Sherbrooke, Québec, Canada, pp. 121–55.

Piotte, M., Bossanyi, F., Perreault, F. and Jolicoeur, C. (1995) Characterization of poly (naphthalene sulfonate) salts by ion-pair chromatography and ultrafiltration. *Journal of Chromatography A*, **704**, 377–85.

Pistilli, M.F., Rau, G. and Cechner, R. (1984) The variability of condensed silica fume from a Canadian source and its influence on the properties of Portland cement concrete. *Cement, Concrete, and Aggregates*, **6**(1), 33–7.

Pistilli, M.F., Winterstein, R., and Cechner, R. (1984) The uniformity and influence of silica fume from a US source on the properties of Portland cement concrete. *Cement, Concrete, and Aggregates*, **6**(2), 120–24.

Ramachandran, V.S., Beaudoin, J.J. and Shilua, Z. (1989) Control of slump loss in superplasticized concrete. *Materials and Structures*, **22**, 107–11.

Ranc, R. (1990) Interactions entre les Réducteurs d'Eau-Plastifiants et les Ciments. *Ciment, Bétons, Plâtres, Chaux*, No. 782, pp. 19–20.

Regourd, M. (1978) Cristallisation et réactivité de l'aluminate tricalcique dans les ciments Portland. *Il Cemento*, No. 3, 323–35.

Regourd, M. (1983) Pozzolanic activity of condensed silica fume, in *Condensed Silica Fume*, Les Éditions de l'Université de Sherbrooke, Université de Sherbrooke, Québec, Canada, ISBN 2-7622-0016-4, pp. 20–4.

Regourd, M., Hornain, H. and Mortureux, B. (1978) Influence de la Granularité des Ciments sur leur Cinétique d'Hydratation. *Ciments, Bétons, Plâtres, Chaux*, No. 712, May–June, pp. 137–40.

Rixom, M.R. and Mailvaganam, N.P. (1986) *Chemical Admixtures for Concrete*, E and FN Spon, London and New York, 306 pp.

Rosenberg, A.M. and Gaidis, J.M. (1989) A new mineral admixture for high-strength concrete. *Concrete International*, **11**(4), April, 31–6.

Ryell, J. and Bickley, J.A. (1987) Scotia Plaza: high-strength concrete for tall buildings, in *Utilization of High-Strength Concrete*, published by Tapir, N-7034 Trondheim NTH, Norway, ISBN 82-519-0797-7, pp. 641–53.

Sellevold, E. (1987) *The Function of Condensed Silica Fume in High-Strength Concrete*, Symposium on Utilization of HSC, Trondheim, Norway, June, ISBN 82-519-0797-7, pp. 39–50.

Sellevold, E.J. and Nilsen, T. (1987) *Condensed Silica Fume in Concrete: A World Review*, Supplementary Cementing Materials for Concrete, Canadian Government Publishing Centre, Supply and Services Canada Ltd., Ottawa, Canada, K1A 0S9, ISBN 0-660-12550-1, pp. 167–243.

References

Tagnit-Hamou, A., Baalbaki, M. and Aïtcin, P.-C. (1992) *Calcium-Sulfate Optimization in Low Water/Cement Ratio Concretes for Rheological Purposes*, 9th International Congress on the Chemistry of Cement, New Delhi, India, Vol. 5, pp. 21–5.

Tucker, G.R. (1938) *Concrete and Hydraulic Cement*, US Patent 2 141 569, December, 5 pp.

Uchikawa, H. (1986) Effect of blending component on hydration and structure formation. Principal Report of the 8th International Congress on the Chemistry of Cement, Rio de Janeiro, Brazil, *Journal of Research of the Onoda Cement Company*, **28**(115), September, 77 pp.

Uchikawa, H. (1994) *Hydration of Cement and Structure Formation and Properties of Cement Paste in the Presence of Organic Admixture*, Conference in Tribute to Micheline Moranville Regourd, Edited by Concrete Canada, Faculty of Applied Science, University of Sherbrooke, Sherbrooke, Québec, Canada, J1K 2R1, 55 pp.

Uchikawa, H., Uchida, S. and Okamura, T. (1987) *The Influence of Blending Components on the Hydration of Cement Minerals and Cement*, Review of the 41st General Meeting, The Cement Association of Japan, ISBN4-88175-002-X, pp. 36–9.

Uchikawa, H., Uchida, S. and Okamura, T. (1990) Influence of fineness and particle size distribution of cement on fluidity of fresh cement paste, mortar and concrete. *Journal of Research of the Onoda Cement Company*,. **42**(122), 75–84.

Uchikawa, H., Hanehara, S. and Sawaki, D. (1997) The role of steric repulsive force in the dispersion of cement particles in fresh paste prepared with organic admixture. *Cement and Concrete Research*, **27**(1), January, 37–50.

Uchikawa, H., Sone, T. and Sawaki, D. (1997) A comparative study of the characters and performances of chemical admixtures of Japanese and Canadian origin, Part 1. *World Cement Research and Development*, **28**(2), February, 70–76.

Uchikawa, H., Hanehara, S., Shirosaka, T. and Sawaki, D. (1992) Effect of admixture on hydration of cement, adsorptive behavior of admixture and fluidity and setting of fresh cement paste. *Cement and Concrete Research*, **22**(6), 1115–29.

Van Damme, H. (1994) Et si le chatelier s'était trompé? Pour une physico-chimie-mécanique des liants hydrauliques et des géomatériaux. *Annales des Ponts et Chaussées*, No. 71, 30–41.

Vernet, C. (1995) Mécanismes chimiques d'interactions ciment-adjuvants, *CTG Spa.Guerville Service Physico-Chimie du Ciment*, Janvier, 10 pp.

Vernet, C. and Cadoret, G. (1992) Monitoring of the chemical and mechanical changes in high-performance concretes during the first days, in *High-Performance Concrete: From Material to Structure* (ed. Y. Malier), E and FN Spon, London, ISBN 0419 17600 4, pp. 145–59.

Vernet, C. and Noworyta, G. (1992) *Reactivity Test for Fine Silicas and Pozzolanic Mineral Additives*, 9th International Congress on Chemistry of Cement, New-Delhi, India, Vol. III, pp. 79–85.

Vichot, A. (1990) Les polyméthylenenaphtylsufonates: modificateurs de la rhéologie. PhD Thesis, Université Pierre et Marie Curie, Paris VI, 151 pp.

Young, J.F. (1983) *Slump Loss and Retempering of Superplasticized Concrete – Final Report*, Civil Engineering Studies – Illinois Cooperative Highway and Transportation, Series 200 UILU-ENG.83-2006, ISSN 0069-4274, University of Illinois at Urbana Champaign, Urbana, Illinois.

CHAPTER 7

Materials selection

7.1 INTRODUCTION

High-performance concrete is prepared through a careful selection of each of its ingredients (Perenchio, 1973; Aïtcin, 1980). It is very difficult to gain the last MPa of compressive strength of a particular concrete mixture, or the 1 hour workability to place it safely and uniformly in the field, but it is so easy to lose them. The performance and quality of each ingredient become critical at a certain point as the targeted strength increases, but there are some issues that are more critical than others. Certain issues have a much stronger impact on the economics of high-performance concrete, and determine whether or not it will be competitive not only against steel but also against usual concrete. These are the issues that will be addressed in this chapter.

At present, the expression 'high-performance' covers a wide range of strengths. In order to facilitate the discussion of the importance of the different factors that have to be considered when designing and making high-performance concretes, these concretes will be divided into five different classes, each defined by compressive strength values corresponding to what can be considered as technological gaps in the present state of the art.

If a 50 to 60 MPa high-performance concrete has in fact only a slightly improved strength compared with a usual concrete, this is certainly not the case for a 120 MPa concrete. Therefore, we will discuss the selection of materials for each class. The proposed limits used to define these five classes should not be considered as absolutes, but rather as average limits that vary somewhat from one place to another.

7.2 DIFFERENT CLASSES OF HIGH-PERFORMANCE CONCRETE

The division of high-performance concrete into five classes is not as arbitrary as it seems at first glance, but derives rather from a combination of experience and the present state of the art. This classification might become codified in the near future as progress takes place in our comprehension of the different phenomena involved in making high-performance concrete; the author is very comfortable with it as this book is being written. The high-strength range has been divided into five

Materials selection

Table 7.1 Different classes of high-performance concrete

Compressive strength (MPa)	50	75	100	125	150
High-performance concrete class	I	II	III	IV	V

classes corresponding in each case to a 25 MPa increment. Class I represents high-performance concrete having a compressive strength of between 50 and 75 MPa, class II between 75 and 100 MPa, class III between 100 and 125 MPa, class IV between 125 and 150 MPa, and class V above 150 MPa, as shown in Table 7.1.

To be somewhat more precise, these compressive strengths correspond to average values obtained at 28 days on 100×200 mm cylindrical specimens cured under the standard conditions used to cure usual concrete. These are not specified strengths or design strengths, for which the standard deviation of the concrete production has to be taken into account.

7.3 MATERIALS SELECTION

As previously stated, when selecting the materials to make high-performance concrete some choices are more critical than others; this is why the selection of Portland cement will be considered first, even in cases where it is certain that one or two other supplementary cementitious materials will be used in conjunction with Portland cement. We will then discuss the selection of the superplasticizer, with or without a set retarder, because experience shows that it is the cement–superplasticizer combination that has to be optimized first. When these crucial choices have been made, the use of one or two supplementary cementitious materials will be discussed. The selection of aggregates will then be discussed because aggregate quality can become a critical issue as the targeted compressive strength increases. It will be seen that, at the present time, in some areas, aggregate performance constitutes the limiting factor when high-performance concrete compressive strength is to be increased. In these cases, for economic reasons, the concrete compressive strength cannot be increased as much as desired.

7.3.1 Selection of the cement

The first choice to be made when making high-performance concrete is definitely that of the cement, even when one or two supplementary cementitious materials will be used, because the performance of the cement in terms of rheology and strength becomes a crucial issue as the targeted compressive strength increases. Although it is not too difficult to make a class I high-performance concrete with most present cements, some of them definitely cannot be used to make class II high-performance concrete. On the other hand, few cement types can be used to make class IV or V high-performance concrete.

Different brands of a given ASTM type of cement do not perform in the same way when making high-performance concrete. Some perform very well in terms of final strength, but very poorly in terms of rheological behaviour. It is very difficult to maintain their workability long enough to place them in the field economically and satisfactorily with a high degree of reliability and uniformity. Others perform very well in terms of their rheology; their slump loss within the first 1 or 2 hours is minimal, or can be managed very easily by redosing with superplasticizer in the field. However, they perform very poorly in terms of compressive strength, and their use does not permit making concrete stronger than class I.

As has already been mentioned in the previous chapter, and will be discussed in more detail in this one, some corrective measures can be taken to try to improve this situation. However, such corrective measures will only partially help to solve the basic poor performance of the cement when it is used at a very low water/binder ratio. This is why the selection of the Portland cement has to be made carefully.

In practice, the selection of the superplasticizer is also very important, because in the end it is the best cement–superplasticizer combination that has to be selected. However, the selection of the superplasticizer is less critical because it is almost always easier and not very expensive to change the superplasticizer brand. Superplasticizers are sold at almost the same price everywhere, and their transportation cost is not a critical issue.

When Portland cement hydration was described in Chapter 6, it was shown that the initial stiffening of the interstitial phase is controlled by the addition of a certain amount of calcium sulfate to Portland cement clinker prior to final grinding. At present, this amount of calcium sulfate is generally optimized to satisfy the normal requirements of Portland cement in terms of initial setting time and compressive strength, and present standards (such as those of the ASTM) require that this be done on a standard paste or mortar with a water/cement ratio of about 0.50.

It is now well established that the rheological properties of Portland cements complying with such standards can be quite different when these cements are used in low water/cement ratio mixtures in the presence of a high dosage of superplasticizer. With some cements, it is very difficult to avoid rapid stiffening in very low water/cement ratio mixtures, even with an increased superplasticizer dosage or redosing, while in other cases the control of the slump is easy for more than 1 hour.

The situation is about the same when looking at high-performance concrete from the point of view of strength. The standard strength performance of a given Portland cement, measured using 50 mm mortar cubes, does not always correlate well with the actual strength that can be

Materials selection

reached when the cement is used at a very low water/cement ratio. This difference in the strength of the cement between usual concrete and high-performance concrete is most probably related to minute differences in the composition of the Portland cement that are almost completely masked by the relatively high water/cement ratio used when carrying out the standard tests. Experience shows that these minute differences in the composition of the cement can make a large difference when the water/binder ratio is lowered to the 0.20 to 0.35 range, generally used to make classes II to IV high-performance concrete.

When reviewing the relevant properties of the basic ingredients of high-performance concrete we have seen that, while the silicate phase plays a key role in the development of strength, it is the interstitial phases that have the key role in determining the rheological characteristics of the low water/binder ratio mixtures. Thus the final performance of the cement when making high-performance concrete will depend on the way in which the rheological behaviour and strength development can be simultaneously optimized.

First of all, let us look at the optimum fineness of the cement. There are conflicting requirements: from a strength point of view, the finer the cement the better, because more of the silicate phase will rapidly and thoroughly come into contact with water; but from a rheological point of view, the finer the cement the more reactive it will be, because more of the silicate and interstitial phases will be in contact with water, and so more ettringite will be formed and more $C-S-H$ will develop rapidly on the surface of the silicate phases (Nawa, Eguchi and Okkubo, 1991).

In present Portland cements, the interstitial phase composition and content are principally dictated by durability concerns in terms of chemical attacks and by manufacturing requirements. The amount of interstitial phase must not be too low or too high in order to optimize the steady flow of clinker out of the rotary kiln. To address the first point, cement producers have only to vary the Al/Ca ratio in their raw meal. As far as the second point is concerned, the most efficient way to produce Portland cement clinker in a modern rotary kiln is to keep the interstitial phase content almost constant, at around 14 to 16%. As a rule for usual concrete, there is no special reason to take account of the morphological form of the C_3A in this interstitial phase, which is quite important from a rheological point of view in high-performance concrete.

Regourd (1978) has shown that the morphology of the C_3A in Portland cement is governed essentially by the amount of alkalis trapped in the C_3A. If the amount of Na_2O is lower than 2.4%, C_3A retains its cubic structure (Regourd, 1978). When the Na_2O content of C_3A lies between 2.4 and 3.8%, the C_3A becomes partially orthorhombic, and at about 3.8% it is orthorhombic. Finally, when the Na_2O content is greater than 5.3%, the C_3A becomes triclinic, though this last situation never occurs in normal Portland cement. This discussion deals only with Na_2O content, but it applies also to K_2O. Therefore, in order to control the morphology of the C_3A during clinker manufacture, it is important to control the amount of alkalis trapped in the C_3A. This can be done by keeping the right balance between the alkalis available in the raw meal and the sulfur available in the kiln. This sulfur can come from the raw materials, but generally speaking it is brought into the kiln by the fuel.

Alkalis, particularly potassium, are volatilized in the clinkering zone and combine with the SO_3 to form one of the three following sulfates: K_2SO_4, Na_2SO_4 or $(K, Na)_2SO_4$. Therefore, if there is enough SO_3 in the clinkering zone, the alkalis are preferentially trapped in one of these three sulfates, and very little remains to be trapped in the C_3A. Here the C_3A will be essentially cubic. When manufacturing Portland cement, the ratio between sulfur and alkalis is usually expressed in terms of the degree of sulfurization of the clinker, as follows:

$$DS = \frac{SO_3 \times 100}{1.292\, Na_2O + 0.85\, K_2O}$$

When dealing with usual concrete, this is not a critical parameter when producing Portland cement. However, from a high-performance concrete point of view, it is an important characteristic of the clinker since it governs not only the reactivity of the C_3A but also the type of ettringite and the rate at which it will be formed, which will further control the hydration of C_3A.

As seen previously, in spite of the fact that cubic C_3A is more reactive than the orthorhombic form, it is easier to control its reaction with SO_4^{--} ions, especially in the presence of superplasticizer molecules. Also, when Portland cement clinkers are observed under a scanning electron microscope, it is seen that frequently the alkali sulfates have a tendency to precipitate just close to the interstitial phase in the form of very small crystals. Thus when the cement comes into contact with water, the alkali sulfates go rapidly into solution because they are easily soluble (Nawa, Eguchi and Fukaya, 1989).

In so far as the rheology of a particular cement is governed by the control of the C_3A through ettringite formation, the more cubic the C_3A of a particular cement, the easier it will be to control its rheology (Vernet and Noworyta, 1992). Since, in order to have a clinker with cubic C_3A, most of the alkalis have to be combined in the form of alkali sulfates in the clinkering zone, the degree of sulfurization of the clinker becomes a very important parameter in controlling the rheology of a particular cement that is intended for use in high-performance concrete.

Obviously, in any case, the lower the amount of C_3A, the easier the control of the cement rheology. Thus when looking for a cement to be used in high-performance concrete, one should choose a cement containing as little C_3A as possible, this C_3A being preferably cubic, or at least a mixture of the cubic and orthorhombic forms in which the cubic form is predominant. This cement should contain the right amount of soluble sulfates, not only the right amount of SO_3, in order to control rapidly and efficiently the formation of ettringite. Finally, from a strength point of view, this cement should be finely ground and contain a fair amount of C_3S, but not too much from a rheological point of view.

In fact, when the requirements for the five standardized ASTM types of cement are examined, it is seen that none has the characteristics of the ideal cement for high-performance applications. Type IIs and V are satisfactory in terms of the C_3A content, but usually they are not finely ground and their C_3S content is kept low in order to lower the heat of hydration. Type III cements are satisfactory in terms of C_3S content, but not in terms of C_3A content and fineness. Finally, Type I cements have the right fineness but their C_3A contents can be too high in some cases.

As stated previously, present cement standards, developed for usual concrete applications, lead to the manufacture of Portland cements that can be either not too far from the previously defined ideal cement or very far from it. This explains why, when looking at the literature, there is no agreement about the type of Portland cement to be used in high-performance concrete. In some cases, a Type I cement is used, in other cases a Type II, in yet others a Type III, and in some cases a so-called Type I/II cement. Sometimes, the cement used is also referred to as a 'special' Portland cement.

Very often the words used to characterize the cement are, deliberately or not, vague or even misleading. In one particular case, where the cement was said to be Type I, the author is aware of the fact that this particular cement was made from a clinker meeting the requirements of both Type I and Type II Portland cement, ground to a fineness of 440 m^2/kg, a fineness that is far from that of a typical Type I cement.

We can thus see that in this respect the literature is not of great help when the type and brand of Portland cement have to be selected. It is better to go back to the basic considerations already discussed to select the best cement available (or at least the least poor one) to achieve the workability and strength level desired. In general, as the requirements for strength and rheology are somewhat conflicting, the selection of the cement will be the result of a compromise. But how can these conflicting conditions be translated into practical terms?

It is easy to check the Blaine specific surface area of a given cement and to have a rough idea of its C_3A content using the Bogue formula. But it is not easy to find the polymorphic form of C_3A and the amount, nature and solubility rate of the sulfates present in the cement. In this last case the knowledge of the total amount of SO_3 in the cement is of little help because it does not tell anything about the form in which this SO_3 is combined with calcium or alkalis in the cement (Zhang and Odler, 1996a,b).

An X-ray diffractogram for bulk Portland cement is not very useful either because most of the desired information about the polymorphic form of C_3A and the types of sulfate existing in the cement cannot be

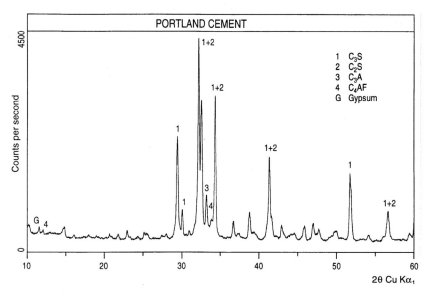

Fig. 7.1 X-ray diffractogram of a Portland cement.

deduced from such diffractograms. The peaks corresponding to this minor component, which represents less than 10%, are simply too small compared with the very strong signals given by the C_3S and C_2S, which represent about 80% of the mass of Portland cement, as seen in Figure 7.1.

However, if Portland cement is first submitted to a SAL (salicylic acid) treatment, which dissolves the silicate phases, the X-ray diffractogram of the remaining part of the cement contains only the peaks of the interstitial phase, of the sulfates, of the limestone filler, if any, and of other minor constituents such as periclase (MgO), and insoluble silica, as seen in Figure 7.2. From such a diffractogram, it is possible to see whether the C_3A is cubic or orthorhombic, or a mixture of both, because C_3A now represents the major constituent of the powder subjected to X-rays. It is also possible to determine the nature of the C_4AF and the calcium sulfate present in the cement. It is definitely possible to see, at least qualitatively, whether the calcium sulfate is present in the form of gypsum, anhydrite, hemihydrate or any combination of these. Unfortunately, it is very difficult, if not impossible, to determine quantitatively the relative proportions of these three forms of calcium sulfate using an X-ray diffractogram.

A SAL treatment is not particularly difficult to carry out and is not beyond the analytical capabilities of a well-equipped cement plant or of a university laboratory. It is recommended that it should be part of a routine test for high-performance concrete applications.

If it is not possible, in a given area, to get this information on C_3A and sulfates from the cement supplier or a nearby laboratory, an indirect

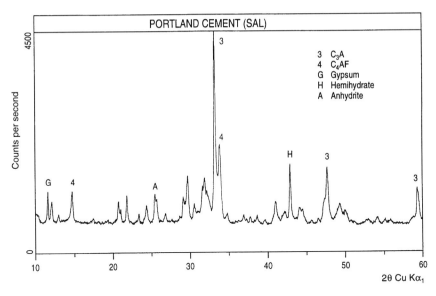

Fig. 7.2 X-ray diffractogram of a Portland cement after a SAL treatment.

measurement of the rheological reactivity of the cement has to be carried out using either a mini-slump method or any kind of grout viscosity measurement, as explained later in this chapter.

As a result of the rheological study, if the available Portland cements are found not to behave satisfactorily, use of a supplementary cementitious material should seriously be considered to try to improve the situation, before considering either the use of a retarder to try to control the slump loss problem or the 'importation' of a more efficient Portland cement. However, in such situations it could be difficult to produce economically a high-performance concrete of a class higher than class I.

With present Portland cements it is usually possible to find at least one cement for class II high-performance concrete (75 to 100 MPa), but the choice shrinks drastically when a class III high-performance concrete (100 to 125 MPa) has to be made. In such cases, the simultaneous use of one or two supplementary cementitious materials is of help.

When making class IV and V high-performance concrete the amount of mixing water has to be decreased to around 120 to 130 l/m^3. Thus if the total cementitious material has to be kept below 550 kg/m^3 for rheological reasons, it is only in a very few exceptional circumstances that it is at present possible to produce a class IV or V high-performance concrete.

In summary, the control of cement quality is quite simple. It consists primarily of measuring the Blaine fineness and making an X-ray diffractogram on a SAL-treated sample of the cement and an indirect measurement of the rheological reactivity of the cement in the presence of the superplasticizer (see later).

7.3.2 Selection of the superplasticizer

The selection of a good and efficient superplasticizer is also crucial when making high-performance concrete, because not all superplasticizer types and brands react in the same way with a particular cement (Aïtcin and Baalbaki, 1994; Uchikawa, Sone and Sawaki, 1997). Experience has shown that not all commercial superplasticizers have the same efficiency in dispersing the cement particles within the mix, in reducing the amount of mixing water, and in controlling the rheology of very low water/binder ratio mixtures during the first hour after the contact between the cement and the water (Daimon and Roy, 1978; Lahalih, Absi-Halabi and Ali, 1988; Singh and Singh, 1989).

As in the case of the cement, this situation is partially due to the fact that the present acceptance requirements for superplasticizers were essentially developed at a time when they were used primarily to fluidify usual concrete, or as high-range water reducers in usual concrete. These conditions of use are far from those prevailing in high-performance mixtures, so compatibility problems can sometimes be faced when using a cement and a superplasticizer that are each in full compliance with their respective acceptance standards. This compatibility problem can almost always be linked to the kinetics of ettringite formation, which depends on the type, amount and reactivity of the interstitial phase, on the type and solubility of the sulfates present in the ground Portland cement and on the initial reactivity of C_3S.

In the following, the selection of a particular type of superplasticizer, of a particular brand within this type and of the method of introduction of the superplasticizer into the mixture will be discussed.

Commercial superplasticizers can be broadly classified into four categories according to the chemical nature of their base, as seen in the previous chapter:

- polycondensate of formaldehyde and melamine sulfonate, also referred to as melamine sulfonate;
- polycondensate of formaldehyde and naphthalene sulfonate, also referred to as naphthalene sulfonate;
- lignosulfonate-based;
- polyacrylates (the market share of this last type has been low up to now, and so they will not be discussed further here).

It must be mentioned that, quite often, commercial formulations contain other molecules mixed with these basic molecules in order to 'improve' their efficiency under particular conditions. For example, certain commercial naphthalene superplasticizers can contain some lignosulfonate. This combination is attractive from an economic point of view for an admixture company, because the cost of lignosulfonate is lower than that of pure naphthalene superplasticizers, and sometimes leads to economies for the concrete producer as well, because it can help to solve the slump loss problem by the slight retarding action of the lignosulfonate. Some concrete producers also maintain that using some lignosulfonate mixed with a naphthalene superplasticizer results in some reduction in the total amount of superplasticizer needed to achieve the required workability.

(a) Melamine superplasticizers

Until recently, the production of melamine superplasticizers was covered by a patent so that practically only one product was available on the market. Now, however, melamine superplasticizers can be produced by any manufacturer. Thus, in future, melamine superplasticizers are likely to become available under different brand names, and so their quality and performance might be less uniform.

Melamine superplasticizers are sold as a clear liquid containing generally only 22% solid particles, but recently formulations containing up to 40% solids have been proposed. Melamine superplasticizers are generally sold in the form of their sodium salt.

Melamine superplasticizers have been used extensively to produce high-performance concrete in Europe as well as in North America (when high-performance concretes were developed in the 1970s and early 1980s). When melamine users are asked why they prefer to use this type of superplasticizer, they give a variety of answers:

- melamine superplasticizers do not delay concrete setting as much as naphthalene ones;
- as their solid content (in their 22% solid formulation) is only one-half that of a naphthalene superplasticizer, any accidental overdosage is not as critical as in the case of a naphthalene;
- melamine superplasticizers entrap less air than naphthalene ones;
- it is easier to get a stable and adequate entrained air system with melamine superplasticizers;
- melamine superplasticizers' quality and performance are very constant (though this may not be true in the future);
- melamine superplasticizers do not give a very slight beige coloration to high-performance architectural concrete made with white cement;
- melamine superplasticizers trap fewer air bubbles on the surfaces of precast elements than naphthalene superplasticizers (though in Scandinavia some precasters maintain that the contrary is true);
- the quality of the service and the reliability of the product are better than in the case of naphthalene superplasticizers;
- as melamines were the first superplasticizer used, and as they still give good results, there is no reason to change.

(b) Naphthalene superplasticizers

Although a naphthalene superplasticizer was patented as early as 1938 (Tucker, 1938), superplasticizers began to be used in the concrete industry only in the late 1960s, and the manufacture of naphthalene superplasticizers is no longer protected by a patent. This explains why a great number of brands have been, and still are, produced.

Naphthalene superplasticizers are sold as a brown liquid, with the total amount of solid particles generally between 40 and 42%. They are also available in solid form as a brownish powder. Both the liquid- and solid-form naphthalene superplasticizers are available as sodium or calcium salts, but more often as sodium salt.

There are special applications in which acceptance requirements make the use of the calcium salt mandatory. This is the case for some nuclear applications of reinforced or prestressed concrete, where even trace quantities of chloride ions are forbidden in order to protect the steel from corrosion. As the soda used to neutralize the sulfonic acid is made from sodium chloride, some chloride ions are always introduced during the neutralization stage. This is not the case for the calcium salt, for which the neutralization is carried out with hydrated lime that does not contain any chloride.

A second type of acceptance requirement that can lead to the use of a calcium salt rather than a sodium salt is when potentially alkali-reactive aggregates are used to make concrete. In such a case, any means of lowering the alkali content of the mix is strongly favoured, so some engineers specify a calcium salt rather than a sodium salt.

In high-strength concrete applications outside the two preceding situations, the sodium salt has been more frequently used because it is the one that is produced most. Sodium salt is usually produced rather than calcium salt because it is easier and less expensive to produce it. The long filtration process necessary when lime is used to neutralize the sulfonic acid is not necessary when soda is used.

Other salts of naphthalene superplasticizers have been studied by researchers but they have not been extensively used for high-performance concrete applications.

Naphthalene superplasticizers have been used predominantly and almost everywhere to produce high-performance concrete. When naphthalene users are asked why they use this type of superplasticizer rather than a melamine one they give answers such as:

- naphthalene superplasticizers have a higher solids content, so they are more cost-effective to obtain a certain degree of workability;
- it is easier to control the rheology of high-performance concrete because of the slight set delay that occurs when they are used;
- naphthalene superplasticizers are less expensive; because they can be obtained from different manufacturers, the bargaining power is better;
- the quality of the service and the reliability of some particular brands are excellent;

- here too, some high-performance concrete producers admit candidly that they started with naphthalene and stay with it because they still achieve good results.

(c) Lignosulfonate-based superplasticizers

When water reducers were discussed in Chapter 6 it was noted that the lignosulfonate dosage could not be increased beyond a certain limit because of secondary effects due to the presence of some impurities in the industrial waste liquor used to prepare lignosulfonates. It was also noted that the removal of these impurities could be too costly for the economic production of a water reducer. However, if the lignosulfonate-bearing liquor is refined further so that the final product can be sold as a superplasticizer, it becomes economically viable to remove most of the remaining sugars and surfactants.

Lignosulfonate superplasticizers are rarely used individually for high-performance concrete applications, except perhaps in some class I high-performance concrete. They are rather used in conjunction with melamine or naphthalene superplasticizers. Some concrete producers like to introduce a lignosulfonate-based superplasticizer at the beginning of the mixing, and then to use a melamine or naphthalene superplasticizer at the end of the mixing or when the slump of the high-performance concrete has to be adjusted in the field.

Some admixture producers also sell superplasticizers that are a mixture of naphthalene and lignosulfonate. From a colour point of view, it is hard to notice any difference between naphthalene and lignosulfonate superplasticizers because both have almost the same brown colour and have about the same viscosity and solids content. In order to see whether a brown superplasticizer is a pure naphthalene or a mixture of naphthalene and lignosulfonate, the best test is an infra-red spectrograph, as shown in Figure 7.3.

(d) Quality control of superplasticizers

It is difficult, long and costly to determine the detailed structure of any polymer in general, and of superplasticizers in particular. Superplasticizer manufacturers generally provide a long list of characteristics of their products which can be verified easily by any well-equipped chemical laboratory, but the problem is that many of these data do not provide an idea of the actual structure of the polymer. They are also not very instructive about the actual efficiency of the superplasticizer or its compatibility with a particular cement. However, from a practical point of view, there are several analyses that can be performed in order to establish the quality and uniformity of a superplasticizer; these include the total solids content and residual content.

An infra-red spectrograph can be obtained to check the composition and uniformity of a superplasticizer (Khorami and Aïtcin, 1989). The infra-red spectrograph can act as a fingerprint of a particular superplasticizer because it contains some minute signals related to particular impurities of the basic materials used to make the superplasticizer.

The determination of the average molecular weight of the superplasticizer molecule and of the ratio of α and β positions in the case of

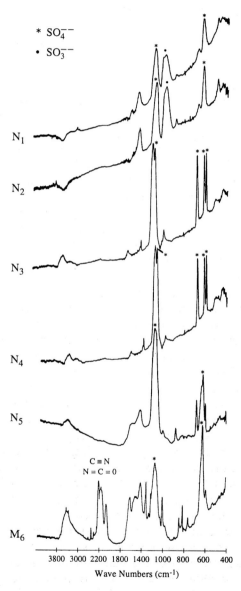

Fig. 7.3 N_1–N_5, infra-red spectrograph of a naphthalene superplasticizer; M_6, lignosulfonate superplasticizer and a mixed superplasticizer.

naphthalene superplasticizers is beyond the capability of a usual industrial chemical laboratory. These kinds of measurement can be made using liquid phase chromatography for the molecular weight, and by nuclear magnetic resonance for the α and β positions. Thus as a direct measurement of these two important characteristics of a superplasticizer is not easy, the best way to check the efficiency of a particular superplasticizer is to make a measurement of its rheological efficiency with a particular cement used as a reference (sections 7.3.3(a) and (b)).

7.3.3 Cement/superplasticizer compatibility

As at the present time it is still impossible to know by looking at the data sheets of a particular cement and a particular superplasticizer what kind of rheological behaviour could be expected in low W/B ratio mixtures (Whiting, 1979; Aïtcin, Jolicoeur and MacGregor, 1994; Huynh, 1996), it is necessary to try them and see how they work, owing to the complexity of chemical phenomena involved (as shown in Figure 6.22). As performing concrete trial batches is time-, material- and energy consuming, several methods, involving much smaller amounts of material and easier to implement and repeat, have been developed. These are generally based on studying the rheological behaviour of a grout. When these methods are properly used, it is possible to make a first screening so that efficient and non-efficient combinations can be easily determined. Factorial design experiments can be used to make this kind of study (section 7.4) (Rougeron and Aïtcin, 1994).

However, from personal experience, this is not a totally foolproof method: there are some combinations that work fine with a grout and which do not perform as well in a concrete, and there are combinations giving only a fair behaviour with a grout that perform much better in a concrete because the mixing conditions of the grout and the concrete are not the same. This is why, in spite of the merit of these simplified methods, it is always necessary to do a few trial batches. In this case, too, factorial design experiments can be used to minimize the hard work while maximizing the experimental output. However, from personal experience, it has never been found that any combination that did not work at all with grouts gave a good rheological behaviour in concrete.

Basically, two simplified methods are widely used, the so-called 'minislump' and the Marsh cone methods. The advantage of the minislump method is that it requires less material to be performed, but the grout is evaluated in a rather 'static' behaviour, while in the case of the Marsh cone method, more material is needed and the grout is tested in a more 'dynamic' condition. The use of one of these two simplified methods is a matter of personal preference. The simultaneous use of both methods is interesting because different rheological parameters are predominant in both tests. The personal preference of the author is for the Marsh cone method.

176 *Materials selection*

(a) The minislump method

As indicated by its name, this method consists of making a slump test on a small amount of grout, using the slump cone presented in Figure 7.4 (Kantro, 1980). The usual procedure is the following:

Sample preparation
- Weighing of 200 g of the cement or binder to be tested.
- Weighing of the appropriate amount of water in a 250 ml beaker. Usually the test is performed at a water/cement or water/binder ratio of between 0.35 and 0.45.
- Weighing of the selected amount of superplasticizer, which is added and dispersed in the previous amount of water.
- Hand mixing of the cement and the water with a spatula for 1 minute after starting the chronometer.
- Mixing of the paste for 2 minutes using an electrical mixer (four speed – Braun MR72 Mixer).

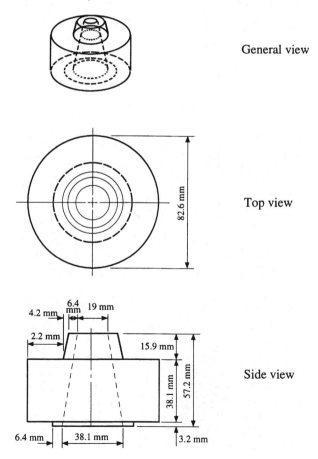

Fig. 7.4 Minicone for minislump test.

Materials selection

- Adjusting the mix temperature to the desired temperature (usually around 20 °C) using a thermoregulated bath, if available. When a thermoregulated bath is not available, the initial temperature of the water has to be adjusted so that after mixing, the paste has the required temperature.

Testing (Figure 7.5)
- A Plexiglas plate is placed on a table, the level of which has been carefully checked.
- The cone is placed at the centre of the plate and after a 15 second hand mixing to homogenize the paste, the minicone is filled with the paste.

Fig. 7.5 Minislump test (courtesy of B. Samet).

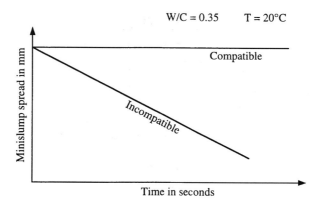

Fig. 7.6 Example of minislump results for a compatible and an incompatible cement/superplasticizer combination.

- Ten strokes are given on top of the minicone before it is raised rapidly so that the paste spreads on the Plexiglas plate.
- The diameter of the spread paste is measured along two perpendicular diameters and the average of these two values is calculated.
- The paste is again put in the beaker and hand mixed for 5 seconds, then the beaker is covered to avoid any desiccation.
- The Plexiglas plate and minicone are cleaned with water and dried for the next test.

Usually the slump is measured at 10, 30, 40, 60, 90 and 120 minutes in order to get an idea of the slump loss of the paste with time.

Figure 7.6 shows the slump loss for typical compatible and non-compatible combinations.

(b) The Marsh cone method

Marsh cones have been used for a while in different industrial sectors to appreciate the fluidity of different types of grout or mud, such as drilling muds in the petroleum industry, injection grouts in rocks or soils and injection grout in prestressed ducts. The principle of the method consists in preparing a grout and measuring how long it takes for a certain volume of the grout to flow through a funnel having a given diameter. The cones that are used can have different geometrical features and the diameter of the funnel can range from 5 mm to 12.5 mm (Figure 7.7).

Different versions of the Marsh cone method can be found in the literature (de Larrard, 1988; de Larrard, Gorse and Puch, 1990; Aïtcin, Jolicoeur and MacGregor, 1994; de Larrard *et al.*, 1996). It is not the intention of the author to discuss the merits of each of these versions of the Marsh cone method, but rather to present the one that has been used at the Université de Sherbrooke over the past few years.

This is a standard plastic cone used by the petroleum industry to measure the flow time of drilling muds. Its total capacity is 1.2 l. A

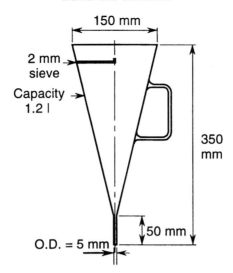

Fig. 7.7 Marsh cone method used at the Université de Sherbrooke.

schematic of this cone is presented in Figure 7.7. This particular cone has been selected because it is commercially available at a low price (less than US$30) and was found convenient for the purpose of studying cement–superplasticizer compatibility.

Sample preparation and flow time measurement

The amount of water, cement and superplasticizer needed to prepare 1.2 l of grout is calculated. Usually the test is performed at a water/cement or water/binder ratio of 0.35 in order to test the cement and superplasticizer under conditions similar to those of the paste of a high-performance concrete. It should be remembered that the amount of material needed for this test is 10 times greater than for the minislump test (around 1.8 kg of cement for the Marsh cone test versus 200 g for the minislump test).

The sample preparation comprises the following steps:

- Weighing of the water and superplasticizer in the container where the mixing will be done.
- Starting the mixer while introducing progressively the amount of cement within 1 minute and 30 seconds (Figure 7.8).
- Stopping the mixing for 15 seconds in order to clean with a spatula the cement glued to the container.
- Mixing for 60 seconds.
- Measuring the temperature.
- Measuring the time it takes to fill a 1 l beaker with the grout (Figure 7.8).
- Placing the grout in a plastic bottle that will be placed on two rollers so that the grout remains homogeneous and also to simulate concrete transportation.

180   Materials selection

- Measuring at different time intervals the flow time up to 60 or 90 minutes. Each time, the temperature of the grout is measured.

Note: In a recent modified version of this test (de Larrard *et al.*, 1996), it has been proposed to make the grout using materials finer than the 2 mm that will be used in the high-performance concrete.

In the following paragraphs, the rheological behaviour of all cement–superplasticizer combinations will be discussed based on the results of a 1.2 l Marsh cone having a 5 mm aperture (4.76 mm ID).

(c) Saturation point

When the flow time is measured with different superplasticizer dosages expressed as a percentage of the solids contained in the superplasticizer to the cement mass at a given time, a curve like that presented in Figure 7.9 is obtained. This curve is composed of two lines having different slopes. The intersection of these two lines corresponds to what is called

 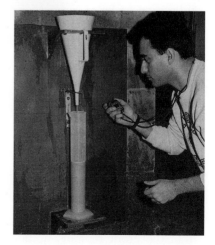

Fig. 7.8 Determination of the Marsh cone flow time (courtesy of M. Baalbaki).

Materials selection

the 'saturation point'. This is the point at which, in the experimental conditions used for the measurement of the flow time, any increase in the dosage of the superplasticizer has no effect on the rheology of the grout. The superplasticizer dosage corresponding to this point is called the saturation dosage, and the flow time, the flow time for the saturation dosage.

From a practical point of view, it is not worth while studying cement–superplasticizer compatibility with a grout that is not too fluid or too thick, because most of the compatibility problems could be hidden. It has been found from experience that it is convenient to adjust the water/binder ratio of the grout so that the 5 minute Marsh cone flow time is between 60 and 90 seconds. Usually, to achieve such a flow time, the water/binder ratio for most Type I Portland cements is between 0.35 and 0.40. However, there are occasions when it is necessary to use a water/binder ratio of between 0.40 and 0.45. In any case, if the water/binder ratio has to be raised to 0.45 to obtain a grout with a 5 minute flow time between 60 and 90 seconds, it is better to look for another cement or another superplasticizer to make a high-performance concrete.

It is important that the temperature of the grout remains within the 20 to 23 °C range at the end of the mixing period in order to reproduce

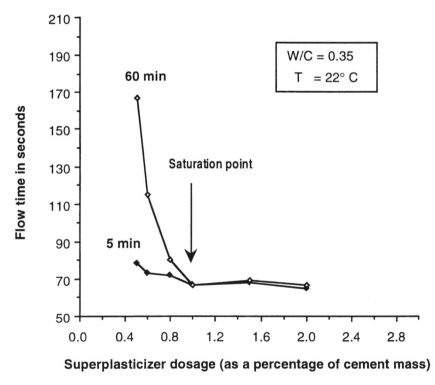

Fig. 7.9 Flow time as a function of the superplasticizer dosage.

normal initial hydration conditions. In concrete, the large mass of aggregates acts as a heat sink for the initial heat of hydration developed during the mixing. As there are no aggregates to reduce the initial temperature in a grout, it may be necessary to cool the mixing water in order to obtain a grout with a final temperature of 20 to 23 °C at the end of the mixing. Experience shows that in some cases it may be necessary to use a mixing water with an initial temperature of between 5 and 10 °C. This initial temperature of the water has to be adjusted to take into account the initial heat developed by the cement when it comes into contact with water and gives an idea of how this cement is reactive.

From a practical point of view, when using the Marsh cone testing method to study the rheological compatibility of a given cement and a given superplasticizer, it is suggested proceeding in the following way:

- A 0.35 water/binder ratio grout is made using a superplasticizer dosage that corresponds to a solid content of around 1% of the cement mass.
- If the measured flow time with a Marsh cone with a 5 mm opening is between 60 and 90 seconds, the 0.35 water/binder ratio is selected to pursue the study of the variation of the flowing time as a function of the superplasticizer dosage. The concentration of the superplasticizer can then be increased and decreased by 0.2% in steps from the initial 1% value.
- If the flow time is greater than 90 seconds, another test should be made, this time using a water/binder ratio of 0.40, and again a 1% dosage of superplasticizer. If the flow time is between 60 and 90 seconds, the experiment is pursued at a water/binder ratio of 0.40, as in the previous case.
- If the Marsh cone flow time is still greater than 100 seconds at a water/binder ratio of 0.40, it is better to use another superplasticizer, or if all the available superplasticizers behave the same with that particular cement, it is better to select another cement.

(d) Checking the consistency of the production of a particular cement or superplasticizer

The saturation dosage can be used to control the consistency of a particular commercial superplasticizer always using the same reference cement, or to control the consistency of a particular Portland cement using the same reference superplasticizer. In both cases, if the cement and the superplasticizer production are well under control, the flowing time should not change by more or less than a few seconds.

At the Civil Engineering Department of the Université de Sherbrooke, an ASTM Type I Portland cement (Type 10 according to CSA standards) has been selected as a reference cement to control the production of a particular naphthalene superplasticizer. When an efficient superplasticizer (i.e. a superplasticizer having 90% β position for the sulfonate group, a degree of sulfonation above 90%, and an average degree of polymerization of 9 to 10 with very few mono-, di- and trimers) is used at a dosage rate of 0.8% in a 0.35 water/binder ratio grout, this grout has a Marsh cone flow time of around 60 seconds (± 2 seconds). When this cement is used, for example, to check the uniformity of a pure naphthalene-based superplasticizer, if the flow time of a production sample of the superplasticizer is greater than this value, this indicates that something has gone out of control during the manufacturing process, for example too many sulfonate groups have been fixed in the α position and not enough in the β position, or the process has produced polymers that are too short and/or have generated too much cross-linking.

This type of test is very simple, very rapid and quite precise, and has never failed to indicate a superplasticizer of lower quality or lower efficiency when it was tested further using liquid phase chromatography and magnetic resonance.

(e)   Different rheological behaviours

Experience shows that cement and superplasticizers that fulfil ASTM standards can present saturation points over a wide range, as shown in Figure 7.10. The dosage at the saturation point can be as low as 0.6% or as high as 2.5%, a 1% value corresponding to a good value for an average cement when the grout is tested at a water/cement ratio of 0.35. The flow time can vary over a large range, from 60 seconds (the flow time of water is 32 seconds, ± 1 second) up to 120 seconds at 5 minutes. In some cases,

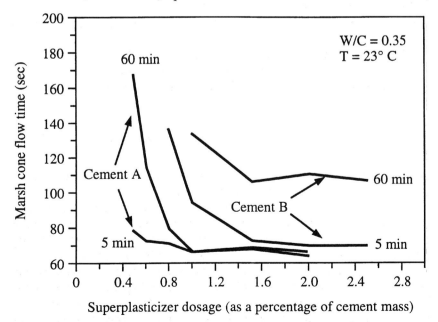

Fig. 7.10 Marsh cone flow time for two different cements.

when the cement and the superplasticizer are not compatible, it is impossible to have any flow-out after 10 to 15 minutes; such a combination is said to be very incompatible. On the other hand, the flow time sometimes remains the same for 1 hour to 1 hour and a half; in such a case the combination is said to be perfectly compatible. However, it must be verified that such compatibility is not linked to any strong retardation by measuring the strength of mortar cubes made with standardized sand and this grout.

It has been found that the most important parameters controlling the rheology of cement and a particular superplasticizer are the following:

- the water/binder ratio at which the test is performed, which should be as close as possible to that used in the actual high-performance concrete;
- the initial temperature of the water used to make the grout;
- the fineness of the cement;
- the phase composition of the Portland cement;
- the amount of calcium sulfate and its rate of dissolution;
- the alkali sulfates present in the cement (Nawa, Eguchi and Fukaya, 1989);
- the efficiency of the mixing.

From a practical point of view, four typical situations can be found when studying the rheological properties of a low water/binder ratio grout using a Marsh cone. Figures 7.11(a) to (d) illustrate these different situations when the superplasticizer dosage is varied.

Figure 7.11(a) represents the case of a fully compatible combination of cement and superplasticizer: the superplasticizer dosage at the point of saturation is low (around 1.0%) and the 60 minute curve is close to the 5 minute one. Thus the increased flow time is maintained for 1 hour. Figure 7.11(b), on the other hand, represents a case of incompatibility: the superplasticizer dosage at the less well-defined saturation point is quite high and the 60 minute curve is much higher than the 5 minute curve. Sometimes, when the incompatibility is more pronounced, the grout stops flowing very rapidly, possibly as early as 15 minutes after the beginning of mixing.

Figures 7.11(c) and 7.11(d) represent intermediate cases. In Figure 7.11(c), the 5 minute curve is similar to the 5 minute curve in Figure 7.11(a) but the 60 minute curve is similar to the 60 minute curve in Figure 7.11(b). In Figure 7.11(d) the 5 minute curve is similar to the 5 minute curve in Figure 7.11(b) and the 60 minute curve has a position relative to the 5 minute curve similar to the situation in Figure 7.11(a).

The typical situations of Figure 7.11 are explained in detail in Lessard et al. (1993), and some solutions mainly involving the addition of an appropriate amount of a set retarder are proposed so that the cement and superplasticizer combinations presented in Figure 7.11(b), 7.11(c) and 7.11(d) behave, in the best cases, as closely as possible to the combination presented in Figure 7.11(a). Unfortunately, it is not always possible to achieve such a solution.

Presently, it is not possible to develop polynaphthalene or polymelamine sulfonates that work with any Portland cement at any water/cement ratio because of the too great variability in the solubility rate of the different calcium sulfates used in commercial cements (Tagnit-Hamou, Baalbaki and Aïtcin, 1992; Tagnit-Hamou and Aïtcin, 1993). However, if attention is paid to this important influence of the nature of calcium sulfate in concretes with a low water/binder ratio, it is possible to manufacture Portland cement that is perfectly compatible with good polynaphthalene or polymelamine superplasticizers. If such an ideal situation does not exist, it is preferable to develop a blended superplasticizer whose composition is adjusted to the particular cement.

(f)   Practical examples

A compatible cement–superplasticizer combination
Figure 7.12 represents the case of an efficient cement–superplasticizer combination. The flow time at 5 minutes is around 60 seconds along a wide range of superplasticizer dosages beginning at a 0.3% dosage. At 60 minutes, the flowing time is still less than 65 seconds (this is a very low 60 minutes flowing time). This figure shows the influence of the fineness

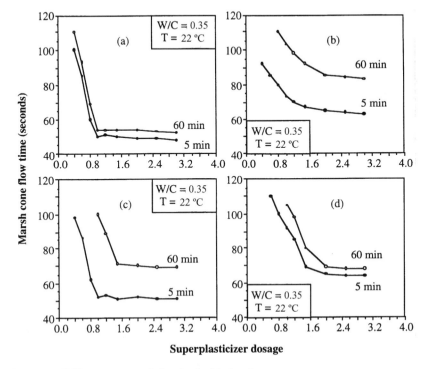

Fig. 7.11 Different types of rheological behaviour.

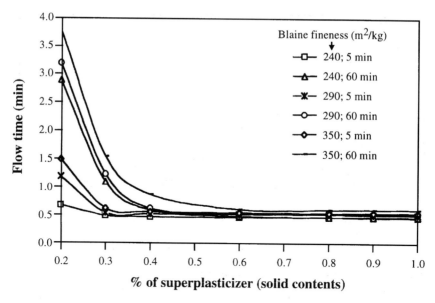

Fig. 7.12 Marsh cone flow time for a compatible type 20 M cement of different fineness and naphthalene superplasticizer.

on the flow time. The higher the fineness, the higher the flow time for low superplasticizer dosage, though beyond the saturation point there is no significant reduction of the flow time whatever the fineness. The chemical and Bogue composition of this Type 20 M cement is given in Table 7.2.

Influence of the cement type

Figure 7.13 shows the Marsh flow time of two different cements produced by the same cement company in the same cement plant. One is the

Table 7.2 Chemical and Bogue composition of different cements tested for Marsh cone flow time

	Type I	Type II (20 M)
CaO	62.4	62.9
SiO_2	20.8	23.1
Al_2O_3	4.0	3.2
Fe_2O_3	3.0	4.7
Alkali equivalent	0.87	0.40
SO_3	3.2	2.4
Loss on ignition	2.7	0.6
C_3S	52	45
C_2S	20	32
C_3A	6.6	1
C_4AF	9.2	14

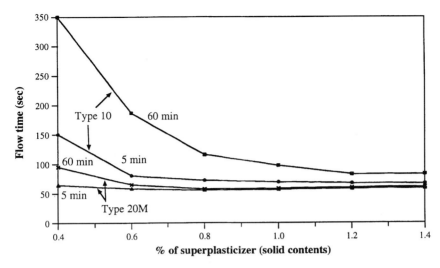

Fig. 7.13 Comparison of Marsh cone flow time of a Type I and II cement produced at the same cement plant.

previous Type I cement (Canadian Type 10) and the other is a Type II cement (Canadian Type 20 M), the compositions of which are given in Table 7.2. It is seen that the modified Type II cement, which has a very low C_3A and C_3S content, has a saturation point corresponding to a much lower superplasticizer dosage and lower flow time; moreover, the flow time at 60 minutes is close to that at 5 minutes, indicating a very low initial reactivity and the potential for good slump retention.

A combination resulting in a significant loss of fluidity during the first hour

A Type I Portland cement was tested with a high-quality superplasticizer. The slump loss as a function of the time is shown in Figure 7.14. The test was performed at a water/binder ratio of 0.35 and 0.8% superplasticizer dosage. The following conclusion can be drawn from this curve: the initial rheological behaviour of the cement up to 45 minutes is not so bad, but between 45 and 60 minutes the concrete experiences a significant slump loss which makes it difficult to use in the field.

The behaviour of this cement at very low water/cement ratios can be improved, for example in two different ways: first by substituting some cement with a supplementary cementitious material and second by using a slight amount of a set retarder.

Improving the rheology of grout using a set retarder

Figure 7.15 represents a case where the addition of a small amount of set retarder significantly improves the rheological behaviour of a grout made with a silica fume blended cement having a water/binder ratio of 0.35.

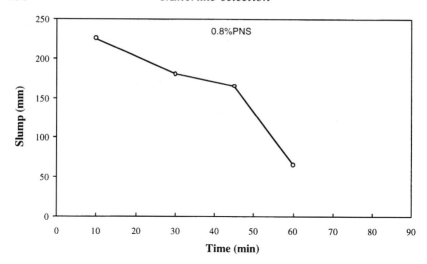

Fig. 7.14 Slump loss of a concrete prepared with a cement presenting a significant slump loss during the first hour.

At 5 minutes, the retarder has no effect on the flowing time, which is excellent (below 60 seconds), but at 60 minutes, the saturation point and the corresponding flow time are drastically lowered, down to 1.0% and 67 seconds, which is excellent compared with 1.2% and about the same

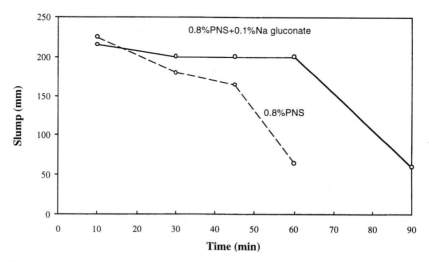

Fig. 7.15 Improving the slump retention with a small amount of retarding agent for the same concrete as in Figure 7.14.

value for the flowing time. The use of the small amount of set retarder resulted in a 50% decrease in the dosage at the saturation point.

7.3.4 Superplasticizer dosage

As has been pointed out, the mixing method has an influence on the saturation point of a grout. Therefore, it is not absolutely certain that the saturation point dosage of a grout corresponds exactly to the optimum dosage of the superplasticizer on the high-performance concrete, whose water/binder ratio can be different from the value used to find the saturation point. However, most of the time it is quite close. Usually, owing to the very strong mixing procedures, the saturation point dosage corresponds to an upper limit of the superplasticizer dosage to be used in the concrete. Therefore, the first concrete trial batch can be designed by introducing only 80% of the superplasticizer dosage of the saturation point and adding the rest of the superplasticizer if needed.

The determination of the optimum dosage of a superplasticizer of a high-performance concrete is not easy. It greatly depends on what is the most critical issue for the particular high-performance concrete to be produced with a particular cement or cementitious system. If the long-term strength issue is the most critical, without any problems on the rheological side, it is preferable to work with the highest amount of superplasticizer and lowest amount of water possible. However, the superplasticizer dosage should never be much higher than the saturation point in order to avoid segregation and excessive set retardation.

If it is the rheological properties that are critical and not the strength, it is best to make the concrete with the highest water/binder ratio and the highest amount of water that can be used to achieve the strength requirements, and to adjust the superplasticizer dosage in order to get the desired rheology.

In an intermediate situation, the most economical balance between the strength requirement (lowest possible water/binder ratio) and the rheological requirement (highest water/binder ratio) has to be found for the particular cement and particular superplasticizer used. When this exercise has to be done for the first time, it might be perceived as long and tedious. However, the most efficient superplasticizer dosage can be found rapidly using the 'three water content approach': the lowest, the highest and an intermediate water/binder ratio, compatible with the strength requirement, are used to make three trial batches containing the same amount of binder but different amounts of water. These trial batches are tested for strength and slump loss.

Another approach consists of making a first trial batch at a water/binder ratio that should provide the targeted strength with a superplasticizer dosage corresponding to 1% and then making the necessary adjustments to this dosage according to the results obtained. Generally speaking, it is possible to find the final water/binder ratio and superplasticizer dosage by the third or fourth trial batch.

(a) Solid or liquid form?

The superplasticizer must be introduced into the concrete in a liquid form in order to obtain the benefit of its full potential as soon as possible. It takes time for a powdered superplasticizer to dissolve in water, so not all of the solids introduced into the mix are fully active during mixing.

(b) Use of a set-retarding agent

If the reactivity of the interstitial phase is critical in controlling the rheology of the concrete, the use of a retarder can help to solve the slump loss problem. It is therefore necessary to find the appropriate balance between the dosage rates of the superplasticizer and the set retarder to avoid any slump loss for the duration of the placing without unduly delaying the strength development of the concrete (Bhatty, 1987).

The early attempts of admixture companies to incorporate a set-retarding agent directly within commercial superplasticizers were unsuccessful because, depending on the characteristics of commercial cements, different behaviours can occur when using a fixed combination of superplasticizer and retarder.

From a practical point of view, the use of a set retarder should be considered only when supplementary cementitious materials are not available or their use has proved to be unsatisfactory to resolve rheological problems.

(c) Delayed addition

Several researchers have shown that the delayed addition of the superplasticizer in the mix can be beneficial from a rheological point of view (Penttala, 1993; Uchikawa, Sawaki and Hanehara, 1995). The slump loss can be reduced significantly when adding the larger part of the superplasticizer at the end of the mixing sequence. Sometimes a lignosulfonate water reducer is added at the beginning of the mixing sequence and the superplasticizer at the end of it. Usually the initial blockage of the C_3A by the calcium sulfate present in the cement is used to explain the beneficial effect of a delayed addition of the superplasticizer. In proceeding with a delayed addition, the superplasticizer molecules are no longer competing with calcium sulfate to combine with C_3A, thus almost all the superplasticizer molecules can be used to fluidify the cement paste.

7.3.5 Selection of the final cementitious system

From experience to date, it is known that high-performance concretes of classes I and II (50 to 100 MPa) can be made using a great variety of cementitious system: pure Portland cement; Portland cement and fly ash; Portland cement and silica fume; Portland cement, slag and silica fume; Portland cement, fly ash and silica fume. However, the literature also shows that almost all class III high-performance concretes (100 to 125 MPa) that have been produced contain silica fume, except for a very few in the lower range of this class which were made with Portland cement only. Up to now, classes IV and V high-performance concretes have all been produced with silica fume. Moreover, it is observed that as concrete compressive strength has increased, fewer and fewer types of cementitious combinations have been used. Thus from a practical point of view, when making class I and II high-performance concrete the use of fly ash or slag should be seriously considered. These are less expensive than cement and usually require less superplasticizer. Hence their use results in savings in terms of both materials cost and easier control of the rheological behaviour of the concrete.

The use of silica fume, on the other hand, generally increases the cost of the mixture. Its use has to be decided on arguments other than economics, such as the ease with which the targeted strength and the control of the rheological behaviour of the fresh concrete can be achieved. However, since for class III high-performance concrete it is almost always necessary to use silica fume in order to achieve the targeted strength, designers should be aware that the decision to design with a 100 MPa concrete rather than a 90 MPa concrete can have a significant impact on the economics of the structure. In North America, with the exception of eastern Canada, the market price of silica fume usually varies between 5 and 10 times the price of Portland cement. This explains why concrete producers as often as possible tend to avoid using silica fume in their class I and II high-performance concretes. This also explains why very often there is a price gap between a 90 MPa concrete and a 100 MPa concrete. The cost of the cementitious system can almost double when making a 100 MPa concrete rather than a 90 MPa one because if the concrete producer has carefully selected a cement and cementitious system, a 90 MPa design strength concrete could be produced without silica fume; but in order to produce a 100 MPa concrete, at present, about 10% silica fume has to be incorporated into the mix.

Up to now, only in a very few cases has the use of supplementary cementitious materials been advocated from solely rheological considerations, but this type of use will certainly increase in the future. Even though one of the major reasons advocated for the incorporation of supplementary cementitious materials has been economics, it has rarely been emphasized that a saving in the cost of the mix also results from the reduction in the amount of superplasticizer necessary to achieve the desired workability. The saving of 5 to 6 l of superplasticizer per cubic metre of concrete, as already reported when using a supplementary cementitious material, is about US$10 per cubic metre of concrete.

In the following, the main characteristics and advantages of the three most important supplementary cementitious materials used in high-performance concrete will be reviewed in order to facilitate the decision as to whether to use a particular supplementary cementitious material, or a combination of two.

7.3.6 Selection of silica fume

(a) Introduction

There is still a myth in the concrete industry that silica fume is necessary to make a high-performance concrete. As has already been pointed out, this is only partially true. Class I and II high-performance concretes (50 to 100 MPa) can be made without silica fume, and in a very limited number of cases class III high-performance concrete (100 to 125 MPa) has been made and delivered in the field without silica fume. However, whenever silica fume is available at a competitive price, its use is recommended because it is then easier to obtain the targeted rheology and strength.

(b) Variability

In the past, some bad experiences have been reported with certain silica fumes, and even 'non-pozzolanic' silica fumes have been described. In the author's opinion, this is due to the fact that in the silicon and silicon alloy industries, the term 'silica fume' is broadly used to identify the very fine powders that are collected in the so-called baghouses. As there are different types of alloys produced, so there are different types of silica fume (Pistilli, Rau and Cechner, 1984; Pistilli, Wintersten and Cechner, 1984; de Larrard, Gorse and Puch, 1990). Since very often a baghouse might serve different furnaces producing different alloys, the collected silica fume is actually a 'cocktail' of different types of silica fume. This last case is not so bad if the different furnaces connected to the same baghouse always produce the same alloys in the same proportions because then the same cocktail is always collected. However, if for market reasons different alloys are produced in different furnaces at different times, the silica fume cocktail varies. In order to avoid such a situation the silica fume user must carefully check the quality of the silica fume source, or buy it from a source that controls such variability (Pistilli, Rau and Cechner, 1984; Pistilli, Wintersten and Cechner, 1984).

In the very few countries that have, to the best of my knowledge, a standard governing the quality of silica fume to be used in concrete or with cement, various requirements in terms of chemical composition and physical properties have been established. These could be used as guidelines to control the quality of any given source of silica fume (Holland, 1995; *ACI Guide for the Use of Silica Fume in Concrete*, 1996).

(c) What form of silica fume to use

At present, silica fume is available in different forms. It can be bought as collected in the baghouse. In this case, its bulk unit weight ranges from 200 to 250 kg/m^3, so the usual 30 tonne cement tanker can transport only around 11 to 12 tonnes of bulk silica fume, making transportation expensive. Bulk silica fume is also available in 15 kg paper bags or in a 1 tonne big-bag. The handling of bulk silica fume is not impossible, but it is not easy.

Silica fume is also offered, more and more commonly, in a so-called 'densified' form. Different processes can be used to densify silica fume, and from the limited comparative data published it seems that densified silica fume performs as well as the non-densified fume. In such cases, the bulk density can vary from 400 to 500 kg/m^3. Of course, it is more economical to transport densified silica fume over longer distances, and densified silica fume presents fewer handling problems. Recently, however, a too strongly densified silica fume has been suspected of being at the origin of localized alkali/silica-type reactions. This took place around some undispersed silica fume hard cores, but up to now the case has not been fully documented.

Silica fume is also available in slurry form, in which the solid content is around 50%, but special equipment is needed to use silica fume in this form.

At the present time silica fume is also available blended directly with Portland cement in Iceland, Canada and France. The percentage of silica fume in the blend varies from 6.5 to 8.5%. If this percentage is expressed as a percentage of the mass of the Portland cement contained in the blended cement, this corresponds to 6.7 to 9.3%, as explained in Chapter 8.

In any case, the choice of the form in which silica fume will be used is usually limited by availability, economics and service considerations; thus these are the determining factors when making this choice (Gjørv, 1991; Cail and Thomas, 1996).

(d) Quality control

The best way to control the quality of bulk silica fume is: (1) to measure its silica, alkali and carbon contents, (2) to take an X-ray diffractogram to check it does not contain too many crystalline parts, as shown in Figure 6.32, (3) to measure its specific surface area by nitrogen adsorption, and (4) to check its pozzolanicity (Vernet and Noworyta, 1992).

Sieve analysis should be performed on bulk silica fume when wood chips have been used to produce it. When using densified silica fume, chemical analysis, X-ray diffractogram and bulk density measurements should be used routinely to control quality. When silica fume is used in a slurry form, the solid content of the slurry should be routinely controlled through chemical analysis and X-ray diffractograms of the solids.

The quality control of a silica fume blended cement should be in terms of its chemical analysis in order to control the silica content. If the chemical analysis of the clinker and the silica fume used to produce that particular blended cement are known, simple proportioning calculations can be used to determine the silica fume content of the blended cement.

It is not appropriate to control the fineness of silica fume blended cement with a Blaine apparatus because it can give misleading results. It is better to evaluate cement fineness by nitrogen adsorption. In such a case it must be remembered that, in general, silica fume has a specific surface area of between 15 000 and 20 000 m^2/kg, while Type I Portland cement generally has a specific surface area of 1200 to 1500 m^2/kg, so a variation in the specific surface area of the silica fume, or of its percentage in the blended cement, can make a significant difference without unduly affecting the strength characteristics of that particular blended cement.

(e) Silica fume dosage

When the use and form of silica fume have been decided, the next step is to decide its dosage. Theoretically, in order to fix all of the potential lime liberated by the hydration of C_3S and C_2S, the silica fume dosage should be 25 to 30%. Such high dosages have occasionally been used in the laboratory but not often in field applications, owing to the high amount of superplasticizer needed. Usually, silica fume has been used in high-performance concrete at a dosage rate of between 3 and 10%.

In fact, from a strength point of view, it can be shown that, at least for usual concrete and for class I high-performance concrete, the strength gains are very significant when silica fume dosage increases from 5 to 10%, but any further addition of silica fume results in a much smaller strength increase. Thus, as the additional cost of this extra silica fume is high and as more superplasticizer will have to be used to disperse the additional silica fume, the payback in terms of $/MPa is less and less attractive.

In spite of the fact that well-documented scientific or economic studies cannot be found in the literature on the optimum dosage of silica fume in the domain of high-performance concrete to the best of my knowledge, I suggest a silica fume dosage of between 8 and 10% of the mass of cement. However, this is not an absolute figure, and local conditions may prove that a different dosage rate can be more efficient in terms of $/MPa.

Usually, the silica fume dosage is expressed as a percentage of the mass of cement used in the particular mixture and not as a percentage of the total cementitious materials, except in the case of a blended cement. In the following, whenever the dosage of silica fume is expressed as a percentage, this percentage will always represent the ratio between the mass of silica fume used in the mixture and the mass of cement or of the total amount of cementitious material (exclusive of silica fume) multiplied by 100.

7.3.7 Selection of fly ash

As stated earlier, the main problem when using fly ash in concrete is that fly ash is a generic term used to define a product that can vary within very broad limits, in spite of the different classifications and acceptance standards developed over the years to subdivide fly ashes into more definite categories. It is not an exaggeration to say that there are as many types of fly ash as there are fly ashes, and that any generalization about fly ash has to be made very carefully (Aïtcin et al., 1986; Swamy, 1993). However, for the sake of simplicity, the ASTM classification dividing fly ashes into two categories according to their chemical composition will be used in the following. Class F includes low-calcium fly ashes, while Class C covers high-calcium fly ashes.

Whenever a fly ash with a good performance record in usual concrete is available its use should be seriously considered for class I high-performance concrete (50 to 75 MPa) because past field experience has shown that there is no problem in meeting the strength requirements. Moreover, the economics speak strongly in favour of the use of fly ash in this case.

The case of class II high-performance concrete (75 to 100 MPa) is not so clear. Very often, 75 MPa seems to be the upper strength limit for many of the fly ash mixes used up to now. There are, however, some cases reported in the literature in which fly ash has been used successfully to make a 90 or even a 100 MPa concrete (Aïtcin and Neville, 1993). In these specific cases, unfortunately, it is difficult to establish what made the fly ash so efficient because of a lack of appropriate data on its physico-chemical characteristics.

In the present state of the art, it does not seem realistic to use fly ash to make class III high-performance concrete (100 to 125 MPa) without the simultaneous use of some silica fume. However, while the strength requirements for class II and even class III high-performance concrete are achievable if silica fume is also used in the mixture, such ternary compositions have not often been used, though this will undoubtedly change in the future.

From reports in the literature, it may be seen that 15% represents the usual average dosage of fly ash in high-performance concrete, but this is not a fixed value: dosages as low as 10% and as high as 30% have also been reported (Mehta and Aïtcin, 1990). Usually, the higher the targeted strength, the lower the dosage because fly ashes are not as reactive as cement or silica fume in a high-strength mixture.

It should be mentioned that a very promising avenue for fly ash in high-performance concrete is currently under development in CANMET laboratories. This is the use of a high content of fly ash coupled with a drastic decrease in the water cementitious ratio which makes it possible to make class I and II high-performance concretes.

(a) Quality control

It is very important to control the quality of a fly ash to be used to make high-performance concrete because it is probably the most variable cementitious material in the mix. The track record of a particular fly ash should be considered with some caution, because it is not certain that the coal burned at present in the power plant is the same as the one burned in the past and also because some power plants add limestone to their coal or lignite in order to control the SO_3 emission; this may be even more common in the future. The reaction of SO_3 with this added limestone results in a higher calcium content of the fly ash, but this is misleading because it is due to an increase in calcium sulfate content and not to calcium trapped in the vitreous phase. This additional calcium sulfate in the fly ash is not harmful *per se* when making high-performance concrete, if it does not modify the cementitious–superplasticizer compatibility drastically. If it does, some serious rheological problems could be foreseen when using this particular fly ash.

The control of the quality and the consistency of a fly ash starts with its chemical composition, especially the SiO_2, Al_2O_3, Fe_2O_3, CaO, alkalis, carbon and SO_3 content.

Next to Blaine fineness, the area must be checked regularly in order to check the variation in grain size distribution and carbon content. Third, it is very important to take routine X-ray diffractograms in order to see if there are any changes in the vitreous state. The more vitreous the fly ash, the better it is: a petrographic examination of a thin section is also very helpful to determine the actual amount of vitreous particles and the mineral composition. For example, the ash from the Mount St Helens explosion was without any pozzolanic value because it was almost totally crystalline. The examination of a thin section also gives a good idea of the shape of the particles and the phase distribution according to the grain size distribution. Unfortunately, this kind of routine testing is rarely performed, even though it can practically eliminate the risk of costly field problems.

7.3.8 Selection of slag

The decision as to whether to use slag clearly depends on its availability at an economic price. Usually, when slag is available, the choice between different slags is quite limited (or may not even exist). At present, from their limited usage in high-performance applications, it appears that slags that perform well with Portland cement in usual concrete also perform well in high-performance concrete. Most of our present experience in high-performance concrete comes from the Toronto area in Canada, where slag has been and continues to be used successfully in the construction of high-rise buildings.

Up to now, slag has always been used in conjunction with silica fume to make class I, II and III high-performance concretes (50 to 125 MPa) and has never been used for class IV and V high-performance concretes (> 125 MPa). This is probably because they were not investigated seriously for these applications, but there is no reason why slags should not be used in the future to make these classes of high-performance concrete.

(a) Dosage rate

As in the cases of silica fume and fly ash, as soon as the decision to use slag to make high-performance concrete has been taken, the next step is to decide the dosage rate. The literature shows that, up to now, slag has been used at a dosage rate of between 15 and 30%. Moreover, in the reported field applications in the Toronto area, this dosage rate is said to vary according to climatic considerations (Ryell and Bickley, 1987). For example, the dosage has been lowered to 15% during winter in order to obtain adequate early strength to strip the forms as early as possible, and increased up to 30% in the summer. However, in the future we may well see slag used in high-performance concretes at a higher dosage ratio. In an experimental field test in a ready-mix concrete plant in Montreal, Baalbaki *et al.* (1992) were able to produce a high-performance concrete with a 91 day compressive strength of 130 MPa containing 60% slag, 30% Portland cement and 10% silica fume.

In general, depending on early strength considerations, the activity of the slag and climatic factors, it is reasonable to use slag at a dosage rate of between 15 and 30% of the cementitious mass in conjunction with silica fume at a dosage rate of 10%.

(b) Quality control

The best way to control the quality of a particular slag is first to check its Blaine fineness, which should be of the order of 450 to 600 m^2/kg, and second to take an X-ray diffractogram in order to see whether its vitreous state varies with time (Figure 6.43). It is usually not necessary to perform a chemical analysis on a routine basis because slag chemical composition does not vary much, but it is useful to be sure that the slag is constant in that respect.

7.3.9 Possible limitations on the use of slag and fly ash in high-performance concrete

In spite of the fact that the use of slag or fly ash to replace part of the Portland cement in high-performance concrete is attractive, this should be done with care, or at lower dosage, in the following cases:

(a) Need for high early strength

Slag and fly ash are not as reactive as Portland cement, so the 24 hour compressive strength of high-performance concrete in which they are incorporated is always lower than when Portland cement is used alone or in combination with silica fume. One way to overcome this lower 24 hour strength is to lower further the water/binder ratio of the high-performance concrete mixture. The saving in superplasticizer will not be as great as anticipated because this water reduction will be obtained through the incorporation of more superplasticizer in the mix, but it could still exist. However, there is always a practical and economic limit to the reduction in the water/binder ratio, and often it is simpler to use less fly ash or less slag in the mix when the 24 hour compressive strength has to be increased.

The author experienced one particular case where a partial substitution of cement by fly ash was beneficial to the very early strength. In that particular case it was not the 24 hour strength that was critical but rather the 12 hour strength because form removal was permitted at 12 hours if the concrete reached a 15 MPa compressive strength. With the materials locally available, the author found that a mix containing 15% fly ash gave a 12 hour compressive strength greater than a mix containing Portland cement only, Portland cement and silica fume, or Portland cement slag and silica fume. However, at 24 hours and beyond 24 hours this mixture resulted in a lower strength than the others, though it was the best one for the 12 hour strength requirement. This high early strength obtained using fly ash was due to the high alkali content of the fly ash. The alkalis acted as a strength accelerator during the very early stage of cement hydration. Limestone filler has also recently been reported to act in the same way (Gutteridge and Dalziel, 1990; Kessal et al., 1996; Nehdi, Mindess and Aïtcin, 1996).

(b) Cold weather concreting

It is well established that during winter in northern countries, lowering the supplementary cementitious material content somewhat gives a higher early strength. The dosage reduction varies, of course, according to the severity of weather conditions.

Materials selection

(c) Freeze–thaw durability

As will be seen in Chapter 17, the freeze–thaw durability of high-performance concrete as tested by ASTM C666 Procedure A is not well established in the case of mixtures containing supplementary cementitious materials with or without entrained air, and conflicting results have been obtained. Therefore, if durability of high-performance concrete to be subjected to repeated freezing and thawing cycles in a saturated state is critical, it is better to be cautious and to test the actual composition to be used for freeze–thaw durability.

(d) Decrease in maximum temperature

The use of slag or fly ash is sometimes advocated to lower the heat development in high-performance casting of massive elements. As has already been reported for usual concrete, slag must be used at a much higher substitution level than 15 to 30% in order to observe a significant decrease in the maximum temperature of the concrete within a massive element because this maximum temperature is not related to the total amount of cement in the mix but rather to the amount of cement that has been hydrated (Bramforth, 1980). This point will be discussed in more detail in Chapter 14, where the properties of fresh concrete and initial heat development are discussed.

7.3.10 Selection of aggregates

As has already been seen, the selection of aggregates must be done carefully because, as the targeted compressive strength increases, the aggregates can become the weakest link, where failure will be initiated under a high stress. Compared with usual concrete, a closer control of aggregate quality with respect to grading and maximum size is necessary, since a primary consideration is to keep the water requirement as low as possible. It should be obvious that only well-graded fine and coarse aggregates should be used.

(a) Fine aggregate

Little research has been done to optimize the fine aggregate characteristics for high-performance concrete applications in spite of the fact that sand characteristics can vary across a wide range (Mack and Leistikow, 1996). Usually, fine aggregates used to make high-performance concrete have a grain size distribution within the limits recommended by the ACI for usual concrete. However, whenever possible, the selected fine aggregate is on the coarse side of these limits, which corresponds to a fineness modulus of 2.7 to 3.0. The use of such a coarse sand is supported by the fact that all high-strength mixtures are rich enough in fine particles because of their high cement or cementitious content, so it is not necessary to use a fine sand from the workability and segregation point of view. Moreover, the use of coarse sand results in a slight decrease in the amount of mixing water necessary for a given workability, which is advantageous from a strength and economic point of view. Finally, the use of a coarse sand results in an easier shearing of the cement paste during mixing.

Generally, there is no particular advantage to using one type of sand rather than another as long as it is clean and free from clay and silt.

In Norway, according to Ronneberg and Sandvik (1990), for the construction of offshore platforms, the fine aggregate grading is achieved through a hydraulic processing unit, where it is separated into eight fractions. The desired grading is achieved by combining these eight fractions.

Natural sand should contain a minimal amount of particles coarser than 5 mm because, very often but not always, these particles are not very strong and can become the weakest link in the concrete.

Partial replacement of natural sand by a manufactured sand made of a strong crushed rock has been proposed but does not seem often to have been put into practice. Moreover, in most ready-mix concrete plants, there is only one bin or silo to store sand. As a usual concrete is simultaneously delivered to the usual customers, concrete using two sands – one for usual concrete, the other one for high-performance concrete – can be difficult. However, if two bins or silos are available for the fine aggregate, then the coarse sand can be used for high-performance concrete.

(b) Crushed stone or gravel

The selection of the coarse aggregate becomes more important as the targeted compressive strength increases and will therefore be discussed in more detail than the selection of the fine aggregate.

Crushed hard and dense rocks, such as limestone, dolomite and igneous rocks of the plutonic type (granite, syenite, diorite, gabbro and diabase), have been successfully used as coarse aggregate in high-performance concrete applications. It is not yet established whether aggregates potentially reactive with the alkali in the cement can be used in high-performance concrete; therefore, it is better to be on the safe side.

The shape of the coarse aggregate is also important from a rheological point of view. During crushing, roughly equidimensional particles (also called cubic particles) should be generated, rather than flat and elongated ones. The latter are weak; they can sometimes be broken with the fingers and tend to produce harsh mixes requiring additional water or superplasticizer to achieve the required workability.

Impact crushers used to be favoured over cone crushers because they produce more cubic particles. However, the author has used cubic aggregates in the range of 5 to 10 mm produced by cone crushers. They were produced by crushing the 14 to 20 mm fraction recovered from the first crushing step. This crushing method increases the production cost of equidimensional particles, but this can largely be compensated by a higher content of coarse aggregate in the mix and by savings in cement and even in the superplasticizer.

From a shape and strength point of view, the best aggregates to make high-performance concrete seem to be glacial gravels or, even better, fluvioglacial gravels because they are generally made of the strongest and hardest part of the rocks crushed by the glacier, and have been thoroughly cleaned by the running water from the melting glacier (Aïtcin, 1988). Every part of the rock that was friable and soft was pulverized under the powerful crushing action of the glacier and these very fine particles were washed away, so the remaining gravel is very clean. This is not the case with glacial gravels of morainic origin. Even when fluvioglacial gravel particles have been transported some distance by running water, they were always transported after the clayey and silty particles, so they were not polished during transportation and therefore can afford a good mechanical bonding through their rough surface. Moreover, as the crushing action of the glacier was very slow, very few fissures or microcracks are present in the individual gravel particles, which is not the case with crushed rock particles blasted before being crushed.

During blasting, the pieces of quarried rock are decompressed and subjected to very high acceleration, so they contain a more or less dense network of microcracks, depending on the texture of the rock. It is well known by granite polishers that blasted granite presents a fissured surface, unlike the blocks extracted in the traditional way using wood wedges or blocks sliced with a steel wire and an abrasive or diamond powder.

Unfortunately, glacial or fluvioglacial gravels are not commonly found. Fluvial gravels are less good than glacial or fluvioglacial ones because they are often composed of particles not as hard and strong as those in glacial gravel. According to the weakest link theory, these softer and weaker particles will be the first to fail, and this will result in a significant decrease in the compressive strength of high-performance concrete.

Moreover, fluvial particles generally have a smooth surface, owing to the polishing action of silt and sand particles entrained by the river. Finally, their surface is not always clean: they can be covered by a very thin layer of clay and/or silt tightly glued to their surface. This thin film results in an increase in water demand during the mixing and in a significant reduction of the bond between the gravel particles and the mortar paste in the hardened concrete, which results in a premature failure. This premature debonding is the cause of the presence of a number of aggregate prints on the failure surface of high-performance concrete and explains the lower strength of concrete when some fluvial gravels are used.

The selection of coarse aggregate must be made after a close examination of its mineralogy and petrography in order to ensure that all the particles are strong enough to avoid premature failure in high-strength concrete.

Aïtcin and Mehta (1990) conducted a laboratory investigation of the influence of aggregate mineralogy on the strength and elastic properties of high-performance concrete. Four different coarse aggregates available in northern California were investigated. One of them was a natural siliceous gravel with round particles and a smooth texture, and the other three were crushed rocks with a rough texture. The rocks consisted of a fine-grained diabase, a limestone and a granite. The maximum size of aggregate was 10 mm except for granite, which was 14 mm. The aggregate particles appeared to be clean, hard, strong and mineralogically uniform, except that the granite contained inclusions of laumonite in the quartz–feldspar matrix. All concrete mixtures had the following proportions: 500 kg/m^3 ASTM Type I Portland cement, 42 kg/m^3 silica fume, 137.5 kg/m^3 water, 10.6 l/m^3 naphthalene sulfonate-type superplasticizer, 675 kg/m^3 natural sand (2.75 fineness modulus) and 1130 kg/m^3 coarse aggregate.

The authors were surprised to find that the concretes containing either granite or siliceous gravel aggregate had a significantly lower strength and elastic modulus than the corresponding concretes containing either diabase or limestone. The average 28 day compressive strength values for concretes made with granite, gravel, limestone and diabase were 85, 92, 97 and 101 MPa, respectively. A probable cause of this behaviour was determined by examination of the failure surfaces of the test specimens. Both the limestone and the diabase concretes showed little evidence of aggregate–cement paste debonding (or a weak transition zone); however, there were frequent occurrences of transgranular fracture with the same shear plane passing through the cement paste and the aggregate. Only the concrete with siliceous gravel showed numerous sites of cement paste–aggregate bond failure, an indication that the aggregate was strong but the transition zone was weak. With granite, although the fracture was essentially of a transgranular type, there was evidence that the aggregate was weak. Since the same shear plane did not pass through the cement paste and the aggregate, it is probable that the aggregate fractured earlier than the cement paste. Mineralogical examination of the aggregate confirmed the presence of inclusions of laumonite, which is known to be unstable in a moist environment.

(c) Selection of the maximum size of coarse aggregate

There is some controversy with regard to the effect of the maximum size of aggregate (MSA) on concrete strength (de Larrard and Belloc, 1992). In concrete technology, it is not uncommon that a change in one variable results in two conflicting effects, and the net result is determined by the more dominant of the two. In usual concrete practice, a small reduction in the water requirement for a given workability can be achieved by an increase in the MSA, for example, from 12 to 25 mm. However, in high-performance concrete the strength gain associated with this increase in the MSA may not be large enough to make up for the strength loss due to the following adverse effects. First, with increasing MSA, the transition zone becomes larger and more heterogeneous, and second, with most rock types, the smaller particles of coarse aggregate are generally stronger than large particles (Aïtcin, 1988). This is because the size reduction process often eliminates internal defects in the aggregate, such as large pores, microcracks and inclusions of soft minerals. Experience shows that it is difficult, but not impossible, to produce class III high-performance concrete using aggregates larger than 25 mm. With most natural aggregates, it seems that, for making high-performance concrete, 10 or 12 mm MSA is probably the safest in the absence of any optimization testing. This does not mean that a 20 mm aggregate cannot be used. When the parent rock (from which the aggregate is derived) is sufficiently strong and homogeneous, 20 or 25 mm MSA can be used without adversely affecting the workability and strength of the concrete.

7.4 FACTORIAL DESIGN FOR OPTIMIZING THE MIX DESIGN OF HIGH-PERFORMANCE CONCRETE

7.4.1 Introduction

The selection of the best available materials to be used to make high-performance concrete is only one step in the production of efficient and economical high-performance concrete. The last steps are, as already seen, the selection of the final cementitious system that will be used and the proportions of the final mixture because modern concrete technology offers many ways to achieve a high-performance concrete having specific properties. At the present time, it must be admitted that the selection of the cementitious materials and the optimization of the composition of a high-performance concrete are more of an art than a science. For example, for a given cementitious system, a given low water/binder ratio can be achieved in different ways: by increasing the amount of binder, by decreasing the amount of water or by doing both at the same time, while in each case adjusting the amount of superplasticizer and retarding agent, if needed, to obtain the required slump for an easy placement, in the most economical way possible.

When optimizing high-performance concrete composition of a given water/binder ratio, it must be remembered that if the amount of mixing water is minimized, the amount of cement is decreased, but at the same time the amount of superplasticizer needed to obtain the required slump is increased. Therefore, what is saved in cement is not necessarily transformed into a global saving in cost at the high-performance concrete level, because this saving in the cement could be absorbed by the extra cost of the superplasticizer needed to be able to decrease the amount of mixing water to save on the cement. Moreover, this higher amount of superplasticizer could significantly delay the early strength of the high-performance concrete, so it will not be interesting for a precaster because the use of such a combination will cause too much of a delay in the re-use of the forms.

At the present time, there is no theoretical approach that can be used to obtain on paper the final answer. Some more or less powerful software packages to get closer to this final answer have been developed, but they give only partial answers; these will be reviewed in the next chapter.

In spite of the information presented in this chapter and the information on all successful high-performance concrete compositions that can be found in the literature, there comes a time when trial batches have to be made using the pre-selected materials in order to be able to produce an economical optimized high-performance concrete fulfilling all the specified requirements.

As it is always a heavy task to develop an optimized composition using trial batches, it is suggested, whenever possible, using a factorial design optimization method. This experimental approach, which was initially developed in the area of agricultural research and then used extensively in chemistry and chemical engineering, is a very powerful tool in concrete technology (Rougeron and Aïtcin, 1994; Kessal et al., 1996). With a minimum of well-planned trial batches, it is possible to explore a large area of compositions in order to find the optimized composition fulfilling the specified requirements. This technique can be used in the laboratory as well as in the field. In spite of the fact that it depends on sophisticated mathematical concepts, its practical use is easy because there are several commercial software packages that allow a very simple, user-friendly and rapid treatment of the experimental data in the area under study. However, this method has its limitations; the software performs very well in making the complicated calculations, but is of no help at all when selecting the limits of the two influencing parameters that will limit the area under study. These limits have to be carefully selected. Second, the parameters, i.e. the high-performance characteristics that will be measured and interpolated in the area studied, have to be selected very carefully like the parameters that will be 'frozen' in the study. When the experimental conditions that will be tested in the area studied are imposed, there are several choices for the kind of factorial design plan.

7.4.2 Selection of the factorial design plan

This analysis technique is based on the statistical analysis of the results obtained from a set of experiments. It gives a lot of information from a

Optimizing the mix design of high-performance concrete

few experiments: the most important factors, their type of influence and the modelling of this influence.

When looking for a linear model, a so-called composite plan can be used. This consists of making concrete batches 1, 2, 3 and 4 (as shown in Figure 7.16) and validating this composite plan by making four additional centre batches, 5, 6, 7 and 8. These four additional trial batches are used to establish the variation of the various responses, assuming that this variation is constant in the whole area under study.

If x is the variable reported in the x-axis and y the variable reported in the y-axis, the answer Z can be expressed as a linear function of x and y in the following manner:

$$Z = a_0 + a_1 x + a_2 y$$

where a_0, a_1, a_2 will be determined so that the experimental points are as close as possible to the plane $Z = g(x, y)$.

When a second-order model is required, the response Z can be written as:

$$Z = a_0 + a_1 x + a_2 y + a_{11} x + a_{22} y + a_{12} x, y$$

Then a star plan corresponding to the previous batches plus batches 9, 10, 11 and 12 has to be done (Figure 7.17). In this case, too, $a_0, a_1, a_2, a_{11}, a_{22}$ and a_{12} will be determined so that the experimental points are as close as possible to the surface $Z = g(x, y)$.

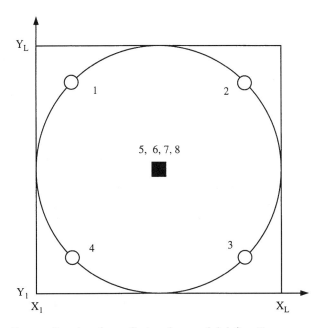

Fig. 7.16 Composite plan for a first-order model (after Rougeron and Aïtcin, 1994). Copyright ASTM. Reprinted with permission.

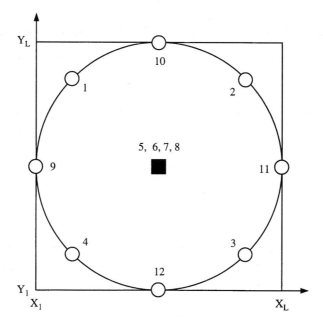

Fig. 7.17 Composite plan for a second-order model (after Rougeron and Aïtcin, 1994). Copyright ASTM. Reprinted with permission.

7.4.3 Sample calculation

The following example related to the optimization of a silica fume, non-air-entrained, high-performance concrete is used to illustrate the usefulness of this technique (Rougeron and Aïtcin, 1994). The cement used is a blended silica fume cement containing 8.5% silica fume. The variables under study are the amount of mixing water in litres per m^3 (w) in the range 120 to 150 l/m^3, and the water/binder ratio (w/b) in the range of 0.22 to 0.34. The areas that are to be studied are the dosage of superplasticizer that has to be used to obtain a 200 ± 20 mm slump and compressive strength at 1, 7 and 28 days.

Iso answer curves are drawn in the area, i.e. curves where the answer has the same numerical value. Moreover, from these answers, the amount of superplasticizer expressed as the percentage of solids (mass of solids contained in the superplasticizer divided by the mass of cement) and the cost of materials used to make a cubic metre of concrete are also deduced. The cement dosage in kg/m^3 is also calculated in the area.

The prices of the materials that have been considered in this study are in Canadian dollars:

Blended silica fume cement $100/t
Coarse and fine aggregate $10/t
Superplasticizer $2/l

However, in order to be more representative of the North American market, a $300/t, $600/t and $1000/t price for silica fume was also

Optimizing the mix design of high-performance concrete

Table 7.3 Composition of the mixes tested in the factorial design plan (Rougeron and Aïtcin, 1994). Copyright ASTM. Reprinted with permission

	W/B	Cement (kg/m^3)	Coarse aggregate (kg/m^3)	Fine aggregate (kg/m^3)	Super.* (l/m^3)	SF** (kg/m^3)	Slump (mm)
1	0.31	395	1100	790	9	34	220
2	0.31	440	1100	720	7	38	210
3	0.25	475	1100	700	11	40	200
4	0.25	520	1100	620	9	45	210
9, 10, 11, 12***	0.28	460	1100	710	9	40	190

*Super. = superplasticizer. **SF = silica fume. ***Average value.

considered. The purpose of this economic study is to see if it is better to use a lower amount of mixing water in order to save some cement, by using a higher amount of superplasticizer. The mix proportion of batches 1, 2, 3, 4, 9, 10, 11 and 12 are given in Table 7.3. The compressive strengths obtained are presented in Table 7.4.

(a) Iso cement dosage curves

Knowing the amount of mixing water and the water/binder ratio, it is easy to represent the iso cement dosage curves in the area studied. In Figure 7.18 it is seen that the lowest cement dosage in the area is 353 kg/m^3 and the highest one is 682 kg/m^3.

(b) Iso dosage curves for the superplasticizer

Iso dosage curves expressed in l/m^3 or as a percentage of the cement mass are represented in Figure 7.19. For example, it is seen that in order to produce concrete with a 200 ± 20 mm slump, a 0.30 W/B high-performance concrete can be made either by using the two limit combinations,

Table 7.4 Compressive strength of the different mixes tested (Rougeron and Aïtcin, 1994). Copyright ASTM. Reprinted with permission

	Average compressive strength, MPa		
	1 day	7 days	28 days
1	27.5	55.2	73.2
2	28.4	56.3	74.2
3	42.3	72.0	96.1
4	46.1	73.1	98.5
5	35.4	63.5	85.0
6	36.2	65.1	88.0
7	24.2	52.1	70.2
8	47.8	76.8	100.3
Average 9, 10, 11, 12	36.2	64.2	86.5

150 l of water and 6.3 l of superplasticizer, or 120 l of water and 10.2 l of superplasticizer. Moreover, the superplasticizer dosage needed to obtain the 220 ± 20 mm slump varied from around 0.7% to around 1.3%. The saturation point for this cement/superplasticizer combination using the Marsh cone method was 1.0%, which in this case corresponds to the dosage needed for the central point of the area.

Iso f'_c curves at 1, 7 and 28 days

Figure 7.20 shows the iso f'_c curves at 1, 7 and 28 days. It is seen that at 1, 7 and 28 days f'_c depends essentially on the W/B ratio value; in particular, the superplasticizer is not causing any delay of the hardening at 1 day.

(c) Iso cost curves

Figure 7.21 presents the iso cost curves in the four cases studied when the silica fume price varies between $100 and $1000/t. At the $100/t price, it is definitely economically advantageous in eastern Canada to use the silica fume blended cement when making a high-performance concrete when the W/B ratio is around 0.30. A parallel study was carried out on non-silica fume high-performance concrete.

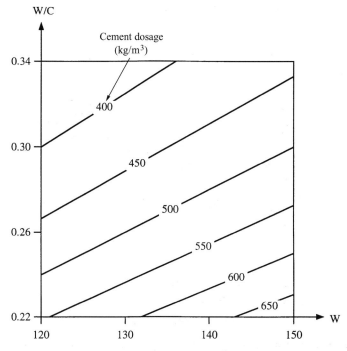

Fig. 7.18 Cement dosages under study (kg/m³) (after Rougeron and Aïtcin, 1994). Copyright ASTM. Reprinted with permission.

Optimizing the mix design of high-performance concrete

Figure 7.21(b) and (c) show that at $300/t price the use of silica fume is still economically advantageous, but that this is no longer true above $600/t. In the latter case there are considerations other than the cost of materials that have to be used to justify the use of silica fume when making high-performance concrete. For example, the use of silica fume reduces the superplasticizer dosage and the risk of delay; it increases the 24 hour and 28 day strength. It should be remembered that it was the slump of all the mixes that was constant and not their ultimate strength. Moreover, silica fume high-performance concrete has a much lower rapid chloride-ion permeability and finally it can be placed more easily because non-silica fume high-performance concretes tend to become viscous and sticky as the W/B ratio decreases.

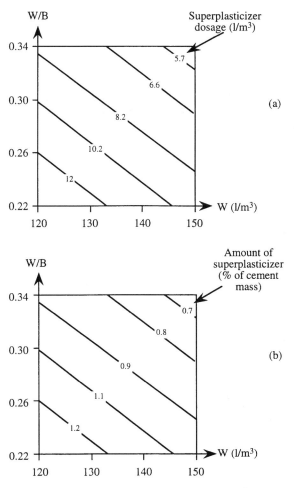

Fig. 7.19 Superplasticizer iso dosages (a) expressed in l/m^3 and (b) expressed as percentage of solids (after Rougeron and Aïtcin, 1994). Copyright ASTM. Reprinted with permission.

Of course, if the cost of silica fume reaches $1000/t, its use has to be justified through considerations other than the cost of materials used to make the high-performance concrete, since the unit price of such a high-performance concrete increases drastically.

7.5 CONCLUDING REMARKS

High-performance concrete has gained wide acceptance during the past decade because it has been proved that, in many locations, it is possible to make reliable and uniform mixes from local sources of materials using conventional concreting practices. High-performance concrete is not made by chance. Combining experience and fundamental work has made the selection of the ingredients for high-performance concrete less empirical, but we have not reached the situation where this selection can be made without at least some laboratory work.

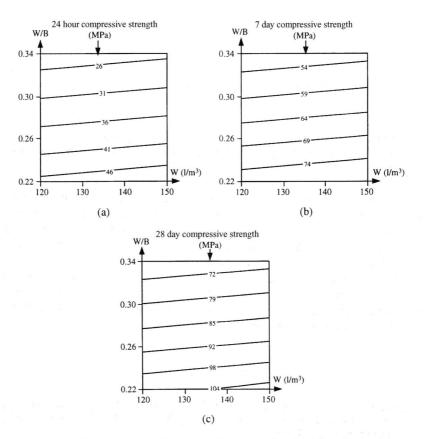

Fig. 7.20 Iso f'_c curves at 24 hours, and 7 and 28 days (after Rougeron and Aïtcin, 1994). Copyright ASTM. Reprinted with permission.

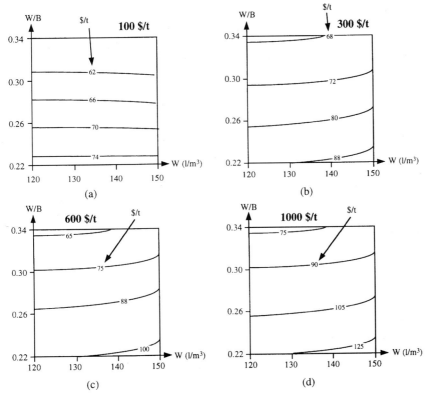

Fig. 7.21 Iso cost graphs (CAN $/m³) for silica fume concretes for different silica fume costs (after Rougeron and Aïtcin, 1994). Copyright ASTM. Reprinted with permission.

It is hoped that, if all the selection criteria developed in this chapter are carefully applied, less work will be necessary to develop an economic high-performance concrete formulation that meets the rheological and strength requirements. However, with the materials available in any given location, some laboratory work will still need to be done.

After considering the selection of the most appropriate materials, we should now discuss proportioning them in order to achieve the desired properties in high-performance concrete. This procedure will be developed in the next chapter.

REFERENCES

ACI Guide for the Use of Silica Fume in Concrete, ACI2342-96. Reported by ACI Committee 234. Available through ACI, PO Box 9094, Farnington Halls, Michigan 48333, 51 pp.

Aïtcin, P.-C. (1980) How to produce high-strength concrete. *Concrete Construction*, **25**(3), March, 222–31.

Aïtcin, P.-C. (1988) From gigapascals to nanometers, in *Engineering Foundation Conference on Advance in Cement Manufacture and Uses* (ed. E. Gartner), American Society of Civil Engineers, Potosi, Mo, pp. 105–30.

Aïtcin, P.-C. and Baalbaki, M. (1994) *Concrete Admixtures – Key Components of Modern Concrete*, Contech 94 International RILEM Workshop on Technology Transfer of the New Trends in Concrete, Barcelona, E & FN Spon, London, ISBN 0-419-20150-5, pp. 33–47.

Aïtcin, P.-C. and Mehta, P.K. (1990) Effect of coarse aggregates characteristics on mechanical properties of high-strength concrete. *ACI Materials Journal*, **87**(2), March, 103–7.

Aïtcin, P.-C. and Neville, A.M. (1993) High-performance concrete demystified. *Concrete International*, **15**(1), January, 21–6.

Aïtcin, P.-C., Jolicoeur, C. and MacGregor, J.G. (1994) Superplasticizers: how they work and why they occasionally don't. *Concrete International*, **16**(5), May, 45–52.

Aïtcin, P.-C., Autefage, F., Carles-Gibergues, A. and Vaquier, A. (1986) *Comparative Study of the Cementitious Properties of Different Fly Ashes*, ACI SP-91, pp. 91–114.

Baalbaki, M., Sarkar, S.L., Aïtcin, P.-C. and Isabelle, H. (1992) *Properties and Microstructure of High-Performance Concretes Containing Silica Fume, Slag and Fly Ash*, ACI SP-132, Vol. 2, May, pp. 921–42.

Bhatty, J.I. (1987) The effect of retarding admixtures on the structural development of continuously sheared cement paste. *International Journal of Cement Composites and Lightweight Concrete*, **9**(3), August, 137–44.

Bramforth, P.B. (1980) In situ measurement of the effect of partial portland cement replacement using either fly ash or ground granulated blast furnace slag on the performance of mass concrete, *Proceedings of the Institution of Civil Engineers*, **69**(2), September, 777–800.

Cail, K. and Thomas, H. (1996) *Development and Field Applications of Silica Fume Concrete in Canada*, CANMET-ACI Intensive Course on Fly Ash, Slag, Silica Fume, Other Pozzolanic Materials and Superplasticizers in Concrete, Ottawa, Canada, April, 21 pp.

Daimon, M. and Roy, D.M. (1978) Rheological properties of cement mixes: 1. Methods, preliminary experiments, and adsorption studies. *Cement and Concrete Research*, **8**, 753–64.

de Larrard, F. (1988) Formulation et propriétés des bétons à très hautes performances. PhD Thesis, École Nationale des Ponts et Chaussées, July 1987, published as Research Report No. 149 of Laboratoires des Ponts et Chaussées, March, 335 pp.

de Larrard, F. and Belloc, A. (1992) Are small aggregates really better than coarser ones for making high-strength concrete? *Cement, Concrete, and Aggregates*, **14**(1), Summer, 62–6.

de Larrard, F., Gorse, J.F. and Puch, C. (1990) Efficacités comparées de diverses fumées de silice comme additif dans les bétons à hautes performances. *Bulletin de Liaison des Laboratoires des Ponts et Chaussées*, No. 168, July–August, pp. 97–105.

de Larrard, F., Bosc, F., Catherine, C. and Deflorenne, F. (1996) La nouvelle méthode des coulis de l'AFREM pour la formulation des bétons à hautes performances. *Bulletin de Liaison des Laboratoires des Ponts et Chaussées*, No. 202, March–April, pp. 61–9.

Gjørv, O.E. (1991) *Norwegian Experience with Condensed Silica Fume in Concrete*, CANMET/ACI International Workshop on the Use of Silica Fume in Concrete, Washington DC, April, pp. 49–63.

Gutteridge, W.A. and Dalziel, J.A. (1990) Filler cement: the effect of the secondary component on the hydration of Portland cement, Part II, fine hydraulic binders. *Cement and Concrete Research*, **20**, 853–61.

Holland, T.C. (1995) *Specification for Silica Fume for Use in Concrete*, ACI SP-154, pp. 607–39.
Huynh, H.T. (1996) La compatibilité ciment-superplastifiant dans les bétons à hautes performances – synthèse bibliographique. *Bulletin de Liaison des Laboratoires des Ponts et Chaussées*, No. 206, November–December, pp. 63–73.
Kantro, D.L. (1980) Influence of water-reducing admixtures on properties of cement paste – a miniature slump test. *Cement, Concrete, and Aggregates*, **2**(2), Winter, 95–108.
Kessal, M., Nkinamubanzi, P.-C., Tagnit-Hamou, A. and Aïtcin, P.-C. (1996) Improving initial strength of a concrete made with Type 20 M cement. *Cement, Concrete, and Aggregates*, CCAGDP, **18**(1), June, 49–54.
Khorami, J. and Aïtcin, P.-C. (1989) *Physico-chemical Characterization of Superplasticizers*, ACI SP-119, pp. 117–31.
Lahalih, S.M., Absi-Halabi, M. and Ali, M.A. (1988) Effect of polymerization conditions of sulfonated-melamine formaldehyde superplasticizers on concrete. *Cement and Concrete Research*, **18**, 513–31.
Lessard, M., Gendreau, M., Baalbaki, M. and Pigeon, M. (1993) Formulation d'un béton à hautes performances à air entraîné. *Bulletin de Liaison des Laboratoires des Ponts et Chaussées*, No. 188, November–December, pp. 41–51.
Mack, W.W. and Leistikow, E. (1996) The sand of the world. *Scientific American*, **275**(2), August, 62–7.
Mehta, P.K. and Aïtcin, P.-C. (1990) *Microstructural Basis of the Selection of Materials and Mix proportions for High-Strength Concrete*, ACI SP-121, p. 265–86.
Nawa, T., Eguchi, H. and Fukaya, Y. (1989) *Effect of Alkali Sulfate on the Rheological Behavior of Cement Paste Containing a Superplasticizer*, ACI SP-119, pp. 405–24.
Nawa, T., Eguchi, H. and Okkubo, M. (1991) Effect of fineness of cement on the fluidity of cement paste and mortar, *Transactions of JSCE*, **13**, 199–213.
Nehdi, M., Mindess, S. and Aïtcin, P.-C. (1996) Optimization of high-strength limestone filler cement mortars. *Cement and Concrete Research*, **26**(6), 883–93.
Penttala, U.E. (1993) Effects of delayed dosage of superplasticizer on high-performance concrete, in *Proceedings of the International Conference on High-Strength Concrete*, Lillehammer, (ed. I. Holland and E. Sellevold), Norwegian Concrete Association, Oslo, ISBN 82-91341-00-1, pp. 874–81.
Perenchio, W.F. (1973) *An Evaluation of Some of the Factors Involved in Producing Very High-Strength Concrete*, Portland Cement Association Research and Development Bulletin RD014, 7 pp.
Pistilli, M.F., Rau, G. and Cechner, R. (1984) The variability of condensed silica fume from a Canadian source and its influence on the properties of Portland cement concrete. *Cement, Concrete, and Aggregates*, **6**(1), 33–7.
Pistilli, M.F., Wintersten, R. and Cechner, R. (1984) The uniformity and influence of silica fume from a US source on the properties of Portland cement concrete. *Cement, Concrete, and Aggregates*, **6**(2), 120–24.
Regourd, M. (1978) Cristallisation et réactivité de l'aluminate tricalcique dans les ciments Portlands. *Il Cimento*, No. 3, pp. 323–35.
Ronneberg, H. and Sandvik, M. (1990) High-strength concrete for North Sea platforms. *Concrete International*, **12**(1), January, 29–34.
Rougeron, P. and Aïtcin, P.-C. (1994) Optimization of the composition of a high-performance concrete. *Cement, Concrete, and Aggregates*, **16**(2), December, 115–24.
Ryell, J. and Bickley, J.A. (1987) Scotia-Plaza: high strength concrete for tall buildings, in *Proceedings of the Symposium on Utilization of High-Strength Concrete*, Stavenger, (ed. I. Holland *et al.*) Tapir, N-7034 Trondheim NTH, Norway, ISBN 82-519-0797-7, pp. 641–53.

Singh, N.B. and Singh, A.C. (1989) Effect of melment on the hydration of white Portland cement. *Cement and Concrete Research*, **19**, 547–53.

Swamy, R.N. (1993) Fly-ash and slag: standards and specifications. Help or hindrance? *Materials and Structure*, **26**(164), December, 600–613.

Tagnit-Hamou, A. and Aïtcin, P.-C. (1993) Cement and superplasticizer compatibility. *World Cement*, **24**(8), August, 38–42.

Tagnit-Hamou, A., Baalbaki, M. and Aïtcin, P.-C. (1992) *Calcium-Sulfate Optimization in Low Water/Cement Ratio Concretes for Rheological Purposes*, 9th International Congress on the Chemistry of Cement, New Delhi, India, November, Vol. 5, pp. 21–5.

Tucker, G.R. (1938) *Concrete and Hydraulic Cement*, US Patent 2 141 569, December, 5 pp.

Uchikawa, H., Sawaki, D. and Hanehara, S. (1995) Influence of kind and added timing of organic admixture on the composition, structure and properties of fresh cement paste. *Cement and Concrete Research*, **25**(2), 353–64.

Uchikawa, H., Sone, T. and Sawaki, D. (1997) A comparative study of the characters and performances of chemical admixtures of Japanese and Canadian origin, Part 2. *World Cement Research and Development*, **28**(3), March, 70–76.

Vernet, C. and Noworyta, G. (1992) Interaction des adjuvants avec l'hydratation du C_3A: points de vue chimique et rhéologique. Personal communication, 56 pp.

Whiting, D. (1979) *Effects of High-Range Water Reducers on Some Properties of Fresh and Hardened Concretes, Portland Cement Association Research and Development Bulletin* RD061.01T, 15 pp.

Zhang, M.H. and Odler, I. (1996a) Investigations on high SO_3 Portland clinkers and cements I. Clinker synthesis and cement preparation. *Cement and Concrete Research*, **26**(9), 1307–13.

Zhang, M.H. and Odler, I. (1996b) Investigations on high SO_3 Portland clinkers and cements II. Properties of cements. *Cement and Concrete Research*, **26**(9), 1315–24.

CHAPTER 8

High-performance mix design methods

8.1 BACKGROUND

The objective of any mixture proportioning method is to determine an appropriate and economical combination of concrete constituents that can be used for a first trial batch to produce a concrete that is close to that which can achieve a good balance between the various desired properties of the concrete at the lowest possible cost. It will always be difficult to develop a theoretical mix design method that can be used universally with any combination of Portland cement, supplementary cementitious materials, any aggregates and any admixture because in spite of the fact that all the components of a concrete must fulfil some standardized acceptance criteria, these criteria are too broad; moreover, to a certain extent the same properties of fresh and hardened concrete can be achieved in different ways from the same materials. This situation must be perceived as an advantage, because in two different locations the same concrete properties can be achieved differently using non-expensive locally available materials (Aïtcin and Neville, 1993). A mixture proportioning method only provides a starting mix design that will have to be more or less modified to meet the desired concrete characteristics. In spite of the fact that mix proportioning is still something of an art, it is unquestionable that some essential scientific principles can be used as a base for mix calculations.

It is interesting to see in the present literature a renewal of interest in mix design and mix proportioning (Day, 1996; Ganju, 1996; de Larrard and Sedran, 1996; Popovics and Popovics, 1996). In fact this renewal of interest only traduces the limitations of the present mix proportioning methods that have been used for usual concrete without any problem for many years. As long as usual concrete was essentially a mixture of Portland cement, water, aggregates and sometimes entrained air, these methods could be used with a very good predictive value to design a concrete mixture having a given slump and 28 day compressive strength. This is no longer the case because:

- the water/cement or water/binder range of modern concretes has been drastically extended towards very low values thanks to the use of superplasticizers;

- modern concrete very often contains one or several supplementary cementitious materials that are in some cases replacing a significant amount of cement;
- modern concrete sometimes contains silica fume which drastically changes the properties of fresh and hardened concrete;
- the slump can be adjusted by using a superplasticizer instead of water without altering the water/cement or water/binder ratio.

Concrete is becoming a more complex material than a simple mixture of cement, water and aggregates and it is more and more difficult to predict concrete properties theoretically in spite of the fact that the use of computers facilitates complex calculations.

Before discussing the mix design method used at the Université de Sherbrooke it has been judged interesting to look briefly at ACI 221-1 Standard Practice for Selecting Proportions for Normal, Heavyweight and Mass Concrete from which it is derived, and also to define precisely the various material characteristics that will be used to perform the calculations necessary to determine the mix proportions as well as the terminology used. Those who are familiar with these characteristics and the terminology could skip sections 8.2 and 8.3 and go directly to section 8.4, where the mix design method is presented.

8.2 ACI 211-1 STANDARD PRACTICE FOR SELECTING PROPORTIONS FOR NORMAL, HEAVYWEIGHT AND MASS CONCRETE

Current North American design procedures for proportioning usual concrete depend on the W/C ratio to achieve a required strength and on the water content to obtain a desired slump. No special consideration is given to the quality of cement and aggregate, or to the type and dosage of supplementary cementitious materials and chemical admixtures that are common in high-performance concrete.

ACI 211-1 Standard Practice for Selecting Proportions for Normal, Heavyweight and Mass Concrete offers a comprehensive procedure for proportioning normal weight concrete of a maximum specified compressive strength of 40 MPa and a maximum slump of 180 mm. The obtained mixture components do not include any supplementary cementitious materials or admixtures, except for an air-entraining admixture. This procedure is applicable to aggregate with a wide range of mineralogical and granulometric properties. It essentially assumes that the W/C ratio and the amount of entrained air are the only parameters affecting strength, and that concrete slump is affected by the maximum size of the coarse aggregate, the amount of mixing water and the presence or absence of entrained air.

Where a set of proper values for the water/cement ratio and the amount of mixing water has been selected, the so-called absolute volume method is applied to calculate the mix proportions, i.e. the proportions of mass of the different ingredients are transformed into volumic proportions, or vice versa using the very simple relationship:

$$\text{Absolute volume} = \frac{\text{Mass}}{\text{Specific gravity}}$$

This method is briefly summarized in the following because not only will it be easier to see its limitations when designing the mix proportions of a high-performance concrete, but it will be easier to understand how it can be modified to provide a framework that still can be used when introducing supplementary cementitious materials and superplasticizers, as is usually the case when making high-performance concretes.

Data that are needed to apply the ACI 211-1 procedure include the fineness modulus of the fine aggregate, the dry-rodded unit weight of the coarse aggregate, the specific gravity of the aggregates, which are determined in the laboratory (the specific gravity of the cement is assumed to be 3.15), and the free moisture and absorption capacity of the aggregate. This procedure assumes that the aggregate is well graded. The method consists of the following steps presented in Figure 8.1.

Step 1: Slump selection

Suggested slump values needed to cast concrete for different types of construction are given in a special table. These values can be used if the slump is not specified otherwise.

Step 2: Determination of the maximum size of coarse aggregate (MSA)

Large coarse aggregates have a lower specific surface than small coarse aggregates, hence less mortar is needed to achieve a desired workability when the MSA is high. For usual concrete it is economical to use a large MSA. The MSA should not exceed one-fifth of the narrowest dimension between the sides of forms, one-third of the depth of slab, or three-quarters of the minimum clear spacing between reinforcing bars, bundled bars or tendons. Depending on the type of construction (reinforced or non-reinforced walls, etc.), and on the maximum and minimum dimensions of the structural member, different MSA values are suggested in a Table.

Step 3: Estimation of mixing water and air content

The amount of mixing water is obtained from a table for given MSA and slump values, both when the concrete is air-entrained and when it is non-air-entrained. The amount of water needed to meet the required slump for a given MSA is lower if the concrete is air-entrained, owing to the lubrication effect of the small air bubbles. The method also suggests appropriate air volumes needed for frost resistance for concretes made with different MSA values.

Step 4: Selection of W/C ratio

Depending on the desired compressive strength within the 15 to 40 MPa range, and the required durability (exposure conditions), the W/C ratio can be determined from two tables.

Step 5: Cement content

The mass of cement is calculated by dividing the mass of the free water by the W/C ratio.

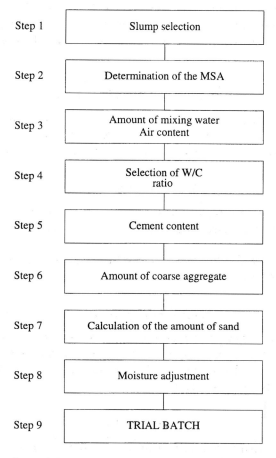

Fig. 8.1 Flow chart of the ACI 211-1 method for mixture proportioning.

Step 6: Coarse aggregate content

The bulk volume of dry-rodded coarse aggregate per unit volume of concrete is determined from a table for a given fineness modulus of sand and a given MSA. This volume is then multiplied by the dry-rodded unit mass of the coarse aggregate to calculate the mass of the coarse aggregate per unit volume of concrete.

Although this method does not differentiate between rounded and crushed aggregates, the effect of angularity of aggregate (which will require more mortar to fill the voids between the aggregates) is expected to be reflected in its lower value of dry-rodded unit mass than that of a rounded aggregate.

Step 7: Fine aggregate content

Volumes of all mixture constituents, except sand, are calculated by dividing each previously determined mass by the specific gravity of the material. The volume of air determined previously should also be added. The total volume of all these ingredients is then deducted from 1 m^3 to obtain the volume of sand. This last value is used to calculate the mass of the sand.

Step 8: Moisture adjustment

The mass of aggregates obtained in this procedure are for aggregate in an SSD state (defined in section 8.3). Therefore their mass, along with that of the mixing water, is adjusted for actual moisture conditions.

Step 9: Trial batches

Trial batches are made, and the mixture proportioning is adjusted to meet the desired physical and mechanical characteristics of the concrete.

If we try to apply this procedure to high-performance concrete there are several drawbacks:

Step 1: Slump selection

The slump of a high-performance concrete is essentially dependent not only on the amount of mixing water but also on the amount of superplasticizer used. The slump can be adjusted to the specific needs by playing with these two values.

Step 2: MSA

The MSA is usually no longer dictated by geometrical considerations. It is also no longer advantageous to select as coarse an aggregate as possible to reduce the amount of mixing water needed to meet a certain slump; rather it is advantageous to select the coarse aggregate as small as possible for placeability considerations and also for concrete strength considerations, as shown in Chapter 7.

Step 3: Estimation of mixing water and air content

As previously stated, in a high-performance concrete the same slump can be achieved using different amounts of mixing water and superplasticizer. The final combination that is usually selected is the one that gives a slump retention appropriate to field conditions. The suggested air content values given for ACI 211-1 are no longer appropriate to make high-performance concrete freeze–thaw resistant; as will be seen later, in Chapter 18, it is rather the spacing factor that is the more relevant parameter in designing a freeze–thaw-resistant high-performance concrete.

Step 4: Selection of the W/C ratio

High-performance concrete most of the time contains one or several supplementary cementitious materials (fly ash, slag, silica fume), so the simple relationship linking the 28 day compressive strength to the water/cement ratio used in ACI 211-1 is in general no longer valid and must be established in each particular case.

Step 5: Coarse aggregate content

This is no longer influenced by the fineness modulus of the sand. The mixture is so rich in paste that, whenever possible, it is better to use a coarse sand.

Moreover, unlike usual concrete, high-performance concrete can have several characteristics that need to be met simultaneously. Among these requirements are low permeability and high durability, high modulus of elasticity, low shrinkage, low creep, high strength, and high and lasting workability. Because of the large number of mixture components used in high-performance concrete and because of the various concrete requirements that can contradict one another, it is very difficult to use a mix design method that gives mix proportions very close to that of the final mixture. Usually it is necessary to make a large number of trial mixtures in order to select the desired combination of materials that should be used to make a high-performance concrete. However, a good mixture proportioning procedure can minimize the number of trial mixtures that are needed to achieve an economical and satisfactory balance between the various desired properties. Therefore, the use of a comprehensive mixture proportioning procedure to produce a high-performance concrete mixture with the minimum amount of trial batches is very desirable. Several methods have been developed and presented in the literature; a few of them are briefly presented in section 8.5.

However, as previously stated before discussing these different methods of mix proportioning, it is very important that the reader becomes familiar with a certain number of definitions in order not only to understand the different steps of these methods, but also to be able to use them properly. The proposed definitions are the ones currently in use in North America.

8.3 DEFINITIONS AND USEFUL FORMULAE

8.3.1 Saturated surface dry state for aggregates

When making concrete either in the laboratory or in the field, the most difficult thing is to keep a tight control of the amount of water actually used when making the mix. It is always easy to weigh a certain amount of cement or aggregates, and to read the number of litres of water passing through a metre. When using a superplasticizer it is not so difficult to calculate the amount of water brought into the mix by the superplasticizer, as will be shown later. What is definitely much more difficult is to keep a tight and constant control of the precise amount of water brought into the mix by the aggregates, especially by the sand, because the water content of the aggregates, and especially that of the sand, can vary considerably. The water content, w_{tot}, of an aggregate is defined as the amount of evaporable water divided by the dry mass of the aggregate and is expressed as a percentage. To measure it, it is only necessary to place a certain amount of wet sand in an oven at 105 °C and to weight it when it has a constant mass. The use of a microwave oven can reduce the time duration of the drying.

Depending on the level of control exercised on the water content of the aggregate, and more particularly of the fine aggregate, a better or poorer quality concrete can be produced. For example, a variation of 1% in the water content of the approximately 800 kg of sand that may be used to produce 1 m^3 of concrete corresponds to a variation of 8 litres of water in the mix. If this variation occurs in a high-performance concrete mix that contains 150 litres of water and 455 kg of cement per cubic metre, which corresponds to a water/cement ratio of 0.33, the water/cement ratio could change from 0.31 to 0.35, which represents a non-negligible difference in terms of slump, compressive strength and permeability. It is therefore very important to be able to control the precise amount of water contained in the aggregates. In order to control the amount of water brought into the mix by the aggregates, it is very important to define a reference state for the aggregates.

By convention in North America this reference state is called the saturated surface dry state or SSD state. This state is defined in ASTM C127 Standards for the Coarse Aggregate and in ASTM C128 Standards

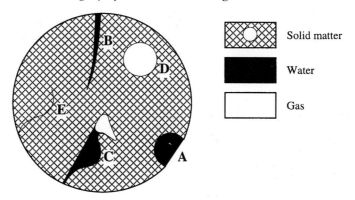

Fig. 8.2 Schematic representation of an aggregate particle in the SSD state.

for the Fine Aggregate. These two standard test methods describe in detail how to measure the SSD state. Very briefly, for the coarse aggregate, the SSD state is obtained by soaking the coarse aggregate in water for 24 hours and then drying it in an oven to constant mass. A coarse aggregate in its SSD state is shown schematically in Figure 8.2. For the fine aggregate, by convention the SSD state is obtained when a small truncated sand cone no longer holds together owing to the capillary forces between the wet sand particles. Figure 8.3 illustrates the determination of the SSD state for a sand and Figure 8.4 the determination of the SSD state for a coarse aggregate.

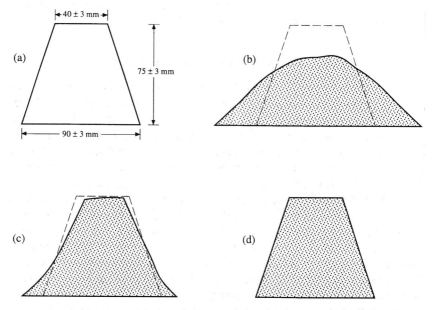

Fig. 8.3 Determination of the SSD state for a sand: (a) the standardized mini-cone used; (b) sand having a water content below its SSD state; (c) sand in an SSD state; (d) sand having a water content above its SSD state.

Definitions and useful formulae 223

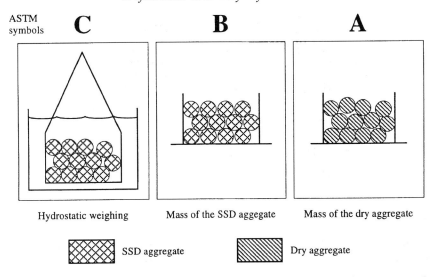

Fig. 8.4 Schematic representation of the measurement of the absorption and SSD specific gravity of a coarse aggregate.

In both cases, the amount of water absorbed in the aggregate when it is in the SSD state, w_{abs}, corresponds to the aggregate absorption. This absorption is expressed as a percentage of the mass of the dry aggregate.

In North America, concrete compositions are always expressed with the coarse and fine aggregates in their SSD states.

8.3.2 Moisture content and water content

The SSD state of an aggregate is very important when calculating or expressing the composition of a particular concrete, because it establishes a clear differentiation between the two types of water typically found in an aggregate. The water absorbed within the aggregate does not contribute either to the slump of the concrete or to its strength because it is not supposed to participate in cement hydration. When the water content of the aggregate is lower than in its SSD state, the aggregate will absorb some water from the mix. This absorption of water by the aggregate increases the rate of slump loss of the concrete. On the other hand, when the water content of an aggregate is higher than in its SSD state, the aggregate will bring water into the mix, as shown in Figure 8.5, if no special correction is made and thus increase the water/binder ratio and increase the slump. Therefore, the amount of mixing water has to be modified in order to keep the water/binder ratio and the slump constant.

The difference between the total water content of an aggregate, w_{tot}, and its water content in the SSD state, w_{abs}, is called the moisture content of the aggregate and is denoted by w_h. The moisture content of an aggregate can

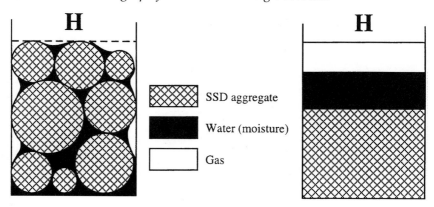

Fig. 8.5 Schematic representation of a wet aggregate.

be negative if the total water content is lower than the water absorption. This occurs frequently in summer for coarse aggregates.

For example, if 1000 kg of coarse aggregate having a water absorption, w_{abs}, of 0.8% is absolutely dry, $w_{tot} = 0$, consequently $w_h = -0.8\%$ and 8 litres of water could be absorbed by the coarse aggregate in the fresh concrete after mixing. This amount of water will undoubtedly have a significant effect on the rate at which this particular concrete will lose its slump during transportation if no water content adjustment is made, because the absorption of these 8 litres of water is not instantaneous.

On the other hand, when the same coarse aggregate has a total water content of 1.8% after a rain shower, $w_h = 1.0\%$ and the 10 litres of extra water brought to the mix by the coarse aggregate drastically increases the amount of mixing water. If no correction of the mix is made, this can have a detrimental effect on the slump, the compressive strength and the permeability.

Therefore in this book the following definitions will be used. The total water content of an aggregate, w_{tot}, is defined as:

$$w_{tot} = \frac{\text{Mass of the wet aggregate} - \text{mass of the dry aggregate}}{\text{Mass of the dry aggregate}} \times 100 \quad (8.1)$$

With the ASTM conventions shown in Figures 8.4 and 8.5:

$$w_{tot} = \frac{H - A}{A} \times 100 \quad (8.2)$$

The absorption of an aggregate, w_{abs}, will correspond to:

$$w_{abs} = \frac{\text{Mass of the SSD aggregate} - \text{mass of the dry aggregate}}{\text{Mass of the dry aggregate}} \times 100 \quad (8.3)$$

or, with the ASTM conventions shown in Figure 8.4:

$$w_{abs} = \frac{B - A}{A} \times 100 \quad (8.4)$$

The moisture content of an aggregate, w_h, is equal to:

$$w_h = w_{tot} - w_{abs} \tag{8.5}$$

Some useful relationships between H and B are:

$$H = B \frac{1 + (w_{tot}/100)}{1 + (w_{abs}/100)} \tag{8.6}$$

$$H = B \frac{1 + [(w_h + w_{abs})/100]}{1 + (w_{abs}/100)} \tag{8.7}$$

$$H \cong \left(1 + \frac{w_h}{100}\right) \tag{8.8}$$

8.3.3 Specific gravity

The specific gravity of an aggregate in the SSD state is called the SSD specific gravity. Figure 8.4 illustrates schematically how to measure the SSD specific gravity of either coarse or fine aggregate.

The SSD specific gravity of an aggregate is equal to:

$$G_{SSD} = \frac{B}{B - C} \tag{8.9}$$

The SSD specific gravity expresses how much denser than water an SSD aggregate is. The application of Archimedes' principle shows that G_{SSD} is the specific gravity that has to be used to calculate exactly the volume occupied by the aggregates in the concrete mix (Aïtcin, 1971).

For Portland cement or any supplementary cementitious material, the specific gravity, G_c, is equal to the mass of the dry material divided by its dried density. The ASTM C188 standard test method explains how to measure this value practically.

$$G_c = \frac{A}{A - C} \tag{8.10}$$

8.3.4 Supplementary cementitious material content

It is convenient to express the amount of supplementary cementitious material and/or filler used when making a high-performance concrete as a percentage of the cement mass, or even to express individually the content of the different supplementary cementitious materials used when making a high-performance concrete. However, the expression 'supplementary cementitious material content' can have two different meanings depending on whether the supplementary cementitious material is incorporated at the concrete plant or has been already incorporated at a cement concrete plant, separately from Portland cement: in this case the term 'content' is only related to the mass of Portland cement that is used to make the concrete. When the supplementary cementitious material is added at the cement plant its content is related to the mass of the blended cement and not only to the cement content.

The following examples illustrate the differences between these two definitions of the expression supplementary cementitious material and/or filler content.

(a) Case 1

A high-performance concrete is made in a concrete plant using 400 kg of Portland cement, 100 kg of fly ash and 40 kg of silica fume that are added separately in the mixer. What is the supplementary cementitious materials content of this concrete?

The supplementary cementitious materials content of this concrete is:

$$\frac{100 + 40}{400} \times 100 = 35\%$$

The fly ash content is $(100/400) \times 100 = 25\%$ and the silica fume content is $(40/400) \times 100 = 10\%$.

(b) Case 2

A high-performance concrete is made using 400 kg of a blended cement containing 7.5% silica fume. What is the actual amount of cement introduced in the mixer?

In this blended cement, Portland cement represents only 92.5% of the blended cement, therefore the actual Portland cement used to make this concrete is:

$$400 \times \frac{92.5}{100} = 370 \text{ kg}$$

As both definitions of the expression 'supplementary cementitious material content' are in use and have their own merits, they will both be used in this book. The reader is asked to remember that this expression has a different meaning when the supplementary cementitious material and/or filler is added at the cement plant to produce a blended cement or at the concrete plant where it is added on top of the amount of cement.

8.3.5 Superplasticizer dosage

The dosage of a superplasticizer can be expressed in different ways. It can be given in litres of commercial solution per cubic metre of concrete. This is the best way to express it at the batching plant. However, this is

Definitions and useful formulae

not the best way to express it in scientific papers or in a book because not all commercial superplasticizers have the same solids content and specific gravity. When making a high-performance concrete, it would be a serious mistake to use the same liquid dosage of superplasticizer with a superplasticizer that has a different specific gravity and solids content. For example, melamine superplasticizers can be found as liquid solutions having a solids content of 22, 33 or 40%. Therefore, it is always better to give the superplasticizer dosage as the amount of solids it contains expressed as a percentage of the mass of cement. This way of expressing the superplasticizer dosage is important when comparing the costs of different commercial superplasticizers. (In fact, it is actually the amount of active solids that should be taken into account and not the total amount of solids, because, first, not all of the solids in a superplasticizer are active dispersing molecules and, second, commercial superplasticizers always contain some residual sulfate. However, for the sake of simplicity, this last distinction will not be made in this book.)

In order to pass from a dosage expressed in litres per cubic metre to a dosage expressed in solids, it is necessary to know the value of the specific gravity of the liquid superplasticizer and its solids content.

(a) Superplasticizer specific gravity

According to Figure 8.6 the specific gravity, G_{sup}, of the superplasticizer is:

$$G_{sup} = \frac{M_{liq}}{V_{liq}} \qquad (8.11)$$

if M_{liq} is measured in grammes and V_{liq} in cubic centimetres.

(b) Solids content

According to Figure 8.6, the solids content, s, of the superplasticizer is:

$$s = \frac{M_{sol}}{M_{liq}} \times 100 \qquad (8.12)$$

Therefore the total solids content M_{sol} contained in a certain volume of superplasticizer having a specific gravity equal to G_{sup} and a solids content equal to s is:

$$M_{sol} = \frac{s \times M_{liq}}{100} = \frac{s \times G_{sup} \times V_{liq}}{100} \qquad (8.13)$$

For example, 6 litres of a melamine superplasticizer having a specific gravity of 1.10 and a solids content of 22% contain $0.22 \times 1.1 \times 6 = 1.45$ kg of solids, while 6 litres of a naphthalene superplasticizer having a specific gravity of 1.21 and a solids content of 42% contain $0.42 \times 1.21 \times 6 = 3.05$ kg of solids.

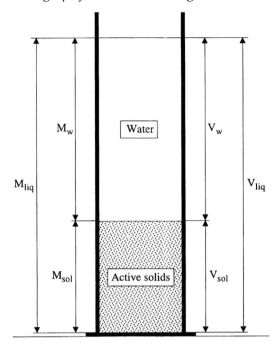

Fig. 8.6 Schematic representation of a superplasticizer.

(c) *Mass of water contained in a certain volume of superplasticizer*

When adding several litres of liquid superplasticizer in a high-performance concrete mixture it is necessary to take into account the amount of water added within the concrete in order to be able to calculate the exact water/binder ratio. This is done in the following way. From Figure 8.6:

$$M_{liq} = M_w + M_{sol} \quad \text{or} \quad M_w = M_{liq} - M_{sol}$$

as from (8.12)

$$M_{liq} = \frac{M_{sol} \times 100}{s} \tag{8.14}$$

Then

$$M_w = \frac{M_{sol} \times 100}{s} - M_{sol}$$

This can be written as

$$M_w = M_{sol}\left(\frac{100}{s} - 1\right) \quad \text{or} \quad M_w = M_{sol}\left(\frac{100 - s}{s}\right) \tag{8.15}$$

Replacing M_{sol} by its value in (8.13):

$$M_w = \frac{V_{liq} \times s \times G_{sup}}{100} \times \frac{100 - s}{s}$$

and finally,

$$M_w = V_{liq} \times G_{sup} \frac{100 - s}{100} \qquad (8.16)$$

when using proper units, g and cm³, and/or kg and l, M_w and V_w are expressed by the same number so,

$$V_w = V_{liq} \times G_{sup} \frac{100 - s}{100} \qquad (8.17)$$

Sample calculation: 8.25 litres of naphthalene superplasticizer with a specific gravity of 1.21 and a solids content of 40% have been used in a concrete in order to obtain the desired slump. What is the volume of water, V_w, that is added to the concrete when using the solution of commercial superplasticizer? According to equation (8.17), the amount of water is:

$$V_w = 8.25 \times 1.21 \frac{100 - 40}{100} = 6.0 \, l/m^3$$

(d) Other useful formulae

If d per cent is the dosage of the solids of a superplasticizer suggested by a manufacturer to obtain a desirable slump in a concrete containing a mass, C, of cementitious material, the volume of liquid superplasticizer, V_{liq}, having a specific gravity G_{sup} and a solids content s can be calculated as follows:

$$M_{sol} = C \times \frac{d}{100} \qquad (8.18)$$

but from (8.12)

$$M_{sol} = \frac{s \times M_{liq}}{100} \qquad (8.19)$$

therefore

$$\frac{s \times M_{liq}}{100} = C \times \frac{d}{100} \qquad (8.20)$$

and replacing M_{liq} by its value deduced from equation (8.11):

$$\frac{s \times G_{sup} \times V_{liq}}{100} = C \times \frac{d}{100}$$

$$V_{liq} = \frac{C \times d}{s \times G_{sup}} \qquad (8.21)$$

(e) Mass of solid particles and volume needed

If C is the total mass of the cementitious materials used in a particular mix and if d per cent is the suggested dosage of solid particles, then the mass M_{sol} of solids needed is:

$$M_{sol} = C \times \frac{d}{100} \tag{8.22}$$

The volume of liquid superplasticizer needed to have M_{sol} of solid particles is calculated as follows.

Replacing in equation (8.11) M_{liq} by its value found from (8.12):

$$V_{liq} = \frac{M_{liq}}{G_{sup}} \quad \text{and} \quad M_{liq} = \frac{M_{sol} \times 100}{s}$$

then

$$V_{liq} = \frac{M_{sol} \times 100}{sup \times G_{sup}} \tag{8.23}$$

(f) Volume of solid particles contained in V_{liq}

From Figure 8.6:

$$V_{sol} = V_{liq} - V_w$$

Replacing V_w by its value given by equation (8.17):

$$V_{sol} = V_{liq} - V_{liq} \times G_{sup} \times \frac{100 - s}{100} \tag{8.24}$$

$$V_{sol} = V_{liq}\left(1 - G_{sup} \times \frac{100 - s}{100}\right) \tag{8.25}$$

(g) Sample calculation

Example 1: Expressing a dosage in l/m³ as a percentage of solids content

A high-performance concrete containing 450 kg of cement per cubic metre of concrete has been made using 7.5 litres of a naphthalene superplasticizer. This naphthalene superplasticizer has a specific gravity of 1.21 and a solids content of 41%. What is the superplasticizer dosage, expressed as the percentage of its solids content, to the mass of cement?

The mass of 7.5 litres of superplasticizer is

$$7.5 \times 1.21 = 9.075 \text{ kg.}$$

The solids content in this mass of superplasticizer is

$$9.075 \times \frac{41}{100} = 3.72 \text{ kg.}$$

The superplasticizer dosage is

$$\frac{3.72}{450} \times 100 = 0.8\%$$

of the mass of cement.

Example 2: Passing from a dosage expressed as a percentage of solids to a dosage expressed in l/m³

In a paper it is stated that a 1.1% superplasticizer dosage has been used in a 0.35 water/cement ratio high-performance concrete containing 425 kg of Portland cement per cubic metre of concrete. A melamine superplasticizer having a specific gravity of 1.15 and a 33% solids content was used in order to reproduce this mix. What is the amount of commercial solution that was used?

The amount of superplasticizer solids is

$$\frac{425 \times 1.1}{100} = 4.675 \text{ kg}.$$

This amount of solids is contained in 4.675/0.33 = 14.17 kg of liquid melamine, which represents 14.17/1.15 = 12.3 l of commercial solution.

8.3.6 Water reducer and air-entraining agent dosages

In these cases, too, there are several ways of expressing the dosages of these admixtures. As in the case of superplasticizer, these dosages can be expressed in terms of litres of the commercial solution per cubic metre of concrete, but with the same drawbacks. In this book, the author has followed the usual way in which water reducer and air-entraining agent dosages are expressed in North America, i.e. in millilitres of commercial solution per 100 kg of cement (or, in US units, in ounces per 100 lb of cement (oz/cwt of cement)). One ounce per 100 lb of cement corresponds to 62.5 ml of admixture per 100 kg of cement, i.e.

$$100 \text{ ml}/100 \text{ kg of cement} = 1.6 \text{ oz/cwt of cement} \quad (8.26)$$

and

$$1 \text{ oz/cwt of cement} = 62.5 \text{ ml}/100 \text{ kg of cement} \quad (8.27)$$

8.3.7 Required compressive strength

Most proportioning methods are based on concrete compressive strength; therefore, it is important to define the exact value of concrete compressive strength that has to be achieved before using any mix proportioning method. In this section we will define the exact meaning of the different types of expression related to compressive strength that are used in practice and propose a clear definition that will be used in this book. The word 'compressive' could be omitted in these definitions, as is usually done in practice.

Specified compressive strength, f'_c, is not used *per se* when calculating a mix design, rather most proposed methods are based on the required average strength values, f'_{cr}. This required average strength value must be calculated from the selected design or specified strength value, the acceptance criteria and the variation observed (or anticipated) in the concrete production.

Design strength or specified strength: f'_c: this is the strength that has been taken into account by the designer in his or her calculations.

Required average strength: f'_{cr}: this is the average strength required to meet the acceptance criteria; the f'_{cr} value depends on the design or specified value, but also on the level of control that is actually achieved in the field and the acceptance criteria.

It is usually assumed that the compressive strength values measured on a specific concrete production follow a normal distribution curve as represented in Figure 8.7(a), where f'_{cr} represents the average value and σ the standard deviation.

Statistically speaking it is totally unacceptable to specify a high-performance concrete by saying that its compressive strength should be always greater than the design strength, f'_c. Concrete specified strength should, rather, be specified by saying that it is accepted that in a limited number of cases, the measured compressive strength can be lower than f'_c. The number of accepted compressive strength values lower than f'_c qualify the acceptance criteria. The dark area, B, under the normal

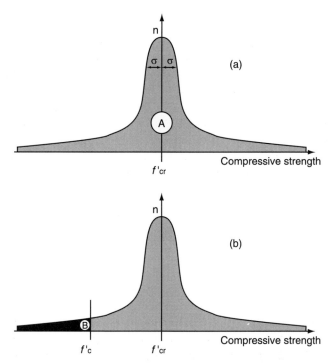

Fig. 8.7 Statistical distribution of compressive strength results.

distribution curve in Figure 8.7(b) expressed as a percentage of the total area, A, under the normal distribution curve in Figure 8.7(a), represents the percentage of tests for which it is accepted that the measured compressive strength is lower than the design strength.

In order to meet the selected acceptance criteria the average concrete compressive strength, f'_{cr}, must be higher than f'_c.

If the standard deviation, σ, of compressive strength values is high, the required strength f'_{cr} needed to achieve the specified strength f'_c has to be much higher than the specified strength. If the acceptance criteria are severe, the required strength f'_{cr} needed to meet the acceptance criteria f'_c has to be much higher.

Very simple formulae can be used to calculate the f'_{cr} value from the f'_c value and from the acceptance criteria. Some of these formulae are presented in the following. For example, ACI 318, Building Code Requirements for Reinforced Concrete and ACI 322, Building Code Requirements for Structural Plain Concrete call for designing concrete using an average compressive strength of field test results (f'_{cr}) that is higher than f'_c in order to reduce the occurrence of strength values lower than f'_c. For usual concrete, the required f'_{cr} should be the larger value given by the following two equations:

$$f'_{cr} = f'_c + 1.34\sigma \tag{8.28}$$

$$f'_{cr} = f'_c + 2.33\sigma - 3 \tag{8.29}$$

where σ is the standard deviation in MPa. Equation 8.28 ensures that there is a 99% probability that the average of all sets of three consecutive compressive strength tests must be equal to or greater than f'_c. The second equation ensures that there is a 99% probability that no single test can have a compressive strength lower than $f'_c - 3$ MPa.

The standard deviation value, σ, can be taken as that determined by the concrete producer from a concrete production made with materials similar to those to be used in the required concrete.

ACI 363 R State-of-the-Art Report on High-Strength Concrete also specifies that the required f'_{cr} (MPa) should be:

$$f'_{cr} = f'_c + 1.34\sigma \tag{8.30}$$

or

$$f'_{cr} = f'_c + 2.33\sigma - 3.5 \tag{8.31}$$

8.4 PROPOSED METHOD

The method that is discussed in the following is related to the calculation of the composition of non-air-entrained high-performance concrete. It can be used to design air-entrained high-performance concrete provided that the strength reduction due to the presence of the air bubble system is taken into account, as explained in Chapter 18 (Lessard *et al.*, 1993; Lessard, Baalbaki and Aïtcin, 1995).

The method itself is very simple: it follows the same approach as ACI 211-1 Standard Practice for Selecting Proportions for Normal, Heavyweight and Mass Concrete. It is a combination of empirical results and mathematical calculations based on the absolute volume method. The water contributed by the superplasticizer is considered as part of the mixing water. A flow chart for this method is presented in Figure 8.8.

The procedure is initiated by selecting five different mix characteristics or materials proportions in the following sequence:

No. 1 – the W/B ratio;
No. 2 – the water content;
No. 3 – the superplasticizer dosage;
No. 4 – the coarse aggregate content;
No. 5 – the entrapped air content (assumed value).

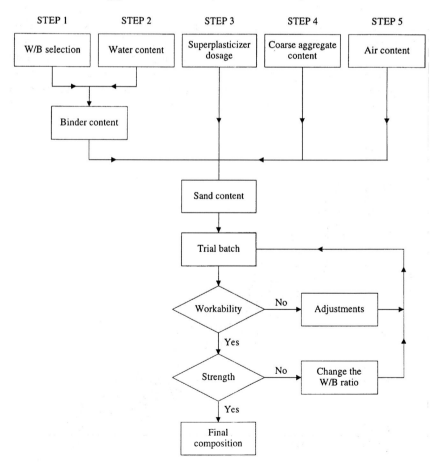

Fig. 8.8 Flow chart of the proposed mix design method.

No. 1. Water/binder ratio

The suggested water/binder ratio can be found from Figure 8.9 for a given 28 day compressive strength (measured on 100 × 200 mm faced cylinders). Owing to variations in the strength efficiency of different supplementary cementitious materials, the curve in Figure 8.9 shows a broad range of water/binder values for a given strength. If the efficiency of the different supplementary cementitious materials is not known from prior experience, the average curve can be used to give an initial estimate of the mix proportions.

No. 2. Water content

One difficult thing when designing high-performance mixtures is to determine the amount of water to be used to achieve a high-performance concrete with a 200 mm slump 1 hour after batching because the workability of the mix is controlled by several factors: the amount of initial water, the 'reactivity of the cement', the amount of superplasticizer and its degree of compatibility with the particular cement. Therefore a 200 mm slump concrete can be achieved when batching the concrete with a low water dosage and a high superplasticizer dosage or with a higher water dosage and a lower superplasticizer dosage. From an economical point of view there is no great difference between the two options, but from a

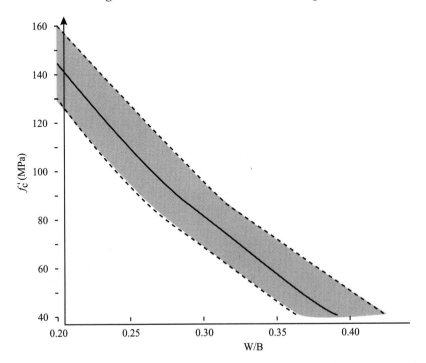

Fig. 8.9 Proposed W/B vs compressive strength relationship.

Saturation point	0.6	0.8	1.0	1.2	1.4	per cent
Water dosage	120 to 125	125 to 135	135 to 145	145 to 155	155 to 165	l/m

Fig. 8.10 Determination of the minimum water dosage.

rheological point of view the difference can be significant, depending on the rheological 'reactivity of the cement' and the efficiency of the superplasticizer. Owing to differences in fineness, phase composition, phase reactivity, composition and solubility of the calcium sulfate of the cements, the minimum amount of water required to achieve a high-performance concrete with a 200 mm slump varies to a large extent. If the amount of mixing water selected is very low, the mix can rapidly become sticky and as a high amount of superplasticizer has to be used to achieve this high slump, some retardation can be expected.

The best way to find the best combination of mixing water and superplasticizer dosage is by carrying out a factorial design experiment (Rougeron and Aïtcin, 1994), but this method is not always practical, therefore a simplified approach based on the concept of the saturation point is given in Figure 8.10. To design a very safe mix, $5 \, l/m^3$ of water can be added to the values presented in Figure 8.10. If the saturation point of the superplasticizer is not known, it is suggested starting with a water content of $145 \, l/m^3$.

No. 3. Superplasticizer dosage

The superplasticizer dosage can be deduced from the dosage at the saturation point. If the saturation point is not known, it is suggested starting with a trial dosage of 1.0%.

No. 4. Coarse aggregate content

The coarse aggregate content can be found from Figure 8.11 as a function of the typical particle shape. If there is any doubt about the shape of the coarse aggregate or if its shape is not known, a content of $1000 \, kg/m^3$ of coarse aggregate can be used to start with.

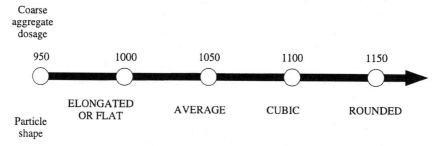

Fig. 8.11 Coarse aggregate content.

No. 5. Air content

For high-performance concretes that are to be used in non-freezing environments, theoretically there is no need for entrained air, so the only air that will be present in the mix is entrapped air, the volume of which depends partly on the mix proportions. However, in order to improve concrete handling placeability and finishability the author strongly suggests use of an amount of entrained air.

When making high-performance concretes with very low water/binder ratios it has been observed that not every cement–superplasticizer combination entraps the same amount of air. Moreover, some concrete mixers tend to entrap more air than others. From experience it has been found that it is difficult to achieve less than 1% entrapped air and that in the worst case the entrapped air contents can be as high as 3%. Therefore, the author suggests using 1.5% as an initial estimate of entrapped air content, and then adjusting it on the basis of the result obtained with the trial mix.

8.4.1 Mix design sheet

All the calculations needed to find the mix proportions are presented on a single sheet (Figure 8.12). This mix design sheet is divided into two parts. In the upper part, the specified properties of the mix are reported, along with the characteristics of all of the ingredients that will be used. This part of the mix design sheet must be completed before any calculations are made because all of these data are essential for the calculations that follow. If some of the physical properties required to make the calculations are not known, it is necessary to assume reasonable values for them, based on the best information available.

The following symbols and abbreviations will be used:

G_c specific gravity of the cement or cementitious material;
G_{SSD} aggregate specific gravity in saturated surface dry condition;
w_{abs} absorbed water in the aggregate in per cent;
w_{tot} total water content of the aggregate in per cent;
w_h moisture content of the aggregate in per cent: $w_h = w_{tot} - w_{abs}$;
G_{sup} specific gravity of the liquid superplasticizer;
s total solid content of the superplasticizer in per cent;
M_{sol} mass of solids in the superplasticizer;
d superplasticizer dosage as a percentage of the mass of solids in comparison to the total mass of cementitious materials;
V_{liq} volume of liquid superplasticizer;
V_w volume of water in the liquid superplasticizer;
V_{sol} volume of solids in the superplasticizer;
W mass of water in kg per cubic metre of concrete;
B mass of binder in kg per cubic metre.

High-performance mix design methods

In order to facilitate the corrections that have to be made to the water dosage in order to take into account the water contained in the superplasticizer, the relevant expressions are shown in a separate section in the bottom of the upper part of the mix design sheet, with space for the important factors to be filled in after calculation.

The lower part of the mix design sheet is in the form of a table in which all of the boxes are numbered in the order in which they have to be filled in. This table is divided into six columns, numbered at the top. In the first column initial data and calculations are reported. In column 2

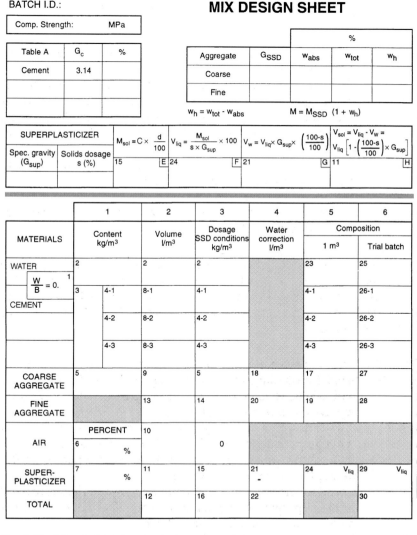

Fig. 8.12 Mix design sheet.

Proposed method 239

the volume of fine aggregate is calculated. In column 3 the SSD proportions of the mix are presented. In column 4 the different water corrections that have to be made are calculated, as explained in Figure 8.12. In column 5 the proportions of the mix using the actual raw materials are given, and in column 6 the proportions of the trial batch can be calculated. The detailed calculations that have to be made to fill this design sheet are explained box by box before doing a sample calculation.

(a) Mix design calculations

Box 1   Record the water/binder ratio found from Figure 8.9.
Box 2   Record the amount of water required, selected from Figure 8.10, and report it in columns 1, 2 and 3 where box 2 appears.
Box 3   From the values appearing in boxes 1 and 2 calculate the necessary mass of binder.
Boxes 4-1, 4-2, 4-3   Calculate the mass of each of the different cementitious materials according to the cementitious composition selected appearing in Table A on the top part of the mix design sheet and report it in columns 1, 3 and 5 where these boxes are found.
Box 5   Fill in the mass of coarse aggregate, given by Figure 8.11 and report it in columns 1 and 3 in box 5.
Box 6   Record the assumed air content.
Box 7   Record the amount of superplasticizer needed, obtained from the saturation point value.

At this stage, the only missing information is the mass of fine aggregate in the mix. This value is found by the absolute volume method, i.e. the volumes of all the ingredients already selected are calculated in order to find the volume of fine aggregate needed to make 1 m^3 of concrete. This is done in column 2.

Boxes 8-1, 8-2, 8-3   The volumes of the different cementitious materials are calculated by dividing their masses, appearing in boxes 4-1, 4-2 and 4-3, by their respective specific gravities that have been reported in the top part of the mix design sheet.
Box 9   The volume of coarse aggregate is calculated by dividing its mass (appearing in box 5) by its SSD specific gravity.
Box 10   The volume of entrapped air, in l/m^3, is obtained by multiplying the air content (box 6) by 10.
Box 11   Calculate the volume V_{sol} (the volume of the solids contained in the superplasticizer) using the formulae presented in the middle part of the mix design sheet.
Box 12   The total mix volume (1000 l/m^3) appears here.

Box 13 The volume of the fine aggregate in l/m³ is calculated by subtracting the volumes of all of the other ingredients (boxes 2, 8, 9, 10 and 11) from 1000.

The mass of fine aggregate and the unit mass of concrete can now be calculated. This is done in column 3.

Box 14 The mass of fine aggregate is calculated by multiplying its volume appearing in box 13 by its SSD specific gravity.
Box 15 The mass of solid in the superplasticizer M_{sol} is reported here.
Box 16 All the masses appearing in column 3 are summed here to give the unit mass of the concrete.

Up to now the aggregate contents have been calculated on an SSD basis.

Next, the water dosage has to be corrected, taking into account the actual total water contents of the aggregates and the water brought into the mix by the liquid superplasticizer. These corrections are made in columns 4 and 5, in boxes 18, 20 and 21 with the following arbitrary sign convention: if aggregate brings water to the mix (i.e. if its moisture content is greater than its absorption with SSD state), its water content has to be subtracted from the water dosage, hence a negative sign will be used in the corresponding box, whereas a positive sign will be used when the aggregate will absorb some water from the mix.

Box 17 Multiply the SSD mass of coarse aggregate by $(1 + w_h/100)$.
Box 18 Subtract the value in box 17 from that in box 5 and enter the result here.
Box 19 Calculate the SSD mass of the fine aggregate.
Box 20 Subtract the value in box 19 from that in box 14 and enter the result here.
Box 21 Write the amount of water brought to the mix by the superplasticizer from box G; the negative sign already appears in this box.
Box 22 Add algebraically all the water corrections.

The final composition of 1 m³ of concrete with the wet aggregates is now calculated in column 5.

Box 23 Add (taking note of its sign) the water correction appearing in box 22 to the volume of water appearing in box 2.
Box 24 Enter the superplasticizer dosage, V_{liq}, from box F.

> The trial batch composition can be calculated in column 6. Each number appearing in column 5 has to be multiplied by a factor, f, equal to the desired mass of the trial batch in kg, divided by the mass in box 16. Factor f can also be calculated on a volume basis. If the trial batch has to have a certain

Proposed method

volume, each number appearing in column 5 has to be multiplied by a factor corresponding to the volume of the trial batch in litres divided by 1000.

Boxes 25 to 29
These are calculated by multiplying the values in column 5 by the factor f.

Box 30
The mass of the trial batch is calculated by adding the masses of the different concrete ingredients appearing in boxes 25 to 29. To check the calculation, multiply box 16 by f; the result should be the same as in box 30.

(b) Sample calculation

Suppose that a 100 MPa concrete has to be made with:

- a Type I Portland cement;
- a naphthalene-type superplasticizer with a total solids content of 40% and specific gravity of 1.21;
- a dolomitic limestone having the following characteristics: D_{max} = 10 mm, G_{SSD} = 2.80, w_{abs} = 0.8% and w_{tot} = 0%, and the shape of the particles can be described as between average and cubic;
- a siliceous natural sand of the following characteristics: G_{SSD} = 2.65, w_{abs} = 1.2% and w_{tot} = 3.5%;
- silica fume at 10% replacement (of total cementitious material) is to be used; its specific gravity is 2.20;
- the dosage of solids superplasticizer at the saturation point is 1.0%.

Filling in the upper part of the mix design sheet:
All of the values characterizing the materials to be used are reported in the upper part of the mix design sheet and the moisture contents of the coarse and fine aggregate are calculated.

(c) Calculations

It is suggested at this stage that an enlarged photocopy of Figure 8.12 is made and that it is filled in as the different calculations are made.

Water/binder ratio
From Figure 8.9 it is seen that in order to achieve a 100 MPa compressive strength the water/binder ratio should be between 0.25 and 0.30. As there are no previous data relating to the strength potential of the particular binder used, let us write 0.27 in box 1.

Water content
From Figure 8.10 we find that the water dosage for a saturation point of 1.0% should be between 135 and 145 l/m^3. Let us write a dosage of 140 l/m^3 for this trial batch in box 2.

Binder content
The binder content is equal to:

$$B = \frac{140}{0.27} = 518.5 \text{ kg}$$

Let us round that value to 520 kg/m³ (box 3). This amount of binder will be composed of 10% silica fume, i.e. 52 kg. Let us take 50 kg to work with round numbers (box 4-2).

The cement content will then be $520 - 50 = 470 \text{ kg/m}^3$ (box 4-1).

The content of coarse aggregate, 1075 kg/m³, given by Figure 8.11, should be written in box 5.

Let us assume a 1.5% volume of entrapped air (box 6).

The dosage of superplasticizer is equal to 1% for the saturation point value (box 7).

At the present time, column 1 has been filled. Before we start filling column 2, let us calculate the mass and volume values related to the superplasticizer in boxes E, F, G and H, and report them in the box number appearing in the upper left part of these boxes.

M_{sol} is equal to $520 \times (1/100) = 5.2$ kg; let us round this value to 5.0 when filling the design sheet (box E and 15).

V_{liq} is equal to $5.2/[(40/100) \times 1.21] = 10.74 \text{ l/m}^3$; let us round this value to 10.7 (box F and 24).

V_w is equal to $10.74 \times 1.21 \times (100 - 40)/100 = 7.8 \text{ l/m}^3$; let us write 8.0 in box G and 21.

V_{sol} is equal to $10.74 - 7.80 = 2.94 \text{ l/m}^3$; let us round this value to 3.0 l/m³ in box H and 11.

Now we can go back to the lower part of the table and start to fill in the missing values in column 2.

The volume of the cement is equal to:

$$\frac{470}{3.14} = 149.7 \text{ l/m}^3;$$

let us write 150 in box 8-1.

The volume of silica fume is:

$$\frac{50}{2.2} = 22.7 \text{ l/m}^3;$$

let us write 23 in box 8-2.

The volume of coarse aggregate is:

$$\frac{1075}{2.80} = 383.9 \text{ l/m}^3;$$

let us write 384 in box 9.

The volume of entrapped air is:

$$1.5 \times 10 = 15 \, l/m^3;$$

let us write 15 in box 10.

The volume of solids in the superplasticizer, 3.0, has already been written in box 11.

The sum of all the numbers appearing in column 2 is:

$$140 + 150 + 23 + 384 + 15 + 3 = 715;$$

let us write 715 in box 12.

The volume of the sand will be:

$$1000 - 715 = 285 \, l/m^3;$$

this is written in box 13.

Now that the volume of sand is known, box 14 in column 3 can be filled in.

The SSD mass of the sand is:

$$285 \times 2.65 = 755 \, kg/m^3;$$

let us write 755 in box 14.

Adding the values appearing in column 3, the unit mass of this concrete can be calculated:

$$140 + 470 + 50 + 1075 + 755 + 5 = 2495 \, kg/m^3;$$

let us write 2495 kg/m³ in box 16.

As the aggregates to be used are not in the SSD condition, and as the 10.8 l/m³ of superplasticizer will bring some water to the mix, water corrections have to be made taking into account the sign convention already proposed. This exercise is carried out in columns 4 and 5.

As the coarse aggregate is dry, it will absorb a certain amount of water from the mix. The mass M_c of dry coarse aggregate to be weighed is $M_c = 1075[1 - (0.8/100)] = 1066 \, kg$ (box 17). This dry coarse aggregate will absorb $1075 - 1066 = 9 \, kg$ of water; hence +9 is written in box 18.

As the fine aggregate is wet a mass greater than 755 kg must be weighed, and the water added has to be subtracted from the total amount of mixing water, as $w_h = 2.3\%$.

$$M_f = 755\left(1 + \frac{2.3}{100}\right) = 772 \, kg$$

is written in box 19.

This mass of fine aggregate will bring to the mix:

$$772 - 755 = 17 \, l/m^3$$

of water. Let us write -17 in box 20.

The correction to be applied to the water content is found by adding algebraically the numbers appearing in column 4:

$$+9 - 17 - 8 = -16 \, l/m^3;$$

let us write -16 in box 22.

This gives the final composition of 1 m³ of concrete with aggregates. The necessary volume of mixing water to be measured is:

$$140 - 16 = 124 \, l/m^3 \quad \text{(box 23)}$$

Suppose that to test the concrete, the following specimens are needed:

- 3 100 × 200 mm cylinders for tests at 1, 7, 28 and 91 days in compression;
- 3 150 × 300 mm cylinders for tests at 28 days in compression;
- 3 150 × 300 mm cylinders for tests for elastic modulus at 28 days;
- 3 100 × 100 × 400 mm beams for tests for modulus of rupture at 7 and 28 days, plus one spare for a total of 7.

A slump test, an air content test and a unit mass test will be done on the fresh concrete. Except for the air content test, the concrete used for these tests will be recovered.

Knowing that:

- a 100 × 200 mm specimen weighs about 4 kg;
- a 150 × 300 mm specimen weighs about 13 kg;
- a 100 × 100 × 400 mm prism weighs about 10 kg;
- an air test needs 15 kg;

the amount of concrete to make this trial batch can be calculated. Assuming 10% extra materials to compensate for losses:

	Specimen			Air	Total
	100 × 200 mm	150 × 300 mm	100 × 100 × 400 mm		
Number	12	6	7	1	
Mass needed (kg)	48	78	70	15	211

Assuming a loss of 10% it will be necessary to mix 232 kg of concrete which represents $232/2495 = 0.093$ of a cubic metre.

All the numbers in column 5 have to be multiplied by this factor to find the mass of each material to be weighed to make the trial batch:

Mixing water	$124 \times 0.093 = 11.5 \, 3 \rightarrow 11.5$	box 25
Cement	$470 \times 0.093 = 43.7 \, kg$	box 26-1
Silica fume	$50 \times 0.093 = 4.7 \, kg$	box 26-2
Coarse aggregate	$1066 \times 0.093 = 99.1 \, kg \rightarrow 99 \, kg$	box 27
Fine aggregate	$772 \times 0.093 = 71.80 \, kg \rightarrow 72 \, kg$	box 28
Superplasticizer	$10.7 \times 0.093 = 0.995 \, l \rightarrow 1 \, l$	box 29

Proposed method

In order to check the final composition, the values appearing in boxes 25 to 28 are added:

$$11.5 + 43.7 + 4.7 + 99 + 72 = 230.9 = 231 \text{ kg}$$

231 is written in box 30 which is very close to the 232 kg value calculated earlier.

All these values appear in Figure 8.13.

MIX DESIGN SHEET

Reference: Sample calculation

Comp. Strength:		100 MPa
Table A	G_c	%
Cement	3.14	90
S.F.	2.2	10

				%	
Aggregate	G_{SSD}	w_{abs}	w_{tot}		w_h
Coarse	2.80	0.8	0.0		-0.8
Fine	2.65	1.2	3.5		2.3

$w_h = w_{tot} - w_{abs}$ $M = M_{SSD}(1 + w_h)$

SUPERPLASTICIZER		$M_{sol} = C \times \dfrac{d}{100}$	$V_{liq} = \dfrac{M_{sol}}{s \times G_{sup}} \times 100$	$V_w = V_{liq} \times G_{sup} \times \left(\dfrac{100-s}{100}\right)$	$V_{sol} = V_{liq} - V_w =$ $V_{liq}\left[1 - \left(\dfrac{100-s}{100}\right) \times G_{sup}\right]$	
Spec. gravity (G_{sup})	Solids dosage s (%)	15	E 24	F 21	G 11	H
1.21	40	5	10.7	8.0	3.0	

MATERIALS		1 Content kg/m³	2 Volume l/m³	3 Dosage SSD conditions kg/m³	4 Water correction l/m³	5 Composition 1 m³	6 Trial batch
WATER	2	140	2 140	2 140		23 124	25 11.5
$\dfrac{W}{B} = 0.27$ CEMENT	3	4-1 470	8-1 150	4-1 470		4-1 470	26-1 43.7
SILICA FUME	520	4-2 50	8-2 23	4-2 50		4-2 50	26-2 4.7
		4-3 —	8-3 —	4-3 —		4-3 —	26-3 —
COARSE AGGREGATE	5	1075	9 384	5 1075	18 + 9	17 1066	27 99
FINE AGGREGATE			13 285	14 755	20 - 17	19 772	28 72
AIR	6	PERCENT 1.5 %	10 15	0			
SUPER- PLASTICIZER	7	1.0 %	11	15 5	21 - 8	24 10.7 V_{liq}	29 1 V_{liq}
TOTAL			12 715	16 2495	22 - 16		30 231

Fig. 8.13 Sample mix calculation.

8.4.2 From trial batch proportions to 1 m³ composition (SSD conditions)

When a satisfactory high-performance concrete mix having the right consistency (initial and final slump) has been obtained after different adjustments, it is necessary to go back to the composition of 1 m³ of concrete in an SSD condition. The mass of the mixing water added in the mixer is known, but the actual water present in the mixture is not, because the aggregates are not likely to be in their SSD state and some water is 'hidden' in the liquid superplasticizer; therefore, the water/binder ratio of the mix is not known. Thus it is necessary to calculate the so-called 'water corrections' to be able to know precisely the amount of water present in the mixture. This is done according to the very simple flow chart presented in Figure 8.14. In order to make all of these calculations systematically, a special calculation sheet entitled 'From trial batch proportions to 1 m³ composition' is presented in Figure 8.15. It is divided into three parts. In the upper part the binder and aggregate characteristics needed to make all of the calculations have to be reported in special boxes. All the expressions used to make these calculations are also presented there. In the middle part several boxes are related to the superplasticizer.

The lower part of the calculation sheet is in the format of a five column table.

(a) Mix calculation

In column 1 the actual masses of the materials used in the trial batch are reported in boxes 1 to 5, as well as the measured air content in box 6.

In column 2 the SSD mass of the coarse and fine aggregates are reported in boxes 7 and 8 and the mass of water they contain are reported in boxes 9 and 10 in column 3. Also in column 3 the mass of water added to the mix by the superplasticizer is recorded in box 11 and the total water correction is calculated and recorded in box 12. The same previous sign convention is used to calculate the actual effective amount of mixing water.

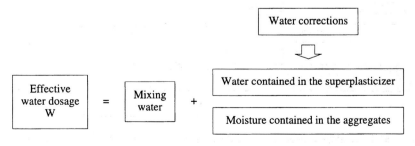

Fig. 8.14 Calculation of the effective water dosage, W, of a trial batch.

Proposed method 247

As the air content is known as a percentage of the mix volume, it is necessary to calculate the volume in litres of each ingredient in order to know exactly the volume of concrete made in the trial batch. This is done in column 4, in boxes 13 to 17. The total of these volumes is recorded in box 18. The actual volume of the mix is then recorded in box 19, taking into account the measured air content of the concrete of the trial batch.

The trial batch represents a certain fraction of a cubic metre; thus in order to calculate the composition of 1 m³ of this concrete it is necessary

FROM TRIAL BATCH PROPORTIONS TO 1 m³ COMPOSITION

CEMENT	G_c = 3.14	$V_{mix} = \dfrac{V_s + V_w}{1 - a/100}$ V_s volume of solids V_w volume of water a air content %	Aggregate	G_{SSD}	w_{abs}	% w_{tot}	w_h
			Coarse				
			Fine				

$w_h = w_{tot} - w_{abs}$ $M = M_{SSD}(1 + w_h)$

SUPERPLASTICIZER		$V_w = V_{liq} \times G_{sup} \times \left(\dfrac{100-s}{100}\right)$	$V_{sol} = V_{liq}\left[1 - \left(\dfrac{100-s}{100}\right) \times G_{sup}\right]$	$M_{sol} = V_{liq} \times G_{sup} \times \dfrac{s}{100}$
Spec. gravity (G_{sup})	Solids content s (%)		G	H	I

MATERIALS	1 Used	2 SSD conditions	3 Water correction I	4 Volume l	5 Dosage SSD conditions kg/m³
WATER	1 l			13	21
CEMENT	2-1 kg			14-1	22-1
	2-2 kg			14-2	22-2
	2-3 kg			14-3	22-3
COARSE AGGREGATE	3 kg	7 kg	9	15	23
FINE AGGREGATE	4 kg	8 kg	10	16	24
SUPER-PLASTICIZER	5 l		11	17 Solids	25-1 l/m³ 25-2 kg of solids
AIR	PERCENT 6 %			18 Volume of solids + water	6 %
		TOTAL	12 Water correction	19	26
			20 Mult. factor		

$\dfrac{W}{C} = 0.$ (27, 28)

Fig. 8.15 Design sheet to calculate 1 m³ composition from trial batch proportions.

to calculate how many times this trial batch is contained in 1 m³ or 1000 l of concrete. 1000 is divided by the volume of the trial batch appearing in box 19. Let us call this value f. The different masses of materials used to make the trial batch will be multiplied by f in order to have the composition of 1 m³ of concrete (SSD conditions).

Water content

The volume of water appearing in box 13 is multiplied by f and recorded in box 21. It should be remembered that the amount of water contained in the superplasticizer is included in this value.

Binder content

The values appearing in boxes 2-1, 2-2 and 2-3 are multiplied by f and reported in boxes 22-1, 22-2 and 22-3.

Aggregate proportions

The values appearing in boxes 7 and 8 are multiplied by f and reported in boxes 23 and 24.

Superplasticizer dosage

The superplasticizer dosage is calculated by multiplying the value appearing in box 5 in column 1 by f and recorded in box 25.

Unit mass of the concrete

This is calculated by adding the numbers appearing in boxes 21, 22-1, 22-2, 22-3, 24 and 25-2 and reported in box 26.

The total amount of supplementary cementitious materials used is noted in box 27 and the water/binder ratio calculated in box 28.

(b) *Sample calculation*

Suppose that a trial batch with an adequate consistency and adequate initial and final slumps has been made using the following quantities of materials.

Water	Cement	Fly ash	Aggregate		Super-plasticizer
			Coarse	Fine	
12 l	45 kg	5 kg	100 kg	70 kg	0.9 l

Proposed method

The air content of this trial batch was 1.5%. The materials used to make this trial batch had the following properties:

Aggregate	G_{SSD}	w_{abs}	w_{tot}
Coarse	2.75	1.0	0
Fine	2.65	1.0	3.9

The fly ash used has a specific gravity of 2.50. The superplasticizer is naphthalene-based with a specific gravity, G_{sup}, of 1.21 and a solid content of 42%.

What is the composition of 1 m³ of such concrete?

(c) Calculations

Let us start by writing 12 l, 45 kg, 5 kg, 100 kg, 70 kg, 0.9 l and 1.5% in the appropriate boxes in column 1.

The SSD mass of coarse aggregate is

$$\frac{100}{1 + (-1/100)} = 101 \text{ kg} \quad \text{(box 7)}$$

The SSD mass of fine aggregate is

$$\frac{70}{1 + 2.9/100} = 68 \text{ kg} \quad \text{(box 8)}$$

The mass of water absorbed by the coarse aggregate is equal to -1 kg. Let us write -1 in box 9, while the amount of water added to the mix by the fine aggregate is $70 - 68 = 2$ kg. Let us write 2 in box 10. The amount of water contained in the superplasticizer is equal to:

$$V_w = 0.9 \times 1.21 \times \frac{100 - 42}{100} = 0.63 \text{ l},$$

to be written in box 11.

The total water correction to be made is $-1.0 + 2.0 + 0.63 = +1.63$ (box 12).

The actual volume of effective water in the mix was $12 + 1.63 = 13.63$ l (box 13).

Let us now calculate the volume of concrete made in this trial batch:

The volume of cement used is

$$\frac{45}{3.14} = 14.33 \text{ l} \quad \text{(box 14.1)}$$

The volume of fly ash used is

$$\frac{5}{2.5} = 2.0 \text{ l} \quad \text{(box 14.2)}$$

The volume of coarse aggregate is

$$\frac{101}{2.75} = 36.73 \text{ l} \quad \text{(box 15)}$$

The volume of fine aggregate is

$$\frac{68}{2.65} = 25.66 \text{ l} \quad \text{(box 16)}$$

The volume of the solids in the superplasticizer is

$$V_{sol} = 0.9 \left[1 - 1.21 \frac{(100 - 42)}{100} \right] = 0.27 \text{ l} \quad \text{(box 17)}$$

The sum of these volumes represents 98.5% of the total volume of the trial batch:

$$13.63 + 14.33 + 2.0 + 36.73 + 25.66 + 0.27 = 92.62 \text{ l} \quad \text{(box 18)}$$

Therefore the actual volume of concrete in this trial batch was:

$$\frac{92.62}{1 - 1.5/100} = 94.03 \text{ l} \text{ of concrete (box 19)}$$

In order to make 1 m³ of this concrete, the proportions of the materials will have to be multiplied by:

$$\frac{1000}{94.03} = 10.63 \quad \text{(box 20)}$$

The composition of 1 m³ of concrete is:

Water	13.63 × 10.63 = 144.88	→ 145 l/m³	(box 21)
Cement	45 × 10.63 = 478.35	→ 480 kg/m³	(box 22-1)
Fly ash	5 × 10.63 = 53.15	→ 53 kg/m³	(box 22-2)
Coarse aggregate	101 × 10.63 = 1073.63	→ 1075 kg/m³	(box 23)
Fine aggregate	68 × 10.63 = 722.84	→ 725 kg/m³	(box 24)
Superplasticizer	0.9 × 10.63 = 9.567	→ 9.6 l/m³	(box 25-1)

The mass of the solids is

$$9.6 \times 1.21 \times 0.42 = 4.88 \text{ kg/m}^3 \rightarrow 5 \text{ kg} \quad \text{(box 25-2)}$$

Proposed method

The unit mass of this concrete is:

$$145 + 480 + 53 + 1075 + 725 + 5 = 2483 \text{ kg/m}^3 \quad \text{(box 26)}.$$

The binder mass is: $480 + 53 = 533$ (box 27).
The actual water/binder ratio is:

$$\frac{145}{533} = 0.27 \quad \text{(box 28)}.$$

All these values are found in Figure 8.16.

FROM TRIAL BATCH PROPORTIONS TO 1 m³ COMPOSITION

	G_c				%		
CEMENT	3.14	$V_{mix} = \frac{V_s + V_w}{1 - a/100}$	Aggregate	G_{SSD}	w_{abs}	w_{tot}	w_h
Fly Ash	2.50	V_s volume of solids	Coarse	2.75	1.0	0	-1
		V_w volume of water	Fine	2.65	1.0	3.9	+2.9
		a air content %					

$w_h = w_{tot} - w_{abs}$ $M = M_{SSD}(1 + w_h)$

SUPERPLASTICIZER		$V_w = V_{liq} \times G_{sup} \times \left(\frac{100-s}{100}\right)$	$V_{sol} = V_{liq}\left[1 - \left(\frac{100-s}{100}\right) \times G_{sup}\right]$		$M_{sol} = V_{liq} \times G_{sup} \times \frac{s}{100}$	
Spec. gravity (G_{sup})	Solids content s (%)	$V_w =$	G	$V_{sol} =$	H	I
1.21	42	$0.9 \times 1.21 \times \left(\frac{58}{100}\right) = 0.6$ l		$0.9\left[1 - \left(\frac{58}{100}\right) \times 1.21\right] = 0.27$ l	$M_{sol} = 9.6 \times 1.21 \times \frac{42}{100} = 5$ kg	

MATERIALS	1 Used	2 SSD conditions	3 Water correction l	4 Volume l	5 Dosage SSD conditions kg/m³	
WATER	1: 12 l		13: 13.63	21: 145		28: W/B = 0.27
CEMENT	2-1: 45 kg		14-1: 14.33	22-1: 480		
FLY ASH	2-2: 5 kg		14-2: 2.00	22-2: 53		27: 533
	2-3: — kg		14-3: —	22-3: —		
COARSE AGGREGATE	3: 100 kg	7: 101 kg	9: -1	15: 36.73	23: 1075	
FINE AGGREGATE	4: 70 kg	8: 68 kg	10: +2	16: 25.66	24: 725	
SUPER-PLASTICIZER	5: 0.9 l		11: +0.63	17 Solids: 0.27	25-1: 9.6 l/m³ 25-2: 5 kg of solids	
AIR	PERCENT 6: 1.5 %			18 Volume of solids + water: 92.62	6: 1.5 %	
	TOTAL		12 Water correction: +1.63	19: 94.03	26: 2483	
				20 Mult. factor: f = 10.63		

Fig. 8.16 From trial batch proportions to 1 m³ composition.

8.4.3 Batch composition

When a high-performance concrete is to be batched, the proportions obtained from the trial mixer or given by the laboratory are always for SSD aggregate conditions. As the aggregates contained in the bins at the batching plant are not usually under SSD conditions, the amount of mixing water has to be adjusted. The content of water added by the superplasticizer also has to be subtracted from the mixing water content.

(a) Mix calculation

In order to facilitate these calculations, another calculation sheet is proposed and presented in Figure 8.17.

It is also divided into two parts: in the upper part, all the characteristics needed to make the water corrections have to be recorded in their specific boxes. Other information related to the type and brand of cement and superplasticizer, the water/binder ratio and the targeted compressive strength can also be reported in this part of the calculation sheet.

The bottom part of the calculation sheet is a four column table. In column 1 the results of the trial batch or the values given by the laboratory are reported (boxes 1 to 5). The mass of the solids, M_{sol}, contained in the superplasticizer is calculated using the formula given in the upper part and written in box 2-1. Summing these values, the unit mass of concrete is found (box 6).

The mass of water, V_w, added to the mix by the superplasticizer is calculated and recorded in box 7.

The mass of wet coarse aggregate, M, to be weighed is equal to $M_{SSD}(1 = w_h)$; it is recorded in box 8 and the mass of water to be added or subtracted to the water content is recorded in box 9.

The same is done for the fine aggregate in boxes 10 and 11.

The same previous sign convention is used to calculate water corrections.

The values in column 2 are added and the result recorded in box 12.

The value recorded in box 12 is added or subtracted to the water content appearing in box 1 in order to calculate the mass of mixing water necessary to make 1 m³ of the desired concrete from the aggregates that are in the bins. This value is reported in box 13.

Now, the masses of the different materials needed to make X m³ of concrete can be calculated by multiplying by X all the values appearing in column 3. This is done in boxes 14 to 18.

The values appearing in boxes 14, 15-1, 16-1, 16-2, 16-3, 17 and 18 are added together and this sum is recorded in box 19.

Multiplying by X the value appearing in box 6 it is possible to check the value recorded in box 19.

Proposed method

(b) Sample calculation

A concrete batch plant is to produce 8 m³ of a high-performance concrete on the basis of the following SSD composition:

W/C	Water	Cement	Aggregate (kg)		Superplasticizer	
			Coarse	Fine	Liquid	Solid
0.29	130	450	1050	750	8 l	4 kg

BATCHING SHEET

BINDER	
TYPE	BRAND

Aggregates			Supplier	W_{tot}	W_h	W_{abs}
Coarse	$\varnothing_{max} =$	mm				
Fine	mf =					

$W_h = W_{tot} - W_{abs}$ $M = M_{SSD}(1 + h)$

COMPRESSIVE STRENGTH	MPa
W/C	0.

SUPERPLASTICIZER		$V_w = V_{liq} \times G_{sup} \times \left(\dfrac{100 - s}{100}\right)$
Specific gravity	Solids content s (%)	
		$M_{sol} = V_{liq} \times G_{sup} \times \dfrac{s}{100}$
TYPE	BRAND	

MATERIALS	1 SSD conditions 1 m³	2 Water correction l/m³	3 Wet conditions 1 m³	4 Wet conditions ____ m³
WATER	1		13	14
SUPER-PLASTICIZER	2 2-1	7	2	15-1 15-2
CEMENT	3-1		3-1	16-1
	3-2		3-2	16-2
	3-3		3-3	16-3
COARSE AGGREGATE	4	9	8	17
FINE AGGREGATE	5	11	10	18
TOTAL	6	12		19

Fig. 8.17 Batching sheet.

254 High-performance mix design methods

The aggregates having the following water contents are in the bins:

Coarse aggregate $w_{tot} = 0.8\%$, $w_{abs} = 0.8\%$
Fine aggregate $w_{tot} = 3.0\%$, $w_{abs} = 1.0\%$

The superplasticizer is a naphthalene superplasticizer containing 42% solids and having a specific gravity of 1.21.

What are the mass of materials that must be weighed to make 8 m³ of concrete?

(c) Calculations

Let us write the appropriate values in (boxes 1 to 5) in column 1. The mass of the solids contained in the superplasticizer is equal to $M_{sol} = 8 \times 1.21 \times 0.42 = 4.06$ kg. Let us write 4 in box 2-1. The unit mass of this concrete is equal to the sum of all the values appearing in this column, i.e. $130 + 450 + 4 + 1050 + 750 = 2384$ (box 6).

The superplasticizer is also adding some water to the mix:

$$V_w = 8 \times 1.21 \times \frac{100 - 42}{100} = 5.61 \, l$$

of water, which is rounded off to 6 l in (box 7).

The number 6 is preceded by sign – using the sign convention already used in other calculation sheets.

The mass of coarse aggregate to be weighed is 1050 kg. Because the coarse aggregate is already in its SSD state (box 8) there is no water correction in this particular case and 0 is recorded in box 9.

In order to get 750 kg of SSD sand, the batcher will have to weigh out $750(1 + 2/100) = 765$ kg of wet sand (box 10). The mass of water included in the wet sand is equal to $765 - 750 = 15$ kg, and -15 is written in box 11.

Summing the values now appearing in column 2: $-6 + 0 - 15 = -21$ (box 12).

21 l of water have to be subtracted from the amount of mixing water appearing in box 1. The mass of mixing water to be measured is: $130 - 21 = 109$ (box 13).

In order to produce the 8 m³ of concrete the values appearing in column 3 are multiplied by 8 and reported in boxes 14, 15-1, 16, 17 and 18. In order to get the value of box 15.2, the volume of superplasticizer has to be multiplied by 1.21: $64 \times 1.21 = 77$.

The concrete load will be equal to: $872 + 77 + 3600 + 8400 + 6120 = 19\,069$ kg (box 19).

All these values are found in Figure 8.18.

Verification

If we multiply the unit mass appearing in box 6 by 8, we obtain 19 072, which is close enough to the value appearing in box 19.

BATCHING SHEET

BINDER	
TYPE	BRAND
Type I	Local A

COMPRESSIVE STRENGTH	— MPa
W/C	0.29

Aggregates			Supplier	W_{tot}	W_{abs}	W_h
Coarse	$\varnothing_{max} =$	mm	Local B	0.8	0.8	0
Fine	mf =		Local C	3.0	1.0	2

$W_h = W_{tot} - W_{abs}$ $M = M_{SSD}(1 + h)$

SUPERPLASTICIZER	
Specific gravity	Solids content s (%)
1.21	42
TYPE	BRAND
Naphthalene	Local C

$$V_w = V_{liq} \times G_{sup} \times \left(\frac{100 - s}{100}\right)$$

$$M_{sol} = V_{liq} \times G_{sup} \times \frac{s}{100}$$

MATERIALS	1 SSD conditions 1 m³	2 Water correction l/m³	3 Wet conditions 1 m³	4 Wet conditions x = 8 m³
WATER	1 130		13 109	14 872 l
SUPER-PLASTICIZER	2: 8 l / 2-1: 4 kg	7 - 6	2 8 l	15-1: 64 l / 15-2: 77
CEMENT	3-1 450		3-1 450	16-1 3600
	3-2 —		3-2 —	16-2 —
	3-3 —		3-3 —	16-3 —
COARSE AGGREGATE	4 1050	9 0	8 1050	17 8400
FINE AGGREGATE	5 750	11 - 15	10 765	18 6120
TOTAL	6 2384	12 - 21		19 19 069

Fig. 8.18 Sample of filled batching sheet.

8.4.4 Limitations of the proposed method

As has been shown in the previous paragraphs, in the present state of the art, the successful fabrication of high-performance concrete depends on a combination of empirical rules derived from experience, laboratory work and a great dose of common sense and observation. It can thus be said that the proposed method is a transition between an art and a science.

Applying step by step the materials selection criteria and the mix design method proposed in this book should help one to find the best, or at least the most acceptable, locally available materials and to select their proportions in order to make the best (or close to the best) high-performance concrete in terms of rheology, strength and economy that can be made from the materials available in a given area to meet the expectations of the designer. However, improvements to this proposed method can always be made.

The most important factor in the mix design of a high-performance concrete, much more than in the case of usual concrete mixes, is the selection of materials. A blind application of the proposed mix design method (or of any other method) does not guarantee success because, in high-performance concretes, all concrete ingredients are working at, or near, their critical limits. Any high-performance concrete has a weak link within it, and when it is tested, failure always initiates at this weakest link and then propagates through the stronger parts. The key to success in designing a high-performance concrete mix is to find materials in which the weakest link is as strong as required to meet the performance requirements, while the stronger parts are not so strong that they lead to an unnecessary expenditure.

The mix design method proposed in this chapter is based on experience gained through the years. Like all mix design methods, it should be used only as a guide. In general, the calculated proportions should provide a mix having almost the desired characteristics. However, if the cement is not suitable, if the aggregates are not strong enough, if the superplasticizer is not efficient enough, if the selected cement/superplasticizer combination is not fully compatible or if some other unexpected factors intervene, the concrete may not reach the desired level of performance. On the other hand, with the application of common sense and with careful observation of the rheology, the surface failure and some other critical characteristics, it should generally not be too difficult to find out why the concrete did not achieve the design performance level that was desired.

For example, when the desired compressive strength is not obtained, and the fracture surface shows a number of intragranular fractures, the aggregates can be blamed, and a 'stronger' aggregate has to be found.

If the fracture surface shows considerable debonding between the coarse aggregate and the hydrated cement paste, the coarse aggregate used has too smooth a surface or is simply too dirty, and therefore a coarse aggregate with a rougher or cleaner surface must be used.

If the fracture surface passes almost entirely through the hydrated cement paste around the aggregates, a stronger concrete can be made with the same aggregates by lowering the water/binder ratio further.

If the compressive strength does not increase when the water/binder ratio decreases, this indicates that the aggregate strength or cement aggregate bonding controls the failure rather than the water/binder ratio, and so a stronger concrete may be possible using the same cement but a stronger aggregate.

Other mix design methods

If the concrete does not have the desired slump, the superplasticizer dosage is not high enough and must be increased, or else the water dosage has to be increased as well as the cement content in order to keep the same water/binder ratio.

If the concrete experiences a rapid slump loss, the cement is perhaps more reactive than expected and the water dosage has to be increased, or the superplasticizer is particularly inefficient with the selected cement, so its dosage has to be increased or another brand of superplasticizer has to be used.

If the workability of the mix is inadequate, a poor shape or gradation of the coarse aggregate, or the incompatibility of the particular cement-superplasticizer combination can be blamed. In the first case, the amount of coarse aggregate has to be decreased, and in the second case, either the cement or the superplasticizer (or both) should be changed.

The method that has been presented is related to non-air-entrained high-performance concrete. In Chapter 18 it will be seen how to design an air-entrained high-performance concrete (Aïtcin and Lessard, 1994).

8.5 OTHER MIX DESIGN METHODS

Several other methods have been proposed to calculate high-performance mixture proportions and in some cases computerized versions of these methods are commercially available (Welch, 1962; Hughes, 1964; Blick, Petersen and Winter, 1974; Peterman and Carrasquillo, 1986; Haug and Sandvick, 1988; Addis and Alexander, 1990; de Larrard, 1990; Mehta and Aïtcin, 1990; Domone and Soutsos, 1994; Gutiérrez and Cánovas, 1996). In order to show the diversity of the different approaches, the author has chosen to discuss briefly three of them with which he is most familiar; this does not mean than the others are not as valuable as those presented here.

The three methods discussed are that proposed by the ACI 363 Committee on high-strength concrete; that proposed by de Larrard in 1990, which is available in a computerized version known as BETONLAB and which is widely used in France (de Larrard, Gillet and Canitrot, 1996; Sedran and de Larrard, 1996); and the simplified method presented by Mehta and Aïtcin (1990).

8.5.1 Method suggested in ACI 363 Committee on high-strength concrete

The steps of this procedure are as follows:

Step 1: Slump and required strength selection

A table suggests slump values for concretes made with superplasticizers and for those without superplasticizer. The first value of the slump is 25 to 50 mm for the concrete before adding the superplasticizers in order to ensure that sufficient water is used in the concrete.

Step 2: Selection of the maximum size of the coarse aggregate (MSA)

The method suggests using coarse aggregate with an MSA of 19 or 25 mm for concrete made with f'_c lower than 65 MPa and 10 or 13 mm for concrete made with f'_c greater than 85 MPa. The method allows the use of coarse aggregate with an MSA of 25 mm with f'_c between 65 and 85 MPa when the aggregate is of a high quality.

As in the case of usual concrete, the MSA should not exceed one-fifth of the narrowest dimension between the sides of forms, one-third of the depth of slab, or three-quarters of the minimum clear spacing between reinforcing bars, bundled bars or tendons.

Step 3: Selection of coarse aggregate content

This method suggests that the optimum content of coarse aggregate, expressed as a percentage of dry-rodded unit weight (DRUW), can be 0.65, 0.68, 0.72 and 0.75 for nominal size aggregate of 10, 13, 20 and 25 mm, respectively. The DRUW is measured according to ASTM Standard C29 Standard Test Method for Unit Weight and Voids in Aggregate. These values are given for concrete made with a sand of fineness modulus 2.5 to 3.2. The dry weight of coarse aggregate can then be calculated from the following formula: mass of coarse aggregate, dry = (% DRUW) × (DRUW).

Step 4: Estimation of free water and air content

A table gives estimates for the required water content and resulting entrapped air content for concretes made with coarse aggregates of various nominal sizes. These estimated water contents are given for a fine aggregate having a 35% void ratio. If this value is different from 35%, then the water content obtained from the table should be adjusted by adding or subtracting 4.8 kg/m^3 for every 1% increase or decrease in sand air void.

Step 5: Selection of W/B ratio

Two tables suggest W/B values for concretes made with and without superplasticizer, respectively, to meet the specified 28 and 56 day compressive strength. These values are based upon the MSA and f'_c of the concrete.

Step 6: Cement content

The mass of cement is calculated by dividing the mass of the free water by the W/B ratio.

Step 7: First trial mixture with cement

The first mixture to be evaluated can be batched using cement and no other cementitious materials. The sand content is then calculated using the absolute volume method described in the previous method.

Step 8: Other trial mixtures with partial cement replacements

At least two different cementitious material contents are suggested to replace some of the cement mass to produce other trial mixtures that can be batched and evaluated. Maximum cement replacement limits are suggested for fly ash and blast-furnace slag; no limits for silica fume are suggested because this method is valid for a maximum f'_c of 85 MPa. These limits are 15 to 25% of the mass of cement for Class F fly ash, 20 to 35% of mass of cement for Class C fly ash, and 30 to 50% of the mass of cement for blast-furnace slag. Again, the unit mass of sand is calculated using the absolute volume method.

Step 9: Trial batches

The mass of aggregates, along with that of the mixing water, are adjusted for actual moisture conditions, and trial batches are made using concretes made with no cement replacement and others using fly ash or blast-furnace slag. The concretes are then adjusted to meet the desired physical and mechanical characteristics.

8.5.2 de Larrard method (de Larrard, 1990)

This method is based on two semi-empirical mix design tools. The strength of the concrete is predicted by a formula which is in fact an extension of the original Feret's formula (Feret, 1892) where a limited number of mix design parameters are to be used:

$$f_c = \frac{Kg \times R_c}{\left[1 + \dfrac{3.1 \times w/c}{1.4 - 0.4\exp(-11\,s/c)}\right]^2}$$

where

f_c = the compressive strength of concrete cylinders at 28 days;
w, c, s = the mass of water, cement and silica fume for a unit volume of fresh concrete, respectively;
Kg = a parameter depending on the aggregate type (a value of 4.91 applies to common river gravels);
R_c = the strength of cement at 28 days (e.g. the strength of ISO mortar containing three parts of sand for each part of cement and one-half part of water).

The workability is assumed to be closely related to the viscosity of the mix, which is computed using the Farris model. In a mix containing n classes of monodispersed grains of size $d_i > d_i + 1$, the viscosity of the suspension is assumed to be:

$$\eta = \eta_0 H\left(\frac{\Phi_1}{\Phi_1 + \cdots \Phi_n + \Phi_0}\right) H\left(\frac{\Phi_2}{\Phi_2 + \cdots \Phi_n + \Phi_0}\right) \cdots H\left(\frac{\Phi_n}{\Phi_n + \Phi_0}\right)$$

where

Φ_i = the volume occupied by the i-class in a unit volume of mix;
Φ_0 = the volume of water;
η_0 = the viscosity of water;
H = a function representing the variation of the relative viscosity of a monodispersed suspension as a function of its solid concentration.

The idea behind using the Farris model is that each granular class has the same interaction with the mix containing the liquid phase plus the finer classes as with a homogeneous fluid.

From these tools the following assumptions are made:

1. The strength of a concrete made of a given set of components is mainly controlled by the nature of the binding paste.
2. The workability of a concrete, where strength grading is fixed, is assumed to be a combination of two terms: the first depending on the concentration of the binding paste and the second on paste fluidity.

The main idea of the method is to perform as many tests as possible on grouts for rheological tests and mortars for mechanical tests.

The first step consists of proportioning a control concrete containing a large amount of superplasticizer with the amount of cement corresponding to the lowest water demand possible. This amount of water is adjusted to obtain the right workability as measured with a dynamic apparatus.

The second step consists of measuring the flow time of the paste of this control concrete using a Marsh cone.

The third and fourth steps involve the adjustment of the binder composition and of the amount of superplasticizer until the flowing time does not decrease anymore. This amount of superplasticizer is said to correspond to the saturation value.

The fifth step involves the adjustment of the water content to obtain the same flowability as in the control mix.

The sixth step consists of following the variation of the flow time with time. If the flow time increases too much, a retarding agent should be added.

The seventh step corresponds to the prediction of the strength of the high-performance concrete using Feret's formula or the measurement of the compressive strength of different mortars.

Other mix design methods 261

In the eighth step, a high-performance concrete is made and its composition is slightly modified in order to get the targeted workability and strength values.

A sample calculation can be found in de Larrard (1990). A computerized programme has been developed from this method and is currently used in France under the trade name of BETONLAB (de Larrard and Sedran, 1994; de Larrard, Hu and Sedran, 1995; de Larrard, Gillet and Canitrot, 1996; Sedran and de Larrard, 1996).

8.5.3 Mehta and Aïtcin simplified method

Mehta and Aïtcin (1990) proposed a simplified mixture proportioning procedure that is applicable for normal weight concrete with compressive strength values of between 60 and 120 MPa. The method is suitable for coarse aggregates having a maximum size of between 10 and 15 mm and slump values of between 200 and 250 mm. It assumes that non-air-entrained high-performance concrete has an entrapped air volume of 2% which can be increased to 5 to 6% when the concrete is air-entrained. The optimum volume of aggregate is suggested to be 65% of the volume of the high-performance concrete.

The steps of this procedure are as follows:

Step 1: Strength determination

A table lists five grades of concrete with average 28 day compressive strength ranging from 65 to 120 MPa.

Step 2: Water content

The maximum size of the coarse aggregate and slump values are not considered here for selecting the water content since only 10 to 15 mm maximum size are considered and because the desired slump (200 to 250 mm) can be achieved by controlling the dosage of superplasticizer. The water content is specified for different strength levels.

Step 3: Selection of the binder

The volume of the binding paste is assumed to be 35% of the total concrete volume. The volumes of the air content (entrapped or entrained) and mixing water are subtracted from the total volume of the cement paste to calculate the remaining volume of the binder. The binder is then assumed to be one of the following three combinations:

Option 1. 100% Portland cement to be used when absolutely necessary.
Option 2. 75% Portland cement and 25% fly ash or blast-furnace slag by volume.
Option 3. 75% Portland cement, 15% fly ash, and 10% silica fume by volume.

A table lists the volume of each fraction of binder for each strength grade.

Step 4: Selection of aggregate content

The total aggregate volume is equal to 65% of the concrete volume. For strength grades A, B, C, D and E the volume ratios of fine to coarse aggregates are suggested to be 2.00:3.00, 1.95:3.05, 1.90:3.10, 1.85:3.15 and 1.80:3.20, respectively.

Step 5: Batch weight calculation

The weights per unit volume of concrete can be calculated using the volume fractions of the concrete and the specific gravity values of each of the concrete constituents. Usual specific gravity values for Portland cement, Type C fly ash, blast-furnace slag and silica fume are 3.14, 2.5, 2.9 and 2.1, respectively. Those for natural siliceous sand, normal-weight gravel or crushed rocks can be taken as 2.65 and 2.70, respectively. A table lists the calculated mixture proportions of each concrete type and strength grade suggested in this method.

Step 6: Superplasticizer content

For the first trial mixture, the use of a total of 1% superplasticizer solid content of binder is suggested. The mass and volume of a superplasticizer solution are then calculated taking into account the percentage of solids in the solution and the specific gravity of the superplasticizer (for naphthalene superplasticizer a typical value of 1.2 is suggested).

Step 7: Moisture adjustment

The volume of the water included in the superplasticizer is calculated and subtracted from the amount of initial mixing water. Similarly, the mass of aggregate and water are adjusted for moisture conditions and the amount of mixing water adjusted accordingly.

Step 8: Adjustment of trial batch

Because of the many assumptions made in selecting a mixture proportioning, usually the first trial mixture will have to be modified to meet the desired workability and strength criteria. The aggregate type, proportions of sand to aggregate, type and dosage of superplasticizer, type and combination of supplementary cementitious materials, and the air content of the concrete can be adjusted in a series of trial batches to optimize the mixture proportioning.

REFERENCES

ACI 211.1-91 (1993) *Standard Practice for Selecting Proportions for Normal, Heavyweight and Mass Concrete, ACI Manual of Concrete Practice,* Part 1, ISSN 0065-7875, 38 pp.

ACI Committee 211 (1993) *Guide for Selecting Proportions for High-Strength Concrete with Portland Cement and Fly Ash* (ACI 211.4R-93), American Concrete Institute, Detroit.

ACI 363 R-92 (1993) *State-of-the-Art Report on High-Strength Concrete, ACI Manual of Concrete Practice,* Part 1, Materials and General Properties of Concrete, 55 pp.

Addis, B.J. and Alexander, M.G. (1990) *A Method of Proportioning Trial Mixes for High-Strength Concrete,* ACI SP-121, pp. 287–308.

Aïtcin, P.-C. (1971) Density and porosity measurements of solids. *ASTM Journal of Materials,* **6**(2), 282–94.

Aïtcin, P.-C. and Lessard, M. (1994) Canadian experience with air-entrained high-performance concrete. *Concrete International,* **16**(9), October, 35–8.

Aïtcin, P.-C. and Neville, A. (1993) High-performance concrete demystified. *Concrete International,* **15**(1), January, 21–6.

ASTM C29/C29M (1993) *Standard Test Method for Unit Weight and Voids in Aggregate, Annual Book of ASTM Standards,* Section 4 Construction, Vol. 04.02 Concrete and Aggregates, pp. 1–4.

ASTM C127 (1993) *Test for Specific Gravity and Absorption of Coarse Aggregate, Annual Book of ASTM Standards,* Section 4 Construction, Vol. 04.02 Concrete and Aggregates, pp. 65–8.

ASTM C128 (1993) *Test for Specific Gravity and Absorption of Fine Aggregate, Annual Book of ASTM Standards,* Section 4 Construction, Vol. 04.02 Concrete and Aggregates, pp. 70–74.

ASTM C188 (1995) *Standard Test Method for Density of Hydraulic Cement, Annual Book of ASTM Standards,* Section 4 Construction, Vol. 04.01 Cement; Lime; Gypsum, pp. 156–7.

Blick, R.L., Peterson, C.F. and Winter, M.E. (1974) *Proportioning and Controlling High-Strength Concrete,* ACI-SP-46, pp. 141–63.

Day, K.W. (1996) Computer control of concrete proportions. *Concrete International,* **18**(12), December, 48–53.

de Larrard, F. (1990) A method for proportioning high-strength concrete mixtures. *Cement Concrete and Aggregates,* **12**(1), Summer, 47–52.

de Larrard, F. and Sedran, T. (1994) Optimization of ultra-high-performance concrete by the use of a packing model. *Cement and Concrete Research,* **24**(6), 997–1009.

de Larrard, F. and Sedran, T. (1996) Computer-aided mix designs: predicting final results. *Concrete International,* **18**(12), December, 38–41.

de Larrard, F., Hu, C. and Sedran, T. (1995) Best packing and specified rheology: two key concepts in high-performance concrete mixture proportioning, in *Adam Neville Symposium on Concrete Technology* (ed. M. Malhotra), CANMET, Ottawa, pp. 109–27.

de Larrard, F., Gillet, G. and Canitrot, B. (1996) Preliminary HPC mix design study for the Grand Viaduc de Millau: an example of LCPC's approach, in *Proceedings of the 4th International Symposium on the Utilization of High-Strength/High-Performance Concrete,* BHP 96, Vol. 3 (eds F. de Larrard and R. Lacroix), Presses de l'ENPC, Paris, ISBN 2-85978-259-1, pp. 1323–32.

Domone, L.J. and Soutsos, M.N. (1994) An approach to the proportioning of high-strength concrete mixes. *Concrete International,* **16**(10), October, 26–31.

Féret, R. (1892) Sur la compacité des mortiers hydrauliques. Mémoires et documents relatifs à l'art des constructions et au service de l'ingénieur. *Annales des Ponts et Chaussées,* Vol. 4, 2nd semester, pp. 5–161.

Ganju, T.N. (1996) A method for designing concrete trial mixes. *Concrete International*, **18**(12), December, 35–8.

Gutiérrez, P.A. and Cánovas, M.F. (1996) High-performance concrete: requirements for constituent materials and mix proportioning. *ACI Materials Journal*, **93**(3), May–June, 233–41.

Haug, A.K. and Sandvik, M. (1988) *Mix Design and Strength Data for Concrete Platforms in the North Sea.* ACI SP-109, pp. 495–524.

Hughes, B.P. (1964) Mix design for high-quality concrete using crushed rock aggregates. *Journal of the British Granite and Whinstone Federation* (16 Berkeley Street, London, W1), **4**(1), Spring, 20 pp.

Lessard, M., Baalbaki, M. and Aïtcin, P.-C. (1995) Mix design of air-entrained high-performance concrete, in *Concrete Under Severe Conditions – Environment and loading*, Vol. 2 (eds K. Sakai, N. Banthia and O.E. Gjørv), E & FN Spon, London, ISBN 0-419-19860-1, pp. 1025–31.

Lessard, M., Gendreau, M., Baalbaki, M., Pigeon, M. and Aïtcin, P.-C. (1993) Formulation d'un béton à hautes performances à air entraîné. *Bulletin de Liaison des Laboratoires des Ponts et Chaussées*, No. 188, Nov.-Déc., Réf. 3982, pp. 41–51.

Mehta, P.K. and Aïtcin, P.-C. (1990) Principles underlying production of high-performance concrete. *Cement, Concrete and Aggregates*, **12**(2), Winter, 70–78.

Peterman, M.B. and Carrasquillo, R.L. (1986) *Production of High-Strength Concrete*, Noyes Publications, Park Ridge, New Jersey, USA, ISBN 0-8155-1057-8, 278 pp.

Popovics, S. and Popovics, J.S. (1996) Novel aspects in computerization of concrete proportioning. *Concrete International*, **18**(12), December, 54–8.

Rougeron, P. and Aïtcin, P.-C. (1994) Optimization of the composition of a high-performance concrete. *Cement, Concrete and Aggregates*, **16**(2), 115–24.

Sedran, T. and de Larrard, F. (1996) René-LCPC: A software to optimize the mix-design of high-performance concrete, in *Proceedings of the 4th International Symposium on the Utilization of High-Strength/High-Performance Concrete*, BHP 96, Vol. 1 (eds F. de Larrard and R. Lacroix), Presses de l'ENPC, Paris, ISBN 2-85978-258-3, pp. 169–78.

Welch, G.B. (1962) Adjustment of high-strength concrete mixes. *Constructional Review*, **35**(8), August, 27–30.

CHAPTER 9

Producing high-performance concrete

9.1 INTRODUCTION

We have seen in the preceding chapter that making high-performance concrete starts with the careful selection of the materials that are to be used. The materials that are presently used to make usual concrete with a compressive strength in the 20 to 40 MPa range may be inappropriate as soon as the water/binder ratio is lowered in order to achieve higher strength. It has also been shown how the selection of the materials used to make a high-performance concrete becomes more and more critical as the water/binder ratio is decreased further and compressive strength increases, but usually a concrete producer still has some choices to make.

On the other hand, when considering the next steps in the use of high-performance concrete – which are producing, delivering and placing high-performance concrete – there is almost no choice: high-performance concrete must be produced, delivered and placed in the same way as usual concrete because its present share of the concrete market is too small to justify a special batching plant or delivery and finishing equipment, except perhaps in special projects, where a new batching plant has to be bought. There is definitely no payback at the present time for any capital investment in modifying or constructing any new piece of equipment to produce, deliver and place high-performance concrete. At the most, in places where the high-performance concrete market is significant, some ready-mix plants occupying strategic locations can be partially devoted to the high-performance concrete market, but in any case the present market for high-performance concrete is not large enough for anyone to make a living by selling only high-performance concrete. Therefore, most of the time high-performance concrete has to be made in a ready-mix plant that is simultaneously producing usual concrete for regular customers, and it has to be transported and placed with the same equipment that is used for usual concrete delivery. By doing this, the extra cost of high-performance concrete will come from the extra cost of the materials used to make it, essentially the cost of the extra cement and superplasticizer, the cost of the more stringent quality control programme that has to be implemented to check not only the quality of the raw materials used, but also the quality of the fresh and hardened high-performance concrete, and finally the expenses related to the cooling or heating of the mix when applicable.

Producing high-performance concrete in a regular ready-mix plant in parallel with normal strength concrete can influence the materials selection process, owing to limitations in storage facilities and the availability of silos and bins. Depending on the number of silos available to store cement and supplementary cementitious materials, or bins to store fine and coarse aggregates, it will be more or less easy to make high-performance concrete with the best available materials in a given area. Of course, if the high-performance concrete producer is producing a large volume of usual concrete, he or she has some bargaining power with suppliers to get their best products at a competitive price. The producer can ask, for example, for a special or modified cement, and the cement supplier will be happy to deliver such a cement, in order to keep such a good customer who is producing a large volume of concrete.

The use of a supplementary cementitious material to make high-performance concrete can also be dictated by the storage facilities of the ready-mix plant. In this respect, there is hope, because in the future more and more ready-mix plants will be using one or two supplementary cementitious materials in their regular production of usual concrete or different blended cements containing some supplementary cementitious materials.

If there are two bins to store fine aggregate, one can be devoted to a coarser sand for high-performance concrete, but if there is only one, high-performance concrete will have to be produced with the sand used to make usual concrete.

It is usually easier to use a particular coarse aggregate to make high-performance concrete. On the one hand, almost all ready-mix plants have more than one bin to store coarse aggregate, and on the other hand, high-performance concrete is usually made with a strong and small-sized coarse aggregate. Therefore, the bin usually used to store the fine fraction of the coarse aggregate can be used to handle the selected small-sized preferred aggregate. This also explains why high-performance concretes are often made with only one small-sized coarse aggregate.

Experience shows, however, that it is not always easy to produce high-performance and usual concrete simultaneously. In some cases (the Two Union Square building (Howard and Leatham, 1989) and Jacques-Cartier bridge in Sherbrooke (Blais *et al.*, 1996)), it was found particularly convenient and economical for the contractor to cast high-performance concrete during the night. During a night production, it is also possible to get more flexibility in the use of the aggregate bins.

In the case of the construction of Two Union Square in Seattle, USA, delivering during the night the 120 MPa high-performance concrete according to a tight schedule was much easier than in the daily traffic jams during rush hours. It allowed the placing crew to pump the 120 MPa concrete without any schedule problems in four big steel pipes (Ralston and Korman, 1989). The author has been told that in some peak night shifts 1000 to 1200 m^3 of high-performance and usual concretes could be placed. Moreover, in order to buy peace with the community living around the ready-mix plant for the great disturbance due to noise produced by the delivery trucks during night placings, the contractor and the ready-mix producer offered to develop, at their own expense, a playground for the children of the community. The payback of this extra expense was rapidly recovered due to the speed of all concreting operations in the nightly deliveries.

In the case of the reconstruction of the Jacques-Cartier bridge in Sherbrooke (Blais et al., 1996), the main problem faced by the contractor was the too low production rate of the specified high-performance concrete at the ready-mix plant. This particular ready-mix plant usually had a peak production rate of 60 m^3 per hour of usual concrete, but it could not produce more than 30 m^3 per hour of high-performance concrete, owing to the longer mixing sequence necessary to achieve a good spacing factor of the air-entrained high-performance concrete, as will be explained later on. Therefore, 10 full hours were necessary to complete the usual scheduled pours of 300 m^3 of concrete during the night. On one occasion, to gain some time on the overall construction schedule, a smaller pour of 100 m^3 was scheduled during the daytime and it took 8 hours for the eight men of the placing crew to place the concrete. Most of the time, the crew was awaiting the next truck delivery because the ready-mix producer was simultaneously producing usual concrete for its usual customers. Placing high-performance concrete at night is also very advantageous for early curing and heat development, as will be seen later.

9.2 PREPARATION BEFORE MIXING

Concrete mixing, and more critically high-performance concrete mixing, does not start with the mixing of the ingredients in the concrete mixer, but rather with the control of the quality of the materials used. Although providing good quality control is important in making usual concrete, it is a crucial step in the manufacturing of high-performance concrete. The lower the water/binder ratio and higher the targeted strength, and the more difficult the job conditions, the more important is quality control. It was shown in Chapter 7 that the selection of the materials to be used in making high-performance concrete is the result of quite an elaborate process. Therefore, it is essential to be sure that the actual materials used to make the high-performance concrete have the same properties as the ones chosen during the selection process (Howard and Leathan, 1989; Keck and Casey, 1991).

It is particularly important to check carefully each delivery of superplasticizer, using a reference cement and any one of the simple tests proposed in Chapter 7. It is also important to test the cement–superplasticizer compatibility as frequently as possible using a reference superplasticizer and the same simple tests.

The author is aware of a particular concrete producer who performs a rheological test before the unloading of each cement tank he receives; it takes less than 10 minutes. As the cement supplier knows that the cement is carefully checked by his customer, and how it is checked, the supplier performs the same test before the cement load leaves the cement plant. Since such a control of the rheological behaviour of the cement has been implemented, not a single cement load has been refused and no cases of flash set, false set or any cement–superplasticizer compatibility problems have occurred. This quality control programme has been also implemented by the superplasticizer supplier, with the same success. The story does not tell who gets the cement and superplasticizer loads that don't meet the requirements!

When the quality of the cement and the superplasticizer is well under control, the remaining critical parameters that have to be checked are the grain size distribution and the shape of the coarse aggregate, as well as the grain size distribution of the sand and its water content.

It is important to check carefully the grain size distribution of the coarse aggregate, and more particularly the shape of its particles, since large quarries often have more than a single production unit. As a result, quarries do not always produce aggregates of the same quality, grain size distribution or shape, even though they are sold at the same price. Moreover, during the summer season, when aggregates are in high demand, some quarries increase their production using second-class equipment which is not always adequate for producing high-quality aggregate for concrete.

It is also important to measure carefully the water content of the sand and to adjust the sand content as well as the amount of mixing water introduced during batching. Checking very carefully the water content of the sand and the coarse aggregate is also crucial in winter time, when, in northern countries such as Canada, the aggregates are heated with steam. A great deal of effort goes into obtaining every bit of strength from the concrete, and hence it is a waste not to account for the extra amount of water contained in the wet sand if it is not carefully checked. Measuring the water content of the sand is sometimes carried out for controlling the quality of usual concrete; however, when making high-performance concrete it is absolutely imperative that this is done. Moreover, it is also essential to check that there is no washing water left in the transit mixer before loading high-performance concrete, as this is frequently done by truck drivers.

If there are supplementary cementitious materials that are to be used to produce high-performance concrete, they must be carefully controlled with the same attention as for the previous materials.

Mixing

It is only when the concrete producer is sure that his or her aggregate bins, cement and supplementary cementitious materials silos, and admixture tanks contain the high-quality materials selected and paid for that the production of high-performance concrete can be started. Any lack of quality in one of the ingredients used to make high-performance concrete will rapidly cause a problem, because when making high-performance concrete the safety margin is not large. Moreover, the mix proportioning cannot accommodate too much variation from its optimum intended state because all the ingredients are working close to their limit.

9.3 MIXING

As stated previously, high-performance concretes are produced in the same way as usual concretes, using the same production equipment, except that the mixing sequence is usually longer. Obviously, all the equipment used to weigh and batch concrete ingredients must be accurate. For example, weighing devices must be calibrated regularly because it is essential that the carefully selected and controlled materials be weighed precisely in order consistently to obtain the targeted strength and workability. High-performance concrete mixtures are very sensitive to any variation in their proportions, especially in water content. An increase of 3 to 5 l of mixing water per cubic metre of concrete can also represent a decrease of 10 to 20 MPa in the compressive strength.

The superplasticizer dispensing device must also be very accurate because any variation in the superplasticizer dosage can result in a slump loss problem or in a segregation problem or a retardation problem.

High-performance concretes of classes I and II, with f'_c between 50 and 100 MPa, have been produced successfully in dry-batch plants. In the case of Scotia Plaza (Ryell and Bickley, 1987), the average compressive strength measured at 91 days from 142 samplings was 93.6 MPa, even though the specified characteristic strength was 70 MPa. In that particular case, the transit mixers were loaded in two steps in order to obtain a very homogeneous mixture. The field results were excellent in spite of the use of so simple a batching method. The standard deviation measured on the 142 sets of tests was 6.8 MPa, which represents a coefficient of variation of 7.3%. However, it is easier to produce high-performance concrete in a ready-mix plant equipped with a central mixer (Woodhead, 1993; Hoff and Elimov, 1995). This mixer can be of the tilt type or the horizontal pan type, with or without counter current-mixing paddles. In the case of the Hibernia project (Hoff and Elimov, 1995), concrete was produced in two twin horizontal mixers, each having a capacity of 2.5 m^3. In the case of the fixed link between Prince Edward Island and New Brunswick in Canada a 6 m^3 tilt mixer was used.

Mixing times are usually longer for high-performance concretes than for usual concretes, but it is difficult to give specific rules. The mixing time has to be adjusted on a case-by-case basis. Mixing is optimized so that any further increase in mixing time does not increase the homogeneity or the workability of the concrete (Ronneberg and Sandvik, 1990; Hoff and Elimov, 1995; Blais *et al.*, 1996). There is still one point of controversy in the domain of mixing, which is when and how the superplasticizer has to be introduced into the mix. There are presently three schools of thought, each pretending that their approach is better and more efficient than the other two. These approaches are briefly described along with their advantages and inconveniences:

- First approach: All of the superplasticizer is added at the same time to the mixture. The supporters of this approach claim that this is the easiest way to achieve simple and full control of the superplasticizer dosage. It is the simplest, it reduces mixing time and increases the output of the high-performance concrete rate, but it can cost more because the superplasticizer dosage necessary to achieve the desired slump in the field is somewhat higher because some of the superplasticizer molecules can be used to block C_3A hydration rather than calcium sulfate.
- Second approach: About two-thirds of the superplasticizer is added to the mixture at the beginning of the mixing, and the last third at the very end of the mixing period (Ronneberg and Sandvik, 1990). The supporters of this mode of introduction (also called the 'double introduction mode') say that this procedure results in some savings in the total amount of superplasticizer needed to reach a certain workability. They say that it is better to let very early ettringite be formed from the sulfate of the cement rather than have some superplasticizer molecules consumed during early batching on some reactive sites of the cement particles. Even though this point can be easily verified in the laboratory, where quite long mixing times of about 10 minutes are usually used, it is not clear that it is still true in industry when the batching times last less than 2 to 3 minutes.
- Third approach: Part of the superplasticizer is added during the batching so that the high-performance concrete leaves the batching plant with a slump of around 100 mm and, when it arrives at the jobsite it still has a slump of at least 50 mm. The remaining superplasticizer is then added to acquire the targeted slump. The supporters of this approach claim that the transportation of a liquid concrete (such as the high-performance concrete obtained with the first two approaches) is problematic (danger of spills whenever the concrete transit mixer brakes) or even dangerous when the transit mixer arrives on uneven terrain at the jobsite. However, the supporters of the first two approaches argue that in order to achieve proper mixing at the jobsite, the transit mixer cannot receive a full load when it leaves the batching

plant. They also argue that the final adjustment of the slump at the jobsite is not always done under the best controlled conditions.

When this last approach is taken, it is convenient to dilute the delayed part of the high solid content superplasticizer in an equal amount of mixing water, which should be subtracted from the initial amount of mixing water during batching.

Whatever the approach that is followed to introduce the superplasticizer, it should be mentioned that the slump of the high-performance concrete should not be higher than 230 mm. High-performance concretes with higher slump values are prone to segregate, unless their composition has been adjusted accordingly.

Moreover, in any case water retempering should not be permitted (Cook, 1989; Keck and Casey, 1991; Aïtcin and Lessard, 1994; Lessard, Baalbaki and Aïtcin, 1995; Blais et al., 1996).

9.4 CONTROLLING THE WORKABILITY OF HIGH-PERFORMANCE CONCRETE

The slump test is the test used to characterize the workability of usual concrete. It has been also used up to now to characterize high-performance concrete workability, though its validity has been questioned. Other methods of evaluation of concrete workability have been proposed, but they are often not very practical for field applications (Hu and de Larrard, 1995; Hu, de Larrard and Gjørv, 1995; Hu et al., 1996).

From a theoretical point of view, it has been proposed that concrete workability be characterized by two parameters: the internal shearing resistance and the plastic viscosity. Recently, Hu et al. (1996) made interesting and useful contributions to the rheology of high-performance concrete. These authors tried to develop a transportable (in a small van) apparatus that can be used in the field. However, it is difficult to foresee in the short run that this apparatus will be substituted for the conventional slump test for measuring concrete workability owing to its high cost.

What makes the measurement of the slump of a high-performance concrete somewhat difficult to evaluate precisely is the fact that very low W/B mixtures tend to be very cohesive, if not sticky, and in some cases at least quite viscous, so the concrete cone collapses progressively and does not seem to stop collapsing. Often, after a while, the concrete spreads out beyond the standard plywood base on which the slump test is performed, so it is difficult to make a precise measurement of the slump and give a fair value for it.

It has always been difficult to characterize the workability and the slump of flowing concrete, and the problem is the same for high-performance concrete. For usual concrete it seems that a flow measurement works quite well and perhaps this measurement should be adapted for high-performance concrete. The test is simple and inexpensive, it can be performed easily in the field and it has proven effective at least for flowing concrete; therefore, it seems likely that it can be adapted for testing high-performance concrete in the field.

9.5 SEGREGATION

The causes of segregation are many, from the presence of some washing water left within the ready-mix drum by the truck driver, to an error in the water or superplasticizer dosage or a sudden increase in the water content of the sand. Segregation of a high-performance concrete mixture can be produced by an accidental increase in the superplasticizer dosage far beyond the saturation point of that particular cement for the particular selected water/binder ratio.

Segregation is not an easy phenomenon to study from either an experimental or a theoretical point of view, but what is sure is that it occurs and that it can harm the quality of concrete. The risk of segregation must be minimized. Generally speaking, increasing the viscosity of the concrete enhances its stability and hence reduces segregation. This can be achieved by reducing the content of mixing water and/or superplasticizer, by adding more silica fume or by adding viscosity-enhancing admixtures, such as colloidal antiwashout admixtures.

From a practical point of view, if no special measures to avoid segregation are taken, it is not recommended using a slump higher than 220 mm, but when optimizing very carefully the composition of concrete, a slump as high as 250 mm has been successfully used (Hoff and Elimov, 1995).

9.6 CONTROLLING THE TEMPERATURE OF FRESH CONCRETE

The control of the temperature of fresh concrete is very important in the case of high-performance concrete, because temperature has major effects on its rheology. If the temperature of the concrete just after mixing is too high, say above 25 °C, hydration is accelerated and it can be difficult to maintain the mix in a workable condition to ensure proper delivery and placing, except if the mix composition is modified to take into account this high initial temperature. Moreover, when the temperature of the concrete is too high, it can be difficult to keep a close control over the entrained air for air-entrained mixes.

As will be shown in Chapter 14, the initial temperature of the fresh concrete is a very important factor in determining the maximum temperature reached within the different structural elements.

On the other hand, if the mix is too cold, say below 10 °C, it must be remembered that liquid superplasticizers are less effective in dispersing

Controlling the temperature of fresh concrete

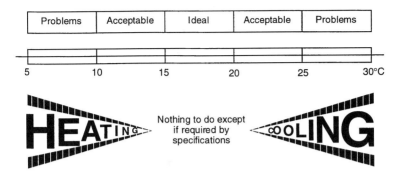

Fig. 9.1 Controlling the temperature of fresh high-performance concrete.

cement particles because their viscosity increases drastically as their temperature decreases. Moreover, as low temperature slows down hydration, the early strength of high-performance concrete may not increase quickly enough, which could result in a costly slowdown of the whole construction process. This situation is even worse when supplementary cementitious materials are used as partial replacements of Portland cement. Figure 9.1 represents schematically the different means that have to be used in order to improve the situation. In the following we will examine some technical means that can be used to adjust the temperature of the mix in order to avoid such problems.

The ideal temperature of a fresh high-performance concrete when it is delivered is between 15 and 20 °C, as in the case of usual concrete, as shown in Figure 9.1. Serious corrective action must be taken if the temperature of the fresh concrete falls below 10 °C or is higher than 25 °C in order to avoid serious rheological problems. In between these limits, it is always beneficial to take action in order to bring the temperature of the fresh concrete within the 15 to 20 °C range. **The author strongly recommends when it is possible that such a temperature range be specified, indicating clearly the minimum and maximum temperature of the fresh concrete that is acceptable.**

9.6.1 Too cold a mix: increasing the temperature of fresh concrete

This situation occurs in northern countries during winter, when it is particularly important to deliver a concrete that can develop enough strength to permit removal of the forms as rapidly as possible or to maintain an economical slipforming rate. It is quite easy to overcome such a situation, because it is easy to heat high-performance concrete using the same techniques used to heat usual concrete, i.e. the water and the aggregates can be heated. In spite of the fact that high-performance concrete contains less water than usual concrete, heating water should be used as a first attempt to increase the temperature of the mix. Ready-mix concrete plants are already equipped to do this. If this is not enough, the heating of the water is carried out simultaneously with the heating of the aggregates with steam, but as said previously, a tight control of the water content of the sand and the coarse aggregate is imperative.

During transportation and placing, concrete loses some heat but not much. In any case this temperature decrease during transportation can easily be reduced, and corrections taken at the batching plant so that the high-performance concrete arrives at the jobsite with a temperature within specifications. As soon as the concrete is placed in the forms, it is not generally necessary to take any special precautions regarding temperature, because when hydration starts it develops enough heat to protect the concrete from the detrimental effects of too low a temperature as far as strength increase is concerned. One case that can be problematic is when a too cold, retarded concrete is placed in steel forms when the ambient temperature is very low. In such a case the concrete just behind the forms can freeze. If the metallic forms are insulated or protected with covers, there should not be any problem. A second case that can be problematic is when a slab is cast in winter when the ambient temperature is low. The temperature of the top of the slab can decrease during the first hours after placing and before hydration starts, delaying the rate of hydration (Lessard *et al.*, 1993, 1994).

9.6.2 Too hot a mix: cooling down the temperature of fresh concrete

The temperature of fresh concrete, especially during summer, must be decreased in order to fulfil the maximum temperature requirement of the specifications (Woodhead, 1993; Hoff and Elimov, 1995; Blais *et al.*, 1996). This was also the case for the high-performance concrete used to build Scotia Plaza in Toronto (Ryell and Bickley, 1987), where the imposed maximum temperature was 18 °C. Most of the high-performance concrete bridges built in Québec up to now required a maximum delivery temperature of 20 °C in summer.

In the case of Scotia Plaza, midsummer conditions for the materials prior to mixing were the following: cement 43 °C, slag 36 °C, fine aggregate 32 °C, 10 mm coarse aggregate 32 °C and water 12 °C, so it was absolutely necessary to cool down the temperature of the fresh concrete produced. This was done by adding some liquid nitrogen, as shown in Figure 9.2, which resulted in a cost increase of $35/m^3 (Canadian $).

Usually, whenever a concrete producer wants to decrease the temperature of fresh concrete, the first attempt made is to use chilled water. If the temperature of the fresh concrete is just over 20 °C, cooling down the aggregate with water can be sufficient to maintain the temperature of the concrete at around 20 °C. If the temperature has to be decreased more than that, it is recommended using some crushed ice (not ice cubes) instead of water to reduce the temperature of the mix. An

Controlling the temperature of fresh concrete

Fig. 9.2 Liquid nitrogen cooling of the high-performance concrete used to build Scotia Plaza in Toronto.

adequate mass of water can be replaced by an equivalent mass of crushed ice that will melt rapidly and reduce the temperature of the fresh concrete; ice cubes do not melt fast enough, so there would be a lack of mixing water when the concrete is batched. This method is perfectly applicable to high-performance concrete; it is possible to replace any fraction of the mixing water by an equivalent mass of crushed ice without creating any workability problems. This method was used in two projects in Québec: the Montée St-Rémi overpass construction, which involved an overnight non-stop pour of 450 m^3 of concrete on 23 June 1993, when the maximum daily temperature recorded was 28 °C (Lessard, Baalbaki and Aïtcin, 1995), and during the Jacques-Cartier bridge deck reconstruction in Sherbrooke, where the 1000 m^3 of high-performance concrete out of the 1800 m^3 of the total project were cast in August and September (Blais et al., 1996).

In both cases the use of crushed ice made it possible to maintain the temperature of the fresh concrete below 20 °C without any problems. In both cases, as the ambient temperature was decreasing during the night, the amount of ice added decreased from 60 to 20 kg per cubic metre of concrete for an average of 40 kg of ice per cubic metre of concrete. The average temperature of the Montée St-Rémi high-performance concrete was 17 °C for 74 deliveries and that for the Jacques-Cartier bridge was 18 °C for 300 deliveries. As previously mentioned, in both cases the cooling of concrete using crushed ice cost a little less than $10/m^3 of concrete (Canadian $).

One more advantage of cooling high-performance concrete is that it has been experienced that a decrease in temperature improves the workability, prolongs the high workability life of the concrete and improves pumpability (Haug and Sandvik, 1988).

However, in this effort to lower the temperature of the fresh high-performance concrete as much as possible, it should be mentioned that decreasing it too much, below 10 °C, could have a harmful effect, because as the temperature of the freshly mixed concrete decreases, so too do the dispersing and water reducing effects of the superplasticizer. At 5 °C the naphthalene superplasticizers are very viscous and are on the verge of crystallization, which they do at -4 °C. Therefore, sometimes, particularly in the winter season, a minimum temperature has to be specified in order to prevent any slump loss problems or any freezing damage due to slow hardening. For example, in winter during the construction of Scotia Plaza in Toronto, Portneuf bridge and the Jacques-Cartier bridge deck in Sherbrooke in Canada, it was specified that the concrete should have a temperature of at least 23 °C. Such a temperature was obtained, as in the case of all usual concretes delivered in Canada during winter, by heating the water and the aggregates. For the Portneuf bridge, built in November 1992 when the maximum ambient temperature was 3 °C, the average temperature of the 27 deliveries of high-performance concrete was 22 °C with a standard deviation of 2.5 °C.

It should also be mentioned that keeping a tight control of the delivery temperature of an air-entrained high-performance concrete helps in controlling the amount of entrained air and the compressive strength of the delivered concrete owing to the strong influence of the amount of air on compressive strength. As an illustration of a successful production of high-performance concrete, the production of the concrete used to build the Jacques-Cartier bridge deck in Sherbrooke in 1995 is described in section 9.8 (Blais *et al.*, 1996).

9.7 PRODUCING AIR-ENTRAINED HIGH-PERFORMANCE CONCRETES

The production of air-entrained concrete implies the use of an air-entraining agent and this complicates the process of adjusting the dosage of the different admixtures because entrained air affects concrete workability and compressive strength (Ronneberg and Sandvik, 1990).

Based on the experience gained at the Université de Sherbrooke through different field projects, it has been found that it is advantageous to develop the mix according to the four-step method presented here (Lessard *et al.*, 1993; Lessard, Baalbaki and Aïtcin, 1995):

Step 1: Evaluation of the cement–superplasticizer compatibility on grouts using the minislump or Marsh cone method.

Step 2: Achieving an increased compressive strength on a non-air-entrained concrete.

Step 3: Developing an adequate bubble system in the preceding concrete.

Step 4: Verifying the spacing factor or the freeze–thaw and scaling resistance of the air-entrained concrete produced.

Of course, step 1 and step 2 are the two steps that are used when designing any non-air-entrained concrete mixture.

The first question that has to be solved before going further with this method is which strength has to be targeted when making the non-air-entrained high-performance concrete?

Based on the experience obtained during the construction of Portneuf bridge (Aïtcin and Lessard, 1994), Montée St-Rémi viaduct (Lessard, Baalbaki and Aïtcin, 1995) and Sherbrooke Jacques-Cartier bridge (Blais et al., 1996), the following rule of thumb was found valid: a 1% variation in air content when the air content is between 4 and 7% results in a variation in the opposite direction of the compressive strength of 5%. Therefore if a represents the average air content necessary to obtain the right spacing factor and p represents the amount of entrapped air found in a non-air-entrained high-performance concrete, the compressive strength must be increased by $5(a - p)/100$. The required compressive strength in the non-air-entrained concrete, f'_{na}, is then expressed as a function of the air entrained compressive strength, f'_a:

$$f'_{na} = f'_a \left[1 + \frac{5}{100}(a - p) \right]$$

As an example, suppose that the necessary air content is $a = 6\%$ and that $p = 2\%$. In this case it will be necessary to make a non-air-entrained concrete having a compressive strength 20% higher than that of the air-entrained high-performance concrete that has to be made.

Of course it will be necessary to increase this strength to take into account the variability of the concrete production in order to obtain the average strength that the concrete producer will have to maintain. For example, if the acceptance criterion is that the average of three consecutive tests be higher than the specified strength 99% of the time, the average strength of the non-air-entrained concrete, f'_{cr}, that has to be produced can be calculated as follows:

$$f'_{cr} = f'_{na} + \frac{2.33\,\sigma}{\sqrt{3}}$$

where σ represents the standard deviation of the production.

In order to find rapidly the exact dosage of the air-entraining admixture when proceeding to step 3 the author uses the 'three dosages method'. Three mixes are made, with one purposely having a low air-entraining admixture dosage, one the dosage recommended by the manufacturer and the third with a purposely higher dosage so as to be sure that the targeted value of entrained air or spacing factor will be between the ones obtained in two of the three trial batches. It is always safer to interpolate rather than extrapolate. The dosage found is used in a fourth batch to verify the validity of the interpolation and to fine-tune the superplasticizer dosage, because owing to the presence of the entrained air, the workability of the high-performance concrete may have changed.

Samples of these four concretes having different air contents and different spacing factors can be submitted to ASTM C666 Procedure A, Test Method for Resistance of Concrete to Rapid Freezing and Thawing, and ASTM C672 Test Method for Scaling Resistance of Concrete Surfaces Exposed to Deicing Chemicals, to find the necessary air content and spacing factors to meet the requirements of these two standards.

9.8 CASE STUDIES

9.8.1 Production of the concrete used to build the Jacques-Cartier bridge deck in Sherbrooke (Blais *et al.*, 1996)

This bridge deck was demolished and reconstructed between 5 August and 26 November 1995. In August, the maximum ambient temperature during the day that preceded the first pour was 28 °C and during the last November pour it was − 5 °C. In spite of such a wide range of ambient temperatures, it was possible to keep a tight control of the delivered concrete temperature.

(a) Concrete specifications

The specifications of the Ministry of Transportation of Québec for the high-performance concrete to be used were:

- maximum water/binder ratio of 0.35;
- 28 day design strength of 60 MPa;
- minimum compressive strength of 10 MPa at 24 hours;
- silica fume blended cement (7.5 to 8.5% silica fume);
- minimum cement content of 340 kg/m^3;
- maximum size of the coarse aggregate 20 mm;
- the concrete had to incorporate a set retarder;
- slump 180 ± 40 mm;
- air content 4 to 7%;
- average spacing factor not greater than 230 μm without any value exceeding 260 μm.

(b) Composition of the high-performance concrete used

The composition of the high-performance concrete used to reconstruct the bridge deck is presented in Table 9.1.

Case studies

Table 9.1 Composition of the high-performance concrete used to rebuild the Jacques-Cartier bridge in Sherbrooke (Blais et al., 1996)

Water/binder ratio	0.32	
Water*	90 ⎫ 150	
Ice (in summer conditions)	60 ⎭	
Silica fume blended cement	470	kg/m³
Sand (SSD)	740	
Coarse aggregate (SSD)	1 050	
Superplasticizer	5.0	
Air-entraining agent	0.315	l/m³
Water reducer**	1.4	

*Including the water of the superplasticizer. **Used as a retarder.

(c) Mixing sequence

The mixing sequence presented in Figure 9.3 was found to be satisfactory for producing a high-performance concrete meeting the specifications. It was not obtained through any optimization process but rather by modifying the sequence used at the plant until satisfactory results were obtained. The mixing sequence was decomposed in the following steps:

1. Loading of the aggregates and water (0 to 20 seconds);
2. Addition of the air-entraining agent and mixing to develop a satisfactory air bubble system and stabilizing it (20 to 80 seconds);
3. Introduction and mixing of the cement and stabilizing it (80 to 140 seconds).
4. Addition of the water reducer to act as a slight retarder (140 to 210 seconds).
5. Addition of the superplasticizer and final mixing (200 to 310 seconds).

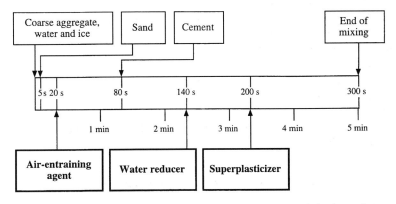

Fig. 9.3 Mixing sequence used to make the air-entrained high-performance concrete of the Jacques-Cartier bridge in Sherbrooke.

Table 9.2 Results of the quality control of the air-entrained high-performance concrete used to rebuild the deck of the Jacques-Cartier bridge in Sherbrooke (Blais et al., 1996)

$n = 56$	Mean value	Standard deviation
Slump	190 mm	25 mm
Air content	5.4%	0.7%
Unit mass	2350 kg/m^3	20 kg/m^3
Temperature	18 °C	1.5 °C
f'_c at 28 days	76.3 MPa	4.1 MPa

The selected mixing sequence and quality control worked well. Of the 300 deliveries, only three were rejected because they did not comply with the specifications: one because the slump was too high, one because the air content was too high (the water reducer valve leaked) and one because big chunks of prehydrated cement were found in the truck mixer when it started its unloading. Table 9.2 presents the average properties and standard deviation of 56 deliveries that were tested by the students and technicians of the Université de Sherbrooke. The acceptance strength criterion was that the average of three consecutive compressive strength tests could be lower than the 60 MPa specified strength only 1% of the time; it can be calculated that the specified compressive strength provided was 70.8 MPa, which was well above the Ministry of Transportation requirement.

(d) Economic considerations

In spite of this successful production of an air-entrained high-performance concrete under the different ambient conditions that were experienced throughout the project, the author is convinced that the mixing sequence could have been improved from an economic point of view. A water reducer was used by the concrete producer for two reasons: first, to retard the mix slightly, and secondly to save about $1/m^3 by replacing 1 litre of superplasticizer by 1 litre of water reducer.

Such a modified mixing sequence represented a saving of 1800 dollars for the concrete producer for the whole project.

Now let us look at the contractor side. The simultaneous addition of a retarder when the cement was added instead of the use of the water reducer could have saved 60 seconds of mixing time and resulted in a higher placing rate for the contractor. The placing crew was able to place more than 30 m^3 per hour and was always waiting for the placing of the next delivery. A rough estimate of the cost of these idle periods of the placing crew was estimated by the contractor at the end of the project to have been around $10/m^3 of placed concrete. Therefore, the contractor could have saved $18 000 for the whole project if the concrete producer had adopted a modified mixing sequence 1 minute shorter.

Case studies   281

The concrete producer was happy to save 2000 dollars but the contractor overspent 18 000 dollars because the final consequence of the use of the water reducer instead of the retarder was not evaluated in a global manner. This poor financial assessment is a direct consequence of the too fractionated nature of the construction industry. The contractor should have paid the concrete producer 2000 dollars for the extra superplasticizer in order to save 16 000 dollars.

9.8.2 Production of a high-performance concrete in a dry-batch plant

(a) Portneuf bridge (Lessard et al., 1993)

The following mixing sequence was used:

- 6 litres of water were used to dilute the admixtures in order to facilitate their hand introduction into the truck mixer;
- half the amount of superplasticizer diluted in an equal volume of water was introduced into the empty drum before the introduction of the aggregates;

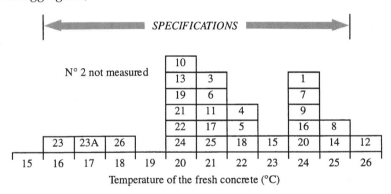

Fig. 9.4 Temperature of the delivered concretes during Portneuf bridge construction.

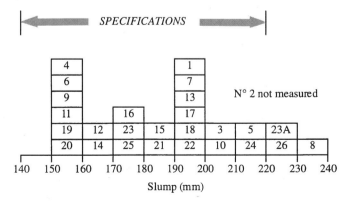

Fig. 9.5 Slump of the delivered concretes during Portneuf bridge construction.

- the aggregates were then added, followed by the cement and the rest of the mixing water;
- the second half of the superplasticizer, the air-entraining agent and retarder were added after the introduction of all the mixing water.

It must be emphasized that this mixing sequence was not the fruit of an optimization process, but it had the merit of working and of producing a well-controlled air-entrained high-performance concrete, as shown in Figures 9.4 to 9.7, which contain the results of a statistical analysis performed on the concrete delivered for this specific project.

(b) *Scotia Plaza building (Ryell and Bickley, 1987)*

The silica fume was used in an as-produced form and batched in the same way as cement. In order to avoid the formation of large lumps during mixing, about half the mix water was introduced in the truck before the introduction of any solid. By adopting such a loading

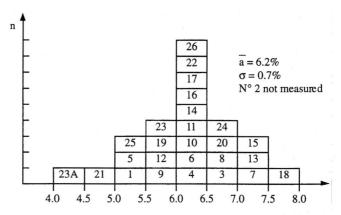

Fig. 9.6 Air content of the delivered concretes during Portneuf bridge construction.

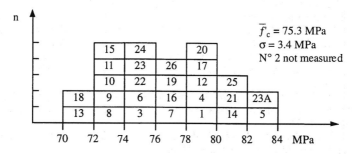

Fig. 9.7 28 day compressive strength of the delivered concretes during Portneuf bridge construction.

sequence very few ballings occurred and no deleterious effect was found on either the strength of standard specimens or the finishing of the concrete in the structure.

REFERENCES

Aïtcin, P.-C. and Lessard, M. (1994) Canadian experience with air-entrained high-performance concrete. *Concrete International*, **16**(10), October, 35–8.

ASTM C666 (1993) *Standard Test Method for Resistance of Concrete to Rapid Freezing and Thawing*, in *1993 Annual Book of ASTM Standards*, Section 4, Construction, Vol. 04.02 Concrete and Aggregates, ISBN 0-8031-1923-2, pp. 326–31.

ASTM C672 (1993) *Standard Test Method for Scaling Resistance of Concrete Surface Exposed to Deicing Chemicals*, in *1993 Annual Book of ASTM Standards*, Section 4, Construction, Vol. 04.02 Concrete and Aggregates, ISBN 0-8031-1923-2, pp. 345–7.

Blais, F.A., Dallaire, E., Lessard, M. and Aïtcin, P.-C. (1996) *The Reconstruction of the Bridge Deck of the Jacques Cartier Bridge in Sherbrooke (Québec) using High-Performance Concrete*, 30th Annual Meeting of the Canadian Society for Civil Engineering, Edmonton, Alberta, May, ISBN 0-921-303-60, pp. 501–7.

Cook, J.E. (1989) 10,000 psi concrete. *Concrete International*, **11**(10), October, 67–75.

Haug, A.K. and Sandvik, M. (1988) *Mix Design and Strength Data for Concrete Platforms in the North Sea*, ACI SP-109, pp. 495–524.

Hoff, G.G. and Elimov, R. (1995) *Concrete Production for the Hibernia Platform*, Second CANMET International Symposium on Advances in Concrete Technology – Supplementary Papers, Las Vegas, 11–14 June, pp. 717–39.

Howard, N.L. and Leatham, D.H. (1989) The production and delivery of high-strength concrete. *Concrete International*, **11**(4), April, 26–30.

Hu, C. and de Larrard, F. (1995) The rheology of fresh high-performance concrete. *Cement and Concrete Research*, **26**(2), February, 283–94.

Hu, C., de Larrard, F. and Gjørv, O. (1995) Rheological testing and modeling of fresh high performance concrete. *Materials and Structure*, **28**(175), January/February, 1–7.

Hu, C., de Larrard, F., Sedran, T., Boulay, C., Bosc, F. and Deflorenne, F. (1996) Validation of BTRHEOM, the new rheometer for soft-to-fluid concrete. *Materials and Structure*, **29**(194), December, 620–31.

Keck, R. and Casey, K. (1991) A tower of strength. *Concrete International*, **13**(3), 23–5.

Lessard, M., Baalbaki, M. and Aïtcin, P.-C. (1995) Mix Design of Air Entrained, High-Performance Concrete, in *Concrete Under Severe Conditions*, Vol. 2 (eds K. Sakai, N. Banthia and D.E. Gjørv), E & FN Spon, London, pp. 1025–34.

Lessard, M., Gendreau, M., Baalbaki, M., Pigeon, M. and Aïtcin, P.-C. (1993) Formulation d'un béton à hautes performances à air entraîné. *Bulletin de Liaison des Laboratoires des Ponts et Chaussées*, No. 188, November–December, pp. 41–51.

Lessard, M., Dallaire, E., Blouin, D. and Aïtcin, P.-C. (1994) High-performance concrete speeds reconstruction for McDonald's. *Concrete International*, **16**(9), September, 47–50.

Ralston, M. and Korman, R. (1989) Put that in your life and cure it. *Engineering News Record*, **22**(7), 16 February, 44–53.

Ronneberg, H. and Sandvik, M. (1990) High strength concrete for North Sea platforms. *Concrete International*, **12**(1), January, 29–34.

Ryell, J. and Bickley, J.A. (1987) Scotia Plaza: high-strength concrete for tall buildings, in *Proceedings of the Symposium on Utilization of High Strength Concrete*, Stavanger (ed. I. Holland *et al.*), Tapir, N-7034 Trondheim NTH, Norway, ISBN 82-519-0797-7, pp. 641–53.

Woodhead, H.R. (1993) Hibernia offshore oil platform. *Concrete International*, **15**(12), December, 23–30.

CHAPTER 10

Preparation for concreting: what to do, how to do it and when to do it

10.1 INTRODUCTION

The delivery of a usual concrete is carried out on a routine basis without any special preparation from the concrete producer or the inspection laboratory. However, it would be an error to proceed in the same way with a high-performance concrete. For some years to come, the production and delivery of high-performance concrete will necessitate much more attention from the concrete producer as well as from the inspection laboratory. At the same time it will require refinements and improvements to be made to conventional quality control and related test procedures.

Successful inspection and testing is already a must for usual concrete; it is surely a more critical process in the case of high-performance concrete, because high-performance concrete is usually used in high-performance structures and a large degree of confidence must be achieved through the quality control, inspection and testing processes. It is pointless specifying high-performance properties for the fresh and hardened concretes if the tests conducted cannot tell clearly if these performance criteria have actually been achieved. Answers, or at least assurances, about their achievement should also meet the needs of the contractor in terms of schedule and economy (Randall and Foot, 1989; Simons, 1989).

Therefore, those individuals or organizations involved in quality assurance, inspection and testing must know the unique characteristics of high-performance concrete in order to do a proper job. They must be convinced that high-performance concrete inspection requires improvements to be made to conventional quality control and test procedures in order to ensure its safe and optimum use. In any case, the proper evaluation of high-performance concrete must be carried out through realistic and meaningful quality control, inspection and testing procedures as part of a well-defined quality assurance programme. This can be done through communication, planning, team work and prequalification (Hughes, 1996). As an example of the kind of co-operation that is necessary, the development of the specification for the first air-entrained high-performance concrete bridge built in Canada in 1992, the Portneuf bridge, will be described.

Despite all the research efforts by Canadian investigators in the area of high-performance concrete, in 1992 high-performance concrete had been used in only a very limited number of indoor construction applications. Moreover, as everywhere, links between the scientific community, engineers, architects and owners were limited, and to complicate matters further the liability of each party had to be considered when innovating with such a new material. For example, an engineer or an architect would not offer an owner solutions that had yet to be tried and tested, but a researcher cannot demonstrate his research results without building experimental structures. In October 1990 a reflection group called the 'Projet Voies Nouvelles du Béton' was launched in Montreal to tackle this situation by having experimental projects built to push high-performance concrete technology ahead and to make its use commonplace. The founding partners included owners, researchers, consulting engineers, research laboratories and centres (including the Network of Centres of Excellence on High Performance Concrete), materials companies, building contractors and organizations that wanted to contribute to the development of new applications for high-performance concrete.

The specifications for the concrete to be used for the Portneuf bridge were discussed during two meetings of the technical committee of the Projet Voies Nouvelles du Béton. It was first agreed that the 60 MPa concrete required by the designers should contain silica fume (between 6 and 10% of the cement mass). Silica fume enhances mechanical strength and also reduces chloride ion permeability, a particularly important property in this case, considering that the concrete slab would not be protected (at least for the first few years) by any type of membrane. In fact, in 1997 it had still not been protected. It was, however, suggested at that time that epoxy-coated rebars be used for the top mat of the slab reinforcement.

Although it was considered that, for such a concrete, an air-void spacing factor of about 300 μm would be sufficient as a protection against freezing and thawing cycles and scaling from deicer salts, it was nevertheless decided that it would be simpler to rely on the usual requirements of the Canadian code CAN A23.1 for usual concrete, which states that the average of all spacing factor measurements should not exceed 230 μm and no single test result should exceed 260 μm.

The slump was fixed at 180 ± 40 mm, and it was decided to allow only two superplasticizer additions on the site. The maximum fresh concrete temperature was fixed at 25 °C (at the discharge of the truck or at the discharge end of the pump, if any), and a maximum period of 2 hours was allowed between the beginning of the batching operations and the final discharge on site.

To ensure an adequate early strength (i.e. in order to verify that hydration would not be exceedingly retarded), a 10 MPa strength at 24 hours was required.

Curing is particularly important for silica fume concrete, since it usually exhibits very little bleeding and can thus easily be subjected to plastic shrinkage. This is also especially important for reinforced concrete slabs exposed to deicer salt applications. It was thus required that a curing compound be applied immediately after the finishing, and that water curing also be used for 72 hours. Furthermore, it was decided that the formwork should not be stripped before 7 days.

Considering the fact that this would probably be the concrete producer's first experience with high-performance concrete, the committee suggested running a field test before casting the retaining walls and the top slab.

Finally, the committee agreed that 100 mm diameter cylinders would be used for the compressive strength tests.

The construction of the bridge was a total success in spite of the very cold conditions that were faced during the casting of the deck. On this late October 1992 day, the maximum temperature was 3 °C and -4.5 °C was the minimum temperature during the first night, so that at 9.00 a.m. the next day a minimum temperature of -1.5 °C was recorded within the concrete. The concrete temperature stayed below freezing for at least 2 hours, but core testing showed that the strength of the concrete had not been affected. Twenty-seven out of the 28 deliveries were fully tested (one load was missed). The results obtained on the fresh and hardened concrete are given in Table 10.1, which shows that the selected specifications were adequate and perfectly feasible (Aïtcin and Lessard, 1994).

10.2 PRECONSTRUCTION MEETING

In order to be sure that there will not be any lack of communication, it is essential to schedule a preconstruction meeting where all the project participants will establish their respective objectives, discuss the inspection and testing programme and procedures, and clarify contract requirements. This advance understanding will minimize future arguments and will give all members of the construction team an opportunity

Table 10.1 Field tests for the fresh and hardened concrete used to build Portneuf bridge (Aïtcin and Lessard, 1994). Reproduced by permission of ACI

$n = 27$	Mean value	Standard deviation	Maximum	Minimum
Concrete temperature	22 °C	2.5 °C	26 °C	16 °C
Slump	180 mm	25 mm	230 mm	150 mm
Air content: fresh concrete	6.2%	0.8%		
hardened concrete	5.4%	0.8%		
Spacing factor	190 µm	33 µm		
Compressive strength ($n = 23$)	75.3 MPa	3.5 MPa		

to participate in the quality process. During this meeting it is especially important to review the precise procedure an inspector will have to follow when he or she finds that a particular load of concrete does not comply with the contract requirement. It will also be essential to determine who will decide to add more superplasticizer to improve the high-performance concrete workability, if necessary. Site sampling and site curing procedures will have to be well defined, as well as transportation of the specimens from the site to the testing laboratory.

Prior to the start of the job it is imperative to establish a correlation curve to determine the relationship between extracted cores and specimens cast for acceptance tests, so that if coring becomes necessary, this relationship has been established and accepted by all the parties involved in the argument.

This preconstruction meeting will not be a waste of time and money, but rather will result in minimizing the so costly arguments that may develop when construction proceeds.

10.3 PREQUALIFICATION TEST PROGRAMME

This preconstruction meeting will be even more fruitful if it precedes a prequalification test programme. This new approach for the selection of the concrete supplier was successfully developed over the years in Ontario in Canada, and has been extended to the selection of the testing laboratory that is to undertake the inspection and control of the high-performance concrete (Bickley, 1993).

A specification can be written correctly stating the requirements of the designer and the criteria to be met, but the ability of the local ready-mix concrete industry to deliver these criteria may lag behind the requirements for the structure. A prequalification test programme is a way of determining the ability of the industry to deliver a high-performance concrete and also a way to apply pressure on suppliers to enhance their capabilities to do so.

The critical event of such a prequalification test programme consists of the delivery of a full-scale trial mix under the same conditions as the ones that will prevail during the job. It can cost the concrete producer up to $7000 to $10 000 to participate in such a prequalification test programme, but it has been well accepted by serious contenders because it makes the bidding process much fairer. Only those that have been proven to be able to deliver a concrete fulfilling all the specifications are allowed to bid, not only for the high-performance concrete used in the particular structure, but also for all the usual concrete used to build the structure, which is particularly interesting.

This new 'philosophy' developed for the selection of the concrete supplier appears at first glance to be in complete contradiction to the universally adopted 'lowest bidder' philosophy. However, there have been a few cases in the past in which this 'lowest bidder' philosophy created dramatic and very costly situations during the construction of a high-rise building, where it happened that the concrete producer was simply not able to make and deliver the high-performance concrete specified.

It is my intention to describe briefly how this kind of prequalification programme works and to report on more details for two particular examples.

First of all, any prequalification programme must start as an open competition. All the concrete producers or contractors operating within the area are invited to participate in the prequalification test programme, for which they receive a pre-contract package. In the first particular case described below, about 18 concrete producers were invited to participate in such a prequalification programme. Of these, only three opted to compete and were found to be able to deliver a concrete fulfilling almost all of the requirements. Only these three concrete suppliers were allowed to bid for the delivery of all the concrete involved in the total project, around 90 000 m^3, of which 26 000 m^3 was high-performance concrete.

Thus the owner was sure to have some reasonable competition to get a good price, and at the same time a good quality. Moreover, the three 'serious' concrete producers were assured of bidding only against fair competitors. What concrete producer will not sharpen its pencils to get a 100 000 m^3 order?

10.3.1 Prequalification test programme for the construction of Bay-Adelaide Centre in Toronto, Canada

The minimum requirement for the testing consisted of the production and testing of several loads of high-performance concrete and the casting of 2 m^3 for heat development and coring from such concrete. The properties and the requirements to be tested were:

- slump on discharge (200 ± 20 mm);
- temperature on discharge (18 °C max., 21 °C in cold-weather conditions);
- compressive strength of 100 × 200 mm specimens;

Age	12 hours	1 day	7 day	28 day	56 day	91 day
f'_c in MPa	15	30	65	80	90	95

- compressive strength of 100 × 200 mm cores horizontally drilled from the 2 m^3 specimen and tested at the same ages as the standard specimens indicated above: 85% of cylinder compressive strength at the same age;

- 150 × 300 mm specimens for 91 day static modulus of elasticity: 45 GPa minimum;
- temperature development at different points in the 2 m^3 specimen over a period of 7 days after casting (moderate without any more precision).

The purpose of casting a 2 m^3 monolith was to reproduce the essential cross-sectional dimensions of the largest components in the structure. It provided an opportunity to check if the mixes placed and compacted reasonably well, and provided data on the thermal characteristics of the concrete when cast in such large elements. Finally, it provided an opportunity to make a correlation between core strength and specimen strength (Mak *et al.*, 1990; Haque, Gopalan and Ho, 1991; Carino, Knab and Clifton, 1992; Leshchinsky, 1992).

10.3.2 Prequalification test programme for the 20 Mile Creek air-entrained high-performance concrete bridge on Highway 20 (Bickley, 1996)

A committee of representatives of the Ministry of Transportation of Ontario and Concrete Canada developed the criteria and documents needed to build a 60 MPa air-entrained concrete in the replacement of a small bridge at 20 Mile Creek.

All prospective bidders were required to attend a pre-bidding meeting where the specific differences in the use of high-performance concrete were covered. The following points were particularly covered: concrete materials and mix design, concrete handling, construction features, placement of a trial slab, curing requirements, sampling and testing, and concerns. The specified trial slab was 30 m long and 23 m wide. It had to be placed, finished and cured using the plant and methods proposed by the contractor.

Ice was used to keep the delivery temperature of the concrete below the 25 °C maximum specified. A double layer of prehumidified wet burlap was applied just after finishing; it was covered for 7 days with a plastic sheet. After 7 days the slab was uncracked, owing to this mode of curing.

Concrete strength was excellent, averaging 77.1 MPa at 28 days; the air void parameters were determined on core samples taken after pumping and on a companion cylinder taken before pumping. Both measurements of the spacing factor were satisfactory to meet the 230 µm average value, but some increase of the spacing factor was noticed following the pumping.

The construction of the trial slab allowed the team to fine-tune the mix design, the placing and the curing of the high-performance concrete so that a fully satisfactory bridge could be built.

10.3.3 Pilot test

In order to accept the high-performance concrete composition submitted by a contractor, the Ministry of Transportation of Québec demand that a pilot test be included in its quotation.

Since high-performance concrete is not a widely used concrete by contractors, a pilot test is a precautionary measure that permits validation of the high-performance concrete formula submitted by the contractor chosen for a given project. The pilot test is done to assure the client that the high-performance concrete specifications are satisfied. The pilot test is mandatory in Québec.

The pilot test has to be done on the construction site under the exact conditions that the casting operations will take place. This means that the equipment used, the high-performance concrete mix composition, the curing conditions, the sampling operations – all the activities related to the high-performance concrete specified in the quotation – must be done and tested during the pilot test. It is the breaking-in period before the regular casting period.

During the pilot test, a sampling programme is done in order to check if the high-performance concrete specified characteristics are met. The pilot test must be done at least 28 days prior to the actual casting activities, since the acceptance of the submitted high-performance concrete mix design is conditional on the test results obtained from the sampled high-performance concrete during the pilot test. The approval of the high-performance concrete mix design is given only if these results are satisfactory.

For the Jacques-Cartier bridge project in Sherbrooke, the pilot test took place on 5 August 1995 (Blais, personal communication, 1995). The casting period started at 0.00 hours and ended at 5.30 a.m. The maximum ambient temperature was 23 °C and the minimum ambient temperature was 19 °C. Half of the approach slab over the south abutment was used for this pilot test. The volume of high-performance concrete cast during the test was 26.5 m^3. It took many hours to cast this small volume of high-performance concrete because for most of the participants it was the first time they had produced, used and placed high-performance concrete. Even if the concrete producer had done some testing to check his mix design proportions prior to the pilot test, when it came to producing the high-performance concrete in the same field conditions, he needed to fine-tune and synchronize its batching procedure. Also, the contractor's staff had to adapt to the special casting equipment used to finish the bridge deck and to the curing procedure.

Initially, the contractors were reluctant to use the actual motorized vibrating screed that would be used for the project since it involved mobilizing a large staff for such a small volume of concrete. Afterwards, everybody involved in the project realized how essential this pilot test was and understood its necessity. The problems encountered during the

Table 10.2 Properties of the fresh high-performance concrete during the pilot test conducted during the construction of the Jacques-Cartier bridge deck in Sherbrooke

	$\bar{x}$	σ
Slump (mm)	195	25
Air content (%)	4.9	0.7
Unit weight (kg/m³)	2380	20
Temperature (°C)	18	0.5

pilot test were solved before the actual casting period. Therefore, a reproducible high-performance concrete was delivered on the construction site throughout the whole project, even when the climatic conditions varied (the minimum ambient temperature varied from 21 °C to − 4 °C during the project).

The results obtained during the pilot test on the fresh and hardened high-performance concrete are shown in Tables 10.2 and 10.3 respectively.

10.4 QUALITY CONTROL AT THE PLANT

As stated by J.A. Bickley (personal communication, 1997):

> the quality control of high-performance concrete should be the sole responsibility of the concrete supplier up to the time of discharge at site. This should be clear in the contract specification which should state in detail all pre-construction and subsequent submittal requirements. By specifying and insisting on comprehensive on-going submissions of data, the concrete supplier and his material suppliers are made to show evidence of consistent quality control programmes.

Chapter 9 emphasized the importance of checking that all the properties of the materials that will actually be used to make the high-performance concrete have the characteristics for which the materials were selected. It was also pointed out that:

1. all the batching equipment must be in good order and its calibrations regularly checked;

Table 10.3 Characteristics of the hardened high-performance concrete sampled during the pilot test conducted during the construction of the Jacques-Cartier bridge deck in Sherbrooke

	$\bar{x}$	σ
f'_c at 28 d (MPa)	75.0	1.2
Air content (%)	4.9	0.9
Spacing factor, $\bar{L}$ (µm)	250	25

2. it is advantageous to make some trial batches and mock-ups and to deliver them under actual field conditions in order to fine-tune their characteristics;
3. it is always interesting at this stage deliberately to deliver an under-slump concrete in order to have to make and define precisely at the jobsite the superplasticizer corrections necessary to restore the slump without any risk of segregation or undue retardation.

It is also important is to check the quality of each load of fresh concrete for slump, temperature, air content and unit mass before the transit mixer leaves the ready-mix plant. This can be done in less than 10 minutes by a well-trained technician devoted full-time to this essential task. If the slump, temperature, air content and unit mass are out of the bracketed values that ensure the delivery of fresh concrete meeting the job specification, immediate action can be taken to correct this particular situation. If the concrete is too far from the targeted objective, it can be transformed into a 30 to 40 MPa concrete by adding some more water and coarse and fine aggregate. Immediate action can also be taken at the batching plant to correct the situation without having to wait for 1 hour to have some feedback from the testing team at the jobsite.

Quality control at the ready-mix plant is an inexpensive way to minimize field arguments, which are always costly to solve. John A. Bickley reported that the removal from a structure of one defective load of concrete in a particular column has been estimated to have cost $250 000: 'a costly reminder of the failure in quality control' (Bickley, personal communication, 1997).

10.5 QUALITY CONTROL AT THE JOBSITE

It is no use specifying high-performance concrete if the tests carried out do not provide some assurance of the achievement of those specifications. Moreover, the tests should satisfy the needs of the contractor in meeting schedules, safety and economy.

It is not enough to control very strictly the quality of the delivered high-performance concrete if its placing and curing are not carried out with the same quality standards, because what is important is that the concrete that is placed has the desired characteristics, not the concrete that is delivered at the jobsite prior to its placing. Therefore, the jobsite inspection crew must pay close attention to placement restrictions, transit difficulties and field curing requirements well ahead of any particular delivery. In this case, good planning and teamwork can solve any foreseen difficulties, leaving the jobsite inspection team with the unforeseen ones which will occur in any event.

If the testing laboratory has not been selected after a prequalification test programme, in which it has demonstrated without any doubt that it can correctly test high-performance concrete, but rather as the cheapest one, it is better for the concrete producer to control in parallel the quality of the concrete he is delivering. This duplication is not a waste of money. The concrete producer can follow day by day the quality of his high-performance concrete production, improve the quality, reduce the number of field problems and optimize the concrete proportioning so as to produce safely the concrete that is required, nothing less, nothing more. The producer can systematically delay by 24 hours the testing of his own test specimens, so as to be able in case of argument to get them tested the next day by an independent agency that will verify whether or not the concrete was satisfactory. Quality control always pays in the long run.

10.6 TESTING

The testing procedures themselves will be discussed in detail in Chapter 15. Here the basic precautions that have to be taken when sampling is carried out and the frequency of sampling will be discussed, since high-performance concrete is more sensitive to testing variables than is usual concrete.

Tests of air content, unit mass, slump and temperature should be made on the very first truckloads to establish that the concrete meets all the specifications. Thereafter, these tests should be performed on a random basis or on a sequential basis. On the Jacques-Cartier bridge deck reconstruction the three first concrete loads of each pour were tested. If they were all well within the specifications, the next testing was performed on the sixth load, so as to proceed faster with the rate of placing of the concrete. If a particular tested load was found to be only marginally good, the following two loads were tested until the concrete properties fell well within the requirements three times in a row. At any time the concrete inspector could also decide to check the characteristics of any load that did not look good to him.

No more than eight specimens can be taken at a time by a well-trained technician unless sufficient test personnel and facilities are available. A minimum of two specimens per test age, with two held in reserve, is recommended to make a proper control.

What should be the frequency of the testing?

During Scotia Plaza construction, since this was the first time that a 70 MPa high-performance concrete had been used in the Toronto area, every concrete load was sampled, so that the final average compressive strength and standard deviation were calculated from the tests of 145 samples of three specimens at a time. However, testing every load of concrete can be very expensive and also unnecessary if the concrete producer is a 'serious' one. The ACI 363 Committee on high-strength concrete recommends a full sampling for every 75 m^3 delivery.

The results obtained from these samples should be used to improve the general quality of the production and to optimize the composition of

Evaluation of quality

the high-performance concrete. Considerable information can be obtained through a well-planned testing programme if the test results are correctly interpreted using quality control charts and related statistical analysis, such as the one found in different papers (Day, 1979; Novokshchenov and Allum, 1992) and even in a book (Day, 1995).

10.6.1 Sampling

It might be surprising to see a short paragraph about concrete sampling in a high-performance concrete book, since sampling concrete is so well known (or supposedly well known). Sampling seems to be so simple that it is generally believed that taking standard specimens does not need special skills. However, in the case of high-performance concrete, sampling must be done carefully in order to avoid any discrepancies between specimens. Moreover, after their casting, the specimens must be stored on an **absolutely horizontal surface** because of the high slump of high-performance concrete. Their tops must be carefully levelled so that their capping or facing, depending on the way they will be tested later on, will be as easy as possible. In Chapter 15 it will be seen that a lack of care at this level can be costly in terms of sample preparation.

The moulds used to take the specimens must be carefully checked for ovality and for perpendicularity. Ovality can reduce the surface over which the load is applied and decrease the ultimate load at which the specimen fails. Moreover, when the mould axis is not perpendicular to the two ends, an extra volume of capping compound will be necessary to compensate for this situation, with the risk that the compression is not applied uniaxially but rather in a shearing mode. Finally, if the top and bottom faces are not parallel, the capping compound will have to compensate for this, with here, too, the risk that the mode of loading will not be uniaxial and will lead to a shearing failure. Moreover, the mould should be as rigid as possible in order to ensure proper and reproducible compaction. All these points will be discussed in detail in Chapter 15.

As far as the sampling is concerned, all other sampling requirements usually enforced for usual concrete have to be used with high-performance concrete. For example, under no circumstances should the technician use another sample to 'top off' the test specimen. If the sample is too small, the concrete must be wasted and another sample should be taken.

10.7 EVALUATION OF QUALITY

The first consideration in evaluating the quality control procedure when testing usual concrete is determining whether or not the distribution of compressive strength test results follows a normal frequency distribution curve. This can no longer be true in the case of high-performance

Table 10.4 Standards of concrete control for usual concretes (ACI 214). Reproduced by permission of ACI

Class of operation	Coefficient of variation (%)				
	Excellent	Very good	Good	Fair	Poor
General construction	< 8	8–10	10–12	12–15	> 15
Laboratory trial batches	< 4	4–6	6–8	8–10	> 10

concrete: as the targeted compressive strength increases, a skewed distribution may prevail because the mean strength is approaching a limit.

In the ACI 214 specification, five overall levels of quality control of usual concrete have been defined in terms of the coefficient of variation of test specimens at the specified age, as shown in Table 10.4. In the case of high-performance concrete, quality control categories based on standard deviations may be misleading; for example, a standard deviation of 5 MPa, which is poor for a 30 MPa concrete, is quite good for a 70 MPa concrete. Recently, the ACI 363 Committee on high-strength concrete recommended that the coefficient of variation be used as a better indicator of high-performance concrete consistency. Table 10.5 presents the values that are suggested as a standard of quality control.

In the current edition of ACI 214, an attempt is made to relate the numerical values of the standard deviation to descriptive evaluations of the quality of the usual concrete production. A standard deviation of less than 2.8 MPa represents an excellent degree of control, whereas a standard deviation greater than 4.8 MPa represents poor control. In the case of high-performance concrete, defining quality control categories based on absolute dispersion may be misleading since standard deviations greater than 4.8 MPa are not uncommon for 70 MPa concrete.

For practical comparisons, the coefficient of variation is a more useful tool for measuring the dispersion of concrete strength than the standard

Table 10.5 Standards of concrete control for high-performance concretes according to ACI 363. Reproduced by permission of ACI

	Standards of concrete control. Specified strength over 5000 psi				
	Coefficient of variation for different control standards (%)				
Class of operation	Excellent	Very good	Good	Fair	Poor
Overall variation					
General construction testing	< 7.0	7.0–9.0	9.0–11.0	11.0–14.0	> 14.0
Laboratory trial batches	< 3.5	3.5–4.5	4.5–5.5	5.5–7.0	> 7.0
Within-test variation					
Field control testing	< 3.0	3.0–4.0	4.0–5.0	5.0–6.0	> 6.0
Laboratory trail batches	< 2.0	2.0–3.0	3.0–4.0	4.0–5.0	> 5.0

deviation. The coefficient of variation is an average strength. It has been suggested (Anderson, 1985; Cook, 1989) that the coefficient of variation must be used because this value is less affected by the magnitude of the strengths obtained and is more useful in comparing the degree of control for a wide range of strength levels.

These standards represent the average for 28 day specimens computed from a large number of tests. It is felt that the higher-strength concrete has a lower coefficient of variation than usual structural concrete, not because of the strength level but because a much higher degree of control is maintained in the production and testing of high-performance concrete. Also, it has been established that the coefficient of variation decreases for any given concrete with increasing strength owing to age. In other words, the increase in variability at early ages and the decrease in variability at later ages are due more to the nature of concrete than to the degree of control exercised during the production and testing of the concrete.

10.8 CONCLUDING REMARKS

Now that it has been shown that high-performance concrete must be treated with more care than usual concrete, that everything has been checked at the concrete plant and at the jobsite, and that the placing method has been approved, it is still not time to start the actual construction, not before being sure that the delivering, placing, controlling and curing of the high-performance concrete has been well treated. Curing high-performance concrete is so important that a separate chapter is devoted to it.

REFERENCES

ACI 363R (1993) *State-of-the-Art Report on High-Strength Concrete, ACI Manual of Concrete Practice*, ACI, Detroit, 55 pp.

ACI 214 (1993) *Recommended Practice for Evaluation of Strength Test Results of Concrete, ACI Manual of Concrete Practice*, Part 2 – Construction Practices and Inspection Pavements, ISSN 0065-7075, pp. 214-1–214-14.

Aïtcin, P.-C. and Lessard, M. (1994) Canadian experience with air-entrained high-performance concrete. *Concrete International*, **16**(10), October, 35–8.

Anderson, F. D. (1985) *Statistical Controls for High-Strength Concrete*, ACI SP-87, pp. 71–82.

Bickley, J.A. (1993) Prequalification requirements for the supply and testing of very high strength concrete. *Concrete International*, **15**(2), February, 62–4.

Bickley, J.A. (1996) High-performance concrete bridge for Ontario Ministry of Transportation: contract 95-39. *Newsletter of the High-Performance Concrete Network of Centres of Excellence*, **1**(2), May, 1–2.

Carino, N.J., Knab, L.I. and Clifton, J.R. (1992) *Applicability of the Maturity Method to High-Performance Concrete*, NRSTIR 4519 United Department of Commerce, Technology, Administration, National Institute of Standards and Technology, Structure Division, Gaithersburg, MD 20899, May, 62 pp.

Cook, J.E. (1989) 10 000 psi concrete. *Concrete International*, **11**(10), October, 67–75.

Day, K.W. (1979) *Quality Control of 55 MPa Concrete for Collins Place Project, Melbourne, Australia*, presented at the 1979 Annual Convention of the American Concrete Institute in Milwaukee, 18–23 March, 18 pp.

Day, K.W. (1995) *Concrete Mix Design, Quality, Control and Specification*, E & FN Spon, London, ISBN 0-419-18190-3, 350 pp.

Haque, M.N., Gopalan, M.K. and Ho, D.W.S. (1991) Estimation of in-situ strength of concrete. *Cement and Concrete Research*, **21**, 1103–10.

Hughes, Y. (1996) *Control of HPC in the Field*, Personal communication.

Leshchinsky, A.M. (1992) Non-destructive testing of concrete strength: statistical control. *Materials and Structures*, **25**, 70–78.

Mak, S.L., Allard, M.M., Ho, D.W.S. and Darwall, L.P. (1990) *In-situ Strength of High-Strength Concrete*, Report No 4/90, Monash University Australia Civil Engineering Research Report, ISBN 07326 0181, 90 pp.

Novokshchenov, V. and Allum, D. (1992) Monitoring concrete operations with the CQC Report. *Concrete International*, **14**(5), May, 51–7.

Randall, V. and Foot, K. (1989) High-strength concrete for Pacific First Center. *Concrete International*, **11**(4), April, 14–16.

Simons, B.P. (1989) Getting what was asked for high-strength concrete. *Concrete International*, **11**(10), October, 64–6.

CHAPTER 11

Delivering, placing and controlling high-performance concrete

11.1 HIGH-PERFORMANCE CONCRETE TRANSPORTATION

Up to now, high-performance concrete has always been transported in the same way as usual concrete using transit truck mixers. However, it can also be transported with stationary trucks with or without agitators. The main problem faced during concrete transportation is slump loss and air-void stability in the case of air-entrained high-performance concrete. This problem has been discussed in terms of cement/superplasticizer compatibility in Chapter 7 but not in terms of duration and ambient temperature during concrete delivery.

Delivering high-performance concrete during the rush hour in large urban areas is risky owing to unpredictable traffic delays, as mentioned in Chapter 4. In order to benefit from a smooth delivery schedule, the transportation of the high-performance concrete used for the construction of the Two Union Square building in Seattle and two bridges in Québec was carried out at night (Howard and Leatham, 1989; Lessard, Gendreau and Gagné, 1993; Blais *et al.*, 1996). A night delivery schedule can also be beneficial for the control of the temperature of the fresh concrete when the ambient temperature is lower.

Most of the time the use of a small dosage of a set retarder helps to maintain the slump during the 60 to 90 minutes that are necessary to place the high-performance concrete (Haug and Sandvik, 1988; Ronneberg and Sandvik, 1990; Aïtcin and Lessard, 1994). In some cases concrete producers prefer to use a high dosage of water reducer to retard the setting slightly (Hoff and Elimov, 1995; Blais *et al.*, 1996).

Moreover, in order to avoid overspill of a high-slump, high-performance concrete from drum mixers, the batching of the truck should be limited to 75% of the capacity.

11.2 FINAL ADJUSTMENT OF THE SLUMP PRIOR TO CONCRETE PLACING

If for any reason, in spite of all the precautions that are undertaken in making and transporting the concrete, the delivered high-performance concrete does not have the right slump upon arrival at the jobsite, the concrete should not be retempered with water. Instead a superplasticizer must be incorporated to attain the desired slump (Aïtcin and Lessard, 1994; Blais *et al.*, 1996). It is, however, quite difficult to give a rule of thumb that would permit an evaluation of the precise amount of superplasticizer needed to increase the slump, such as so many litres per 100 kg of cement to increase the slump by so many millimetres. The best way to make the necessary adjustments is to carry out preliminary tests prior to starting of the actual delivery by purposely making a high-performance concrete that does not have the right slump and by finding, not under pressure and improvisation, the right amount of superplasticizer to be added. In any case, this experimental concrete would not be wasted because it could always find a use at the jobsite. For example, it could be recycled for the casting of a conventional structural concrete element. Making such a pilot test prior to the actual delivery is a low-cost solution to the slump adjustment problem because the problem is then solved in a quiet situation, as opposed to the very hot situation that prevails when all of the contractor's placing crew is in place and wasted time represents a great deal of expense. In any case, if the mix design has been well adjusted to the jobsite conditions, the addition of superplasticizer in the field should not be more than 1 to 2 litres per cubic metre of concrete (Aïtcin and Lessard 1994).

It has been suggested that at the jobsite, high-performance concrete should be redosed with a melamine-based superplasticizer rather than with a naphthalene-based one for the following reasons (Ronneberg and Sandvik, 1990):

- melamine-based superplasticizers, when used at high dosage rates, are significantly less retarding than naphthalene-based superplasticizers;
- as melamine-based superplasticizers generally have a 20% solids content, any accidental overdosage is less critical for segregation;
- as melamine-based superplasticizers usually have only a 20% solids content, they do not need to be diluted in an equal volume of water to be properly dispersed in the mix;
- as melamine-based superplasticizers contain more water than equivalent volumes of naphthalene-based superplasticizers, this water helps to recover some of the lost slump. (Note from the author: but it also increases the W/B ratio.)

However, from a practical point of view concrete producers do not like to work with two different types of superplasticizer, one at the plant and one in the field. It is important to note from a colour point of view that working with these two types of superplasticizer is foolproof: naphthalene superplasticizers are brown, while melamine ones are pale milky white.

11.3 PLACING HIGH-PERFORMANCE CONCRETE

High-performance concrete has been placed using pumping lines, cranes, buckets and conveyor belts. Each placement method has its own specific advantages and disadvantages, depending on the field conditions. Mixture ingredients, site accessibility and location, required capacity and casting rate, time of placement and weather conditions can dictate the use of one method rather than another (Hover, 1995).

11.3.1 Pumping

Pumping concrete is more and more favoured by the construction industry, and this applies also to high-performance concrete. The rules developed for usual concrete still apply for high-performance concrete: the aggregate grading is an essential parameter, particularly that of sand, in order to obtain a cohesive mix, which is necessary for high-pressure pumping (Haug and Sandvik, 1988). The addition of a small amount of silica fume, 1 to 3% by mass of cement, has been found to improve pumpability (Ronneberg and Sandvik, 1990).

When the mix is well designed, it can be pumped as high as 277 m, as in the case of Scotia Plaza in Toronto (Ryell and Bickley, 1987; Page, 1989), or even higher: 295 m, as in the case of 311 South Wacker Drive Tower in Chicago (Page, 1990). Other successful examples of placement of high-performance concrete by pumping are presented in Page's paper (1990).

There is, however, one point of controversy related to pumping of air-entrained high-performance concrete. This concerns the modification of the characteristics of the air-bubble system and the related freeze–thaw and scaling durability. When pumping air-entrained high-performance concrete having a slump greater than 200 mm, a significant drop in air content is often observed. Such a drop in air content is accompanied by a significant increase in the spacing factor and a decrease in the specific surface area, as reported by Aïtcin and Pigeon (1996). It does not seem that the modification of the configuration of the pumping line can help in solving this problem, as shown in Figure 11.1 (Lessard, Baalbaki and Aïtcin, 1995).

This modification of the spacing factor causes a problem in Canada because Canadian standard A23.1 specifies that in order to be freeze–thaw resistant, a concrete (only usual concretes were considered when this standard was written and reapproved) must have an average spacing factor, $\bar{L}$, lower than 230 μm, with no individual values higher than 260 μm. This is not easy to achieve when pumping high-slump, air-entrained concrete; consequently, up to now, the Ministry of Transportation of Québec, for example, is forbidding pumping to place air-entrained high-performance concrete for bridge construction, in spite of the fact that it has been proved that high-performance concrete with

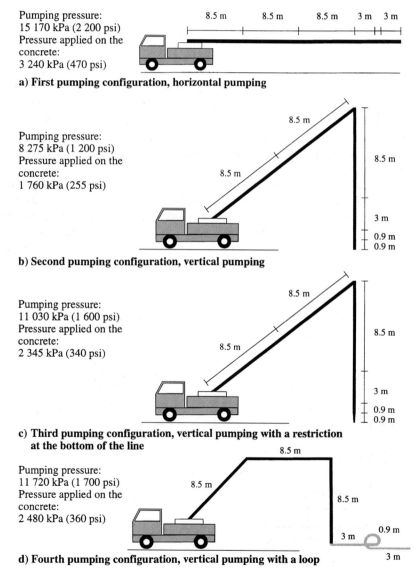

Fig. 11.1 Pumping configurations studied by Lessard, Baalbaki and Aïtcin (1995).

spacing factors of the order of 400 μm can be durable to freezing and thawing when tested according to ASTM C666 (Aïtcin, 1996).

In a recent paper, the author requested the CSA Committee A23.1 to introduce a waiver for freeze–thaw and scaling durability so that either the $\bar{L} = 230$ μm or the ASTM C666, Procedure A test result could be employed to determine if an air-entrained high-performance concrete could be considered as durable to freezing and thawing and to scaling (Aïtcin, 1996).

11.3.2 Vibrating

It is commonly believed that because high-performance concrete usually has a high slump, it does not need to be vibrated when placed in the forms; however, this is not true. Owing to their sticky and more cohesive consistency, high-performance concretes tend to entrap large air pockets and bubbles that must be eliminated by internal or external vibration. The usual rules of consolidation that are used for usual concrete still apply for high-performance concrete.

During a recent full-scale experiment, a 1000 × 1000 × 2000 mm column was cast with a 120 MPa concrete (Miao et al., 1993). This column was later sawed into three pieces and extensively cored in order to evaluate any strength differences within the column. In total, 24 cores were taken in each of the three sections of the column (Figure 11.2). It

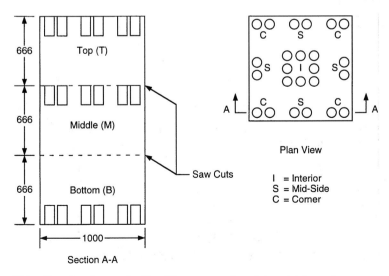

Fig. 11.2 Cored columns built with one usual and two high-performance concretes (Miao et al., 1993).

was found that the compressive strength of the concrete was quite uniform through the whole column.

11.3.3 Finishing concrete slabs

Usually concrete finishers do not like their first experience with high-performance concrete because they have often not been told that finishing a high-performance concrete is quite different from finishing a usual concrete surface. Usually they are unhappy with the stickiness of the mix, they are waiting for bleeding water that never comes, they see the concrete surface experiencing a rapid drying shrinkage, and then they find that it is very difficult to close the cracks appearing at the surface of the concrete because the high-performance concrete surface is quite difficult to work with. However, if the placement of high-performance concrete is well planned and well explained to the placing crew, it can be handled smoothly. For flat surfaces, striking non-air-entrained concrete is quite painful and yields bad results; therefore the use of a vibrating screed is essential. After the use of a vibrating screed, the minor defects that remain at the surface of the high-performance concrete have to be worked with a magnesium trowel or a darby immediately in order to take advantage of the improved workability of the top layer of the slab (Figure 11.3) (Blais *et al.*, 1996). It has recently been observed that a trowel finish just after the use of the vibrating screed also improved the position of the coarse aggregate in the top layer of the slab.

The placement and finishing of high-performance concrete can be enhanced by entraining some air in the concrete. It has already been advocated that entrained air helps the finishing and improves the appearance of vertical surfaces in offshore platforms (Haug and Sandvik, 1988; Ronneberg and Sandvik, 1990; Hoff and Elimov, 1995). Such an improvement can be much more significant for flat surfaces (Lessard *et al.*, 1994; Blais *et al.*, 1996). Therefore, it is the opinion of the author that in the future high-performance concrete will contain a small amount of entrained air voluntarily added to facilitate its placement and finishing and not specifically to make it resistant to freezing and thawing. But the question is: how much air must be entrained to improve the placing and finishing of high-performance concrete? The author thinks that an addition of 2 to 3% of very small bubbles on top of the 1.5 to 2% of naturally entrapped air drastically improves the placing and finishing of high-performance concrete. Of course, the water/binder ratio has to be decreased slightly to recover the 10 to 15 MPa that can be lost owing to the presence of this 2 to 3% of air (this rule of thumb has already been mentioned in Chapter 9). The reduction of the water/binder ratio is simple and not so costly as having difficulties in placing and finishing where the crew cannot maintain the required placing and finishing rate because of the sticky consistency of the concrete. Of course, the lower the

Placing High-Performance Concrete

Fig. 11.3 Finishing of the Jacques-Cartier bridge deck in Sherbrooke.

water/binder ratio (less than 0.30), the more difficult the placing and finishing.

When placing the air-entrained high-performance concrete in the second sidewalk entrance at the McDonald's restaurant in Sherbrooke, the placing crew was happier than when they placed the non-air-entrained one in the first entrance (Lessard *et al.*, 1994). The result is that the quality of the finish of the surface of the entrance cast with the non-air-entrained concrete does not look as good as that of the air-entrained one. Moreover, owing to the highly viscous nature of the concrete at low shear rates, the placing crew was able to shape the fresh concrete in the desired 3-D profile between the entrance and the plain sidewalk (Figure 11.4).

11.4 SPECIAL CONSTRUCTION METHODS

11.4.1 Mushrooming in column construction

Some special attention has to be paid when placing high-performance concrete over column and shear wall locations before casting floor slabs with a lower-strength concrete, to ensure the continuity of high-performance concrete, as shown in Figure 11.5. It is essential to extend the cap far enough from this location and to pay some attention to the overlap of the two concretes in order to avoid cold joint formation between the two (Moreno, 1990). The use of a set-retarding agent should be considered in these special locations in order to be sure that the high-performance mushroom is plastic enough to blend well with the usual concrete of the slab.

11.4.2 Jumping forms

Jumping forms have been used where high-performance concrete is employed when the presence of a cold joint is not critical from a

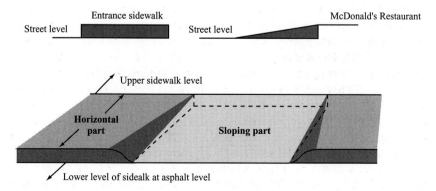

Fig. 11.4 Sidewalk profile shaped in the fresh high-performance air-entrained concrete at the McDonald's restaurant in Sherbrooke.

Special construction methods

Fig. 11.5 Mushrooming technique in the case of columns. The columns are made of a darker high-performance silica fume concrete, while the slabs are made of a paler non-silica fume concrete. A darker area is seen at the base of each column on each floor. Courtesy of J. Moreno.

construction point of view, such as in the case of the construction of Scotia Plaza (Figure 4.6(a)) (Ryell and Bickley, 1987). In this case, the forms were raised every 15 hours. In order to be sure that the concrete was strong enough, an elaborate testing programme based on pullout strength on inserts was developed to the satisfaction of the owner, the contractor and the quality control laboratory.

11.4.3 Slipforming

If no cold joints can be tolerated within the structure, it is necessary to proceed by slipforming. This technique has been developed and extensively used in offshore platform construction in Norway (Haug and Sandvik, 1988; Ronneberg and Sandvik, 1990) and was recently used successfully during the construction of the Hibernia offshore platform (Hoff and Elimov 1995) (Figure 11.6). In this last case, the challenge was quite tough, not only because of the field conditions and the ambient temperature that varied between $-25\,°C$ and $+25\,°C$, but also owing to the special characteristics of the high-performance mix used to build this platform: the W/B ratio was 0.31, and the air content had to be between 3.5 and 6% to secure a spacing factor below 300 μm so that the mix could still pass the ASTM C666 Procedure A freeze–thaw test after pumping. Half the volume of the coarse aggregate was lightweight aggregate to secure the desired buoyancy of the platform, and the

308 *Delivering, placing and controlling*

pumping of this concrete could be as long as 800 m, with a maximum height of 60 m and a downward drop of 60 m.

When slipforming is used in an offshore platform, it comprises three decks, as shown in Figure 11.6: the buggy deck, the working deck and the trailing deck. The set-up can be 'winterized' in order to protect the

Fig. 11.6 Slipforming.

concrete long enough (about 3 days), so that when it is exposed to very low temperatures most of its hardening has occurred.

11.4.4 Roller-compacted high-performance concrete

Recently, in Québec, Canada, high-performance concrete has started to be placed using the roller-compacted technique. Several large slabs having areas from 10 000 m^2 to 87 000 m^2 have been built in different industrial yards in order to facilitate material handling in particularly harsh conditions: smelters, pulp and paper plants, scrap iron plants, etc. Usually two layers of 150 mm and 0 mm slump high-performance concretes are laid using an asphalt paver. The high-performance concrete can also be prepared in an asphalt plant. The two layers are compacted using asphalt vibrating rollers. Such pavement sections are not reinforced and contain no joints. The high-strength slabs have hard surfaces and perform quite well under the very severe service conditions. This new type of high-performance concrete is presented in more detail in section 18.7.

In the Domtar pulp and paper plant near Sherbrooke, the 87 000 m^2 yard, which is equivalent to 16 adjacent football or soccer fields, was cast within one-and-a-half months with a reduced staff using asphalt paving standard equipment.

11.5 CONCLUSION

It is important to take great care when placing and finishing high-performance concrete, because such concrete is not finished in the same way as usual concrete. In spite of the usual high slump of such concrete, it still must be internally or externally vibrated to facilitate placement and enhance performance. In flat slabs, vibrators should not be used to displace the concrete into place. High-performance concrete should not be over-vibrated, to avoid segregation and local bleeding. High-performance concrete slabs should be finished using vibrating screeds, though a trowel finish should be applied as soon as the vibrating screed has passed. It is then very important to apply a temporary curing compound or provide fog misting as soon as the surface is finished, as will be seen in the next chapter, to avoid the rapid development of drying shrinkage near the exposed surface.

REFERENCES

Aïtcin, P.-C. (1996) The pumping of HPC – satisfactory results. *Newsletter of the High-Performance Concrete Network of Centres of Excellence*, **4**(2), May, 4.

Aïtcin, P.-C. and Lessard, M. (1994) Canadian experience with air-entrained high-performance concrete. *Concrete International*, **16**(10), October, 35–8.

Aïtcin, P.-C. and Pigeon, M. (1996) *Freezing and Thawing Durability of High-Performance Concrete*, Proceedings of the Annual Meeting of Concrete Canada, August, pp. 9–15.

Blais, F.A., Dallaire, E., Lessard, M. and Aïtcin, P.-C. (1996) *The Reconstruction of the Bridge Deck of the Jacques Cartier Bridge in Sherbrooke (Québec), using High-Performance Concrete*, 30th Annual Meeting of the Canadian Society for Civil Engineering, Edmonton, Alberta, May, ISBN 0-921-303-60-0, pp. 501–7.

Haug, A.K. and Sandvik, M. (1988) *Mix Design and Strength Data for Concrete Platforms in the North Sea*, ACI SP-109, pp. 495–524.

Hoff, G.C. and Elimov, R. (1995) *Concrete Production for the Hibernia Platform*, Second CANMET International Symposium on Advances in Concrete Technology – Supplementary Papers, Las Vegas, 11–14 June, pp. 717–39.

Hover, K. (1995) Investigating effects of concrete handling on air content. *Concrete Construction*, **40**(3), September, 745–9.

Howard, N.L. and Leatham, D.M. (1989) The production and delivery of high-strength concrete. *Concrete International*, **11**(4), April, 26–30.

Lessard, M., Gendreau, M. and Gagné, R. (1993) *Statistical Analysis of the Production of a 75 MPa Air-Entrained Concrete*. 3rd International Symposium on High-Performance Concrete, Lillehammer, Norway, ISBN 82-91-341-00-1, pp. 793–80.

Lessard, M., Baalbaki, M. and Aïtcin, P.-C. (1995) *Effect of Pumping on Air Characteristics of Conventional Concrete*, Transportation Research Record, Transportation Research Board, Washington, DC 20418, ISBN 0-309-05904-6, pp. 9–14.

Lessard, M., Dallaire, E., Blouin, D. and Aïtcin, P.-C. (1994) High-performance concrete speeds reconstruction for McDonald's. *Concrete International*, **16**(9), September, 47–50.

Miao, B., Aïtcin, P.-C., Cook, W.D. and Mitchell, D. (1993) Influence of concrete strength on in-situ properties of large columns. *ACI Materials Journal*, **90**(3), May–June, 214–19.

Moreno, J. (1990) 225 W. Wacker Drive. *Concrete International*, **12**(1), January, 35–9.

Page, K.M. (1989) Profiles in concrete pumping. *Concrete International*, **11**(10), October, 28–30.

Page, K.M. (1990) Pumping high-strength concrete. *Concrete International*, **12**(1), January, 26–8.

Ronneberg, A. and Sandvik, M. (1990) High-strength concrete for North Sea platforms. *Concrete International*, **12**(1), January, 29–34.

Ryell, J. and Bickley, J.A. (1987) Scotia-Plaza: high-strength concrete for tall buildings, in *Proceedings of the Symposium on Utilization of High-Strength Concrete*, Stavanger (ed. I. Holland et al.), Tapir-N-7034 Trondheim, NTH, Norway, ISBN 82-519-0797-7, pp. 641–53.

CHAPTER 12

Curing high-performance concrete to minimize shrinkage

12.1 INTRODUCTION

Usual concrete must be cured. No one has any doubts about it. But this does not mean that concrete is always well cured, if it is cured at all. For usual concretes, water curing is necessary to ensure the highest degree of hydration possible in order to obtain the highest strength and the lowest permeability (Neville, 1995). Uncured concrete dries more or less rapidly, depending on its water/binder ratio, since it never even approaches its full strength and durability potential. Early curing is always better than late curing, and in the case of usual concretes late curing is better than no curing at all.

There are different ways to water-cure concrete: fog misting, wet burlap, water ponding, water spraying and curing membranes. It is not the purpose of this chapter to review usual concrete curing techniques; they are extensively treated in many books, manuals and specifications. Rather the purpose is to show that:

- high-performance concrete has to be water cured as early as possible;
- late curing is of practically no value, but is still better than no curing at all;
- total shrinkage can be drastically reduced by appropriate early water curing;
- no water curing at all can be catastrophic.

The need for high-performance concrete to be cured is still a controversial subject (Aïtcin and Neville, 1993). Some say that high-performance concrete must be cured like any usual concrete. Others say that owing to its very dense microstructure, high-performance concrete does not need any curing at all. Among those who realize the need for a proper curing, there is another controversy: how long does high-performance concrete need to be cured? The answer to this controversial subject is not unique;

it depends ... First, it depends on the type of curing involved: is it the immediate curing just after placing that is in question, or is it rather the long-term curing of the hardened concrete? We will examine both of these in order to define some rules that should be adapted to particular field conditions.

12.2 THE IMPORTANCE OF APPROPRIATE CURING

No one will argue that it is absolutely necessary to protect the top surface of the deck of a bridge built with a high-performance concrete from drying shrinkage and early drying. Nobody will argue for long that it is very useful to wet-cure the top surface of a column made of high-performance concrete that is exposed to drying for 7 days or more after being cast. But what should be done in other cases? Common sense or entrenched positions are not satisfactory; it is absolutely necessary to understand what is happening when the hydration reaction develops in a high-performance concrete (as well as in a usual concrete, of course) in order to take the appropriate measures to reduce as much as possible the shrinkages that will develop in any uncured concrete.

The concrete skin is always the first line of attack, if there is an attack, and its weakening due to the lack of appropriate curing can be critical for the durability of the structure. So it is always important to evaluate the risk taken in weakening the concrete skin when there is improper curing. In any case, too much curing is always better than no curing at all, but it must also be recognized that proper curing is not always practical if it has not been well planned before the start of construction. When it has been planned well ahead of schedule, there is never a problem in carrying out proper curing, because it only involves adding an almost free material to concrete: water.

Curing is carried out for two reasons: to hydrate as much as possible the cement present in the mix and, what is sometimes forgotten, to minimize shrinkage.

12.3 DIFFERENT TYPES OF SHRINKAGE

Concrete shrinkage is both a very simple phenomenon in its manifestation – a decrease in the apparent volume of the concrete – and quite a complex one when the causes of the phenomenon have to be understood. It would be better to speak of concrete **shrinkages** because the measured shrinkage corresponds to a combination of several elementary shrinkages (Aïtcin, Neville and Acker, 1997):

- plastic shrinkage, which develops at the surface of a fresh concrete subjected to drying (Wittmann, 1976);
- autogenous shrinkage (also called self-desiccation or chemical shrinkage), which can develop when cement hydrates (Tazawa, Myazawa and Kasai, 1995);

- drying shrinkage, which follows the loss of water in a hardened concrete owing to evaporation of its internal water;
- thermal shrinkage, which follows a decrease in the temperature of the concrete;
- carbonation shrinkage.

In order to understand the origin and the main causes of each of these elementary shrinkages, it is essential to understand the hydration reaction and its physical, thermodynamic and mechanical consequences. Then, from a practical point of view, it should be possible to take the appropriate measures to be sure that the negative effects of shrinkage can be minimized. In fact, it is well known that usual concrete does not shrink at all if it is cured under water. Shrinkage is not an unavoidable phenomenon; it is rather the consequence of a lack of adequate curing or cessation of proper curing. The time at which water curing is stopped, the water/binder ratio value and the amount and type of aggregate will determine in large part the final shrinkage of a concrete.

12.4 THE HYDRATION REACTION AND ITS CONSEQUENCES

The expression 'hydration reaction of Portland cement' is not very scientific. It conveys only one fact: when Portland cement is put in contact with water, the cement paste starts to harden after a while following the reaction between the water and the constituents of Portland cement. This chemical reaction results in the release of heat and a contraction of the solid volume. Therefore, the strength gain of any hydrated cement paste is always associated with the release of heat and a contraction of the solid volume, and vice versa. In order to illustrate this fact, we may say, to borrow an image, that the cement paste, or more generally concrete, evolves within the 'Bermuda Triangle': strength–heat–volumetric contraction. This triangle has been schematically represented in Figure 12.1. It is impossible for any concrete to gain strength without exhibiting these

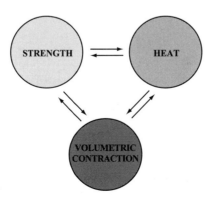

Fig. 12.1 The 'Bermuda Triangle' of concrete.

two concomitant phenomena, which cause so many worries to engineers and with which concrete users have to learn to live. However, upon further reflection, it is not so bad that concrete evolves within such a triangle, because it is possible to live with these two parasitic phenomena. Concrete would not be the most widely used material in the world if it were a swelling material with a temperature decrease during hardening.

The terms 'hydration reaction' and 'heat of hydration' are therefore very useful terms that describe a physical reality, but this simplicity of vocabulary can be misleading since Portland cement is not a well-defined pure material (as shown in Chapter 6) but rather a multiphasic material, the composition of which can vary over a wide range. This simple vocabulary in fact integrates the development of a series of different complex chemical reactions that are all exothermic and therefore contribute to increasing the temperature of concrete.

In order to explain schematically those properties of concrete in which we are interested, Portland cement can be considered to be composed, to a first approximation, of essentially five different phases: C_2S and C_3S, C_3A, C_4AF and calcium sulfate (which is added to control the setting and hardening of the cement). In this analysis, the so-called impurities of Portland cement (alkali sulfates, free lime, unreacted silica, periclase, etc.) are usually neglected. When we look more closely at the role of these impurities in the hydration reaction, however, we see that their role is not always so negligible.

When observing the hydration of pure phases, the hydration of the two silicates results in the formation on the one hand of a single calcium silicate hydrate that is written in an abbreviated form as C–S–H, and on the other hand of portlandite, $Ca(OH)_2$. The C_3A in the presence of calcium sulfate and water is transformed into ettringite, $C_3A \cdot 3CaSO_4 \cdot 32H_2O$ and later on to monosulfoaluminate $C_3A \cdot CaSO_4 \cdot 12H_2O$ when there is no more calcium sulfate, as already shown in Chapter 6. As for the C_4AF, it always hydrates like C_3A but more slowly.

According to the respective proportions of each of these phases, the amount of water used, the specific surface area of the cement, the initial temperature of the concrete and the ambient temperature, more or less heat and more or less strength will be developed in the concrete. Therefore, a very simple means of following the evolution of the strength of the concrete is to follow the evolution of the temperature of the concrete, and vice versa: this is the principle behind maturitymeters.

Portland cement hydration is also accompanied by a reduction of the solid volume because the hydration of the dicalcium and tricalcium silicates that constitute the essential part of Portland cement and the formation of ettringite are accompanied by a reduction of the solid volume. In fact, if in order to hydrate a volume A of cement, it is necessary to use a volume B of water, to form a volume C of hydrate, it is always seen that $C < A + B$.

Even though researchers do not agree on the respective values of A, B and C, it is usually accepted that the reduction in the solid volume that is due to hydration is of the order of 10%, as was found around 100 years ago by Le Chatelier (1904). This volumetric contraction has some very important practical consequences on the hydration reaction.

In the case of the volumetric contraction, the bad habit has been developed of saying that this contraction of the solid volume is due to self-desiccation. This is only true when a sample of concrete is isolated from any external influence by sealing it in a perfectly tight envelope. Only in such a case does the volumetric contraction of the solid volume develop shrinkage when the very fine porosity created by cement hydration drains water from the coarser capillaries and thus results in the formation of menisci within the capillary system. The result of chemical contraction in this case is the same as if the concrete had dried; this is why the expression 'self-desiccation' is used to describe this phenomenon. It happens in any concrete whatever the W/B ratio in the absence of water curing.

However, if a sample of concrete is cured under water as soon as the hydration reaction starts to develop, the very fine porosity that is created by the volumetric contraction is immediately filled with water drained from coarse capillaries, but at the same time coarse capillaries drain water from the external source, as long as this porosity is connected to the external source of water. Therefore, in this case, the term self-desiccation is not at all appropriate to describe what is happening in the concrete during its hardening. The more appropriate expression is rather 'chemical contraction of the solid volume'. Chemical contraction is a general phenomenon that may or may not result in self-desiccation, depending on the way concrete is cured.

The evolution of concrete within the 'Bermuda Triangle' has very important practical implications. The more a concrete gains strength, the more it develops heat and the more its solid volume contracts, or, if a concrete can still gain some strength, it will develop a certain amount of heat during this strength gain, and its solid volume will decrease.

We will examine successively in the following sections what is happening at the apex of this famous Bermuda Triangle in order to understand better what is in fact the hydration reaction, and what are the consequences of its development on the development of shrinkage.

12.4.1 Strength

It is quite easy to determine experimentally that concrete gains strength with time, even when it is not so well cured, but it is still quite difficult at the present time to explain the profound causes of this strength increase (Van Damme, 1994). Thanks to the progress in electron microscopy it is now possible to follow all of the phase changes that can be observed at the micrometre level during concrete hardening. It is well known that it is the formation of the calcium silicate hydrate created during the hydration of dicalcium and tricalcium silicates that is essentially responsible for the strength gain of concrete, but it must be admitted that at the present time the exact composition and structure of this hydrated calcium silicate are not well known, so that prudently it is still written in the form of C–S–H. From a morphological point of view it is well accepted that C–S–H has a layered structure, but this structure is not as well defined and well known as the layered structures of two other closely related hydrated silicates: kaolinite (hydrated aluminium silicate) and serpentine or chrysotile asbestos (hydrated magnesium silicate). Moreover, it is also not well known how the different C–S–H crystallites are 'welded' together finally to give strength to concrete; this has been extensively discussed by Van Damme in a recent paper (Van Damme, 1994).

However, as has already been stated, when there is a strength gain between two ages, there is necessarily on the one hand some heat release that may or may not be manifested by an increase in temperature, according to the thermodynamic conditions in which this heat release is developed, and on the other hand a reduction of the solid volume that may or may not be accompanied by a contraction of the apparent volume, according to the curing conditions. When measuring the compressive strength of a perfectly sealed concrete sample cured in an adiabatic manner, or more simply when monitoring its temperature, it is also possible to follow the evolution of the volumetric contraction of the hydrates, as can be seen in Figure 12.2.

12.4.2 Heat

Portland cement hydration is always accompanied by a heat release. This heat release generally results in an increase in the temperature of the concrete, though this temperature increase depends on a great number of

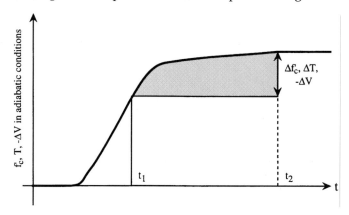

Fig. 12.2 Evolution of compressive strength, temperature and volumetric contraction of a concrete sample under adiabatic conditions.

factors, such as the cement dosage, the type of cement used, the initial temperature of the concrete, the nature of the aggregates, the ambient temperature, the thermodynamic conditions of curing, and the geometry and shape of the element in which the concrete is cast. The two extreme cases are: theoretical isothermal curing (constant temperature) and theoretical adiabatic curing (without any heat exchange with the exterior). Practically, concrete curing is neither isothermal nor adiabatic, it depends upon the particular conditions in which the hydration reaction is taking place.

From a practical point of view, there is a period of time, more or less long, during which there is no heat release. This is followed by an increase in the temperature which may be more or less rapid and more or less intense, during which the temperature of the concrete increases, and finally this is followed by a more or less long period during which the concrete returns to ambient temperature. This temperature evolution is schematically shown in Figure 12.3 and will be useful when discussing the most appropriate way to cure concrete in order to attenuate the effects of its volumetric changes.

12.4.3 Volumetric contraction

(a) Apparent volume and solid volume

It is convenient to define simply and clearly what is the difference between the apparent volume and the solid volume. The apparent volume of a material is the volume that is seen, without any consideration of the internal structure or porosity of the material. A prismatic

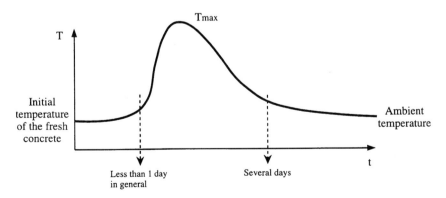

Fig. 12.3 Typical curve showing the evolution of the temperature of concrete in a structural element.

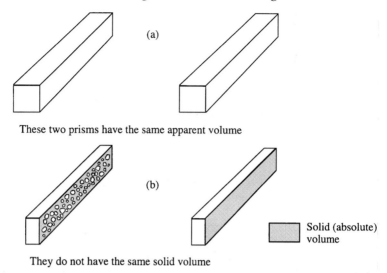

Fig. 12.4 Notion of apparent and solid volumes.

sample of concrete that has the external dimensions 300 × 300 × 750 mm has an apparent volume of 67 500 000 mm³ or 67.5 l, and a cylinder of 100 × 200 mm has an apparent volume of 1 570 000 mm³ or 1.57 l.

The solid volume or the absolute volume of a material corresponds to the part of the apparent volume that is really occupied by solid matter. The two prisms that have the same apparent volume in Figure 12.4(a) do not have the same solid volume, as shown in Figure 12.4(b). For example, concrete samples with different air contents can have the same apparent volume but not the same solid volume. Their mass will be a function of the value of the solid volume.

(b) Volumetric changes of concrete (apparent volume)

Like most other materials, the apparent volume of concrete changes due to the effect of temperature: concrete volume increases when it is heated and contracts when it is cooled, but concrete displays other volumetric changes that are particular to concrete: its solid volume decreases when cement hydrates and its apparent volume decreases when it is cured in air.

Each of these three volumetric changes can be the cause of different forms of shrinkage that have to be added algebraically in order to obtain the total shrinkage of a concrete.

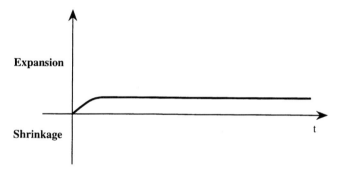

Fig. 12.5 Typical curve showing the evolution of the apparent volume of a concrete sample continuously cured in water.

(c) *Chemical contraction (absolute volume)*

When Portland cement hydrates, the absolute volume of the hydrates that are formed within the concrete is smaller than the sum of the solid volume of the cement and the volume of water that combined. This contraction of the solid volume develops as long as the hydration reaction goes on. Most people believe that this contraction of the absolute volume is almost automatically accompanied by a contraction of the apparent volume. This is not always the case: when usual concrete is cured continuously under water, its apparent volume does not decrease, but, rather has a tendency to increase, at least at the beginning of the water curing, as can be seen in Figure 12.5. The **solid volume** of usual concrete cured under water definitely **decreases**, but its **apparent volume increases**. This apparent contradiction is explained by considering the expansion of fresh concrete created by the growth of ettringite and portlandite crystals that are formed when concrete starts to harden or by the swelling of C–S–H crystals owing to water adsorption, as happens with clay materials. As usual concrete at this time is still a soft material with a very weak cohesion, the growth of these crystals pushes other solid particles apart and results in an increase of the apparent volume. As soon as the tensile strength of usual concrete is greater than the pressure created by the growth of the crystals that are formed, the apparent volume of usual concrete stops increasing.

Therefore, when usual concrete is cured under water it does not present any apparent volume contraction and shrinkage although its solid volume decreases.

Unfortunately for concrete, human beings are not mermaids, and do not live under water. That is why most concrete structures are built in air and will certainly be exposed to dry conditions one day. After that time, external drying has to be added to self-desiccation. Drying of concrete is always accompanied by a certain shrinkage, during which the apparent volume of concrete decreases; this shrinkage is called 'drying shrinkage'.

(d) The crucial role of the menisci in concrete capillaries in apparent volume changes

Let us see how drying and self-desiccation develop within concrete. Concrete drying is a result of the evaporation of a part of the water contained within the network of concrete capillaries that are connected to the surface, following an imbalance between the relative humidity of the ambient air and of the capillaries. Of course, it is the water that is contained in the very coarse pores and capillaries located near the surface of the concrete that leaves the concrete first, because the capillary forces are weaker there. However, as water continues to leave the concrete, menisci develop in finer and finer capillaries, and therefore the capillary forces that are developed within the concrete become stronger and stronger (Wittmann, 1968; Buil, 1979; Baron, 1982). But the stronger the capillary forces become, the more strongly the water is retained within the concrete and the more difficult it is for it to evaporate. This explains the shape of the mass loss versus time curves of concrete samples that are drying (Figure 12.6) (L'Hermite, 1978). The shape of these curves is a function of the total volume of the pores, the diameter of the pores, their degree of connectivity and their shape, the dimensions of the sample of concrete and the dryness of the ambient air.

The capillary forces that are developed within capillaries are inversely proportional to the diameter of these capillaries. As long as these capillary forces are lower than the tensile strength of the concrete, the concrete contracts in an elastic manner. If these capillary forces are greater than the tensile strength of the concrete, the paste cracks. However, **as long as concrete is water-cured there are no empty capillaries formed during hydration because they are filled with water; therefore no menisci are formed and no autogenous shrinkage develops within the concrete and no cracks appear.** It is only when the water/binder ratio is

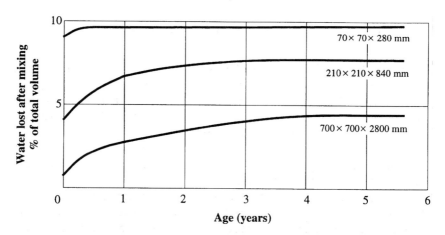

Fig. 12.6 Water lost from prisms of various sizes (relative humidity: 55%).

very low that some pores can be disconnected from the external source of water during the hydration process and some autogenous shrinkage can develop in spite of a permanent water curing.

The hydration reaction stops, of course, when there is no more water or no more cement to be hydrated or, as has been shown by Powers (1947, 1958), when the partial pressure of water in the capillaries decreases too much. At that time, the attraction that the unhydrated cement grains are exerting on the water is not strong enough to attract water that is retained in the capillaries by capillary forces, and the hydration reaction stops even though there are still unhydrated cement grains and water within the concrete.

(e) Essential difference between self-desiccation and drying

It has been seen that in some cases chemical contraction can result in the creation of menisci within concrete capillaries and a porosity filled with water vapour, as when concrete is drying. However, there exists a fundamental difference between these two phenomena, the result of which is the same: the appearance of menisci within the capillary system.

Self-desiccation develops progressively within the mass of a non-water-cured concrete in a homogeneous manner in so far as cement grains are homogeneously dispersed within the mass of concrete. On the other hand, drying is a localized phenomenon that starts to develop at the surface of concrete, where evaporation can take place. This phenomenon progresses within concrete more or less rapidly, depending on the compactness of the microstructure of the concrete and the dryness of the ambient air. Therefore, tensile strength gradients are developed at the skin of the concrete by drying, whereas in the mass of the concrete self-desiccation develops in a homogeneous manner.

(f) From the volumetric changes of the hydrated cement paste to the shrinkage of concrete

Self-desiccation and drying develop within the hydrated cement paste, which is only one component of concrete, the others being the aggregates which usually have a constant volume. The variation of the apparent volume of hardened concrete (which is called shrinkage) is thus related to the volumetric change of the hydrated cement paste and to the thermal volumetric change that is developed within the hydrated cement paste and the aggregates. The thermal shrinkage of the paste is always greater than that of the aggregates.

Of course, each of the elementary volumetric changes contributes to the global or total shrinkage of the hardened concrete, and therefore it is sometimes useful to divide the total shrinkage into autogenous shrinkage, drying shrinkage and thermal shrinkage.

Autogenous and drying shrinkage are linked to the corresponding volumetric changes of the hydrated cement paste, but these shrinkages will not develop freely, owing to the relative rigidity of the coarse aggregate skeleton, which will oppose the contraction of the apparent volume of the concrete (Alexander, 1996). Therefore the shrinkage of a hardened concrete is always smaller than that of a hydrated paste of the same water/binder ratio. It is well known that one easy way to reduce concrete total shrinkage is to increase the coarse aggregate content. However, it must be emphasized that if the total shrinkage of concrete is reduced by incorporating more coarse aggregate in the mix for the same volume of paste, there will be more microcracks within the paste because the paste is more restrained from shrinking and the shrinking of the paste has nothing to do with the amount of coarse aggregate within the concrete.

The coarse aggregate distribution is not homogeneous in a concrete structural element; the concrete just behind the form, the skin concrete, is richer in mortar and paste owing to the so-called 'wall effect'. The skin concrete will therefore develop more shrinkage and will develop it more freely, which can result in the development of larger cracks than in the mass of the concrete, where the coarse aggregate skeleton opposes the development of large fissures.

The shrinkage of the skin of a high-performance concrete can be greater than that of a usual concrete because this skin is richer in paste (Sicard et al., 1992; Sicard, Cubaynes and Pons, 1996), making water curing more imperative. Of course, the longer the water curing, the better.

12.5 CONCRETE SHRINKAGE

The shrinkage of hardened concrete that is not water-cured corresponds to the addition of its thermal shrinkage (if any), its autogenous shrinkage and its drying shrinkage (carbonation shrinkage that can develop thereafter is not considered here). We can write:

$$\varepsilon_{tot} = \varepsilon_{thermal} + \varepsilon_{autogenous} + \varepsilon_{drying}$$

When calculating ε_{total} we consider that $\varepsilon_{thermal}$ is positive when the concrete is cooling and that $\varepsilon_{thermal}$ is negative when the concrete temperature is increasing. Autogenous shrinkage can be equal to 0 if the concrete is cured under water and drying shrinkage can be equal to 0 if the concrete is covered by an impervious film so that none of its water molecules can leave it. However, autogenous and drying shrinkage cannot both be equal to 0 at the same time, except if the concrete is always cured under water.

12.5.1 Shrinkage of thermal origin

Like any other material, concrete can be subjected to volumetric changes of thermal origin. However, where concrete behaves differently from other construction materials is that while it is gaining strength its temperature increases, and therefore the mechanical links that are developed within the concrete during its hardening are formed at a temperature that is not necessarily the one that it will have during its service life. Moreover, these links are created at a temperature that evolves during cement hydration, as was schematically shown in Figure 12.3. After the maximum temperature is reached, thermal losses are finally greater than the amount of heat liberated by hydration, so that concrete starts to cool, to return to ambient temperature.

The physical laws that govern these heat exchanges within concrete are well known and it is quite easy to develop models in order to predict the variation of the temperature of a concrete cast in a specific structural element exposed to specific ambient conditions when it cools down.

12.5.2 How to reduce autogenous and drying shrinkage and its effects by appropriate curing of high-performance concrete

It has been seen that if concrete is cured under water it does not undergo any shrinkage. However, in practice concrete structures cannot be cured forever by sprinkling some water on them. Therefore, one day concrete curing is stopped and some autogenous shrinkage and drying shrinkage start to develop. In order to reduce the shrinkage of high-performance concrete, it is only necessary to delay as long as possible the time when water curing is stopped. The question is, how long is long enough? We will see that it is difficult to give a precise answer to such a question, but a good knowledge of the way cement hydrates and how each of the elementary shrinkages develops helps, in each particular case, in selecting curing conditions that will result in a reduced shrinkage when water curing stops.

Therefore, it is necessary to cure high-performance concrete with water as soon as it starts to hydrate and for as long as possible. The more rapidly cement hydrates, the more rapidly it must be cured, and the more critical it is to cure concrete. Of course, during this curing under water there will not be any drying shrinkage.

The curves showing the temperature increase of concrete under adiabatic conditions in Figure 12.7 can be used to determine when to stop water curing so that autogenous shrinkage remains within acceptable limits and can be developed within the concrete without causing any cracking because the tensile strength of the concrete is high enough. When water curing is stopped self-desiccation develops.

In order to eliminate drying shrinkage it is only necessary to prevent the water contained in high-performance concrete from evaporating (Bissonnette, Pierre and Pigeon, 1997). Therefore, it is necessary to cover concrete with an absolutely impervious membrane so that not a single water molecule can be lost. But this high-performance concrete can still develop some autogenous shrinkage as long as some cement particles

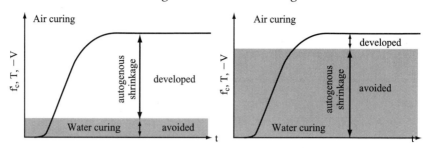

Fig. 12.7 Induced self-desiccation shrinkage as a function of the duration of water curing.

continue to hydrate or until the partial pressure of water vapour within the capillaries drops below 80%, as found by Powers (1947).

We are not used to painting concrete with a completely impervious film, but it is necessary to evaluate the cost of such an additional curing treatment and compare it with the increased durability of the structure when dealing with high-performance concrete. If an economic analysis shows that the development of drying shrinkage does not decrease the durability of a structure in a significant manner, it is not necessary to paint a high-performance concrete with an impervious film, and it will be more economical to let the high-performance concrete dry. If, on the contrary, an economic study shows that it is better to invest in painting the high-performance concrete with an impervious membrane, it will have to be done. After all, we are used to seeing steel structures painted in order to delay their rusting, so why not paint the high-performance concrete in order to avoid its drying?

Coming back to the curve representing the increase in the temperature of a high-performance concrete in a structural element, it is easy now to see in Figure 12.8 how to select the most appropriate curing conditions during high-performance concrete hardening so that plastic, autogenous and drying shrinkage can be minimized during the hydration process.

In Figure 12.8 it can be seen that the best way to protect a high-performance concrete against shrinkage is to start by protecting it using a temporary curing membrane or fog misting, or ponding, as long as it is plastic. Then the high-performance concrete must be cured by spraying water on it or fog misting or ponding, as long as the hydration reaction is progressing, and finally the high-performance concrete must be covered by an impervious film as soon as water curing is stopped.

12.6 WHY AUTOGENOUS SHRINKAGE IS MORE IMPORTANT IN HIGH-PERFORMANCE CONCRETE THAN IN USUAL CONCRETE

Every chemical contraction of hydrated cement paste results in a 10% decrease in its absolute volume. Therefore, the following simplistic

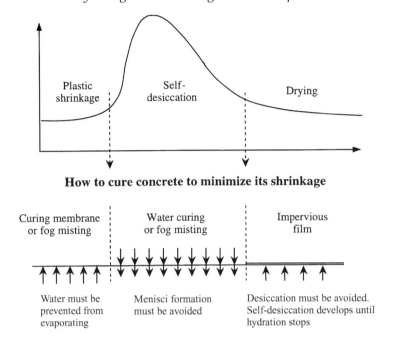

Fig. 12.8 The most appropriate curing regimes during the course of the hydration reaction.

reasoning can be made: as high-performance concretes have a higher cement content than usual concretes, high-performance concretes shrink more. However, such reasoning is absolutely wrong. In fact, as in the case of usual concretes, if a high-performance concrete is continuously cured under water as soon as hydration of the cement starts, a slight increase in the apparent volume is observed in the early stage of hydration. Then, no further changes in the apparent volume of the concrete are observed except when the water/binder ratio is so low that the pore system can become disconnected when hydration proceeds. Therefore, high-performance concrete shrinkage is essentially linked to the curing or absence of curing, and not to its cement content.

It is necessary to go back to the fundamental causes of the different types of shrinkage occurring in any concrete to understand why self-desiccation, if it is out of control, becomes the major cause of shrinkage in any high-performance concrete, which is not true for usual concretes. This difference in the behaviour of high-performance concrete and usual concrete can be explained by the major differences that can be observed at the microstructural level between the two types (Hua, Acker and Erlacher, 1995; Hua, Erlacher and Acker, 1995; Baroghel-Bouny, Godin and Gawsewitch, 1996).

The capillary network existing in a usual concrete becomes coarser and coarser as the water/binder ratio of the concrete is increased. In such a case, if owing to an initial lack of water curing, cement particles have to be hydrated with only the water introduced into the concrete during its batching, it is the water contained in the coarser capillaries that will be drained by the very fine porosity generated by the chemical contraction. This water is easily drained because it is weakly retained in these big capillaries. The tensile forces accompanying the creation of menisci in coarse capillaries are not very strong, and the resulting autogenous shrinkage is very small or negligible. Of course, when the water contained in the coarser capillaries has been drained, the porosity created by the volumetric contraction will drain the water contained in the finer capillaries and the corresponding tensile stresses will increase. But in usual concretes, most of the mixing water is contained in very coarse capillaries that contain a large volume of water, and as there is not very much cement in them, tensile stresses never become high enough to result in significant autogenous shrinkage.

At the surface of a usual concrete, drying occurs rapidly and progresses deeply as this concrete is porous and has coarse capillaries, owing to its high water/binder ratio. Therefore, drying shrinkage rapidly becomes the predominant shrinkage at the surface of any usual concrete, and it governs the total shrinkage of that concrete.

It can be concluded that, generally speaking, chemical contraction of the cement paste develops a very low autogenous shrinkage in usual concretes, so the final shrinkage is mostly due to concrete drying.

In the case of high-performance concrete of low water/binder ratio, in the absence of any water curing, autogenous shrinkage develops rapidly because the water drained by the very fine porosity resulting from volumetric contraction is drained from capillaries that already have a small diameter. Moreover, in the case of high-performance concrete, when the hydration reaction starts it develops very rapidly, so that water is drained rapidly from capillaries that are finer and finer, and high tensile stresses are developed rapidly, which results in the development of a rapid autogenous shrinkage. Therefore, autogenous shrinkage develops very rapidly and intensely in high-performance concrete (Tazawa and Myazawa, 1995). Jensen and Hansen (1995, 1996) have measured autogenous shrinkage as high as 3000 and 1500 microstrains in 0.25 W/C cement paste, while Tazawa and Myazawa (1993) have measured autogenous shrinkage of low water/binder ratio paste as high as 4000 microstrains and concrete autogenous shrinkage as high as 600 microstrains.

At the surface of a high-performance concrete, drying develops more slowly because high-performance concrete porosity is globally very fine, and because it has already been drained by the self-desiccation process. Drying shrinkage therefore develops more slowly in high-performance concrete than in usual concrete.

Therefore, high-performance concrete shrinkage will be essentially influenced by self-desiccation, while in usual concrete shrinkage is essentially due to drying.

It is therefore normal to observe experimentally that when a high-performance concrete is air-dried at a very early age, less than 24 hours, its initial shrinkage is usually greater than that of a usual concrete. Of course, it is also normal to find that the total shrinkage of any usual concrete is a function of the amount of mixing water used when preparing the concrete, which is not the case for high-performance concrete.

12.7 IS THE APPLICATION OF A CURING COMPOUND SUFFICIENT TO MINIMIZE OR ATTENUATE CONCRETE SHRINKAGE?

The answer is, 'it depends'.

If the concrete that is protected by a curing membrane is a concrete with a high water/binder ratio (greater than 0.40), in principle it contains enough water for all of the cement particles to be completely hydrated with the amount of initial mixing water it contains. The higher the water/binder ratio, the better the concrete will be protected against autogenous shrinkage. As soon as the curing membrane loses its efficiency or disappears, the concrete will no longer be protected against drying and drying shrinkage will develop. Drying shrinkage will be the essential part of concrete shrinkage.

If the concrete that is protected by a curing membrane has a water/binder ratio lower than 0.40, self-desiccation will develop very rapidly within the concrete if the concrete is not properly water cured. The lower the water/binder ratio, the higher the autogenous shrinkage. This explains why it is very often found, or almost always found, that a high-performance concrete with a very low water/binder ratio has an initial shrinkage that is greater than that of a concrete with a higher water/binder ratio. In this case, when the curing membrane disappears or loses its efficiency, drying shrinkage will develop as in the preceding case.

Therefore, curing compound membranes that were acceptable to cure concrete with a high water/binder ratio (0.50 to 0.70) are no longer sufficient to cure high-performance concrete with low water/binder ratios. The lower the water/binder ratio, the higher the autogenous shrinkage.

12.8 THE CURING OF HIGH-PERFORMANCE CONCRETE

There are two sure ways to decrease the durability of a reinforced concrete structure: first, the use of low-strength concrete with a high water/binder ratio, and second, the elimination of the curing of concrete whatever its water/binder ratio, so that this concrete shrinks as much as possible and develops as many cracks as possible. When both occur simultaneously, the durability of any concrete structure is shortened dramatically because the penetration of aggressive agents is favoured, so they can attack very rapidly either the concrete or the reinforcing steel rebars or both at the same time.

When using a high-performance concrete with a low water/binder ratio, only the penetration of aggressive agents through the concrete itself is decreased. In order to avoid the penetration of the aggressive agents through cracks, it is absolutely essential to cure high-performance concrete properly to eliminate these cracks as much as possible.

All specifications and textbooks repeat that it is absolutely necessary to cure concrete. In all specifications there are always several paragraphs describing in detail how concrete has to be cured in the field, but unfortunately all too often these specifications are not enforced strictly or are not implemented at all. Contractors have always been very convincing when they claim that it is unavoidable that concrete cracks and that there is nothing they can do about it.

In fact, appropriate curing is always essential to lower as much as possible the final shrinkage of any usual concrete so that it does not crack, but, as we shall see in the following, appropriate curing is much more crucial in the case of high-performance concrete.

The dramatic importance of appropriate curing for high-performance concrete constitutes one fundamental difference between usual concrete and high-performance concrete. In usual concrete, the final shrinkage is essentially of thermal origin and due to drying, while in high-performance concrete, severe shrinkage can develop:

- in the plastic state, owing to the very low bleeding rate of high-performance concrete (Samman, Mirza and Wafa, 1996);
- in the hardened state, owing to self-desiccation following the rapid and intense development of the hydration of cement (Tazawa and Myazawa, 1996),
- in the hardened state, owing to thermal gradients created by the non-homogeneous development of temperature within the mass of concrete during cooling.

In high-performance concrete, drying shrinkage is not so important, because the concrete microstructure is very compact. This is because the water contained in the capillaries is tightly held in very small capillaries and also because the partial pressure of water vapour in the capillaries has already decreased as a result of the self-desiccation process when drying starts to develop.

Therefore, if durable high-performance concrete structures are to be built, it is essential to fight appropriately each of these three elementary shrinkages as they develop. Unfortunately, there is no universal, foolproof way to cure high-performance concrete, although there are always

The curing of high-performance concrete

the same three fundamental mechanisms that reduce the apparent volume of concrete. In some high-performance concrete structural elements, one particular mechanism can take on a particular importance so that it will be responsible for most of the final shrinkage.

In this respect, the author believes that the recent development of admixtures that reduce concrete shrinkage (Tomita, 1992; Balogh, 1996) is very interesting, but he is not sure that the best way to use them is to introduce them within the concrete during its mixing. Would it not be better to use them at the end of the water curing? As far as the author knows, these admixtures are only reducing the tensile stresses and contact angle developed in the menisci within the capillary pores.

As seen previously, the application of a curing compound membrane or of a very brief water curing that is often sufficient to reduce the shrinkage of most usual concretes to a tolerable level is no longer appropriate in the case of high-performance concrete. The more appropriate way to cure high-performance concrete is very simple in principle: high-performance concrete must be water cured as long as possible. However, this is not always easy to implement in all the parts of any structure, and there is always the eternal question: how long is long enough?

It will be seen below, in some particular cases, what can be an appropriate curing regime and how to implement it in order to minimize the major causes of shrinkage, so that a high-performance concrete structural element does not become a structural element built with the most impervious concrete in between two cracks.

12.8.1 Large columns

Let us examine the volumetric changes of a high-performance concrete at three particular points, A, B and C, of a large column (1 × 1 m and 2.5 m high, for example) as they appear in Figure 12.9.

(a) Volumetric changes at A

At A, there is no plastic shrinkage and no drying shrinkage. In the latter case, it can be shown that it will take years, if not a century, before drying

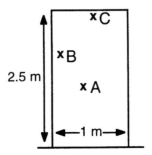

Fig. 12.9 Example of a large column.

reaches A, if it reaches it. The only shrinkages that are developed at A are autogenous and thermal shrinkages. It is at A that the thermal shrinkage will almost all the time be the maximum because it is at A that the maximum temperature will be reached. It is also the point at which the thermal gradient could be high during cooling. If it is not too difficult to predict the thermal shrinkage at A, it is rather difficult to predict theoretically the autogenous shrinkage. However, it is possible to assert that at A the hydration reaction will generally stop when the partial pressure of the water in the pores reaches 80% following the self-desiccation process.

At the centre of the Lavalin Building and the Concordia University library experimental columns, autogenous shrinkage decreased drastically after the fourth day following the placing of concrete, when the concrete temperature was back to ambient temperature (El Hindy et al., 1994; Dallaire, Lessard and Aïtcin, 1996).

As at A, at the centre of the column, autogenous shrinkage is developing while the concrete is experiencing a very rapid increase in tensile strength, and therefore there is no risk of the development of cracks. Moreover, when thermal shrinkage starts to develop, the concrete has usually already developed more than 50% of its final strength, so that there is usually no risk of cracking.

(b) Volumetric changes at B

At B there is also no plastic shrinkage, but in this case drying shrinkage will be added to autogenous shrinkage and thermal shrinkage as soon as the forms are released if no curing is done. Of course, thermal shrinkage at B is much lower than at A because at B the concrete temperature never gets as high as at A. But the nature of the material used to make the form is crucial, because significant thermal gradients can be developed in concrete when it cools. Experience shows that at B thermal gradients are low when plywood forms are used, but not when steel forms are used.

The following very simple rule can be stated: in order to limit the risk of cracking of thermal origin behind the forms when using a high-performance concrete, it is better to use plywood forms or insulated steel forms.

Moreover, in order to limit the effects of self-desiccation at B, it is only necessary to release the forms slightly as soon as the high-performance concrete has started to hydrate (or to increase its temperature) and to let a small amount of water run on the concrete surface just to dampen it. It is better to pursue this water curing for 7 days. After 7 days about 75% of the autogenous shrinkage due to self-desiccation should have been eliminated because the concrete will have reached about 75% of its strength. Moreover, this kind of water curing retards the development of the drying shrinkage until an age when the concrete is stronger.

If drying shrinkage has to be eliminated, it is only necessary to paint the column surface with an impervious film. However, this impervious film will not stop concrete shrinkage totally because the self-desiccation process can continue behind this film until hydration stops, resulting in autogenous shrinkage.

When the only curing given to a high-performance concrete column is to leave the forms on it for 7 days, little has been done to prevent shrinkage. The only thing done has been to delay drying shrinkage until the concrete has reached a higher tensile strength, but nothing has been done to reduce autogenous shrinkage, which is the major cause of shrinkage for a high-performance concrete.

(c) Volumetric changes at C

It is of course at C that high-performance concrete can suffer most from a lack of curing, but it is also the easiest place for an appropriate curing to be implemented. At C, plastic shrinkage, autogenous shrinkage and desiccation shrinkage can develop in a cumulative way while high-performance concrete is not at all strong. The only shrinkage that does not develop at C is thermal shrinkage, because thermal exchange is so easy at C that the high-performance concrete temperature is almost equal to the ambient temperature. Experience also shows that thermal gradients are quite low at C.

In order to prevent high-performance concrete from cracking at C, it is absolutely necessary either to cover the concrete surface with a temporary curing compound to avoid plastic shrinkage or to expose it to fog misting as soon as the placing of the concrete has been finished. Moreover, as soon as the concrete starts to develop some heat, water curing must be applied even when a curing compound has been used, or fog misting must be continued, so that the concrete surface at C is kept continuously wet.

12.8.2 Large beams

By a 'large beam', the author means a beam in which the smaller dimension is greater than 500 mm, so that thermal effects will develop within the mass of the beam, not as high as in the previous example but sufficiently large that thermal shrinkage cannot be considered negligible.

In this case, too, we will consider the volumetric changes at three particular points, A, B and C, as shown in Figure 12.10. Points A and C can be treated as in the preceding case except that in A thermal shrinkage will not be as high.

The more critical zone this time is B, because it is this part of the beam that will be subjected to tensile stress. Moreover, at B the forms are usually maintained as long as possible, so autogenous shrinkage can develop to its maximum if no corrective measures are taken. The release

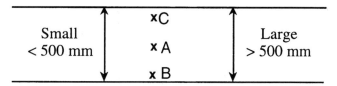

Fig. 12.10 Example of large and small beams.

of the forms and water curing as soon as the high-performance concrete starts to develop its strength can limit the effect of self-desiccation at B.

The material used to make the forms is in this case more important than in the previous one in order to limit thermal gradients. It is better to use plywood forms, or insulated metallic forms if metallic forms have to be used.

12.8.3 Small beams

A small beam is a beam in which the smaller dimension is lower than 500 mm (Figure 12.10). It is a beam in which the temperature of the high-performance concrete will not rise too much above the ambient temperature or above the initial temperature of the concrete, so thermal shrinkage is not a problem at A, B or C.

As in the case of the large beam, it is still at B that the uncontrolled development of autogenous shrinkage can be the most harmful, and the same kind of water treatment can be applied after a slight release of the forms. In this case, too, it is easy to protect the beam from plastic shrinkage at C.

However, drying shrinkage in a small beam is no longer a negligible shrinkage, owing to the high ratio existing between the surface of the beam and the volume of the concrete. It is in this kind of structural element that drying must be minimized by painting the concrete surface with an impervious film as soon as the forms are released.

It is appropriate to remember that in this case while it is still better to use plywood forms or insulated steel forms, this is not as crucial as in the previous cases.

12.8.4 Thin slabs

A thin slab is a slab with a thickness of less than 300 mm (Figure 12.11). Such a structural element represents the case in which a lack of curing

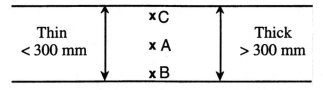

Fig. 12.11 Example of thin and thick slabs.

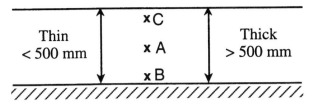

Fig. 12.12 Example of thin and thick slabs on grade.

can be the most dramatic but also the case in which proper curing is the easiest to implement.

In such a slab, there are no thermal problems because the temperatures at A, B and C never get much higher than the high-performance concrete's initial temperature or the ambient temperature. On the other hand, plastic shrinkage is a critical factor at C as well as autogenous and drying shrinkage if the high-performance concrete is not cured properly. Each of the two resulting shrinkages can be fought easily.

A curing compound membrane can be applied on the high-performance concrete just after its placing, or fog misting can be implemented. In order to eliminate autogenous shrinkage, it is necessary to apply water curing as soon as the concrete temperature starts to rise, even in the case where a curing membrane has been applied, or to continue fog misting for at least 7 days.

For a thin high-performance concrete slab on a grade (Figure 12.12), it is particularly important that the lower part of the slab is not in contact with a source of humidity. In such a case, the lower part of the slab which is in contact with an external source of water hardens in ideal curing conditions and does not develop any shrinkage. On the other hand, autogenous shrinkage and drying shrinkage develop very rapidly at the top of the slab, so the thin slab curls. A draining layer of crushed stone or gravel and/or the use of a plastic membrane below the slab is recommended to avoid curling; it is also recommended that a smaller joint spacing is used when a slab on a grade is not reinforced.

12.8.5 Thick slabs

A thick slab is, of course, a slab thicker than 300 mm (Figure 12.11). While in the case of a thin slab there are practically no thermal problems, this is no longer the case in thick slabs because the temperature at A can rise quite high, and because thermal gradients at A, B and C can be quite high when the concrete is just starting to harden or to cool. Moreover, owing to the temperature difference, the concrete hardens more rapidly at A than at B and even more rapidly than at C.

In order to limit the detrimental effects of early thermal gradients, it is essential to use the coldest concrete possible and to make this concrete with a cement having a low heat of hydration, or to use supplementary cementitious materials. The use of ice to decrease the concrete's initial temperature is recommended (this costs less than 10$/m^3 in Canada). It is also suggested that high-performance concrete is cast during the night when the ambient temperature is lower.

When dealing with thick slabs on a grade it is therefore very important to try to limit thermal shrinkage problems (Figure 12.12). Of course, at C the same precautions to limit the development of the different shrinkages apply here; a 7 day water curing is essential. Decreasing the temperature of the soil near B before concreting is recommended when feasible.

Thick slabs on a grade are less prone to curl, but it is always better to place a drain at the bottom in order to eliminate the risk of curling or to use plastic sheets to avoid contact between the bottom part of the slab and an external source of water.

12.8.6 Other cases

It is not possible here to consider all of the situations that might be faced with the various structural elements made with any kind of high-performance concrete. However, if the previous curing measures that have been recommended in the particular cases treated have been well understood, the appropriate means of reducing shrinkage in the most critical places of any other structural element can be easily devised. Moreover, if the fundamental causes of each particular shrinkage have been understood, defining the most appropriate way to cure a high-performance concrete structural element involves only common sense.

Finally, it can never be emphasized enough that shrinkage is not an unavoidable phenomenon: if usual concrete is always cured under water, it does not usually shrink. Shrinkage development in concrete is the result of a lack of water curing; it has nothing to do with the amount of cement used to make the concrete.

Water curing is important for usual concretes. It is crucial for high-performance concretes.

12.9 HOW TO BE SURE THAT CONCRETE IS PROPERLY CURED IN THE FIELD

The curing of concrete in the field is often neglected, and it is the nightmare of all inspectors to be sure that curing specifications are implemented. Very often this is a lost battle and the contractor is almost always the winner; much of the time concrete is not cured at all or at least not as much as it should be. Concrete must be a very tough construction material in order to be able to survive all of this lack of curing!

It could be quite easy to solve this problem when looking at the way responsibilities are shared in the field and the way concrete is used in the field. The contractor and the laboratory that is controlling the quality of the concrete are paid at the present time from two different budgets, and the sum of these two budgets represents the total cost of the project.

The contractor and the laboratory want to maximize their profits. For the contractor, the easiest way to maximize profits is to lower as much as possible the manpower cost, and in that respect concrete curing represents a gold mine that the contractor exploits as soon as inspection is relaxed. On the other hand, the laboratory that is checking concrete quality is paid by the hour or by the number of tests done, and all owners know that they very often have to moderate overly zealous inspectors who are prone to charge hours for tests. Therefore, as concrete curing is as crucial as checking the quality of concrete when it is delivered, why not pay for it item by item, by the hour? In that case, it is absolutely certain that concrete curing will be the object of zealous attention.

When writing the specification and determining the importance of the budget that will be allocated to the control of the quality of the concrete and its curing, the engineer indirectly will decide on the importance given to the control of the quality of the concrete and to its curing and thus indirectly to the quality of the structure. In fact, in order to build a durable structure, it is necessary not only to check that the concrete has the quality needed, but also that it is cured properly. How many concrete structures built with a good quality concrete have seen their service life dramatically shortened because this quality concrete was badly cured? Our societies are not rich enough to tolerate such a lack of curing which is so costly in terms of the service life of concrete structures.

12.10 CONCLUSION

If water curing is essential to usual concrete, it is crucial to high-performance concrete. If water curing is applied as soon as high-performance concrete temperature begins to rise, most of the autogenous shrinkage can be eliminated. The absolute volumetric contraction accompanying the hydration reactions occurs, but as external water penetrates the concrete as soon as very fine pores are formed owing to chemical contraction, there are no menisci formed within the capillaries and therefore no tensile stresses, so no shrinkage of the apparent volume occurs in spite of the contraction of the absolute volume that is developed.

The best way to cure high-performance concrete is to protect it against plastic shrinkage with a temporary curing membrane or by mist fogging or ponding **as soon as it has been finished**, then to cure it with water or continuing mist fogging, and then to paint it when curing stops in order to avoid the development of any external drying. How long must the water curing be? The longer the better of course, but 7 days seems to be a period long enough to drastically reduce high-performance concrete shrinkage. In any case water curing should never be shorter than 3 days. Depending on the size and the type of the structural element in which the concrete is used, different appropriate curing solutions will have to be implemented.

REFERENCES

Aïtcin, P.-C. and Neville, A.M. (1993) High-performance concrete demystified. *Concrete International*, **15**(1), 21–6

Aïtcin, P.-C., Neville, A. and Acker, P. (1997) Integrated view of shrinkage deformation. *Concrete International*, **19**(9), September, 35–41.

Alexander, M.G. (1996) Aggregates and the deformation properties of concrete. *ACI Materials Journal*, **93**(6), November–December, 569–77.

Balogh, A. (1996) New admixture combats concrete shrinkage. *Concrete Construction*, **41**(7), July, 546–51.

Baroghel-Bouny, V., Godin, J. and Gawsewitch, J. (1996) *Microstructure and Moisture Properties of High-Performance Concrete*, 4th International Symposium on Utilization of High-Strength/High-Performance Concrete, Paris, pp. 451–61.

Baron, J. (1982) *Les Retraits de la Pâte de Ciment, Le Béton Hydraulique*, Presses de l'École Nationale des Ponts et Chaussées, ISBN 2-85978-033-5, pp. 485–501.

Bissonnette, B., Pierre, P. and Pigeon, M. (1997) Influence of key parameters on drying shrinkage of cementituous materials. Accepted for publication in *Cement and Concrete Research*.

Buil, M. (1979) 'Contribution à l'étude du Retrait de la Pâte de Ciment durcissante. Rapport de recherche LPC No 92, ISSN 0222-8394, December, 72 pp.

Dallaire, E., Lessard, M. and Aïtcin, P.-C. (1996) *Ten Year Performance of High-Performance Concrete Used to Build Two Experimental Columns*, ASCE Structures Congress – High Performance Concrete Columns, Chicago, April, 10 pp.

El Hindy, E., Miao, B., Chaallal, O. and Aïtcin, P.-C. (1994) Drying shrinkage of ready-mixed high-performance concrete. *ACI Structural Journal*, **91**(3), May–June, 300–305.

Hua, C., Acker, P. and Erlacher, A. (1995) Retrait d'autodessiccation du ciment. Analyse et modélisation macroscopique. *Bulletin de Liaison des Laboratoires des Ponts et Chaussées*, No. 196, March–April, pp. 79–89.

Hua, C., Erlacher, A. and Acker, P. (1995) Retrait d'autodessiccation du ciment. Analyse et modélisation – II. Modélisation à l'échelle des grains en cours d'hydratation. *Bulletin de Liaison des Laboratoires des Ponts et Chaussées*, No. 196, September–October, pp. 35–41.

Jensen, O.M. and Hansen, P.F. (1995) A dilatometer for measuring autogenous deformation in hardening Portland cement paste. *Materials and Structure*, **28**(181), August/September, 406–9.

Jensen, O.M. and Hansen, P.F. (1996) Autogenous deformation and change of the relative humidity on silica fume – modified cement paste. *ACI Materials Journal*, **93**(6), November–December, 539–43.

Le Chatelier, H. (1904) *Recherches Expérimentales sur la Construction des Mortiers Hydrauliques*. Dunod, Paris, pp. 163–7.

L'Hermite, R.G. (1978) Quelques problémes mal connus de la technologie du béton. *Il Cemento*, **75**, 231–46.

Neville, A.M. (1995) *Properties of Concrete*, 4th edn, John Wiley, New York, and Longman, England, 844 pp.

Powers, T.C. (1947) A discussion of cement hydration in relation to the curing of concrete. *Proceedings of the Highway Research Board*, **27**, 178–88.

Powers, T.C. (1958) Structure and physical properties of hardened Portland cement paste. *Journal of the American Ceramic Society*, **41**, January, 1–6.

Samman, T.A., Mirza, W.H. and Wafa, F.F. (1996) Plastic shrinkage cracking of normal and high-strength concrete: a comparative study. *ACI Materials Journal*, **93**(1), January–February, 36–40.

Sicard, V., Cubaynes, J.-F. and Pons, G. (1996) Modélisation des déformations différées des bétons à hautes performances: relation entre le retrait et le fluage. *Materials and Structures*, **29**, July, 345–53.

Sicard, V., François, R., Ringot, E. and Pons, G. (1992) Influence of creep and shrinkage on cracking in high-strength concrete. *Cement and Concrete Research*, **22**(1), 159–60.

Tazawa, E. and Myazawa, S. (1993) Autogenous shrinkage of concrete and its importance in concrete technology, in *Creep and Shrinkage of Concrete* (eds Z.P. Bazant and I. Carol), E and FN Spon, London, ISBN 0419 18630 1, pp. 159–68.

Tazawa, E. and Myazawa, S. (1995) Experimental study on mechanism of autogenous shrinkage of concrete. *Cement and Concrete Research*, **25**(8), December, 1633–8.

Tazawa, E. and Myazawa, S. (1996) *Influence of Autogenous Shrinkage on Cracking in High-Strength Concrete*, 4th International Symposium on Utilization of High-Strength/High-Performance Concrete, Paris, ISBN 2-85878-258-3, pp. 321–30.

Tazawa, E.I., Myazawa, S. and Kasai, T. (1995) Chemical shrinkage and autogenous shrinkage of hydrating cement paste. *Cement and Concrete Research*, **25**(2), 288–92.

Tomita, R. (1992) A study on the mechanism of drying shrinkage reduction through the use of an organic shrinkage reducing agent. *Concrete Library of JSCE*, No. 19, June, pp. 233–45.

Van Damme, H. (1994) Et si le Chatelier s'était trompé – pour une physico-chimio-mécanique des liants hydrauliques et des géomatériaux. *Annales des Ponts et Chaussées*, No. 71, pp. 30–41.

Wittmann, F.H. (1968) Surface tension shrinkage and strength of hardened cement paste. *Materials and Structure, RILEM*, No. 6, pp. 547–52.

Wittmann, F.H. (1976) On the action of capillary pressure in fresh concrete. *Cement and Concrete Research*, **6**(1) 49–56.

CHAPTER 13

Properties of fresh concrete

13.1 INTRODUCTION

It is important to control the properties of fresh high-performance concrete for two major reasons. First, high-performance concrete must be as easily placed as usual concrete; second, high-performance concrete whose fresh properties are under full control is likely to yield a hardened concrete whose properties are also well controlled. This second point is true for usual concrete, but as the margin of error is much smaller when making high-performance concrete, greater attention has to be paid to the properties of the fresh concrete. The properties of fresh high-performance concrete have to be controlled both at the concrete plant just after mixing and at the jobsite prior to placement. At the plant, unit mass, slump, air content and temperature have to be controlled, while at the jobsite only slump and air content (when air-entrained) should be regularly verified.

The control of the properties of fresh concrete at the batching plant avoids problems during delivery and placement. As previously discussed in Chapters 6 and 7, the rheology of low water/binder ratio mixtures is very sensitive to any changes in the quality of the materials, the formulation of the concrete and the temperature. At this stage, the main problems that are faced are:

- a dramatic and irrecoverable slump loss;
- a strong set retardation that severely delays strength development;
- a dramatic segregation in the highly fluid mixture.

The slump loss problem has been discussed in section 9.4 from a materials point of view. Such a problem should be avoided when a rigorous quality control programme is adopted for the different materials used to make high-performance concrete when received and when the batching process is well controlled. However, slump loss can take place and when it does it is very important to notice it, process it as quickly as possible, identify the causes of the problem and take adequate measures to correct it.

An extensive set retardation can take place when the superplasticizer dosage used is quite high or when a set-retarding agent is used to solve the slump loss problem. An extensive set retardation can be costly, especially in a precast operation when it delays the removal of the forms or the cutting of the pre-tensioned tendons. An over-retardation can also have dramatic consequences when slipforming or when jumping forms are used, because in these cases any delay in the hardening of the concrete definitely results in a slowdown of the construction rate.

Severe segregation is observed from time to time when, for any reason, an error in the dosage of superplasticizer or in the amount of mixing water occurs. The cement paste becomes too fluid and can no longer sustain the aggregates in suspension. Unfortunately, it is very difficult to save such a segregated mixture. In one particular case, experienced by the author, it was the wash water left by the truck driver within his transit mixer that caused a severe segregation.

It is very important to measure systematically the properties of fresh high-performance concrete just after mixing. This is the easiest way to figure out if something has gone wrong during the manufacturing process and to enable quick remedial action. If the properties of the fresh concrete change, it is not always easy to determine whether the problem comes from the materials (changes in their quality, in the water content of the aggregates, etc.), from the batching (inadequate formulation, wrong cement, wrong aggregates, wrong sequence), from the equipment used (blockage of a trap) or from the admixture dispersing system, etc., but it is certain that something has gone wrong somewhere. As high-performance concrete is more sensitive to variations in the mix because the materials are working close to, or at, their maximum potential, any deviation from the optimized characteristics of the materials and their mixing can strongly modify the properties of both the fresh and the hardened concrete.

It is easy, economical and not time-consuming to control the properties of the fresh concrete in the case of high-performance concrete because these properties are measured in the same way as those of usual concrete. They are measured using the well-accepted simple tests that are already used in the construction industry. They do not require any sophisticated or expensive equipment, and they are rapid. So why not take advantage of the situation? Moreover, if these results are treated statistically they provide useful information regarding the efficiency of the materials and of the mixing operation, as shown in Chapter 9. However, all too often these tests are not systematically carried out, because they seem to be so simple that many people believe they cannot provide any useful information. This attitude is unfortunate because often concrete producers do not take advantage of these inexpensive tools that give them an early warning of trouble, or a particular satisfaction in seeing that their high-performance concrete is being produced uniformly.

In this chapter the advantages of controlling the unit mass, the slump and the air content will be reviewed. The set retardation that can occur during the hardening of some high-performance concretes will then be discussed.

13.2 UNIT MASS

The measurement of unit mass does not pose any problem. This simple test should be carried out at the mixing plant; it is less important to evaluate it at the jobsite, where it is more important to measure the air content in the case of air-entrained high-performance concrete.

It should be pointed out that the unit mass of fresh high-performance concrete is somewhat higher than that of usual concrete made from the same materials. High-performance concrete contains more cement and less water. The unit mass of a non-air-entrained high-performance concrete is often close to 2500 kg/m^3, and it is close to 2400 kg/m^3 for an air-entrained mix, which represents roughly a 50 to 100 kg/m^3 increase in the unit mass from the usual values obtained for usual concrete.

The systematic performance of this test eliminates the necessity for frequent measurements of the air content, which take a little longer. If the unit mass is constant, the air content must be constant: a variation of 1% in the air content results in a 25 kg/m^3 difference in the unit mass, which is noticed when measuring unit mass. If the unit mass changes suddenly, it is important to check the air content in order to see whether it is the amount of entrapped or entrained air that has varied. If the air content stays the same, the decrease in unit mass is due to a change in the mix proportions or to an error in the measurement of the unit mass.

Used alone the unit mass cannot be too instructive, but if it is used in conjunction with the other measurements of the properties of the fresh concrete it can confirm whether changes have occurred in the fresh concrete and help to make a rapid diagnosis of the problem.

13.3 SLUMP

13.3.1 Measurement

In spite of the fact that this very simple test is regularly criticized with regard to its scientific and technological value (Punkki, Golaszewski and Gjørv, 1996), it is not going to be tomorrow when it will no longer be used to control high-performance concrete workability.

Many factors affect the slump of concrete, but from a rheological point of view the slump depends essentially on factors related to the aggregate skeleton and to the amount and fluidity of the paste used when mixing the concrete.

The measurement of the slump of usual concrete is easy, because after a few seconds the concrete cone stops collapsing and keeps a stable height which is easy to measure. If properly carried out, the value of the slump is reproducible within ± 10 mm. The measurement of the slump of high-performance concrete is not as easy: the collapse of the concrete cone is progressive and sometimes it is difficult to decide when to measure the height of the slowly collapsing concrete cone. In the case of particularly fluid high-performance concretes it is not even unusual to see the concrete spilling over the edges of the standard plywood board on which the slump test is performed.

A more fundamental approach has been taken by some researchers to study the complexity of the rheological properties of high-performance concrete (Hu, de Larrard and Gjørv, 1995; Punkki, Golaszewski and Gjørv, 1996). These scientific approaches to the problem have already yielded some interesting results, but the techniques used are quite sophisticated (Hu et al., 1996). They work well in the laboratory, but none of them has led to a simple test that can be used in the field to replace the slump test. As for any viscous system, these rheological studies are based on the measurement of the shear strength that is developed within fresh concrete. Even though these studies will probably not have a direct impact on the measurement of slump in the field in the near future, they are very important from a scientific and practical point of view because they provide a powerful tool in the comprehension of the cement/superplasticizer compatibility problem and will surely help cement and admixture manufacturers to develop cements and superplasticizers with improved rheological properties at very low water/binder ratios. All this can be done with cement paste rheology. However, the effect of coarse and fine aggregates is important in concrete evaluation, and problems of pumping stability and slipforming are possible with concrete rheology.

13.3.2 Factors influencing the slump

There are many factors that affect the slump consistency of concrete, but from a rheological point of view they can be classified into two broad categories: those related to the aggregate skeleton and those related to the rheological behaviour of the cement paste itself.

The total amount of aggregate present in the mix, the relative proportions of the coarse and fine aggregates, their respective grain size distribution and the shapes of the aggregate particles are the main factors related to the aggregate skeleton that affect the slump of the concrete, whether it is of the usual or high-performance type.

The rheological behaviour of the cement paste in usual concrete is essentially related to the water/binder ratio. The higher this ratio, the more diluted in water are the fine particles, so physically the water plays a key role in the rheology of the hydrated cement paste when the water/binder ratio is high, say above 0.50. At such a high water/binder ratio, the cement and cementitious particles are so far apart from each other within the paste that their interaction during hydration does not particularly affect the slump of the concrete. The amount of entrained air also drastically affects the slump of a usual concrete, and this is well documented in the literature (ACI 211, 1993; Mehta and Monteiro, 1993; Neville, 1995).

Even when a water reducer is used in usual concrete, its dosage is relatively low (around $1 \, l/m^3$) and its effect on the reduction of water needed to get a certain slump is generally limited to 8 to 10%.

The fluidizing effect of superplasticizers in usual concrete is also well documented in the literature and some admixture companies have promoted, very optimistically, the use of a rule of thumb saying that the addition of so many litres of superplasticizer in a cubic metre of concrete of a given slump could increase the slump by so many millimetres.

As the water/binder ratio decreases and the superplasticizer dosage increases, the situation becomes more and more complex from a rheological point of view, because the water by itself no longer plays the key role in the rheology of the cement paste. The cement and cementitious particles interact physically, and this is affected by their shape, their grain size distribution and their chemical reactivity (Aïtcin, Jolicoeur and MacGregor, 1994; Huynh, 1996). Moreover, the superplasticizer used to deflocculate the cement particles interacts with the hydrating cement particles so that it is now a very complex set of factors that influences the rheology and slump of high-performance concrete. In this respect, it has been shown that the sulfate content of the cement and its dissolution rate play a key role, which is not particularly the case in usual concrete (Tagnit-Hamou and Aïtcin, 1993).

Of course, the lower the water/cement or water/binder ratio, the more complex the situation becomes, so that terms such as 'sticky' and 'thixotropic' are used to describe the consistency of some high-performance concretes.

13.3.3 Improving the rheology of fresh concrete

In spite of the fact that until recently the entrainment of air has usually been avoided in high-performance concrete because in most of its early applications freezing and thawing resistance was not a problem, it is the experience of the author that, as in usual concrete, entrained air drastically improves the rheology of fresh high-performance concrete (Lessard et al., 1994). Of course, a strength loss results from the entrainment of air, but, using very efficient air-entraining agents, it is possible to obtain a multitude of very fine air bubbles that occupy no more than 4 to 4.5% by volume. These bubbles not only improve the freezing and thawing resistance of the high-performance concrete, they also greatly improve the workability of the fresh concrete. In one particular case the author was able to make a concrete having a compressive strength of 100 MPa at 91 days with an air content of 4.5%. The bubble spacing factor was 200 μm. It is the opinion of the author that the entrainment of a small amount of air will be used in the future in certain cases to improve the rheological behaviour of low water/binder ratio systems.

Recently, some Japanese and Canadian researchers have proposed the use of a colloidal agent in high-performance concrete in order to develop successfully self-levelling high-performance concrete that can be placed very easily in high reinforcement-congested areas (Khayat, Gerwick and Hester, 1993; Khayat *et al.*, 1996; Khayat, Manai and Trudel, 1997). Professor K. Khayat at the Université de Sherbrooke is presently developing an 85 MPa fibre-reinforced self-levelling concrete containing 0.6% steel fibres.

13.3.4 Slump loss

Different practical ways to solve the slump loss problem have been discussed in Chapter 7. These mostly consist of using supplementary cementitious materials as a partial replacement for the cement. This point is not well documented from a theoretical point of view, most authors saying that as supplementary cementitious particles are less reactive in the fresh mix than cement particles, their rheological effect is limited to their physical aspect.

From a practical point of view, using supplementary cementitious materials works well to solve the large slump loss problem, and these have frequently been used since the development of high-strength concrete in the 1970s. As seen in Chapter 7, the only limitations to the use of supplementary cementitious materials as a partial replacement for cement in high-performance concrete concern early strength development and freezing and thawing durability. It is not well established whether these mixtures are freeze–thaw durable, or even if the tests used to decide whether these mixtures are durable are appropriate. It is hoped that these points will be clarified in the near future. In such cases some entrained air could be used to solve the problem; it will not hurt from a rheological point of view, though it will result in the loss of some strength.

13.4 AIR CONTENT

13.4.1 Non-air-entrained high-performance concrete

It is not essential to measure repeatedly the air content of a non-air-entrained high-performance concrete, but it is recommended measuring it from time to time as a cross-check for the unit mass. Usually, high-performance concretes can entrap from 1 to 3% of air because the mixtures are usually more sticky than usual concrete. The lower the water/binder ratio, the stickier the high-performance concrete becomes. However, with some cement–superplasticizer combinations, it is possible to make 0.30 water/binder ratio concretes in which the amount of entrapped air is between 1 and 1.5% when the fluidity of the mix is carefully adjusted. However, as the water/binder ratio decreases below 0.30 it becomes difficult to reduce the amount of entrapped air to below 1.5 to 2%. The examination of polished sections of hardened high-performance concrete shows that in such cases the air bubbles are large, having a diameter in the millimetre range, which makes them absolutely inadequate to provide any freeze–thaw protection.

A strange phenomenon has been observed during the mixing of some high-performance concretes, following an overdosage of superplasticizer when a relatively high amount of mixing water was used. The high-performance concretes became very fluid and tended to entrap a large volume of large air bubbles, resulting in what has been called 'bubbling concrete' or the 'champagne effect' (Aïtcin, Jolicoeur and MacGregor, 1994). Large air bubbles tend to spring out of concrete but seem to be regenerated as easily as they tend to disappear during mixing. When placed in the forms, this kind of mix is prone to a severe segregation and is strongly retarded. After hardening, such a mix is found to contain a large volume of very coarse voids, which drastically lower the compressive strength. As it is very difficult to correct this situation when it occurs at the concrete plant just after mixing, it is better to discard the concrete load and check carefully what went wrong with the superplasticizer or water dosage. When this happens at the jobsite, after excessive redosing of superplasticizer to increase slump, the situation is more critical because the contractor is not always ready to wait for the next truck load, especially when it is near the end of the day. Usually the highly flowable mix is placed, and everybody hopes that it will start to harden. In one particular case, the author had to wait 36 hours to see the concrete reach the 20 MPa compressive strength necessary to remove the forms. At 28 days that specified 100 MPa concrete tested 20 MPa less than it should have done.

When compressive strength is of major concern in non-air-entrained high-performance concrete, it is important to keep the amount of entrapped air as low as possible in order to avoid any strength loss. It is so difficult to obtain the last MPa's that any method that can decrease the amount of entrapped air should be considered and exploited to its full potential.

13.4.2 Air-entrained high-performance concrete

In spite of the fact that some researchers pretend that entrained air is not necessary to make high-performance concrete freeze–thaw durable, as will be discussed in more detail in Chapter 18, the use of air-entrained high-performance concretes is recommended in the case of concretes that are to be exposed to freezing and thawing. However, in such cases the spacing factor ensuring freeze–thaw durability is much higher than the one necessary for usual concrete, as discussed in Chapter 18.

In spite of the fact that some researchers and concrete producers pretend that it is difficult to entrain a proper amount of air to get a bubble system with the right spacing factor in the case of high-performance concrete, it has been shown many times that there are already on the market air-entraining agents that perform well in that respect in very low water/binder ratio mixes (Hoff and Elimov, 1995; Bickley, 1996). Pumping air-entrained concrete can be critical to maintaining a low spacing factor (Lessard, Baalbaki and Aïtcin, 1996), as will be discussed in more detail in Chapter 17.

Usually the air-entraining agent dosage has to be significantly increased, but it is possible to entrain 4 to 5% air in a high-performance concrete and still have a bubble spacing factor smaller than 200 μm (Bickley, 1996). As previously pointed out, it should be remembered that entrained air drastically improves the workability of high-performance concrete, and that in many cases having 3 to 4% of entrapped and entrained air helps the placing and finishing of high-performance concrete so much that it is worthwhile losing some MPa. Therefore, if the strength is not a critical issue, as in class I concrete (50 to 75 MPa), the incorporation of a small amount of air should be attempted if the non-air-entrained mixture has poor rheological characteristics.

Above 100 MPa, for class III, IV and V concretes, it is not recommended to entrain any air at all. For class II concrete, as previously stated, it is possible to use a small amount of entrained air to improve the rheological behaviour of high-performance concrete.

13.5 SET RETARDATION

As discussed in Chapters 6 and 7, a secondary effect of a large dosage of superplasticizer is a definite retardation of the setting of the concrete. The higher the specified compressive strength, the higher the superplasticizer dosage and very often the longer the retardation. Retardation longer than 24 hours has been more or less voluntarily experienced in the field, without any particular damage to the ultimate compressive strength of the concrete. However, it should be recognized that this is not a comfortable situation when it happens involuntarily or unexpectedly (Aïtcin et al., 1984).

There are some cases when it is not possible to tolerate such a situation; for example in the case of prestressed elements, of slipforming or when jumping forms have to be raised rapidly. In a precast plant this situation is less dramatic because precast elements are usually heated to accelerate the form removal and the cutting of the tendons. For example, the prestressed box girders that were precast to build the Île de Ré bridge in France, that weighed from 40 to 80 tonnes, were heated so that they could be removed from the forms and transported to a special curing yard within less than 15 hours (Cadoret and Richard, 1992).

It has been shown in the literature that it is possible to decrease the retardation of high-performance concrete somewhat by using a melamine superplasticizer instead of a naphthalene one. However, when the slipforming rate had to be decreased, some naphthalene superplasticizer was mixed with the melamine (Ronneberg and Sandvik, 1990).

In the case of the jumping forms used for the construction of Scotia Plaza, a 15 MPa compressive strength was specified at 12 hours before the forms were removed (Ryell and Bickley, 1987). This stringent strength criterion was met without many problems throughout the construction of the building, in spite of the fact that a naphthalene superplasticizer was used, slag was employed as a partial replacement for cement, and the outside temperature was as low as $-10\,°C$ in the winter. This early strength control was achieved through the use of proper mixing techniques and a close control of the temperature of the mix within the specifications.

When looking for high early strength, in one particular case the author was quite surprised to find that it was a mix containing 15% Class C fly ash as a partial replacement for cement that performed the best. It was possible to get the highest strength at 15 hours, despite the fact that at 24 hours and beyond this particular mix performed not as well as the other combinations tried in that particular case. The high early strength of the fly ash concrete was attributed to the high alkali content of the fly ash (1.5%). The alkalis in the fly ash accelerated the initial setting and hardening of that particular high-performance concrete during the first 15 hours, but resulted in a decrease of the 24 hour and later strength.

If the early strength of a particular high-performance concrete is of concern, it should first be determined whether that particular concrete can be heated after casting, and second if a decrease in the superplasticizer dosage is possible by simultaneously increasing the cement and water content in order to keep the same water/binder ratio. Finally, the use of an accelerator should be considered.

There are also situations in which it is absolutely necessary to retard the setting of a high-performance concrete that displays rapid slump loss. In such a case, if it is not possible to use any supplementary cementitious materials that could help to solve the problem at least partially, a set-retarding agent should be used (Lessard *et al.*, 1993). The dosage of this retarding agent has to be carefully adjusted, taking into account the reactivity of the cement, its fineness, the external temperature, the transportation and placing times, etc. In such a case one should not be suprised if, from time to time, the retardation that is obtained through the use of the set-retarding agent gets out of control.

13.6 CONCLUDING REMARKS

It is important to measure regularly the properties of fresh high-performance concrete because concrete that has constant fresh properties is likely to have constant hardened properties as well. Each of the measured properties does not bear very much information in itself, but the set of values that is obtained can be quite useful.

The slump test is definitely not the most appropriate way to evaluate the workability of high-performance concrete. It is essentially used to monitor the uniformity of concrete delivery. As high-performance concrete is complicated rheologically, slump measurement is not sufficient to evaluate workability in the field to appreciate the ease with which high-performance concrete can be placed and finished.

Entrained air has been found to improve significantly the workability and finishability of high-performance concrete, making low water/binder ratio concrete less sticky and easier to place. The compressive strength decrease accompanying the entrainment of some air in high-performance concrete can be easily recovered by lowering the water/binder ratio slightly. It is the opinion of the author that in the future some high-performance concrete will contain a small amount of entrained air, not for freeze–thaw durability but rather to improve workability.

Slump loss is not always an easy problem to resolve because it can be the consequence of so many factors. The addition of a retarder can be tried. On the other hand, if there is an accidental overdosage of the water and/or superplasticizer, the mix can become too fluid and severe segregation can be obtained which makes it difficult to recover such a mixture.

When a high-performance concrete has to receive an additional dosage of superplasticizer in the field to increase its slump, it is important to avoid the overdosing in order to avoid the effect of bubbling high-performance concrete which will not be possible to recover.

It is the author's opinion that the ideal temperature of freshly mixed high-performance concrete is between 15 and 20 °C. Below a 10 °C ambient temperature, high-performance concrete can be heated to accelerate its set, and above a 25 °C ambient temperature, it must be cooled or strongly retarded to avoid slump loss problems. In between these 10 and 25 °C ambient temperatures, it is a matter of judgement to decide whether the temperature of the high-performance concrete has to be slightly increased or decreased.

REFERENCES

ACI 211 (1993) *Standard Practice for Selecting Proportions for Normal, Heavyweight and Mass Concrete, ACI Manual of Concrete Practice*, Part 1, pp. 1–55.

Aïtcin, P.-C., Jolicoeur, C. and MacGregor, J. (1994) Superplasticizers: how they work and why they occasionally don't. *Concrete International*, **16**(5), May, 45–52.

Aïtcin, P.-C., Bédard, C., Plumat, M. and Haddad, G. (1984) *Very High Strength Cement for Very High Strength Concrete*, Symposium of the Material Research Society, Fall meeting, Boston, November, pp. 201–10.

Bickley, J.A. (1996) High-performance concrete bridge for Ontario Ministry of Transportation: contract 95-39. *Concrete Canada Newsletter*, **1**(2), May, 1–2.

Cadoret, G. and Richard, P. (1992) Full scale use of high-performance concrete in buildings and public works, in *High-Performance Concrete — from Material to Structure* (ed. Y. Malier), E and FN Spon, London, pp. 379–411.

Hoff, G.C. and Elimov, R. (1995) *Production for the Hibernia Platform*, Supplementary papers, 2nd CANMET/ACI International Symposium on Advances in Concrete Technology, Las Vegas, Nevada, June, pp. 717–39.

Hu, C., de Larrard, F. and Gjørv, O. (1995) Rheological testing and modelling of fresh high-performance concrete. *Materials and Structures*, **20**, 1–7.

Hu, C., de Larrard, F., Sedran, T., Boulay, C., Bosc, F. and Deflorenne, F. (1996) Validation of BTRHEOM, the new rheometer for soft-to-fluid concrete. *Materials and Structures*, **29**, 620–31.

Huynh, H.T. (1996) La compatibilité ciment-superplastifiant dans les bétons à hautes performances – synthèse bibliographique. *Bulletin de Liaison des Laboratoires des Ponts et Chaussées*, No. 296, November–December, pp. 63–73.

Khayat, K.H., Gerwick, B.C., Jr. and Hester, H.T. (1993) Self-levelling and stiff consolidated concretes for casting high-performance flat slab in water. *Concrete International*, **15**(8), August, 36–43.

Khayat, K.H., Manai, K. and Trudel, A. (1997) In situ mechanical properties of wall elements cast using highly flowable high-performance concrete. Accepted for publication in *ACI Materials Journal*, **94**(6), November–December, 491–500.

Khayat, K.H., Soneli, M., Yahia, M. and Skaggs, C.D. (1996) Statistical models to predict flowability, washout, resistance and strength of underwater concrete, in *Production Methods and Workability of Concrete, Proceedings of the International RILEM Conference*, June, ISBN 0 419 22070 4, pp. 463–81.

Lessard, M., Baalbaki, M. and Aïtcin, P.-C. (1996) Effect of pumping on air characteristics of conventional concrete. *Transportation Research Record*, **132**, 9–14.

Lessard, M., Dallaire, E., Blouin, D. and Aïtcin, P.-C. (1994) High-performance concrete speeds reconstruction for McDonald's. *Concrete International*, **6**(9), September, 47–50.

Lessard, M., Gendreau, M., Baalbaki, M., Pigeon, M. and Aïtcin, P.-C. (1993) Formulation d'un béton à haute performance à air entraîné. *Bulletin de Liaison des Laboratoires des Ponts et Chaussées*, No. 180, November–December, pp. 41–51.

Mehta, P.K. and Monteiro, P.J.M. (1993) *Concrete-Microstructure, Properties and Materials*, 2nd edn, MacGraw Hill, New York, pp. 309–56.

Neville, A.M. (1995) *Concrete Properties*, 4th edn, Longman, Harlow, England, pp. 182–4.

Punkki, J., Golaszewski, J. and Gjørv, O.E. (1996) Workability loss of high-strength concrete. *ACI Materials Journal*, **93**(5), September–October, 427–31.

Ronneberg, H. and Sandvik, M. (1990) High strength concrete for North Sea platforms. *Concrete International*, **12**(1), January, 29–34.

Ryell, J. and Bickley, J.A. (1987) Scotia Plaza: high-strength concrete for tall buildings, in *Proceedings of the Symposium on Utilization of High-Strength Concrete*, Stavenger (ed. I. Holland *et al.*), Tapir, N-7034 Trondheim, NTH, Norway, ISBN 82-519-0797-7, pp. 641–53.

Tagnit-Hamou, A. and Aïtcin, P.-C. (1993) Cement superplasticizer compatibility. *World Cement*, **24**(8), August, 38–42.

CHAPTER 14

Temperature increase in high-performance concrete

14.1 INTRODUCTION

Portland cement hydration is an exothermic reaction, as shown in Chapters 6 and 12, so that concrete develops heat during its setting and hardening. The more cement particles hydrate within a certain period of time, the more heat is developed during that period of time. As high-performance concrete contains more cement than usual concrete, it is prudent to have some concerns about too high a temperature increase and/or too large thermal gradients within structural elements made of high-performance concrete. Maximum temperatures of about 60 to 70 °C have been recorded in massive structural elements made of high-performance concrete (Aïtcin, Laplante and Bédard, 1985; Burg and Ost, 1992; Cook *et al.*, 1992; Miao *et al.*, 1993a; Lachemi, Lessard and Aïtcin, 1996).

Such a temperature corresponds to a temperature increase of 40 to 50 °C relative to ambient conditions. Such a temperature increase can raise three concerns. First, in a structural element, concrete hydrates in conditions completely different from those that prevail in the small specimens used to control its properties, so the measured properties of the concrete cannot be at all representative of the concrete that actually hardens within the structure. It is well known, for example, that if a usual concrete receives an early curing at a high temperature, a decrease of its long-term compressive strength is observed. Second, as the temperature increase is not uniform through the whole mass of concrete, the maximum temperature reached varies within the structural element so that the concrete does not harden everywhere at the same rate and maximum temperature. Third, too high a temperature increase during initial hardening can generate large thermal gradients within the structural element when the concrete cools down to the ambient temperature later on. During concrete cooling these thermal gradients generate tensile stresses that can result in microcracking if these stresses are greater than the concrete tensile strength at that moment.

As current Type I cements usually contain from 50 to 65% C_3S, 15 to 25% C_2S, 5 to 10% C_3A and 5 to 10% C_4AF, it is obvious that it is the hydration of C_3S that generates most of the initial heat in a high-performance concrete. Of course, the amount of heat liberated at a given time depends on the rate of the hydration reaction, which depends in turn on factors such as the phase composition of the particular Portland cement used, the initial temperature of the high-performance concrete and the action of the admixtures on the development of the hydration reaction. This heat development usually results in a temperature increase of the concrete, except in thin elements exposed to low ambient temperature.

The temperature increase of a high-performance concrete within a structural element also depends on geometrical and thermodynamic factors, such as the shape and size of the structural element, the ambient temperature and the heat exchange rate through the forms and the top surface, which depends on the ambient temperature (Schaller et al., 1992; Lachemi, Lessard and Aïtcin, 1996).

Therefore, the amount of cement that is used in a particular concrete is not the only factor that dictates the maximum temperature increase at a particular location in a structural element; it is rather the amount of cement that has actually been hydrated, until the heat released by hydration equals that dissipated in the environment.

Moreover, in spite of the fact that it is the same concrete that is cast in a structural element, which would imply that the heat release is uniform, the temperature increase is not uniform because heat is not lost at the same rate from all parts of a structural element (Schaller et al., 1992; Lachemi, Lessard and Aïtcin, 1996). Thus the temperature rise is not uniform throughout the whole structural element and results in the development of thermal gradients and non-uniform thermal shrinkage during cooling, which can result in some locations in the development of tensile stresses high enough with respect to the tensile strength of the concrete to cause cracking.

As very few concerns about the heat development in structural elements made of 30 to 40 MPa concrete can be found in the literature, why then should the heat development in high-performance concrete become a very controversial matter?

14.2 COMPARISON OF THE TEMPERATURE INCREASES WITHIN A 35 MPa CONCRETE AND A HIGH-PERFORMANCE CONCRETE

Very surprisingly, as shown by Cook et al. (1992), a 35 MPa concrete having a water/cement ratio of 0.45 and containing 355 kg/m^3 of cement developed as much heat as two high-performance concretes having water/binder ratios of 0.31 and 0.25 and cement and supplementary cementitious material contents of 470 and 540 kg/m^3, respectively, when these three concretes were used to cast three 1 × 1 × 2 m columns. Finite element calculations showed that in the case of these three identical columns the 35 MPa concrete was the only one prone to develop microcracks during cooling owing to the development of thermal gradients, because its tensile strength was developing at a slower rate than that of the two high-performance concretes.

The fact that almost the same maximum temperature increase was observed in the three columns, 68 °C, 68 °C and 63 °C, respectively, indicates that almost the same amount of cement hydrated in the three columns. The heat generated within a concrete is proportional to the amount of cement that hydrates simultaneously and not to the total amount of cement contained in a particular mixture. However, as the water/binder ratio of the 35 MPa concrete was 0.45 instead of 0.31 or 0.25, as in the case of the two high-performance concretes, this meant that in the 35 MPa concrete there was as many cement particles that had been hydrated but that they were not as close to each other as those in the high-performance concretes, so that the 35 MPa concrete developed a lower tensile strength than the high-performance concretes in spite of the fact that the same amount of cement was hydrated.

Cement particles are closer to each other in high-performance concrete; they develop strong interparticle bonds more rapidly, so the early tensile strength of high-performance concrete is greater than that of usual concrete. This increase in the early tensile strength makes high-performance concrete less prone to microcracking due to the thermal gradients that develop during the cooling of the concrete to ambient temperature.

14.3 SOME CONSEQUENCES OF THE TEMPERATURE INCREASE WITHIN A CONCRETE

The temperature rise in structural elements made of usual concrete has not been considered a worrisome issue in the past and, in fact, has been ignored most of the time, except when massive elements have been built. However, when high-performance concretes started to be used, many engineers influenced by the high cement content of high-performance concretes raised some concerns about this temperature rise:

- How much greater was the temperature increase in a high-performance concrete compared with that of a usual concrete?
- Were the strength and more generally all concrete properties altered by this temperature rise?
- Was the strength measured on standard specimens cured at 23 ± 3 °C representative of the strength of the actual concrete in the structural element?
- Were the thermal gradients generated during cooling high enough to generate unacceptable cracking?

- Was thermal shrinkage high enough to create cracking?

Instead of trying to answer all these questions in a quantitative manner, which is impossible, owing to the great diversity of concrete mixtures that are used to make high-performance concrete, of the thermal and boundary conditions and the shape of structural elements, in the following we will rather review the different factors that influence the temperature rise in any concrete structural element that is built using either usual or high-performance concrete and show how this temperature rise can be controlled. It will be the duty of the engineer to look at the particular application of the high-performance concrete with which he or she is dealing and decide if the temperature rise (or decrease) is really of major concern. There are already several finite element software programs on the market that will allow the engineer to make appropriate calculations and justify the implementation of some of the proposed methods to control the temperature rise, so that thermal cracking can be minimized or eliminated when all concrete structural elements finally return to their average ambient working temperature.

14.3.1 Effect of the temperature increase on the compressive strength of high-performance concrete

For usual concrete, a high initial curing temperature favours the rapid growth of outer hydration products during the initial hardening period and results in the rapid build-up of a loose and weak microstructure, as compared with that obtained under curing at lower temperature. The inner products developed later at ambient temperature will never be able to compensate for this loose and weak initial microstructure. This explains why it is important to see whether the initial temperature increase during the early stage of hardening of high-performance concrete has the same effect on the long-term compressive strength as in usual concrete.

If we look closely at the temperature rise within a structural element made of high-performance concrete, we notice that it occurs under completely different conditions than in the case of steam curing of precast elements made of usual concrete. The temperature increase due to hydration does not occur before 12 to 18 hours after the casting of the concrete element. This retardation is due to the high dosage of superplasticizer, which retards the hydration of C_3S. This retardation does not mean that during this time hydration does not develop at all but rather that it proceeds very slowly. During this period of time, which is very crucial from a microstructural point of view, as will be seen in section 14.3.4, very little outer product develops. At the end of this period of low chemical activity, when hydration starts it develops uniformly very rapidly, so the concrete temperature increases very rapidly at almost the same rate everywhere within the whole mass of concrete. The maximum temperature is reached within a matter of a few hours (4 to 8 hours, depending on the geometry of the structural element and ambient conditions). When the maximum temperature has been reached hydration does not stop, but rather proceeds at a slower rate; the heat thus generated is not sufficient to compensate for the heat lost through the forms later on. Depending on the size of the high-performance concrete element and the ambient temperature, it can take from 2 to 5 days for concrete to come back to ambient temperature during this cooling period and later on.

When high-performance concrete returns to ambient temperature, hydration then proceeds by diffusion of the very few remaining free water molecules within a very dense microstructure; it proceeds very slowly and eventually stops.

14.3.2 Inhomogeneity of the temperature increase within a high-performance concrete structural element

Experience has shown that a 5 to 10 °C difference, or in some cases even more, is observed between the maximum temperature reached by high-performance concrete in the centre and at the surface of a massive concrete element, so that curing conditions are not exactly the same within the entire mass of the concrete element. However, this difference in the maximum temperature reached is not large enough to change hydration conditions so drastically that, close to the forms, Portland particles hydrate differently from those at the centre of the element. Therefore the inhomogeneity of the maximum temperature reached within high-performance concrete elements usually has little effect on high-performance concrete properties (Miao et al., 1993a), but this is not true all the time.

14.3.3 Effect of thermal gradients developed during high-performance concrete cooling

When Portland cement starts to hydrate, the heat development is so rapid that it is difficult to see how a large thermal gradient could develop within a massive concrete element during the period in which the temperature is increasing. However, as the temperature of the high-performance concrete approaches its maximum, and later during its cooling to ambient temperature, the temperature difference within the mass of concrete become large enough to create significant thermal gradients. It seems therefore that one should have some concern about the possible generation of microcracks due to the tensile stresses generated within hydrated cement paste that does not cool down at the same rate. In order to prevent such a situation from occurring, thermal insulation of massive high-performance concrete has been specified.

However, it should be clearly pointed out that it is not the absolute value of the thermal gradients that must be of concern, but rather the relative values of the thermal stresses generated by these thermal gradients and the tensile strength of the high-performance concrete at that time. In the case of two large columns ($1 \times 1 \times 2$ m) previously cited, it has been shown by Cook *et al.* (1992) that the tensile strengths of the two high-performance concretes tested were greater than the tensile stress generated by the maximum gradient developed within their mass.

In fact, high-performance concretes rapidly develop a significant tensile strength not only because they contain a large amount of cement that hydrates almost simultaneously, but also, and more importantly, because the cement particles are very close to each other owing to the very low water/binder ratio used. Under such conditions hydration products can bridge very rapidly and very strongly the interparticle gaps and rapidly develop strong bonds.

14.3.4 Effect of the temperature increase on concrete microstructure

It is normal to have some concerns about the type of concrete microstructure that develops when a maximum temperature of 50 to 70 °C, or even higher, is reached within some structural elements (such a temperature corresponds to a temperature increase of 40 to 50 °C relative to ambient conditions). It is well known that when usual concretes are externally heated to such temperatures to increase their early strength, their long-term strength is reduced (Neville, 1995). Moreover, as the temperature increase of high-performance concrete is not uniform through the whole mass of the structural element, the maximum temperature varies and the concrete does not harden everywhere at the same maximum temperature. Finally, too great an increase in temperature during initial hardening of a high-performance concrete structural element can generate significant thermal gradients during cooling, whereas when heat curing is applied to a usual concrete element, the whole mass of concrete cools at about the same rate when the external heat source is removed.

When analysed more closely, the temperature increase of a high-performance concrete structural element is completely different from that which takes place in a usual concrete element that is heated to increase its early compressive strength. Therefore, it is totally incorrect to apply to high-performance concrete what is well known and well documented for usual concrete exposed to an initial heat treatment (Baroghel-Bouny, Godin and Gawsewitch, 1996).

In a usual concrete exposed to low-pressure steam curing 2 to 3 hours after casting, a high initial curing temperature favours the rapid growth of outer hydration products during the initial hardening period and results in the rapid build-up of a loose and open microstructure, as compared with that obtained under curing at ambient temperature. This forced hydration also results in quite a high degree of hydration, so, very little inner hydration product will develop later at ambient temperature. This absence of late inner product development will never be able to compensate for the development of this loose and weak initial microstructure; hence the reduced long-term strength of such concretes (Moranville, 1996).

On the other hand, the temperature rise within a high-performance concrete usually does not occur before 12 to 18 hours after casting, owing to the retardation of C_3S hydration by the superplasticizer. This retardation does not mean that during this time hydration does not proceed at all, but that it proceeds very slowly. Moreover, C_3A and C_4AF hydration also proceeds at a very slow rate, in the presence of a high dosage of superplasticizer, which is very important from a microstructural point of view. In fact, it is rather difficult to find any well-developed ettringite crystals in high-performance concrete (Aïtcin *et al.*, 1987). It has been shown that in the presence of a superplasticizer, ettringite crystallizes as minute crystals having a massive form. It is also difficult to find well-developed Portlandite crystals, in spite of the fact that thermal analysis shows the presence of some hydrated lime in low water/binder ratio pastes (Aïtcin *et al.*, 1987; Moranville-Regourd, 1992).

Therefore, during the temperature rise, very little outer product is formed. The type of hydration product is instead of the inner type, and is formed at almost the same rate within the whole mass of a structural element.

In a high-performance concrete, the maximum temperature is reached within a matter of a few hours after the end of the dormant period (4 to 8 hours) almost everywhere at the same time, so the concrete reaches a more or less uniform high strength at the same time, a strength that is much higher than that of standard cured specimens of the same age.

The development of inner- rather than outer-type hydration product during the initial hardening is a consequence of the initial close packing of the cement particles within high-performance concrete and of the limited volume of mixing water used during the batching owing to the use of a superplasticizer. Therefore, hydration proceeds mostly by a diffusion process of water molecules within the mass of unhydrated cement particles, rather than through a dissolution–precipitation process. When observed under a scanning electron microscope, the hydrated cement paste of a high-performance concrete appears as a homogeneous, very compact, 'amorphous-like' paste.

The rapid development of this very dense microstructure soon stops the hydration process of the Portland cement, because there is less and less free water for hydration and water molecules have more and more difficulty diffusing through this barrier. This explains why the temperature rise slows down and stops rapidly. Of course, during the cooling period some hydration continues to proceed by diffusion, but when a high-performance concrete is finally back to ambient temperature, its compressive strength does not increase very much with time.

The observation of the microstructure of mature or very old high-performance concretes with a scanning electron microscope almost always shows a significant volume of unhydrated cement particles and unreacted supplementary cementitious particles; the lower the water/binder ratio, the higher the amount of unhydrated cement and unreacted supplementary cementitious particles.

It is interesting to look at how hydration proceeds when some supplementary cementitious material has been used in producing the high-performance concrete. It has been observed in the laboratory and in the field that, as in the case of usual concretes, the amount of cement substituted for must be quite high before a significant decrease in the maximum temperature of the high-performance concrete is observed. In fact, when hydration starts, all the mixing water is available for the hydration of the cement particles, and as the heat generated depends on the number of cement particles that are hydrating, it is necessary to reach a high degree of substitution before a significant decrease in the maximum temperature reached by high-performance concrete is seen. However, the rapid temperature rise accelerates the rate at which any pozzolanic reaction proceeds. In fact, some very fine pozzolanic particles may hydrate faster than some coarse cement particles and prevent them from future hydration by fixing part of the mixing water. But in general the pozzolanic particles will have less chance to hydrate than coarse cement particles unless some additional external water is available to hydrate them. This explains why, when the microstructure of a high-performance concrete that contains a supplementary cementitious material is observed under a scanning electron microscope, it is common to observe many unreacted coarse particles of supplementary cementitious material even in the case of silica fume. In high-performance concrete containing supplementary cementitious materials, the hydration conditions can be different from those observed in usual concrete containing pozzolanic materials, so the late testing that is sometimes specified for high-performance concretes containing fly ash or slag can be questioned as it is not certain that these supplementary cementitious particles will have had a chance to take part in a pozzolanic reaction.

14.4 INFLUENCE OF DIFFERENT PARAMETERS ON THE TEMPERATURE INCREASE

A number of different laboratory studies, field experiments and monitoring of high-performance structures have been carried out during the past few years at the Université de Sherbrooke to answer some of these concerns (Aïtcin, Laplante and Bédard, 1985; Cook *et al.*, 1992; Miao *et al.*, 1993a, b; Aïtcin *et al.*, 1994; Lachemi, Bouzoubaâ and Aïtcin, 1996; Lachemi, Lessard and Aïtcin, 1996; Lachemi *et al.*, 1996; Lachemi and Aïtcin, 1997). Some of this work is summarized below.

14.4.1 Influence of the cement dosage

In the particular experiment already reported (Cook et al., 1992) three $1 \times 1 \times 2$ m plain concrete columns were cast using three ready-mix concretes having about 35, 90 and 120 MPa 28 day compressive strengths. The water/binder ratios of these three concretes were 0.45, 0.31 and 0.25, respectively, and their binder contents were 355, 470 and 540 kg/m^3. The maximum temperatures recorded at the geometrical centre of each column were, respectively, 68, 68 and 63 °C, and were reached after 24, 27 and 34 hours, respectively. These maximum temperatures were similar to the 66.5 °C reported in a previous experiment in which two experimental columns were monitored during the construction of a Montreal high-rise building (Aïtcin, Laplante and Bédard, 1985).

The temperatures of the delivered concretes were, respectively, 25, 22 and 16 °C, so the maximum temperature rises above the initial concrete temperature were 43, 46 and 47 °C, respectively: the maximum temperature rise at the centre of each column was almost the same in spite of the great difference in the amount of binder used when making these three concretes. The similarity of the temperature rise in the three columns is in fact a consequence of the crucial influence of the initial temperature of the concrete, of the retarding action of the superplasticizer on the rate of hydration, and of the fact that it is not the total amount of cement contained in a mix that is important but rather the actual amount of cement that hydrates. These factors explain the observed delay in the development of the hydration reaction within the two high-performance concretes, as shown in Figure 14.1.

A finite element model fitted to match the recorded temperature variations was used to analyse the thermal stresses. In this analysis the following assumptions had to be made: (1) the coefficient of thermal expansion was constant throughout the hydration process; and (2) the elastic modulus varied with time according to the following relationship:

$$E'_c = A\sqrt{f'_c(t)}$$

where the value of A was calculated from 28 day cylinder test results and the early compressive strength was given by the equation:

$$f'_c(t) = f'_c(28) \frac{t}{t+B}$$

The value of B was calculated using a regression analysis fitted to measured compressive strengths at 1, 7, 28 and 91 days.

In a first attempt, it was assumed that the thermal effects were applied as short-term loading. In such a case the stresses were overestimated because during the first hours the modulus of elasticity of each concrete column is smaller than that assumed in the analysis and this results in smaller stresses. A more complex analysis was also performed to give a

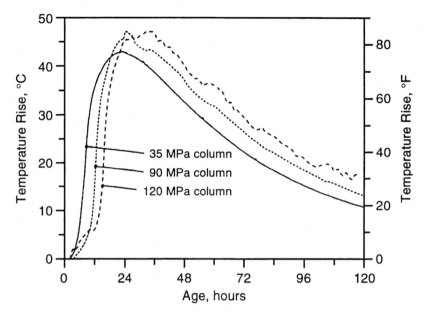

Fig. 14.1 Temperature increase at the centre of each column.

better approximation of the actual behaviour, but in that case creep effects were neglected, so the predicted stresses still represented upper bond solutions.

This analysis demonstrated that the 120 MPa column had the smallest thermal gradient and the smallest thermally induced stress. The ratios between the thermal stress and the tensile strength of all the concretes were also calculated and found to be below 1. Moreover, it was found that the risk of cracking was much higher for the 35 MPa concrete than for the 120 MPa concrete. The ratio between the thermal stress and concrete tensile strength was found to be 0.83, 0.70 and 0.37 for the 35, 90 and 120 MPa concretes, respectively. In fact no cracking was observed on the surface of any of the three columns.

Later on these three 2-m high columns were cut into three equal pieces 0.66 m high, which were cored vertically parallel to the direction of casting, as shown in Figure 11.2 (Miao et al., 1993b). This coring was done at 84 days. The cores were stored in sealed polyethylene bags until testing at 91 days. All the core samples (ϕ-95 mm) and standard cylinders (ϕ-100 mm) were prepared by grinding their ends.

It was found that the 35 MPa column presented larger variations in compressive strength over the height and the cross-sections of the column. A 23% drop between the top and the bottom of the column was measured; this drop was only 10% for the 90 MPa concrete, while the 120 MPa column did not present any significant variation of compressive strength over its full height.

The mean compressive strength of the cores was found to range from 85 to 90% of the compressive strength of standard water-cured specimens, and from 101 to 106% of the compressive strength of the air-cured cylinders for the three columns. These numbers are within the limits usually observed for both usual concretes and high-performance concretes.

Core compressive strengths were compared with those of specimens water-cured and air-cured just after their demoulding at 24 hours and with those of sealed specimens (Aïtcin et al., 1994). After demoulding, the curing under sealed conditions was achieved by wrapping the specimens first with a polyethylene film and then with an aluminium foil. Measurements indicated that the weight loss of these sealed specimens was less than 0.1%. For each of these curing conditions, three specimens were tested at 1, 7, 28, 91 and 365 days. The compressive strengths of the cores and the air-dried specimens were found to be lower than those measured on sealed cured specimens for the three concretes. The lowest strength was obtained with the air-cured specimens, and showed a decrease of 17 to 22%. These results are in good agreement with those presented by Haque, Gopalan and Ho (1991).

Table 14.1 compares the influence of curing on the 91 day and 1 year compressive strength and core compressive strength. Comparing the ratios of the 1 year strengths for water-cured versus sealed conditions, the 35 and 90 MPa concretes show only a 3% increase in the water-cured strength, while the 120 MPa concrete shows a 13% increase. The compressive strength of the water-cured specimens with a very low water/binder ratio concrete was due to the penetration of some water within the specimen and the hydration of unhydrated cement particles (essentially at the periphery of the specimens within the first 20 to 30 mm), which was not the case with the sealed specimens. The different compressive strengths measured can be ranked in the following way: water-cured > sealed > cores > air-dried.

Finally, in order to compare the degree of microcracking in the three concretes, chloride ion permeability tests were performed according to the ASTM 1202 testing procedure on cores taken from the central part of the column. Although this test is usually carried out over 6 hours, it had

Table 14.1 Influence of curing on compressive strength (Khayat, Bickley and Hooton, 1995). Copyright ASTM. Reprinted with permission

	91 days		1 year		
	Cores Water-cured	Cores Air-dried	Air-cured Water-cured	Air-cured Sealed	Water-cured Sealed
35 MPa	0.91	1.04	0.83	0.85	1.03
90 MPa	0.84	1.06	0.79	0.81	1.03
120 MPa	0.86	1.01	0.78	0.87	1.13

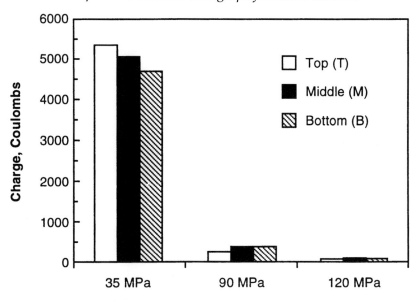

Fig. 14.2 Chloride ion permeability measurements (after 3 hours) taken from core samples (Miao *et al.*, 1993a). Reproduced by permission of ACI.

to be shortened to 3 hours because the 35 MPa concrete had too high a chloride ion permeability and the concrete specimens were heating too much during the test. This shortened time was also used for the 90 and 120 MPa concretes. As shown in Figure 14.2, the two higher-performance concretes had significantly lower chloride ion permeabilities; the higher the strength, the lower this value. The connectivity of the pore system existing in the two high-performance concretes was therefore much smaller than that of the 35 MPa concrete.

14.4.2 Influence of the ambient temperature

The ambient temperature largely dictates the heat exchange rates in the different parts of any concrete structure. As it is rather difficult to monitor the influence of the ambient temperature on structural elements in the laboratory, the effect of three different ambient temperature conditions on three similar structural elements monitored under field conditions will be used to illustrate this point. These three structural elements are three thin slabs made of approximately the same air-entrained high-performance concrete. They are part of the Portneuf bridge deck (Aïtcin and Lessard, 1994; Lachemi *et al.*, 1996), the McDonald's sidewalk (Lessard *et al.*, 1994) and the Montée St-Rémi bridge deck (Lachemi, Lessard and Aïtcin, 1996). In each case the temperature of the high-performance concrete upon its delivery to the site was approximately the same, ranging from 21 °C to 18 °C, though, as shown in Figure 14.3, the temperature at the centre of the three thin concrete elements was markedly influenced by the ambient temperature.

Influence of different parameters 361

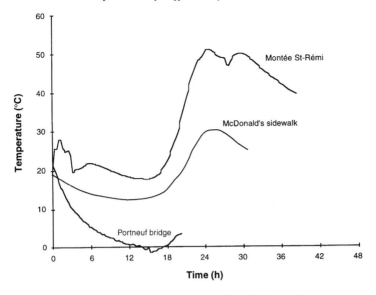

Fig. 14.3 Temperature variation of three similar high-performance concretes exposed to different ambient conditions. Reproduced by permission of ACI.

In the case of Portneuf bridge, where the ambient temperature varied from a maximum of 3 °C to a minimum of − 4.5 °C, insulated blankets were placed over the bridge deck 10 hours after the casting of the concrete. The temperature of the high-performance concrete dropped to − 1.5 °C, 16 hours after casting. The compressive strength of cores taken from the slab at 14 days showed that this very brief freezing temperature did not damage the concrete. This temperature was not low enough to freeze the interstitial water, which at that time was almost saturated with various ions. It must be remembered that, for example, sea water starts to freeze only at − 1.8 °C.

In the case of the McDonald's restaurant sidewalk reconstruction carried out in late autumn, the ambient temperature varied from 9 °C to 22 °C (Lessard *et al.*, 1994), and it is seen that the concrete temperature in the sidewalk was greatly influenced by the ambient temperature. The peak temperature of the concrete was also in this case influenced by the initial temperature of the concrete.

In the case of the Montée St-Rémi bridge deck, cast in summer conditions with a maximum ambient temperature of 33 °C, the peak temperature at a mid-height of the slab was 51.5 °C (Lachemi, Lessard and Aïtcin, 1996). Therefore, the thinner a concrete element, the more the ambient temperature influences its maximum temperature.

14.4.3 Influence of the geometry of the structural element

In the case of the Montée St-Rémi bridge, temperature variations were monitored in three different structural elements: the bridge deck, a

T-beam and a massive girder, which are shown in Figure 14.4 (Lachemi, Lessard and Aïtcin, 1996). Thirty-two thermocouples were installed and the monitoring lasted 7 days, until the centre of the large girder was back to ambient temperature. Figure 14.5 shows the variations of the temperature at the centre of these three elements.

It can be seen that the geometry of the structural element greatly influences the maximum temperature reached at its geometrical centre: 67.5 °C for the massive girder, 63.5 °C for the T-beam and 51.5 °C in the slab, but does not much influence the time at which this maximum is reached: 22 to 24 hours. The heat loss rate is a direct function of the surface/volume ratio of the concrete within the structural element, while

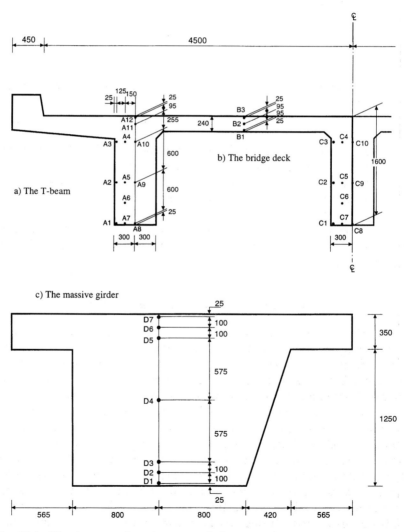

Fig. 14.4 The three monitored structural elements of Monteé St-Rémi viaduct.

Influence of different parameters

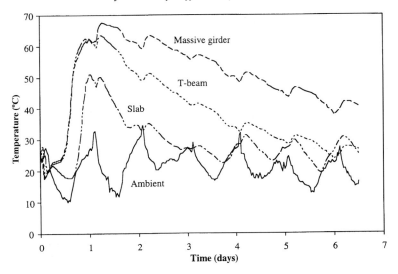

Fig. 14.5 Temperature variations at the centre of the three structural elements under study.

the time at which the maximum temperature is reached depends primarily on the ambient temperature and the concrete characteristics such as the initial temperature, the heat development rate, which depends on the initial temperature, and the influence of the admixture on the heat development rate.

Similar temperature variations were observed in different structural elements during the reconstruction of the Jacques-Cartier bridge deck in Sherbrooke (Blais et al., 1996).

14.4.4 Influence of the nature of the forms

During the reconstruction of the Jacques-Cartier bridge deck in Sherbrooke, temperature variations were monitored in a beam for which the bottom form was a steel beam and the lateral forms were made of 20 mm-thick plywood, as shown in Figure 14.6. The recorded temperature just behind the forms showed that, as expected, heat losses were greater through the steel form than through the plywood forms, generating a lower peak temperature but greater thermal gradients during cooling. Therefore, if the development of too high thermal gradients can damage a high-performance concrete structural element, the use of plywood forms or insulated steel forms is recommended. The use of such forms does not have a major impact on the maximum temperature within the mass of the concrete, but it significantly decreases the thermal gradients generated during cooling owing to its better insulating properties.

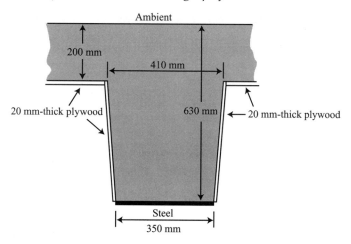

Fig. 14.6 Forms used for the reconstruction of the Jacques-Cartier bridge deck.

14.4.5 Simultaneous influence of fresh concrete and ambient temperature

Using the TEXO module of the CESAR finite element code developed by the Laboratoire Central des Ponts et Chaussées of Paris (Humbert, 1989; Dubouchet, 1992), it has been possible to simulate numerically the temperature rise at the geometrical centre of a massive girder of the Montée St-Rémi bridge for three different temperatures of the fresh high-performance concrete and three different ambient temperatures (Lachemi and Aïtcin, 1997). The numerical model was fitted to match the measured temperature variations during and after the construction of the bridge.

Maximum temperatures and the times at which they were reached could be matched in the model within ± 2 °C and ± 1 hour (Lachemi, Lessard and Aïtcin, 1996). Table 14.2 reproduces the maximum temperature at the centre of the massive girder for the nine temperature configurations under study. This table shows that the temperature of the fresh concrete has a much greater influence on the maximum temperature of a massive element than does the ambient temperature. The temperature difference between the centre of the girder and its top surface was always greater than the temperature difference between the

Table 14.2 Predicted temperatures at the centre of the massive girder (Lachemi and Aïtcin, 1997)

Fresh concrete temperature	Ambient temperature		
	10 °C	20 °C	28 °C
10 °C	58	61	63
18 °C	67	69	70
25 °C	75	76	77

Table 14.3 Temperature difference between the centre of the massive girder and its top and bottom surfaces (Lachemi and Aïtcin, 1997)

Fresh concrete temperature	Surface exposed ambient temperature			20 mm plywood formwork ambient temperature		
	10 °C	20 °C	28 °C	10 °C	20 °C	28 °C
10 °C	18	16	14	17	14	12
18 °C	21	18	16	19	16	14
25 °C	22	19	17	20	17	15

centre of the girder and its lower part that was separated from ambient conditions by the plywood forms, as shown in Table 14.3. Moreover, the temperature difference between the centre of the girder and its top surface increases with the temperature of the fresh concrete for any kind of ambient temperature, and decreases with an increase of the ambient temperature for a given temperature of the fresh concrete.

The smallest temperature difference is obtained when the casting of a 10 °C concrete is carried out when the ambient temperature is 28 °C. In such a case the hydration reaction is slowed down and somewhat retarded owing to the low initial temperature of the concrete. Moreover, the difference between the maximum temperature reached within the concrete and the ambient temperature is minimal because the ambient temperature, which is already high, does not have a significant influence on the maximum temperature reached within the concrete. Therefore, the casting of the massive girder under study in the summer with an 18 °C high-performance concrete generated smaller thermal gradients than if it had been cast under winter conditions.

14.4.6 Concluding remarks

In conclusion it can be said that, generally speaking, the use of a high-performance concrete results in a temperature rise within its mass, except in the case of thin elements cast under winter conditions. The maximum temperature reached within a high-performance concrete does not depend on the amount of cement used to make the concrete but rather on the amount of cement that is actually hydrated when the heat losses become equal to the heat generated by the cement that is hydrating. This maximum temperature also depends on the type of cement, the amount of mixing water, the effect of admixtures, the initial temperature of the concrete, the ambient temperature, the size and the shape of the structural element and the nature of the forms, etc. Therefore the assumption that the heat generated by a high-performance concrete is a direct function of the amount of binder used when making it can be totally misleading.

Moreover, in spite of the homogeneity of the temperature of the high-performance concrete when it is cast, the combined effects of geometrical factors and different thermodynamic boundary conditions result in the development of a non-homogeneous temperature increase within the structural element, which of course generates non-homogeneous thermal shrinkage and different thermal gradients in different parts of the structural element. If in a particular case the temperature increase of the high-performance concrete in a particular structural element can create some thermal cracking problems, several measures can be taken to lower the temperature rise or the thermal gradients so that the high-performance concrete hardens under totally safe thermal conditions (Branco, Mendes and Mirambell, 1992; Bélanger and Shirlaw, 1993). Some of these methods are reviewed in the following section.

14.5 HOW TO CONTROL THE MAXIMUM TEMPERATURE REACHED WITHIN A HIGH-PERFORMANCE CONCRETE STRUCTURAL ELEMENT

There are several ways of controlling the maximum temperature reached within a high-performance concrete. As it is the cement, and particularly its C_3S content, that is responsible for the heat development, then from a logical point of view one should tackle the problem at its source and make high-performance concretes with a cement having a low heat of hydration, and use as far as possible supplementary cementitious material which develops little or no heat of hydration. Unfortunately, at the present time, this logical solution is often not very practical to implement for both economic and technical reasons.

Concrete producers do not always have several silos in which to store different types of cement. The delays in the release of the forms owing to the use of a cement with a low heat of hydration and with too much supplementary cementitious material can be unacceptable from a construction point of view because it is indirectly too costly owing to the delays it introduces into the construction schedule. Therefore in many cases the first practical solution that should be considered to lower the maximum temperature of a high-performance concrete is to decrease its temperature upon delivery.

14.5.1 Decrease of the temperature of the delivered concrete

In this respect high-performance concrete can be treated like any usual concrete because the decrease of the initial temperature of any concrete is a purely thermodynamic problem. Whenever possible, the first thing to do is to use ingredients having the lowest temperature possible. Usually very little can be done about the cement temperature, since decreasing cement temperature is not easy and is very costly when it has to be implemented. Coarse aggregates can be protected from direct exposure to sun, and they can be sprayed with cold water, but it should be understood that this last measure has a very limited impact on the peak temperature within a high-performance concrete. As in the case of usual concrete, the most efficient means of decreasing high-performance concrete temperature is to decrease the temperature of the water (Neville, 1995). Up to now, two different techniques have been used to lower the initial temperature of high-performance concretes: liquid nitrogen cooling and crushed ice.

(a) Liquid nitrogen cooling

Liquid nitrogen cooling has been used successfully in summer to lower the temperature of fresh concrete (Ryell and Bickley, 1987), but it remains an expensive solution. In the case of Scotia Plaza, the use of liquid nitrogen cost $35 CAN/m^3. Moreover, liquid nitrogen is not readily available everywhere, and usually the volume of high-performance concrete is not large enough to justify the construction or rental of storage and dispensing facilities within a ready-mix concrete plant. Liquid nitrogen also embrittles steel and creates some maintenance problems with the blades of the drum mixer. Moreover, the cloud that accompanies the introduction of the liquid nitrogen into the truck mixer can be dangerous from a traffic point of view in the ready-mix plant area. Finally, nitrogen introduction significantly delays concrete delivery.

(b) Use of crushed ice

A more common approach to lowering the initial temperature of a high-performance concrete is to replace part of the mixing water with an equal mass of crushed ice, as is done for usual concrete in summer when the concrete temperature has to be decreased or when making mass concrete. This technique was found to be very efficient in keeping the temperature of the delivered concrete at around 18 °C during the construction of the Montée St-Rémi bridge, when the maximum ambient temperature was 33 °C (Lachemi, Lessard and Aïtcin, 1996), and in August during the reconstruction of the Jacques-Cartier bridge deck in Sherbrooke (Blais et al., 1996). In both cases an average of 40 l of water were replaced by 40 kg of ice. In fact the amount of crushed ice used varied from 80 kg to 20 kg, depending on the ambient temperature conditions. When 80 kg of ice were substituted for 80 l of mixing water this represented more than half the amount of mixing water. However, this crushed ice melted very rapidly, so there were no slump problems at the end of the mixing. In both cases, the extra cost related to the use of ice was evaluated at $9.50 CAN/m^3. This cost included the cost of the ice, the transportation of the ice from Montreal, the rental of the crushing system, the introduction of the ice and the salary of the student hired to check the temperature of the concrete.

14.5.2 Use of a retarder

It has been found that when a superplasticizer is used at a high dosage rate it has a tendency to retard cement hydration, owing to the fixing of a certain number of superplasticizer molecules on C_3S (Simard et al., 1993). Usually this delay in the hydration of Portland cement is accompanied by a decrease in the peak temperature of a few degrees Celsius. Moreover, it has also been found that when retarders are correctly dosed they are more effective in reducing the peak temperature than are superplasticizers, so that in the various field projects undertaken by the Université de Sherbrooke a retarder was systematically introduced within the high-performance concrete (Aïtcin et al., 1987; Blais et al., 1996; Lachemi, Lessard and Aïtcin, 1996). In each case the amount of retarding agent was adjusted so that it did not delay too much Portland cement hardening.

The specific amount of retarder that is added can be easily adjusted by introducing an increasing amount of retarder in different high-performance concretes or grouts having the same water/binder ratio and superplasticizer dosage, and measuring the influence of this retarding agent on the early strength. The selected dosage involves a compromise between the strength and the maximum temperature requirements.

In the case of a particular cement, for instance, it was found from a study conducted on grouts that for a specific superplasticizer the retarder dosage had to be lower than 50 ml/100 kg of cement, while for another superplasticizer type it had to be lower than 85 ml/100 kg of cement. In both cases these two dosages did not result in a significant decrease of the 24 hour compressive strength.

In very hot countries, where the availability of ice can be problematic, the use of a retarder can be an attractive solution to control the maximum temperature reached within a high-performance concrete.

14.5.3 Use of supplementary cementitious materials

As supplementary cementitious materials do not react as rapidly as Portland cement, at first sight it might be thought very advantageous to replace a certain amount of highly reactive Portland cement that develops a significant amount of heat of hydration with a slowly reacting cementitious material that does not. However, it must be emphasized that in high-performance concrete the amount of heat generated during the first 24 hours is related not only to the total amount of cement but also to the volume of mixing water that is available to react with this cement, so that in order to have a significant decrease in the peak temperature of a high-performance concrete, the level of substitution must be of the order of 50% or more, as has been observed for usual concrete (Bramforth, 1980; Baalbaki et al., 1992). Below this level of substitution there is some decrease in the peak temperature, but this decrease is not directly proportional to the level of substitution because the unsubstituted cement has more chance to be hydrated by the mixing water. In fact, even if the water/binder ratio is low, the water/cement ratio of the mixture is high and increases as the substitution level increases, and therefore all of the available water can be used to hydrate the cement.

14.5.4 Use of a cement with a low heat of hydration

Of course, whenever the alternative of using a low heat of hydration cement is possible, it should be implemented. Low-heat cement contains less C_3S and C_3A, the two most reactive phases of Portland cement that are responsible for the early heat development. Moreover, low-heat cements are usually coarsely ground in order to lower their reactivity, so that they initially develop heat at a lower rate. This is not a disadvantage when making high-performance concrete, because high compressive strength can be achieved by having the cement grains closer to each other through a reduction in the water/binder ratio. It is the author's opinion that in the future ASTM Type II cements will be frequently used for high-performance applications. Most probably cement producers will market a Type II cement with different fineness in order to achieve different early strength requirements. Moreover, the use of a cement with a low amount of C_3S and C_3A is very advantageous, both from a rheological point of view, because it results in a significant decrease in the amount of superplasticizer needed to achieve a high slump while slump retention is improved, and because it can result in reduced autogenous shrinkage (see Chapter 12).

14.5.5 Use of hot water and insulated forms or heated and insulated blankets under winter conditions

There are cases when high-performance concrete has to be protected against freezing in winter. The use of hot water, insulated forms or insulated blankets, as is usually done with usual concrete, is very efficient in protecting high-performance concrete exposed to freezing temperatures. Two bridge decks were cast under severe winter conditions without causing any major concreting problems (Lessard *et al.*, 1993; Blais *et al.*, 1996). In both cases, hot water was used during the mixing and insulating blankets were used to protect the top surface of the high-performance concrete deck. In winter conditions, it is good practice to increase the temperature of the delivered concrete in the 20 to 25 °C range and insulate the surface, while in summer it is better to stay in the 15 to 20 °C range.

14.6 HOW TO CONTROL THERMAL GRADIENTS

Thermal gradients are developed within a structural element whenever heat dissipation varies between two points. Thermal gradients can be critical in the case of massive elements and in beams having a minimum dimension greater than 500 mm. In thinner elements, the thermal gradients developed should never be high enough to cause any cracking.

The best way to minimize thermal gradients is to use insulated forms; the worst way is to use steel forms. Finite element calculations can be done to reach a final decision on the appropriateness of using insulated forms or not, in which case the thermal gradient will be lowered but thermal shrinkage will be slightly increased (Miao et al., 1991).

During the construction of the high-performance concrete part of the Pont de Normandie, the girders were heated from the inside in such a way that the heating sources compensated for the non-homogeneous temperature increase that would have occurred within the girder owing to the different thicknesses of its various parts (Acker, personal communication, 1996).

14.7 CONCLUDING REMARKS

The temperature increase observed when using high-performance concrete must be well understood. Surprisingly, very often it is not higher than the too often ignored temperature increases occurring in 30 to 35 MPa concretes. When the origin of this temperature rise is well understood, measures can be taken to control it and its effects on concrete properties. Unfortunately, there is not a simple universal answer to this problem, owing to its complex and numerous origins, such as:

- the great diversity of the mix compositions used when making low water/binder ratio concretes;
- the great variation in initial and ambient conditions;
- the great diversity of the shape and geometry of structural elements.

As is always the case in engineering, there is no single way to approach and solve a particular problem and the final decision is always the responsibility of the engineer. This decision must be taken after studying all the available solutions to reduce the temperature increase, after balancing their advantages, their weaknesses, their economic impact and their effect on the durability of the structure. It is only after such an exercise has been carried out that one can be confident that the use of a high-performance concrete will result in the construction of a structure that is safe, durable and economical.

REFERENCES

Aïtcin, P.-C. and Lessard, M. (1994) Canadian experience with air-entrained, high-performance concrete. *Concrete International*, **16**(10), October, 35–8.

Aïtcin, P.-C., Laplante, P. and Bédard, C. (1985) *Development and Experimental Use of a 90 MPa (13 000 psi) Field Concrete*, ACI SP-87, pp. 51–67.

Aïtcin, P.-C., Sarkar, S.L., Regourd, M. and Hornain, H. (1987) Microstructure of a two-years old very high strength field concrete (100 MPa), in *Proceedings of the Symposium on Utilization of High Strength Concrete*, Stavanger (ed. I. Holland et al.), Tapir, Norway, N-7034 Trondheim, NTH, Norway, ISBN 82-519-0797-7, pp. 99–109.

Aïtcin, P.-C., Miao, B., Cook, W.D. and Mitchell, D. (1994) Effects of size and curing on cylinder compressive strength of normal and high-strength concretes. *ACI Materials Journal*, **91**(4), July–August, 349–54.

Baalbaki, M., Sarkar, S.L., Aïtcin, P.-C. and Isabelle, H. (1992) *Properties and Microstructure of HPC containing Silica Fume, Slag and Fly Ash*, 4th CANMET/ACI International Conference on Fly Ash, Silica Fume, Slag and Natural Pozzolans in Concrete, Istanbul, Turkey, ACI SP-132, Vol. 2, pp. 921–42.

Baroghel-Bouny, V., Godin, J. and Gawsewitch, J. (1996*) Microstructure and Moisture Properties of High-Performance Concrete*, 4th International Symposium on High-Performance Concrete, Paris, May, pp. 451–61.

Bélanger, P.R. and Shirlaw, M.R. (1993) Temperature control in high-strength massive concrete girders. *Concrete International*, **15**(11), 30–32.

Blais, F.A., Dallaire, E., Lessard, M. and Aïtcin, P.-C. (1996) *The Reconstruction of the Bridge Deck of the Jacques-Cartier Bridge in Sherbrooke (Québec), using High-Performance Concrete*, 30th Annual Meeting of the Canadian Society for Civil Engineering, Edmonton, Alberta, May.

Bramforth, P.B. (1980) *In Situ Measurement of the Effect of Partial Portland Cement Replacement using Either Fly Ash or Ground Granulated Blast-Furnace Slag on the Performance of Mass Concrete*, Proceedings of the Institution of Civil Engineers, **69**(2), September, 777–800.

Branco, F.A., Mendes, P.A. and Mirambell, E. (1992) Heat of hydration effects in concrete structures. *ACI Materials Journal*, **89**(2), 139–45.

Burg, R.G. and Ost, B.W. (1992) *Engineering Properties of Commercially Available High-Strength Concretes*, RD 104.01T, Portland Cement Association, Skokie, IL, USA, 55 pp.

Cook, W.D., Miao, B., Aïtcin, P.-C. and Mitchell, D. (1992) Thermal stresses in large high-strength concrete columns. *ACI Materials Journal*, **89**(1), January–February, 61–8.

Dubouchet, A. (1992) Développement d'un pôle de calcul: CESAR-LCPC. *Bulletin de Liaison des Laboratoires des Ponts et Chaussées*, No. 178, March–April, pp. 77–84.

Haque, M.N., Gopalan, M.K. and Ho, D.W.S. (1991) Estimation of in situ strength of concrete. *Cement and Concrete Research*, **21**(6), 1103–10.

Humbert, P. (1989) CESAR-LCPC: un code de calcul par éléments finis. *Bulletin de Liaison des Laboratoires des Ponts et Chaussées*, No. 160, March–April, pp. 112–15.

Khayat, K.H., Bickley, J.A. and Hooton, R.D. (1995) High-strength concrete properties derived from compressive strength values. *Cement, Concrete, and Aggregates*, **17**(2), December, 126–33.

Lachemi, M. and Aïtcin, P.-C. (1997) Influence of ambient and fresh concrete temperatures on the maximum temperature and thermal gradient in a high-performance concrete. *ACI Materials Journal*, **94**(2), March–April, 102–10.

Lachemi, M., Bouzoubaâ, N. and Aïtcin, P.-C. (1996) *Thermally Induced Stresses During Curing in a High-performance Concrete Bridge: Field and Numerical Studies*, ICCE-96, The Second International Conference in Civil Engineering on Computer Applications Research and Practice, Bahrain, April, Vol. 2, pp. 451–7.

Lachemi, M., Lessard, M. and Aïtcin, P.-C. (1996) *Early-Age Temperature Developments in a High-Performance Concrete Viaduct*, ACI SP-167, pp. 149–74.

Lachemi, M., Bois, A.P., Miao, B., Lessard, M. and Aïtcin, P.-C. (1996) First year monitoring of the first air-entrained high-performance bridge in North America. *ACI Structural Journal*, **93**(4), July–August, 379–86.

Lessard, M., Gendreau, M., Baalbaki, M., Pigeon, M. and Aïtcin, P.-C. (1993) Formulation d'un béton à haute performance à air entraîné. *Bulletin de*

Liaison des Laboratoires des Ponts et Chaussées, No. 180, November–December, pp. 41–51.

Lessard, M., Dallaire, E., Blouin, D. and Aïtcin, P.-C. (1994) High-performance concrete speeds reconstruction for McDonald's. *Concrete International*, **16**(9), September, 47–50.

Miao, B., Aïtcin, P.-C., Cook, W.D. and Mitchell, D. (1991) *Effect of Thermal Gradients and Microcracking of Large Structural Elements Made of High Strength Concrete*, 70th Annual Transportation Research Board Meeting, Washington.

Miao, B., Aïtcin, P.-C., Cook, W.D. and Mitchell, D. (1993a) Influence of concrete strength on in situ properties of large columns. *ACI Materials Journal*, **90**(3), May–June, 214–19.

Miao, B., Chaallal, O., Perraton, D. and Aïtcin, P.-C. (1993b) On-site early-age monitoring of high-performance concrete columns. *ACI Materials Journal*, **90**(5), September–October, 415–20.

Moranville-Regourd, M. (1992) Microstructure of high-performance concrete, in *High-Performance Concrete: From Material to Structure* (ed. Y. Malier), E & FN Spon, London, pp. 3–13.

Moranville, M. (1996) *Implications of Curing Temperatures for Durability of Cement Based Systems*, Seminar on Cement for Durable Concrete, Royal Melbourne Institute of Technology, Melbourne, 8 pp.

Neville, A.M. (1995) *Concrete Properties*, 4th edn, Longman, Harlow, England, pp. 359–411.

Ryell, J. and Bickley, J.A. (1987) Scotia Plaza: high-strength concrete for tall buildings, in *Proceedings of the Symposium on Utilization of High-Strength Concrete*, Stavanger, (ed. I. Holland *et al.*), Tapir, N-7034 Trondheim, NTH, Norway, ISBN 82-519-0797-7, pp. 641–54.

Schaller, I., de Larrard, F., Sudret, J.P., Acker, P. and LeRoy, R. (1992) Experimental monitoring of the Joigny Bridge', in *High-Performance Concrete: From Material to Structure* (ed. Y. Malier), E & FN SPON, London, pp. 432–57.

Simard, M.A., Nkinamubanzi, P.-C., Jolicoeur, C., Perraton, D. and Aïtcin, P.-C. (1993) Calorimetry, rheology and compressive strength of superplasticized cement paste. *Cement and Concrete Research*, **23**, 939–50.

CHAPTER 15

Testing high-performance concrete

15.1 INTRODUCTION

As stated by Bickley, Ryell and Read (1990): 'it is of no use specifying high-performance concrete if the tests made do not provide some assurance of its achievement, but at the same time it must be essential that the tests are done properly, which has not been always the case of high-performance concrete'.

At first glance, it seems unnecessary to devote a whole chapter of a book on high-performance concrete to properly defining and detailing ways of carrying out simple tests such as compressive strength and shrinkage measurements. Standard test methods are well established, and since usual concrete has been used extensively for a long time, almost all possible situations should already have been addressed. Moreover, empirical relationships based on a great number of field and laboratory tests have been established over the years, creating a high degree of confidence and comfort when dealing with test data. Finally, building codes that rely on the results of these standard tests and the values given by these empirical relationships have proven to be safe. Therefore, why should present standard test methods and empirical relationships be questioned when dealing with high-performance concrete (Carino, Guthrie and Lagergen, 1994)?

In fact, it was realized from practical experience that some of the standard procedures and relationships used and developed for usual concrete are not well suited for high-performance concrete and that they need to be adjusted or partially modified. Fortunately, rock mechanics deals with quite strong rocks having about the same mechanical properties as high-performance concrete, or even stronger ones, for which standard test methods and procedures have been developed. As with concrete, over the years a large number of empirical relationships have been developed in rock mechanics, and some of these are readily applicable to high-performance concrete. In some respects high-performance concrete can be considered as an artificial rock that is as strong, or even stronger, than many natural rocks. In a number of cases, rock mechanics testing methods and procedures can easily be adapted to the needs of high-performance concrete testing, though this is not always the case.

Fortunately, not all of the present standard tests used to characterize usual concrete need to be modified or adapted. For example, the standard test methods for measuring the modulus of rupture, modulus of elasticity, splitting tensile strength and abrasion resistance of fresh usual concrete can be used directly without modification for testing high-performance concrete, because in these cases the energy and the stresses developed during the testing of high-performance concrete are not drastically different from those developed when testing usual concrete.

There are, however, other cases for which the testing methods usually used for usual concrete should be applied cautiously, or modified, because unstable conditions can develop during the test. This is the case in particular when trying to determine the stress–strain curve of high-performance concrete.

There are other cases in which the standard testing devices presently used to test a particular critical value do not work at all with high-performance concrete, so that new testing procedures, devices and apparatus have to be developed. This is particularly the case for shrinkage and creep measurements. In the latter case the use of springs to develop a constant load does not work owing to the very high constant loads that need to be maintained on the specimens.

Finally, there are cases in which the present testing methods are not suited for testing high-performance concrete and can neither be adapted nor copied from rock mechanics, such as shrinkage and permeability measurements. For example, if we consider the evaluation of the permeability of a particular high-performance concrete, experience shows that it is necessary to measure the apparent permeability of high-performance concrete according to AASHTO T-277 (*Standard Method of Test for Rapid Determination of the Chloride Permeability of Concrete*) or its ASTM equivalent, ASTM C 1202 (*Test Method for Electrical Indication of Concrete's Ability to Resist Chloride Ion Penetration*), until something better can be proposed. People will have to get used to expressing the chloride ion permeability in terms of coulombs, in spite of the fact that few people are familiar with such a permeability scale.

In this chapter, the most appropriate methods of measuring the compressive strength, complete stress–strain curve (especially in the post-peak domain), shrinkage, creep and permeability of high-performance concrete will be discussed in detail.

15.2 COMPRESSIVE STRENGTH MEASUREMENT

At present, in order to evaluate usual concrete compressive strength, 150×300 mm standard cylindrical specimens (or 150 mm cubes) are

Compressive strength measurement

used. The diameter of the specimen can vary slightly from one country to another (Neville, 1995), but it is always around 150 mm and the aspect ratio (height/diameter) is equal to 2.

After 28 days of standard curing in a fog room or in lime-saturated water at 23 °C ± 3 °C, specimens are capped a few hours before testing. This is usually done with a fast hardening sulfur-based material, also called a capping compound. This compound is melted and poured into a special device, ensuring that both caps are parallel to each other and perpendicular to the specimen axis. This sample preparation ensures that the concrete specimen will be tested in uniaxial compression mode and not in a shear mode. For usual concrete, the compressive strength of the capping compound is stronger than that of the tested concrete.

The capped specimen is then placed between the two platens of a testing machine and the load is applied at a defined rate until failure. For example, failure occurs when a load of 350 kN is applied by the testing machine to a 150 × 300 mm specimen of 20 MPa concrete:

$$\frac{\pi \times 0.15^2}{4} \times 20 \times 10^6 \text{ Newtons} \simeq 350 \text{ kN}$$

Of course, this load doubles to 700 kN when concrete compressive strength jumps to 40 MPa. As long as the concrete compressive strength stays between 20 and 40 MPa, the ultimate load that has to be developed by the testing machine fits in well with the loading capacities of conventional testing machines found in most laboratories. In fact, most testing machine manufacturers have been building testing machines with a full capacity of about 1.3 MN.

In order properly to measure the compressive strength of high-performance concrete, it is necessary to see what types of new situation must be faced. In the following paragraphs, we will address, successively, the testing machine limitations, the implications of the specimen end preparation, the specimen position during the testing, and the influence of the shape and size of specimens. Finally, we will address fundamental choices such as at what age and after which curing condition should the specimens be tested.

15.2.1 Influence of the testing machine

(a) Testing limitations due to the capacity of the testing machine

At present, repeatedly testing 40 MPa 150 × 300 mm concrete specimens does not cause any problems. Since it is generally considered that a testing machine should not be used routinely and repeatedly at more than two-thirds of its full load capacity in order to avoid any premature fatigue, a 1300 kN testing machine is able to break 40 MPa concrete specimens repeatedly. Moreover, occasionally using the full capacity of such a testing machine, it is possible to break a 150 × 300 mm standard specimen having a compressive strength of 70 MPa. However, this last compressive strength value is about, or is lower than, most of the high-performance concretes that are currently used in the construction industry and it can be dangerous, and costly in the long run, to test 60 to 70 MPa high-performance concretes repeatedly with a 1.3 MN testing machine. Therefore, present testing machines are not at all suited to testing the lowest level of strength of Class II high-performance concretes.

In order to test high-performance concrete with a compressive strength higher than 50 MPa, there are two options: the first is to buy a new testing machine with a higher full load capacity, and the second is to test high-performance concrete using smaller specimens.

Let us consider the first option. Buying a new testing machine with a greater full load capacity (and rigidity) represents a significant long-term investment for a testing laboratory. In such a case, it is not wise to buy a machine that will be obsolete within 5 to 10 years, which means that if the Class III high-performance concrete market develops, as it should, a testing machine with a full capacity of at least 3.3 MN has to be purchased (if the two-thirds safety rule is considered). The purchase of such a large piece of equipment that will not often be used to its full capacity represents a lot of money for a testing agency.

Buying a new testing machine is an even more costly investment for a university or a research laboratory that wants to explore the feasibility of Class V high-strength concrete with a compressive strength greater than 150 MPa. In this case, the full load capacity of the testing machine will have to be greater than 4 MN (Lessard and Aïtcin, 1992). Such an acquisition requires not only finding a large amount of money, but also a large area with high head space to install such a large testing machine, whose base has to be big enough to dissipate all the energy stored in the testing machine just before the failure of the specimen. In spite of the fact that this first option could represent an 'El Dorado' for testing machine manufacturers, it is rather better to decide to test high-performance concrete (and usual concrete also) on smaller specimens. However, it has to be decided what diameter of specimen should be selected: 125, 100 or 75 mm.

Let us go back to our 1.3 MN standard testing machine that is presently found in most serious testing laboratories, and let us calculate the maximum compressive strength that can be tested in each case using the two-thirds safety rule. A simple calculation shows that for a 1.3 MN testing machine $f'_c = 1.1/D^2$, where f'_c is expressed in MPa and D in mm. If concrete specimens have a diameter of 75 mm, the maximum compressive strength that can be tested safely on a repeated basis is 196 MPa. Similarly, for 100 mm specimens, it is 110 MPa, and 70 MPa for 125 mm specimens. So if the present testing machines have to be used to test high-performance concrete, it is necessary to test high-performance concrete with specimens having a diameter of 100 mm. Occasionally, when used at their full capacity, existing 1.3 MN testing machines could be used to test 165 MPa concrete, which corresponds to a strength that stands in our highest class of high-performance concretes and that has not been used up to now in industrial projects.

However, such a drastic decision as changing the standard diameter of concrete specimens with which high-performance concretes are to be tested cannot be taken exclusively on the basis of trying to increase the life cycle of 1.3 MN testing machines and avoiding the necessity for testing agencies and universities to invest money in a new piece of equipment. The pros and cons of such a change have to be weighed up on a more scientific and technological basis. This will be discussed in more detail in the section dealing with the influence of specimen size on the evaluation of high-performance concrete compressive strength.

(b) Influence of the dimensions of the spherical head

The decision to change the size of standard specimens (which has already been taken by a number of researchers and testing agencies) has a direct consequence on the set-up of the testing machine. This is sometimes forgotten, and might explain some discrepancies found in the literature regarding the compressive strength values of high-performance concrete. In order to comply with ASTM C39, the compressive strength test must be done on a testing machine in which any lack of parallelism between the two ends of a specimen is automatically corrected for by a platen than can rotate (slightly) around a spherical head, the dimension of which varies with the diameter of the specimen. Figures 15.1 and 15.2 represent spherical heads to be used when testing 100×200 and 150×300 mm concrete specimens (Lessard, 1990). Whenever a concrete specimen of a different size is to be tested, the spherical head of the testing machine must be changed, which is not an easy task if the testing machine has not been adapted for that. Figure 15.3 represents a particular set-up used at the Université de Sherbrooke which greatly simplifies the change of the spherical head. With such a set-up, it is easy to change the spherical head in less than 5 minutes without much effort.

If the proper spherical head is not used, the compressive strength measured will be different from what it would have been if the standard head had been used, as shown by Lessard (1990) in Table 15.1. When the standard spherical head for 100 mm specimens was used, 100 mm specimens had a 112 MPa compressive strength, while it was only 108 MPa when measured on 100 mm specimens with the standard spherical head for 150 mm specimens. However, when 150×300 mm specimens were tested on a standard spherical head for 100×200 mm specimens, the compressive strength value obtained on the same concrete was only 89.3 MPa, or 17% lower than it would have been had the right spherical head been used.

Moreover, during this experiment a change in the mode of failure of the concrete specimens was observed when non-standard spherical heads were used. When the 100 × 200 mm and 150 × 300 mm specimens were tested with their appropriate spherical heads, the mode of failure was in the shape of the usual two opposite cones, as shown in Figures 15.4(a) and 15.5. On the other hand, when the 150 × 300 mm specimens were tested with the non-standard spherical head, the failure followed the development of vertical fissures, as shown in Figure 15.4(b). Neville and Brooks (1987) identify this type of failure as characteristic of a splitting test.

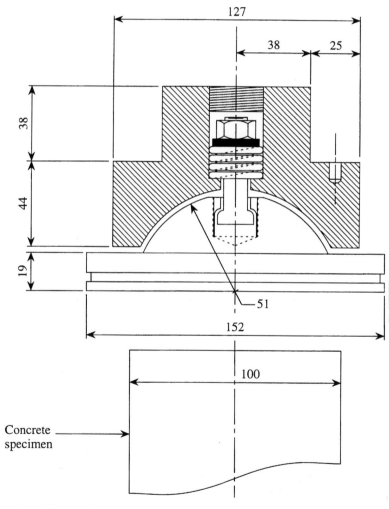

Fig. 15.1 102 mm diameter normalized bearing block for testing 100 × 200 mm specimens (dimensions in mm).

This experiment was repeated with a second concrete, this time using 12 specimens for each measurement. Some of the specimens were tested with their ends ground, and the others with their ends capped with a high-strength capping compound. The results of this second experiment are also reported in Table 15.1. The same decrease in the measured compressive strength was observed when the 150 × 300 mm specimens were tested with the standard spherical head for 100 × 200 mm specimens.

It is the opinion of the author that some reported significant differences in compressive strength values between 100 × 200 and 150 × 300 mm specimens found in the literature could be attributed to the use of an incorrect spherical head.

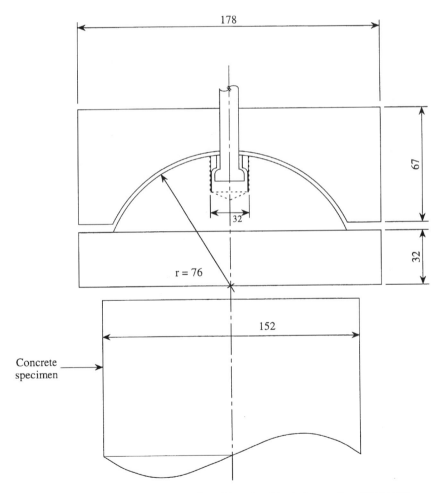

Fig. 15.2 152 mm diameter normalized bearing block for testing 150 × 300 mm specimens (dimension in mm).

Set-up for measuring compressive strength on 150 × 300 m specimens

Set-up for measuring compressive strength on 100 × 200 m specimens

Fig. 15.3 Université de Sherbrooke set-up for easy change of the spherical head.

Table 15.1 Average 28 day compressive strengths for different bearing block diameters

		Average compressive strength (MPa)					
		Capped			Faced		
Concrete identification	Type of bearing	100 × 200 mm cylinders	150 × 300 mm cylinders	$\overline{f'_{c100}}/\overline{f'_{c150}}$ (%)	100 × 200 mm cylinders	150 × 300 mm cylinders	$\overline{f'_{c100}}/\overline{f'_{c150}}$ (%)
C10	⌀ 102 mm	–	–	–	112[a]	89.3[b]	125
	⌀ 152 mm	–	–	–	108[b]	–	104[c]
C11	⌀ 102 mm	118[d]	93[d]	127	119[d]	93.5[d]	127

[a] Average of 3 specimens. [b] Average of 7 specimens. [c] Based on the ratio 112/108. [d] Average of 12 specimens.

 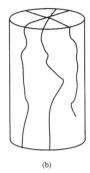

Fig. 15.4 (a) Conical type of rupture (due to shear); (b) splitting type of rupture.

According to ASTM C39-86, when the proper spherical head is used, the failed specimen should be in the shape of two opposite cones because the two ends of the specimen are confined by the platens when the compressive load is applied. When a specimen is loaded with a uniaxial compressive load, its diameter increases in the direction perpendicular to the axial load (Thaulow, 1962), but at the same time frictional stresses develop between the specimen ends and the two platens of the testing machine and prevent any diameter increase in these particular areas where the concrete is confined. These frictional stresses, which tend to counteract the natural horizontal expansion, create a horizontal compressive stress zone on both ends which is responsible for the conical failure shape (Thaulow, 1962). If the spherical head is not large enough to cover the entire specimen end, a part of the end of the specimen is not fully confined (Hester, 1980), a splitting mode of failure develops in this area and the measured compressive strength decreases (Neville, 1995). Thus it is very important to be sure that the right spherical head is used whenever the diameter of the tested specimen changes.

(c) Influence of the rigidity of the testing machine

The rigidity of the testing machine is a subject of major concern for researchers interested in studying the so-called post-peak behaviour of usual concrete. In the case of high-performance concrete, this problem can be even more severe. When the failure of a concrete specimen occurs, all of the elastic energy stored within the columns of the testing machine is suddenly released. The frequent repetition of such a brutal release of energy can deregulate the testing machine and put it out of calibration (Holland, 1987). Moreover, the testing machine must be fixed on a massive base capable of absorbing all of the energy liberated at failure without generating dangerous vibrations around it (Lessard and Aïtcin, 1992).

It is also important to note that when the compression test is carried out in a very stiff testing machine, the specimen shortens as the load increases, but at the same time the deformation of the testing machine is negligible when compared with that of the specimen, and thus the rate at which the load is applied decreases, so the measured compressive strength is decreased (Neville, 1995).

15.2.2 Influence of testing procedures

It is necessary not only to have an appropriate testing machine to perform compression tests, but also to recognize the importance of properly preparing the specimen and of ensuring that the load is applied

Fig. 15.5 Typical conical failures: (a) the coarse aggregates have been crushed; (b) debonding of the coarse particles of the gravel; (c) a 150 MPa failure.

along the axis of the testing machine, so there is no eccentricity between the load and the axis of the specimen.

(a) How to prepare specimen ends

For usual concrete, the strength of regular sulfur capping compounds is high enough to be sure that during the compression test, failure always occurs first within the concrete and not within the sulfur caps. According to CSA A23.2-9C-M90, a 50 mm cube of regular sulfur compound should usually have a compressive strength of not less than 35 MPa. Recently, a high-strength capping compound has been introduced which has a compressive strength of 60 to 70 MPa on 50 mm cubes. Even with such a capping compound, the level of strength reached is still lower than the compressive strength of many high-performance concretes. But it should be noted that this compressive strength, which is measured on 50 mm cubes, is obtained in testing conditions that are far from the actual working conditions of the capping compound within the very thin layer (2 to 3 mm thick) that is squeezed between the specimen and the platens of the testing machine. Within the thin cap, the sulfur compound is compressed in a confined state (like the ends of the concrete specimens) and it is well known that confined materials exhibit a much higher apparent compressive strength than unconfined ones (Lessard and Aïtcin, 1992). It will be seen later that high-strength capping compounds can thus be used to test high-performance concretes with a compressive strength higher than 70 MPa.

In rock mechanics, the problem of the preparation of the ends of the specimens has long since been addressed and solved. They are ground with a grinding machine and the faced specimen is placed directly between the two platens of the testing machine so that the load is transferred directly to the specimen (Vuturi, Lama and Saluja, 1984).

As facing equipment for concrete specimens is not presently available in all testing laboratories (though commercial equipment for this purpose is now being marketed), two other options to solve the end preparation problem can be explored. The first option is to stop testing high-performance concretes using cylindrical specimens and rather test them as cubes cast in rigid moulds having perfectly parallel faces. The consequence of such a drastic change will be discussed in section 15.2.3(a) on the influence of the shape of the specimens on the compressive strength. However, from extensive testing done at the INSA of Toulouse in France and at the Laboratoire Central des Ponts et Chaussées in Paris (de Larrard et al., 1994), it has been found that it is difficult to obtain sufficient parallelism of the lateral faces of cube specimens even though they are cast in moulds made of very thick and rigid steel. The number of shear failures increases drastically as the strength of the concrete increases when high-performance concrete is tested in the form of cubes. It has been found that in order to restore a proper failure, the two faces of the cube that are in contact with the platens of the testing machine have to be faced. So the use of cubes does not eliminate the necessity of facing; this leaves us now with only one option to explore.

The second option is to use a testing device that suppresses the need for any facing of the ends of the specimens, or that can play the role of the capping compound without presenting its strength problem.

In this vein of thought two options have been proposed: the use of neoprene pads (Grygiel and Amsler, 1977; Ozyildirim, 1985; Matsushita, Kawai and Mohri, 1987; Carrasquillo and Carrasquillo, 1988a), which are already used to test normal strength concrete, or the use of a 'sand box', as proposed by Boulay (Boulay, 1989; Boulay and de Larrard, 1993; Boulay, 1996). Experience has shown that neoprene pads work, as long as they are not used more than 5 to 10 times each for testing high-performance concrete. This solution becomes quite expensive in the long run. Moreover, the two metallic ends used to confine the neoprene pads are quite heavy and not very practical to use, and eccentricity problems are difficult to address.

On the other hand, the sand-box option proposed by Boulay is attractive from an operational cost point of view, because the only expense is the purchase of standard Ottawa sand meeting the grain size requirements of ASTM C109-90; however, the testing device is quite expensive (Boulay and de Larrard, 1993; Boulay, 1996). In this testing method, the specimens are seated on a thin sand cushion which compensates for the irregularities of the specimen ends, as shown in Figure 15.6. This testing method, which has been developed at the Laboratoire Central des Ponts

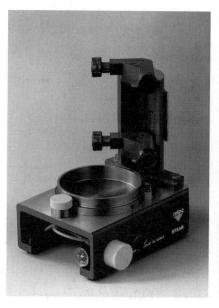

Fig. 15.6 The 'sand box' set-up for testing high-performance concrete.

et Chaussées in Paris, France, seems promising. It has been used extensively to test all the high-performance concrete of the Pont de Normandie. According to Boulay (1989), there is practically no difference between the compressive strength values obtained with the sand-box method and those of faced specimens.

If it is decided not to use either neoprene pads or the sand-box method, there remain only three other options for preparing the ends of concrete specimens: the use of a so-called high-strength capping compound, the facing of both ends of the specimens as in rock mechanics, or the use of a confined capping system, as recently proposed by Johnson and Mirza (1995) and Mirza and Johnson (1996), in which the capping compound is confined in steel rings that can be re-used when the crushing of the concrete sample has been completed.

In order to evaluate the influence of the end preparation of high-performance concrete specimens, Lessard (1990) tested five different concretes on 100×200 mm and 150×300 mm cylindrical specimens. Half of these specimens were capped with a high-strength capping compound and the other half had both ends faced on a modified lathe, shown in Figure 15.7. A total of 203 specimens were tested, 103 capped and 100 faced. From the results presented in Table 15.2(a) and 15.2(b), it may be seen that almost the same compressive strengths were measured in both cases, whatever the diameter of the specimen. The average

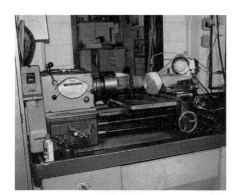

Fig. 15.7 Modified lathe used to face specimen ends.

Table 15.2(a) 91 day compressive strength of capped and ground 100 × 200 mm specimens

		Concrete mix				
		F6	F6	F9	L2[a]	L3[a]
Capped	Number of specimens	10	10	9	12	12
	$\bar{f'_c}$ (MPa)	120	121	129	118	115
	σ (MPa)	3.1	3.8	4.0	5.7	4.8
	$\sigma/\bar{f'_c}$ (%)	2.6	3.1	3.1	4.8	4.2
Ground	Number of specimens	10	9	9	12	12
	$\bar{f'_c}$ (MPa)	121	123	132	119	117
	σ (MPa)	2.3	1.9	2.8	2.7	2.5
	$\sigma/\bar{f'_c}$ (%)	1.9	1.5	2.1	2.3	2.1

[a] 28 day compressive strength.

compressive strength measured on faced specimens was almost systematically higher by 1 to 5 MPa than that of the capped specimens. For an average concrete compressive strength of 119.8 MPa, the faced specimens on average had a 4.8 MPa higher compressive strength, which represents a difference of only ± 1.5%.

Moreover, if we look in Table 15.2(a) and 15.2(b) at the standard deviation values, it may be seen that the standard deviation is always greater for the capped specimens. The average value of the standard deviation is 4.1 MPa for the capped specimens, while it is only 2.5 MPa for the faced specimens. Thus the dispersion of the results is slightly lower for faced specimens. These results tend to confirm those of Moreno (1990) but are contrary to the view that the dispersion of results is larger

Table 15.2(b) 91 day compressive strength of capped and ground 150 × 300 mm specimens

		Concrete mix				
		F6	F6	F9	L2[a]	L3[a]
Capped	Number of specimens	10	9	9	10	12
	$\bar{f'_c}$ (MPa)	115	114	119	93	93
	σ (MPa)	3.1	4.9	4.5	2.8	3.9
	$\sigma/\bar{f'_c}$ (%)	2.7	4.3	3.8	3.0	4.2
Ground	Number of specimens	10	7	9	12	10
	$\bar{f'_c}$ (MPa)	114	117	122	94	94
	σ (MPa)	2.9	3.6	3.5	1.6	1.1
	$\sigma/\bar{f'_c}$ (%)	2.5	3.1	2.9	1.7	1.2

[a] 28 day compressive strength.

when smaller specimens are used (Malhotra, 1976; de Larrard, Torrenti and Rossi, 1988).

So it is apparently not true that the measurement of concrete compressive strength on 100 × 200 mm specimens is not as consistent as that obtained with 150 × 300 mm specimens. It is also incorrect to say that if high-performance concrete has to be tested on 100 mm specimens, it is necessary to test more specimens to get the same degree of reliability as in the case of 150 mm specimens. There is thus no need to increase the capacity of 1.3 MN testing machines in order to obtain a more representative value of the compressive strength of high-performance concrete.

From the results presented here, it can be concluded that high-strength capping compounds, have a compressive strength of 55 to 60 MPa on 50 mm cubes, can be used to test high-performance concretes having a compressive strength as high as 130 MPa, **as long as the caps are not thicker than 2 mm, and are perfectly parallel** (Lessard, 1990). However, the standard deviation will always be significantly higher than if the specimen ends were faced.

As a result of this extensive study, it can be concluded that testing high-performance concrete on 100 × 200 mm specimens produces the same degree of reliability as testing on 150 × 300 mm specimens. Therefore, there is no need to increase the number of specimens tested when testing high-performance concrete with 100 × 200 mm specimens.

As a general rule at the Université de Sherbrooke, high-performance concretes with a compressive strength expected to be higher than 75 MPa are systematically faced (Figure 15.7). Between 50 and 75 MPa either practice can give satisfactory results, with a limited variability in the test results. Facing has one advantage over capping: it can be done at any convenient time prior to the test, as long as faced specimens are replaced in the same curing conditions that they were in before facing.

In two recent papers, Johnson and Mirza (1995) and Mirza and Johnson (1996) present results on the use of a confined capping system for compressive strength testing of high-performance concrete cylinders. In one of the studies, 55 test cylinders 100 × 200 and 150 × 300 mm in size were cast using three different high-performance concretes with compressive strengths ranging from 90 to 110 MPa, this compressive strength (which will be referred to as the 'standard strength') being measured on 150 × 300 mm specimens with their ends ground. Table 15.3 presents the results obtained. It is shown that for the 150 × 300 mm cylinders with confined caps the relative strength is 98% of the standard strength. The results for the 100 × 200 mm cylinders are in close agreement with those of the standard strength, considering the effect of cylinder size. The relative strengths obtained are, respectively, 101% and 105% for confined caps and ground ends. Moreover, the test showed that the coefficient of variation for 20 individual test groups ranged from 0.9 to 5.9% for cylinders with ground ends and from 1.2 to 3.6% for cylinders with confined caps.

Table 15.3 Comparison of the compressive strength of concrete specimens tested with their ends ground or using confined caps, as proposed by Johnson and Mirza (1995)

Mix	Cylinder size (mm)	28 day strength (MPa)		61 day strength (MPa)	
		Ground ends	Ends with confined caps	Ground ends	Ends with confined caps
DM9-1	100 × 200	98.2	94.8	102.6	97.9
	150 × 300	91.6	89.6	95.6	93.8
DM9-2	100 × 200	107.7	105.5	–	–
	150 × 300	106.8	102.6	–	–
DM19-1	100 × 200	96.3	90.3	94.9	94.9
	150 × 300	89.5	90.9	94.7	91.8

Note: Each value represents the average of three tests for most cases.

A patent application for the technique has been made in Canada and the United States under the name of Claude D. Johnson, S. Ali Mirza, Eric Powell and Edith Ramanathan.

(b) Influence of eccentricity

It is known that any eccentricity between the axis of the specimen and that of the testing machine can influence the compressive strength measurement of usual concrete. Neville (1995) has reported that, on usual concretes, an eccentricity of 6 mm does not affect the compressive strength measurement, while one of 12.5 mm leads to a compressive strength decrease of 10%. It should be pointed out that a careful naked-eye observation of the position of the specimen within the testing machine can easily spot a 6 mm eccentricity.

In order to see if this observation also applies to high-performance concrete, Lessard (1990) tested five specimens of usual and high-performance concretes with eccentricities of 0, 4, 6 and 12.5 mm. His results are presented in Table 15.4 and the mode of failure in Figure 15.8.

Table 15.4 28 day compressive strengths for different eccentricities (average of five specimens)

Eccentricity (mm)	Compressive strength (MPa)			
	L4	Type of rupture	L5	Type of rupture
0	29	conical	115	conical
4	30	conical	115	conical
6	29	conical	108	diagonal
12.5	27	diagonal	95	splitting

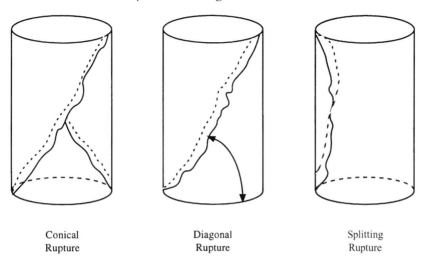

| Conical Rupture | Diagonal Rupture | Splitting Rupture |

Fig. 15.8 Types of rupture observed after eccentricity tests.

When making a statistical analysis of the results with the Student 't' test, Lessard (1990) found that for usual concrete there was no significant difference between the average compressive strengths obtained with 0, 4 and 6 mm eccentricity. The same statistical test indicated, of course, that the average compressive strength value obtained with a 12.5 mm eccentricity was definitely affected by this large eccentricity. For high-performance concrete, the same statistical test placed the limit of eccentricity beyond which the average value is affected at 4 mm.

The results obtained for usual concrete agree well with those published by Neville (1995). However, in the case of high-performance concrete an eccentricity of 6 mm was found to lower the compressive strength by 6%. For high-performance concrete, a 4 mm eccentricity was found to have no effect on the compressive strength value. It should be noted that this level of eccentricity is at the limit of a naked-eye observation for a technician who is performing compressive tests with some care. Table 15.4 shows that eccentricity can influence the mode of failure; the higher the eccentricity and the compressive strength, the sooner the failure mode ceases to be conical.

In spite of the limited nature of this study of the effects of eccentricity on compressive strength, it should be emphasized that great care must be taken to ensure that each specimen of high-performance concrete is tested as well-centred as possible with the axis of the testing machine. To solve this eccentricity problem, a simple plastic guide in use at the Université de Sherbrooke, shown in Figure 15.9, can be used to centre 100 and 150 mm diameter specimens on the testing machine (Figure 15.10).

15.2.3 Influence of the specimen

(a) Influence of the specimen shape

It is not our intention here to renew the controversy on the pros and cons of testing concrete either as cubes or as cylinders. When testing high-performance concrete in the form of cubes, it seems that the parallelism of the two faces on which the two platens of the testing machine apply the load is critical. A lack of parallelism can result in an increased number of shear failures, which tend to lower the compressive strength value (de Larrard *et al.*, 1994). So, to restore the parallelism of the two faces of the cube, it is necessary either to use a capping compound or to grind them; the use of cube specimens does not solve the end preparation problem. In addition, those who work with both cubes and cylinders know well the inconvenience of using cube moulds: they are heavy, take a long time to clean, need careful maintenance and are costly when compared with the present reusable plastic cylindrical moulds.

Fig. 15.9 Plastic guide to eliminate eccentricity.

Fig. 15.10 Testing high-performance concrete.

It must also be remembered that the same concrete will not give the same compressive strengths when tested as cubes and cylinders. The compressive strength measured on cubes is always higher than that obtained on cylinders.

(b) Influence of the specimen mould

Standard test methods for concrete are not restrictive on the material used to make the specimen mould, as long as it is rigid enough and non-absorptive. ASTM C-39 *Standard Test Method for Compressive Strength of Cylindrical Concrete Specimens* does not contain any restrictions on the material used to make the moulds in which the specimens are cast.

However, from field experience it is known that great care must be taken when selecting a particular mould of a given diameter in order to avoid different 'geometrical' problems. For usual concrete, the tolerance on ovalization is 0.5 mm. In the absence of any hard data, the only thing that can be said is that for high-performance concrete it should certainly not be more than that.

Moreover, the perpendicularity between the mould axis and its base is very critical when testing high-performance concrete. On one occasion, the author was confronted with an abnormal series of shear failures during the testing of 100×200 mm specimens that had been cast in disposable thin-walled (inexpensive) plastic moulds. After carefully checking that the capping device was producing specimen ends that were perfectly parallel, it was found that the real problem was the lack of perpendicularity between the bottom of the disposable mould and the axis of the mould. The moulds were improperly made and leaning.

The use of cardboard moulds has also been strongly recommended against for testing high-performance concrete. As early as 1971 Freedman recommended the use of rigid steel moulds to test high-performance concrete because it was noticed that the compressive strength obtained with rigid steel moulds was always higher and had a lower standard deviation than that obtained with disposable cardboard or disposable thin-walled metallic moulds. The strength decrease associated with the use of these non-rigid moulds was attributed to some ovalization of the top end of the mould and, to a lesser degree, to compaction obtained on the concrete cast in the non-rigid thin-walled moulds, even though the concrete received the correct number of strokes of the tamping rod. It should be remembered that at that time high-performance concretes were delivered with a 75 to 100 mm slump in order to reduce as much as possible the amount of mixing water because superplasticizers were not available, and the consolidation of high-performance concrete in the moulds was more critical than for the present, almost flowing, high-performance concretes.

Currently, plastic moulds with rigid walls (6 mm thick) are perfectly adapted for testing high-performance concrete: they are rigid enough, and if well maintained they can be re-used more than 100 times before any abnormal wear or ovalization is noticed. They are light, easy to maintain, easy to clean and easy to demould with compressed air. Using such moulds, Lessard (1990) has been able to reproduce the compressive strength of a 150 MPa concrete within 3.5 MPa when testing nine different specimens.

(c) Influence of the specimen diameter

As has already been discussed, it is better from an economical point of view to test high-performance concrete as 100×200 mm specimens rather than as 150×300 mm specimens, without sacrificing the quality and reproducibility of the results. This is not a revolutionary change in the way concrete should be tested, since some testing agencies and many researchers have already promoted the use of smaller specimens. In the author's opinion, and that of his technicians and graduate students, such a move has some practical advantages: smaller specimens are easier to handle, take up less space in the curing room, are much easier to cap correctly or faster to grind. So why stick to the heavy 150×300 mm specimens if there is absolutely no advantage in doing so, except being conservative (which cannot be considered as an advantage)? However, as most building codes are based on the sacrosanct compressive strength value obtained on 150×300 mm specimens, it is necessary during a transition period to determine exactly what would be the implication of such a change in the size of concrete specimens, particularly in the case of high-performance concrete.

The issue is to know exactly what a 100 MPa compressive strength value obtained on a 100 × 200 mm specimen would be had it been obtained using a 150 × 300 mm specimen. Such a correlation is well documented for usual concrete (Gonnerman, 1925; Malhotra, 1976; Carrasquillo, Nilson and Slate, 1981; Forstie and Schnormeier, 1981; Date and Schnormeier, 1984; Nasser and Kenyon, 1984; Nasser and Al-Manaseer, 1987; Carrasquillo and Carrasquillo, 1988b; Howard and Leatham, 1989; Moreno, 1990), but few comprehensive studies have been done on high-performance concretes and the results are somewhat conflicting.

For example, Carrasquillo and Carrasquillo (1988b) found that 150 × 300 mm specimens gave around 7% higher compressive strength than 100 × 200 mm specimens for 48 to 80 MPa concretes. Moreno (1990) found, on the contrary, a 1% increase in the compressive strength when it was measured on 100 × 200 mm specimens rather than on 150 × 300 mm specimens. Cook (1989) indicated that for a 70 MPa mix design, 100 × 200 mm specimens had a compressive strength approximately 5% higher than 150 × 300 mm specimens.

Lessard (1990) presented results obtained from 10 different high-performance concretes sampled either in the field or in the laboratory in 100 × 200 mm and 150 × 300 mm moulds with compressive strengths ranging from 72 to 126 MPa when tested at 28 and 91 days; all the specimens were faced. Lessard's results are presented in Table 15.5. Each value reported in this table is the average of three individual tests. This represents a total of 108 individual compressive strength tests. In Table 15.6, the different compressive strength results have been expressed as the ratio (expressed as a percentage) of the compressive strength of 100 × 200 mm specimens and that of 150 × 300 mm ones. The maximum value of this ratio was obtained for high-strength concrete No. 3 at 28 days and No. 9 at 91 days, for which it is equal to 110%. The minimum value is 100% (concrete No. 6 at 91 days). The average value of this ratio is 105%, with a standard deviation of 2.8% for the 18 compressive

Table 15.5 Compressive strengths of 100 × 200 mm and 150 × 300 mm specimens

Concrete mix		F1	F2	F3	F4	F5	F6	F7	F8	F9	L1
$\bar{f_c}$ (MPa)	28 days	75	93	97	101	99	110	105	90	114	112
100 × 200 mm	91 days	–	111	104	122	108	117	121	111	126	–
$\bar{f_c}$ (MPa)	28 days	72	91	88	95	95	105	104	86	108	108
150 × 300 mm	91 days	–	105	97	114	106	117	116	103	115	–

Note: Average of three specimens.

Table 15.6 Compressive strength correlation between 100×200 mm and 150×300 mm specimens at 28 and 91 days

Concrete identification		C1	C2	C3	C4	C5	C6	C7	C8	C9	C10
$\dfrac{\bar{f}_{c100}}{\bar{f}_{c150}}$ (%)	28 days	105	102	110[a]	106	105	105	101	105	106	104
	91 days	–	106	108	107	102	100[b]	104	108	110[a]	–

Average value of strength ratios: 105%.
[a]Maximum value of strength ratios. [b]Minimum value of strength ratios.

strength ratios presented in Table 15.6, which is in good agreement with the results obtained by Cook (1989).

Table 15.7 presents the standard deviations obtained for each testing condition. The average standard deviations are quite low and not very different for the 100×200 and 150×300 mm specimens at either 28 or 91 days, but at the same time the average standard deviation is slightly higher at 91 days (3.1%) than at 28 days (1.8%). This can most probably be attributed to the higher strength level obtained at 91 days. Finally, when looking at the values obtained on the standard deviation of the compressive strengths that were obtained on both sizes at both ages, it is difficult to conclude that one size gives more scattered results than the other.

From these results, it can be concluded that testing high-performance concrete as 100×200 mm specimens is as valid as testing it as 150×300 mm specimens. Moreover, the compressive strength measured on the smaller (100×200 mm) specimens is systematically slightly higher (average value of 5%) than that obtained with the larger (150×300 mm) specimens.

As a general rule, in order to avoid any argument, when controlling and/or reporting the compressive strength value of any high-performance concrete, the size of the specimens used to measure it must be clearly specified and/or indicated.

Table 15.7 Coefficients of variation of average compressive strengths

Concrete mix		F1	F2	F3	F4	F5	F6	F7	F8	F9	L1	Average
						100×200 mm						
Coefficient of variation[a] $\sigma/\bar{f}_{c100}$ (%)	28 days	2.0	1.8	1.0	1.2	2.5	0.9	1.0	0.9	1.3	2.4	2.4
	91 days	–	2.6	6.4	2.2	2.9	1.7	3.3	5.0	3.6	–	
						150×300 mm						
Coefficient of variation[a] $\sigma/\bar{f}_{c150}$ (%)	28 days	2.4	2.3	0.8	3.6	0.2	1.0	1.6	4.2	1.4	2.7	2.3
	91 days	–	2.9	3.7	0.5	2.5	4.5	1.0	2.6	3.5	–	

[a]Within three specimens.

15.2.4 Influence of curing

(a) Testing age

Concrete has traditionally always been tested at an age of 28 days, but when the first high-performance concretes were used in the Chicago area in the early 1970s, concrete producers were able to convince designers to use the 91 day rather than the 28 day compressive strength in the design. Several reasons were advanced to support such a rationale. First, high-performance concrete is never loaded to its full capacity at 28 days (nor is usual concrete). Second, at that time high-performance concrete contained fly ashes that did not have time to develop their full cementitious potential within 28 days. Third, the dense concrete still contained enough water after 28 days to hydrate Portland cement and fly ash further, so that it would have been a pity not to take advantage of the strength gain between 28 and 91 days. The cost of these extra MPa's is, of course, very low when compared with the high cost of the 28 day MPa's.

Since that time, it has become a more or less accepted rule to test high-performance concrete at 91 days, in spite of the fact that concrete technology has evolved since then, and the arguments prevailing in the early 1970s are not necessarily valid in the 1990s. It is now time to see whether this practice is still valid and if it is necessary to transform it into a permanent rule, or if it is better to stay with the conventional 28 day testing age.

The strength increase obtained by lengthening the curing period from 28 to 91 days for nine of the ten high-performance concretes presented in Table 15.6 is shown in Table 15.8. The average strength increase is about 14% for the 100 × 200 mm specimens. This represents a significant dollar saving when designing the structure. The variations found in the individual strength increases between 28 and 91 days are mainly due, in the author's opinion, to the wide variety in the composition of the

Table 15.8 Effect of lengthening curing period from 28 to 91 days

Concrete identification	$\bar{f_c}$ 100 × 200 mm		Difference	
	28 d	91 d	MPa	% of 91 d
C1	37.5	44.4	6.5	17
C3	93.4	111	17.6	19
C4	96.7	104	7.3	8
C5	101	122	21	21
C6	99.3	108	8.7	9
C7	110	117	7	6
C8	105	121	16	15
C9	90.0	111	21	23
C10	114	126	12	11

high-performance concretes tested, and the reactivity of their respective cementitious materials.

However, if the 91 day testing age can be scientifically justified for a concrete still having at 28 days an excess of water with regard to the amount of water needed to hydrate all the cement grains fully, it is more difficult to use this argument for a concrete having a water/binder ratio lower than 0.30. Lengthening the standard water-curing period from 28 to 91 days for concrete specimens containing so little water can result in an artificial increase in the compressive strength of the high-performance concrete, because a more or less thick outside rim of concrete in the small test specimen can benefit from an external supply of water which will increase its strength, while the concrete that is placed in the structure in much larger sections will not benefit from such an additional source of strength.

Thus it is the personal opinion of the author that if standard specimens of high-performance concrete are to be wet-cured, as is presently done for usual concrete, it should be good practice to maintain the age of testing of high-performance concrete at 28 days in order to get a fair value of the compressive strength of the concrete that has been cast in the structure. However, if the curing conditions of the specimens are modified as proposed in the next section, then the 91 day age could be used. This opinion is supported by extensive core testing carried out in past years using existing structural elements and experimental full-size columns built with high-performance concrete. These results are presented in section 15.2.5.

(b) How can high-performance concrete specimens be cured?

It is not easy to give a clear answer to the question of how to cure high-performance concrete specimens, because high-performance concrete can harden quite differently within the structure to how it hardens within standard specimens. To the best knowledge of the author, there has not been enough work done that addresses this very simple question.

For example, the temperature increase experienced by the concrete during the first hours or days in the forms is always much higher than that measured in small standard specimens, especially those having a 100 mm diameter. In order to simulate the temperature increase of the concrete within the structure, standard specimens should be stored for the first 24 hours within insulated containers. But what maximum temperature would be reached by these specimens? In fact, within the forms, a 15 to 20 °C variation in the highest temperature reached by the concrete can exist, depending on how far it is from the forms, as will be seen in the next chapter. Aïtcin and Riad (1988) have shown that this temperature increase is not disadvantageous in all circumstances for high-performance concrete, though it is for steam-cured usual concrete.

Moreover, usually the concrete cast in a structure does not get any additional curing water in spite of what might have been written in the specifications under the heading: curing. So, why should standard specimens used to evaluate high-performance concrete compressive strength receive such a benefit? Is it fair that these standard specimens be allowed to increase their strength somewhat from a well-controlled wet curing when the concrete within the structure will instead be exposed only to self-desiccation and drying?

Wet curing has been found to be an adequate mode of curing for usual concrete specimens, but usual concretes always contain large amounts of mixing water, so that one can imagine that if the field concrete is well cured it will be able to develop its strength under such conditions. But is this still true for a high-performance concrete?

One possible way to reproduce similar curing conditions, as far as moisture is concerned, would be to wrap the specimens in an impermeable thin envelope as soon as the water curing is finished in the structure and to cure them at room temperature until the selected age for testing. Almost impermeable plastic films are readily available in any supermarket; it is easy to wrap specimens with them and there is no need to invest in a sophisticated curing room. From the few fully documented studies on this topic it would seem that this is not a bad idea (Lessard, 1990; Khayat, Bickley and Hooton, 1995). But those who do not like such a 'revolutionary' move can stay with the more orthodox 28 day wet-curing period already used for usual concrete, or the well accepted 91 day water curing. The debate is open ...

15.2.5 Core strength versus specimen strength

One question that is often raised concerns the representativity of the compressive strength measured on specimens when compared with the actual compressive strength of the concrete in the structure. It is obvious that the concrete that is tested has not been cured under the same conditions as the concrete of the structure and, more particularly, has not the same thermal history. It is well known that a heat treatment changes the compressive strength of usual concrete and that this technique is commonly used in precast plants, in brick- and block-making plants and in pipe plants, etc.

Usually, an initial heat treatment increases the early compressive strength but results in a 28 day strength decrease when compared with the 28 day compressive strength of a specimen cured at ambient temperature.

In the previous chapter, it was seen that the temperature profile followed by a high-performance concrete during its hardening can be quite different from the usual heat treatment we are used to seeing in the precast plant. Some major differences are that the temperature increase in a high-performance structural element occurs usually after more than 10 hours of dormancy and is quite homogeneous and isotropic, the heat being generated within the concrete itself instead of being applied through an outside medium very shortly (a couple of hours) after the casting of the element. Moreover, the maximum temperature reached within the high-performance concrete is most of the time lower than the maximum temperature reached during the industrial heat treatment applied to usual concrete.

There are not many comprehensive studies that have been done on this subject. One of those that has is that reported in section 11.3.2 (Miao et al., 1993); some others have been published by Mak et al. (1990), Haque and Gopalan (1991), Burg and Ost (1992), Aïtcin et al. (1994) and Bartlett and MacGregor (1994, 1996). All these experimental studies have shown that the compressive strength measured on cores at 28 or 91 days was between 85 and 90% of the compressive strength measured on standard-cured specimens. This value is about the same as that found with usual concrete and which is reported in most codes. Moreover, these studies showed that, in spite of the fact that the thermal history of the cores within the structural element could be somewhat different, the measured compressive strength did not vary very significantly with the core location. Miao et al. (1993) found that the compressive strength values measured on cores taken at three different levels in their two high-performance concrete columns were more homogeneous than in the case of the 35 MPa concrete column.

15.3 STRESS–STRAIN CURVE

Designers are interested in the shape of the post-peak part of the stress–strain curve for their calculations. For usual concrete, researchers have proposed a number of mathematical formulations for this part of the curve, which are supposed to represent appropriately this important characteristic of concrete (and of the testing conditions) (Kaan, Hanson and Capell, 1978).

National codes favour one mathematical formulation rather than another for reasons that are more often nationalistic rather than actually scientific. Thorenfeldt, Tomaszewicz and Jensen (1987) proposed the following equation:

$$\frac{f_c}{f_c'} = \frac{\varepsilon_c}{\varepsilon_c'} \times \frac{n}{n - 1 + (\varepsilon_c/\varepsilon_c')^{nk}}$$

where

f_c is the longitudinal stress;
ε_c is the longitudinal strain;
f_c' is the maximum stress;
ε_c' is the strain corresponding to f_c';
n and k are correction factors.

However, Collins and Porasz (1989), Collins and Mitchell (1991) and Collins, Mitchel and McGregor (1993) suggested that:

$$k = 0.67 + \frac{f'_c}{62} \text{ MPa}$$

$$n = 0.8 + \frac{f'_c}{17} \text{ MPa}$$

Is this formulation valid for all high-performance concretes? How do the stiffness of the testing machine, the testing procedure and the specimen aspect ratio affect the shape of this part of the stress–strain curve for high-performance concrete (Van Gysel and Taerwe, 1996; Wee, Chin and Mansur, 1996)? It should be admitted that at present these questions have received only scant attention from the research community and have not been fully answered for the influence of the stiffness of the testing for usual concrete and *a fortiori* for high-performance concrete (Shah, Gokoz and Ansari, 1981).

Moreover, if we look at what is already known in rock mechanics about the post-peak stress–strain curves of rocks, it is seen that, here too, it would be appropriate to adopt, and perhaps adapt, the rock testing procedures for high-performance concrete. For example, the influence of the stiffness of the testing machine on the post-peak part of the stress–strain curve has already been studied for concrete by Sigvaldson (1966). The aspect ratio of the specimen being tested also plays a key role in the shape of the post-peak stress–strain curve.

A very important point when determining the post-peak part of the stress–strain curve of some high-performance concretes is that it can, in certain cases, have the same shape as that of a Class II rock. This means that just after the ultimate failure, there is a surprising strain regression, i.e. after the ultimate failure occurs, high-performance concrete exhibits for a very short period of time a slight increase in volume. This slight volume increase may be explained in two ways:

1. When the two rough surfaces of a given shear plane start to slide on each other (as shown in Figure 15.11), this generates a slight increase in the apparent volume of the broken specimen.

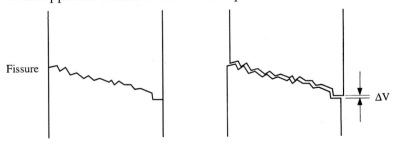

Fig. 15.11 Apparent volume increase when two rough surfaces start to slide on each other.

2. There is an elastic decompression in the two parts of the broken specimen that results in a volume increase. This possible slight volume increase creates a challenging situation as far as the programming of the testing machine is concerned. At present, most testing machines are programmed for a constant strain rate; when such a Class II behaviour occurs during the test, the servo valves no longer respond appropriately and this can lead to the crushing of the specimen and of the whole expensive set-up used to record the stress–strain curve. In such a case, it is better to regulate the testing machine using a linear stress–strain function which will depend on the elastic modulus of the ascending branch of the stress–strain curve, as shown in Figure 15.12.

15.4 SHRINKAGE MEASUREMENT

Concrete shrinkage has received a great deal of attention from researchers and specialists. In North America it is measured according to ASTM C157 *Standard Test Method for Length Change of Hardened Cement Mortar and Concrete,* and this method has been applied to high-performance concrete with more or less success. It is the purpose of this section to show that this standard testing method, that has been developed and successfully used to evaluate usual concrete drying shrinkage, is not pertinent at all to measuring the total shrinkage of any high-performance concrete. Other experimental methods have been proposed but they have not yet been widely used.

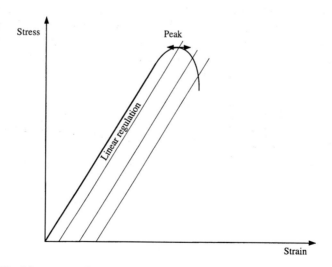

Fig. 15.12 Linear regulation for safe stress–strain curves recording in the descending part.

15.4.1 Present procedure

Prismatic concrete samples having a length L_o and which include two steel pins at the centre of each of their extremities are cast and cured for 24 hours at ambient temperature in a moist cabinet having a relative humidity of not less than 55%, so that the top surface of the sample is not subjected to any external drying (plastic shrinkage and drying shrinkage) during these first 24 hours. The next day the samples are demoulded and L_{ref}, the initial reference length, is measured before applying lime-saturated water curing. Usually after 28 days of lime-saturated water curing the samples are air-cured at 50% relative humidity and constant temperature and their change in length is measured regularly by measuring the length between the two steel pins. This procedure is schematically presented in Figure 15.13.

This standardized procedure has been in existence for many years and has been used successfully to measure the drying shrinkage of usual concrete. But if we look at what is actually happening in the concrete sample from its casting until its demoulding, we find that some autogenous shrinkage can develop in the sample during the first 24 hours, so that Figure 15.14 better represents what actually happens when preparing a specimen for drying shrinkage measurement.

Concrete having a high water/binder ratio, for example higher than 0.50, does not develop any autogenous shrinkage, or so little that it is legitimate to suppose that L_{ref} is the initial length of the sample when cast. Present ASTM C 157 standards preclude that autogenous shrinkage is negligible during the first 24 hours, so there is no problem in supposing that $L_{ref} = L_o$. In any case, this is what has always been done up to now when testing usual concretes.

However, this assumption becomes less and less valid as the water/binder ratio of concrete decreases, because autogenous shrinkage does not wait 24 hours to develop in high-performance concretes. Generally speaking, the lower the water/binder ratio, the faster autogenous shrinkage develops. This is a major difference between usual concrete and

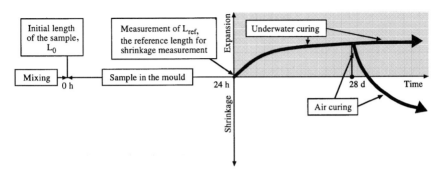

Fig. 15.13 Experimental procedure presently used for drying shrinkage measurement.

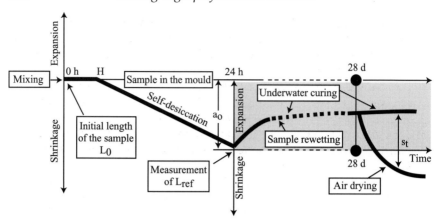

Fig. 15.14 What actually happens when preparing a specimen of high-performance concrete for shrinkage measurement.

high-performance concrete, and Figures 15.13 and 15.14 will be used to illustrate more closely what is happening when the present procedure is used in the case of a usual concrete having a high water/binder ratio, and of a high-performance concrete having a low water/binder ratio.

15.4.2 Shrinkage development in high water/binder concrete

Usually, hydration begins after a dormant period of about 4 hours, and bleed water and the water present in the large capillaries are essentially used to hydrate the cement particles during the first 24 hours, and fill the fine porosity developed by the volumetric contraction of the hydrated cement paste so that no autogenous shrinkage is developed when the concrete remains in the mould when L_{ref} is measured. It is therefore valid to assume that $L_{ref} = L_o$.

As shown in Figure 15.13, when the sample is placed under water after demoulding, it usually experiences a small expansion which is essentially due to rewetting. When after 28 days of water curing, the concrete specimen is air-cured, drying shrinkage develops rapidly. The higher the water/binder ratio, the more rapid and the greater the drying shrinkage.

15.4.3 Shrinkage development in a low water/binder concrete

As shown in Figure 15.14, the beginning of hydration, H, varies widely, usually from 6 to 18 hours, but sometimes it can be sooner or later; the beginning of hydration is essentially determined by the type of cement, and the amounts of superplasticizer and retarder used. Moreover, the water/binder ratio also influences the amount of self-desiccation; other things being equal, the lower the water/binder ratio, the higher the autogenous shrinkage developed during the first 24 hours when using the present recommended procedure. Therefore, when concrete samples are demoulded at 24 hours, the hydration reaction is more or less developed, so the autogenous shrinkage, a_0, developed within the sample at 24 hours, can vary over a wide range.

The autogenous shrinkage is no longer negligible and therefore it is no longer valid to assume that $L_{ref} = L_0$. Then when the concrete sample is placed under water, it begins to expand up to a maximum, owing to its rewetting. But in some cases, the development of further self-desiccation within the concrete, even though it is cured under water, results in a global shrinkage after several hours or days, as shown in Figure 15.15.

Therefore, using present procedures, it is not possible to make any kind of reliable shrinkage measurement in the case of high-performance concrete. That is why the author is proposing a new experimental procedure for the curing of the specimen during the first 24 hours when drying shrinkage has to be measured. This procedure, which is described in section 15.4.6, can be used for any type of concrete.

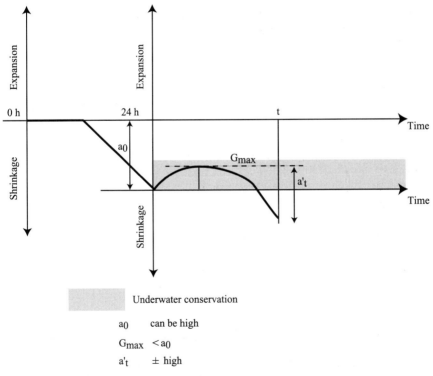

Fig. 15.15 Development of shrinkage in some very low water/binder ratio concretes when using the present standard procedure.

The measurement of self-desiccation shrinkage, if necessary, can be performed as proposed by Jensen and Hansen (1995) or Tazawa and Miyazawa (1996).

15.4.4 Initial mass increase and self-desiccation

When a sample in which some autogenous shrinkage has developed is placed under water, it is observed that its mass increases, because it absorbs external water to fill the empty spaces generated by the self-desiccation process that was developed during the first 24 hours. It is also observed that the sample expands somewhat, as any dry concrete sample usually does when it absorbs water when rewetted. It must be emphasized that the cause of this expansion is not very clear. This expansion is not proportional to the mass increase due to the intake of water; it is only a function of the kind of self-desiccation porosity that was developed during the first 24 hours within the concrete sample. Autogenous shrinkage is linked to the diameter of emptied capillary pores, not to the total volume of pores that were created due to self-desiccation. When the water/binder ratio is high, the diameters of the capillaries that have emptied during the first 24 hours are large, so the menisci present in the concrete create very weak tensile stresses and therefore practically no autogenous shrinkage is developed during this time. This is not the case for low water/binder ratio concretes.

The initial mass gain cannot be used to indicate the amount of autogenous shrinkage developed within the concrete sample, because two concrete samples can present the same mass gain without having the same autogenous shrinkage. The fact that they absorb the same amount of water only proves that the same amount of cement has hydrated in both samples (if all internal empty spaces created by self-desiccation are connected to the external source of water). However, the self-desiccation may not have developed the same type of microstructure and, therefore, the same capillary pore size distribution, which will generate tensile stress within the mass of concrete when menisci develop in the pores.

15.4.5 Initial compressive strength development and self-desiccation

The 24 hour compressive strength also cannot be used to evaluate the progress of self-desiccation in concrete, in spite of the fact that it is directly linked to the amount of cement that has been hydrated during these first 24 hours, because compressive strength is not linked only to the capillary pore size distribution (Jung, 1968). Therefore, if the amount of autogenous shrinkage developed in a concrete sample has to be established, it must be done from direct measures on the sample as it is hydrating, as was performed by Tazawa and Miyazawa (1993, 1995, 1996).

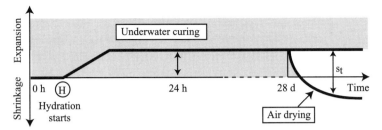

Fig. 15.16 Suggested procedure for drying shrinkage measurement.

When after 28 days of water curing, the concrete sample is air-dried, drying shrinkage begins to develop but at a much lower rate than in the case of a high water/binder ratio concrete.

15.4.6 New procedure for drying shrinkage measurement

In order to avoid any ambiguity for the measurement of the reference length, L_{ref}, it is proposed to cure concrete specimens under water just before hydration starts or the moment they are cast, until they are exposed to air drying, as shown in Figure 15.16. They can be demoulded rapidly at 24 hours, L_{ref} measured, and replaced as soon as possible under water, so that they do not dry superficially, in order to maintain as long as possible the connectivity of the capillary system within the concrete. This is the best procedure to limit the development of any autogenous shrinkage within the concrete, whatever its water/binder ratio.

When a concrete sample is air-cured, the measured drying shrinkage, s_t, develops and increases very rapidly for usual concretes, owing to their large porosity and the large diameter of their coarser capillaries, but it develops and increases very slowly for high-performance concretes, owing to their very low and refined porosity.

If a low water/binder ratio concrete sample is placed under water as soon as it is cast, when the cement starts to hydrate water can be drained from within the concrete to fill the very fine porosity appearing during hydration, as long as the gel pores and capillary pores form a continuous network ensuring the continuity of the water phase. This is why no autogenous shrinkage develops in some low water/binder concrete samples cured under water, at least during the early development of the hydration reaction. But as soon as some capillary pores are no longer connected to the surface of the sample, some menisci appear within these unconnected capillaries and some autogenous shrinkage starts to develop, even though the concrete sample is cured under water. This explains why a curve like the one presented in Figure 15.17 can be obtained with some very low water/binder concretes (Jensen and Hansen, 1995; Tazawa and Miyazawa, 1996).

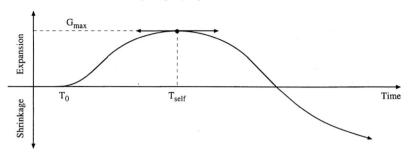

Fig. 15.17 Volumetric variation of a very low water/binder ratio concrete sample (W/B < 0.30) cured under water just after its placing (after Tazawa and Miyazawa, 1996).

It is difficult to be precise about the exact value of the maximum expansion, the time when this maximum is reached and the rate and intensity of the autogenous shrinkage that is developed, because these phenomena are directly linked to the development of the microstructure of the particular concrete. The development of this microstructure depends on the water/binder ratio, the initial and ambient temperatures, the type and fineness of the cement, the effect of admixtures and, more generally, on all factors that have an effect on the development of the hydration reaction. Tazawa and Miyazawa (1993) have developed a formula allowing the prediction of the autogenous shrinkage of a concrete from a linear regression analysis obtained in an experimental study where they studied several types of cement or cementitious materials:

$$\varepsilon_{auto} = -0.012 \times \alpha\, C_3S(t) \times (\%\, C_3S) - 0.070 \times \alpha\, C_2S(t) \times (\%\, C_2S)$$
$$+ 2.256 \times \alpha\, C_3A(t) \times (\%\, C_3A) + 0.859 \times \alpha\, C_4AF(t) \times (\%\, C_4AF)$$

where

ε_{auto} represents autogenous shrinkage;
$\alpha\, i(t)$ represents the hydration degree of phase i at time t;
(% i) represents the percentage of phase i in the cement.

It is surprising to see that in this linear regression a negative sign precedes the two coefficients governing the influence of the two silicate phases, and a positive sign the two coefficients governing the influence of C_3A and C_4AF, despite the fact that ettringite formation is also accompanied by a volume decrease, as was shown by Le Chatelier 100 years ago: volume of ettringite < volume of C_3A + volume of calcium sulfate + volume of water.

The preponderant role of C_3A in the development of autogenous shrinkage is somewhat surprising, because the volume change that is essentially responsible for autogenous shrinkage is due to the volumetric change of the silicate phase. Moreover, the C_3A content is small (usually less than 10%) when compared with the total amount of silicates (usually more than 80%).

However, Eckart, Ludwig and Stark (1995) have recently shown that hydration products resulting from the hydration of C_3S and C_2S generate practically no capillary pores, only gel pores, while C_3A hydration, and C_4AF to a lesser extent, primarily generate capillary pores. It is in these very fine capillary pores that tensile stresses generating autogenous shrinkage are developed.

15.5 CREEP MEASUREMENT

15.5.1 Present sample curing (ASTM C 512)

ASTM standard C 512 *Standard Test Method for Creep of Concrete in Compression* recognizes two types of curing for the sample to be tested for creep: standard curing and basic creep curing, which are schematically represented in Figure 15.18. In the case of standard curing, the specimens should stay in the mould for not less than 20 hours or more than 48 hours, and then be stored under moist conditions until the age of 7 days. After this moist curing, they are air-dried ($23.0 \pm 1.1\,°C$ and RH = $50 \pm 4\%$). Finally, when the creep potential of different concretes is to be compared, the specimens are loaded at 28 days.

15.5.2 Development of different concrete deformations during a 28 day creep measurement

The development of the different deformations during a 28 day creep measurement is schematically represented in Figure 15.19. The value of the autogenous shrinkage, a_0, at the time of demoulding is zero or negligible for high water/binder ratio concretes, but it can be significant for low water/binder ratio concretes. The lower this ratio, the greater the value of a_0.

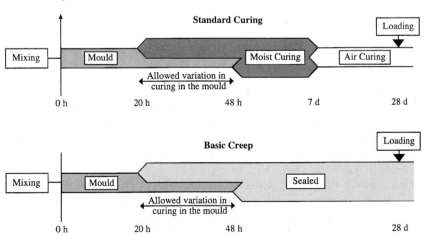

Fig. 15.18 Present ASTM C 512 Standard Test method curing procedure for creep measurement.

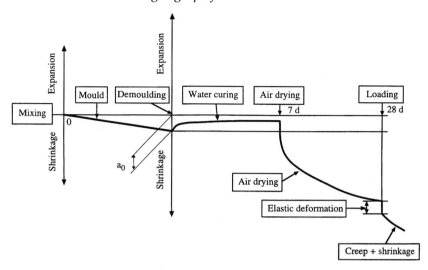

Fig. 15.19 Development of the different deformations in high-performance concrete during a 28 day creep measurement.

When concrete samples are air-dried after 7 days of moist curing, most or all of the autogenous shrinkage has been developed, so that it can be considered that the shrinkage developed is only drying shrinkage. Therefore, this procedure is convenient for both high and low water/binder ratio concretes. However, this is not the case where an early creep measurement at 2 days or less has to be performed.

15.5.3 Deformations occurring in a high water/binder ratio concrete subjected to early loading during a creep test

Figure 15.20 represents the different deformations that are developed in a high water/binder ratio concrete sample subjected to an early loading, as well as those of unloaded companion samples exposed to three different types of curing. The three methods of curing are: underwater curing, sealed curing and air drying. It is important to conduct these three methods of curing simultaneously in order to study the simultaneous action of drying and creep on the concrete of the sample.

First of all, the initial autogenous shrinkage at demoulding, a_0, is zero or negligible. When the concrete sample is moist-cured after its demoulding, it experiences a slight volume increase owing to its rewetting.

When the concrete sample is sealed so that it does not dry due to an external source, the only shrinkage that can develop within the sample is autogenous shrinkage. But for high water/binder ratio concrete the autogenous shrinkage remains negligible if not zero.

When the concrete sample is air-dried, drying shrinkage develops; the higher the water/binder ratio, the faster and the greater the development of this drying shrinkage.

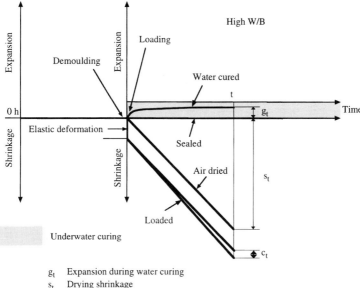

Fig. 15.20 Shrinkage development in a high water/binder ratio concrete when an early creep measurement is made.

By comparing the different deformations of the four specimens, it is possible to measure all of the elementary deformations that are developed within the concrete sample or within a structural element. For example, the covercrete will react rather like the air-dried specimen, while the concrete at the centre of the structural element will react rather like the sealed specimen.

15.5.4 Deformation occurring in a low water/binder ratio concrete subjected to early loading during a creep test

Figure 15.21 represents the different deformations that are developed in a low water/binder ratio concrete sample subjected to an early loading, and those of three unloaded companion samples exposed to the same curing regimes described in section 15.5.3 for usual concrete having a high water/binder ratio.

The main experimental problem in such a case is that the initial autogenous shrinkage, a_0, which develops when the loading is applied is, first of all, no longer negligible and can even represent the major part of the total shrinkage, and, secondly, is highly variable, depending on the time at which the loading is applied with respect to the beginning of the hydration reaction.

As mentioned previously, the time at which the hydration reaction begins depends essentially on the type and fineness of the cement and on the amount of superplasticizer and retarder (if any) that are used.

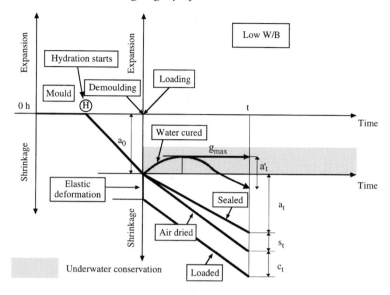

Fig. 15.21 Shrinkage development in a low water/binder ratio concrete when an early creep measurement is made.

a_0	Self-desiccation shrinkage at the demolding (can be very high)
g_{max}	Maximum expansion during water curing ($g_{max} < a_0$)
a'_t	Self-desiccation shrinkage in a water-cured sample (a'_t more or less high)
a_t	Self-desiccation shrinkage developed in a sealed sample (a_t more or less high)
s_t	Drying shrinkage (small)
c_t	Creep

It should also be noted that after the application of the load, the essential part of the shrinkage that develops within the concrete is the autogenous shrinkage, and that drying shrinkage is small. The lower the water/binder ratio, the higher the autogenous shrinkage and the smaller the drying shrinkage, even within the concrete.

15.5.5 Proposed curing procedure before loading a concrete specimen for creep measurement

It is suggested that concrete samples be cured under water just before hydration starts or as soon as they are cast, as in the case of shrinkage measurements, in order to avoid the development of any autogenous shrinkage within the concrete sample before its demoulding and loading. Figure 15.22(a) and (b) represents what is happening in the different companion samples cured in the three different ways. Using such a procedure, there will no longer be any ambiguity in the measurement of the various deformations that are developed within the loaded specimen.

Creep and shrinkage measurements

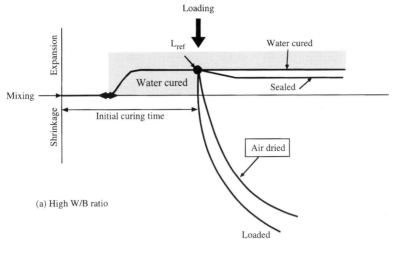

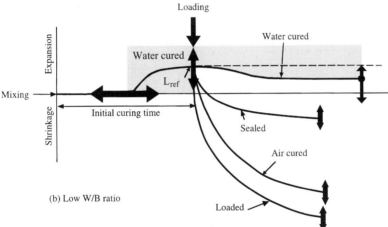

Fig. 15.22 Proposed procedure for creep measurement.

15.6 CONCLUDING REMARKS ON CREEP AND SHRINKAGE MEASUREMENTS

The present procedures and testing methods, which have been developed to measure creep and drying shrinkage in usual concrete with a high water/binder ratio, are no longer appropriate to measure the creep and shrinkage of high-performance concretes with a low water/binder ratio, owing to the complexity of the physico-chemical processes involved in the development of the different forms of shrinkage. In the author's opinion, it would be better to begin to cure under water the concrete samples used to measure concrete shrinkage and creep before the hydration reaction starts, or at least by the time initial setting is achieved. It is only with such curing conditions that a reliable value of

L_{ref} can be measured without the need to use expensive gauges embedded (and lost) in the samples to measure concrete shrinkage. It is a very simple and non-expensive way of achieving a reliable value of L_{ref} from a practical point of view. Finally, it should be emphasized that in high-performance concrete, the predominant component of shrinkage varies greatly according to the location of concrete within the mass of a structural element, as already shown in section 12.8.

15.7 PERMEABILITY MEASUREMENT

The measurement of the water permeability of usual concrete is relatively easy; since the water/binder ratio is high, the concrete is very porous, its capillaries being interconnected. It is the presence of non-reacted water that creates this fully interconnected network of channels through which water can flow when a pressure gradient is applied.

The water permeability of usual concrete decreases as the water/binder ratio decreases until this ratio approaches 0.40. When such a water/binder ratio is reached most permeameters are unable to register any significant water flow through the concrete sample, although a high pressure is applied on the water. This does not mean that there are no longer any interconnected capillaries within the concrete, but rather that the water cannot move through them owing to their fineness (El-Dieb and Hooton, 1995). Darcy's law, on which most permeameters are based, is no longer valid and the electric and capillary forces developed within these fine capillaries are strong enough to counteract the usual pressure gradients used when measuring water permeability (usually $\cong 1$ MPa).

As most high-performance concretes have a water/binder ratio lower than 0.40, it is not possible to evaluate the degree of interconnection of the existing network of fine pores by trying to force water to flow through it. However, it would be dangerous to assume that high-performance concretes are totally impervious, because even though water cannot percolate through this interconnected network, aggressive ions can, because the osmotic pressures can be strong enough to make these ions move more or less easily through the high-performance concrete.

As there is evidence of the existence of such an interconnected network of pores, it is necessary to approach concrete permeability measurements in another way. In this case, too, it is interesting to look at what is done in rock mechanics because one can find the value of the permeability of a rock as impervious as granite in a rock mechanics book.

The water permeability of rocks is usually measured in triaxial cells using convergent or divergent water flow, but such a measurement necessitates quite a sophisticated and expensive piece of equipment and has been tried in very few cases for high-performance concrete (El-Dieb and Hooton, 1995). For rocks, the permeability is more easily evaluated using air permeameters. In the case of rocks the preparation of the sample does not create too many problems: the sample is dried to constant mass in an oven prior to testing, which can be done without much worry in rock because water does not generally interact with the rock microstructure; however, this is not the case with concrete.

Measurement of the air permeability of concrete is not easy, owing to the problems created by the drying of the specimens, especially in the case of high-performance concrete. It has been shown that it is quite a difficult and a very slow process to dry a piece of concrete. If the size of this piece of concrete is decreased in order to facilitate its drying, then it is no longer very representative of a massive concrete element. Moreover, it is difficult or impossible to dry a piece of concrete so as to remove only the free water present in the interconnected network of pores. Recently, Acker (personal communication, 1996) has shown that the interconnected pore network in usual concrete is in fact composed of two interconnected networks of pores, such as the ones that exist in wood. One network is composed of rather large pores, and it is interconnected with a much finer one from which it is very difficult, if not impossible, to remove water. This very fine network can reactivate water flow through concrete after its drying.

Thus, as it is impossible to measure the water permeability and impossible to define precisely the dryness state of high-performance concrete, high-performance concrete permeability will have to be measured in another way. In this particular case the answer for permeability evaluation does not come from rock mechanics but rather from a technique already used in concrete technology, i.e. the AASHTO T-277 *Standard Method of Test for Rapid Determination of the Chloride Permeability of Concrete*, or its ASTM equivalent C 1202 *Test Method for Electrical Indication of Concrete's Ability to Resist Chloride Ion Penetration*.

The name by which this test is known is not perhaps the most appropriate one, but the test itself works quite well for usual concrete and can easily be correlated to the water permeability for usual concrete when it is measurable. This explains why it has been so rapidly adopted by those interested in evaluating the permeability of high-performance concrete (Tsutsumi *et al.*, 1993).

It has been shown, for example, that there is a strong correlation between the water/binder ratio and the chloride ion permeability of high-performance concrete (Mobasher and Mitchell, 1988; Perraton, Aïtcin and Vézina, 1988; Whiting, 1988; Gagné, Aïtcin and Lamothe (1993)). Moreover, this test is simple to perform and the piece of equipment required to measure it is not too expensive. A multicell apparatus is readily available, which simultaneously measures the chloride ion permeability in four to eight samples at a time (Figure 15.23).

The concrete samples used to measure chloride ion permeability are neither too small nor too big: they have a diameter of 95 mm and a length of 50 mm. They are placed between two solutions, one containing

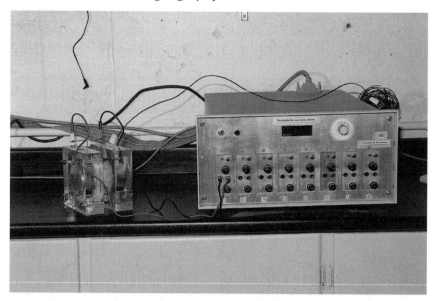

Fig. 15.23 Chloride ion permeability set-up developed at the Université de Sherbrooke.

at the beginning of the test a definite amount of sodium chloride, and the other one containing a definite amount of sodium hydroxide. The test consists of recording the intensity of the current developed through the sample when a voltage of 60 volts is applied over 6 hours. The permeability of concrete is expressed in coulombs, and represents the total amount of electrical charge passed through the sample during these 6 hours. Whiting (1981) established a chloride ion permeability scale that covers all kinds of concrete; it is reproduced in Table 15.9. The particular set-up developed at the Université de Sherbrooke is presented in Figure 15.23.

Experience has shown that concrete for which it is not possible to measure any water flow using a water permeameter can present different degrees of chloride ion permeability (Perraton, Aïtcin and Vézina, 1988).

Table 15.9 Chloride ion permeability based on charge passed

Charge passed (coulombs)	Chloride ion permeability
> 4000	High
2000–4000	Moderate
1000–2000	Low
100–1000	Very low
< 100	Negligible

Table 15.10 Rapid chloride ion permeability ranges for concretes of different water/binder ratios, with and without silica fume

	With silica fume		Without silica fume	
	From	To	From	To
0.60	1800	3500	5500	9000
0.50	1200	5000	4000	6000
0.40	250	1000	2500	4000
0.30	150	500	1800	3000
0.25	100	200	800	2500

First, the fact that a substantial electrical current passes through the sample shows that there actually exists an interconnected network of very fine pores in which the chloride ions (and OH^-) ions can move. These ions are not diffusing through the solid parts of the high-performance concrete but rather moving through a network of pores. Table 15.10 shows the chloride ion permeability of some high-performance concretes and Figure 15.24 the variation of the chloride ion permeability as a function of the water/binder ratios as compiled by Whiting (1984).

15.8 ELASTIC MODULUS MEASUREMENT

The measurement of the elastic modulus of a high-performance concrete can be done in the same way as for usual concrete. It is the purpose of this section to present a very efficient and convenient set-up for strain measurement developed at the University of Berkeley that is now in use in the author's laboratory.

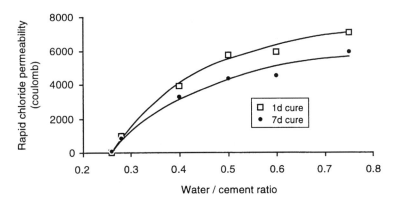

Fig. 15.24 Chloride ion permeability as a function of the water/binder ratio (Whiting, 1984). Reproduced by permission of ACI.

416 Testing high-performance concrete

The use of this set-up reduces considerably the time devoted to the measurement of the elastic modulus because it is not necessary to glue any strain gauge onto the sample. There is no need to prepare the concrete specimen in any way, and it has been checked at Berkeley University and at the Université de Sherbrooke that the results obtained are as good and as reproducible as the ones obtained with expensive strain gauges.

The set-up is composed of two aluminium yokes separated by a distance equal to half the length of the specimen to be tested (Figure 15.25). These two yokes are fixed onto the specimen using three thumbscrews, which are slightly tightened on the specimen. These three thumbscrews are at 120° to each other. Between these two yokes, which

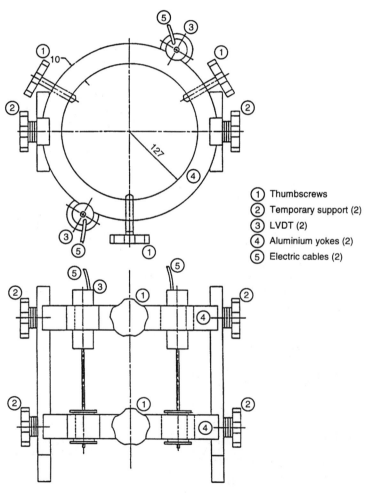

① Thumbscrews
② Temporary support (2)
③ LVDT (2)
④ Aluminium yokes (2)
⑤ Electric cables (2)

Fig. 15.25 The Berkeley modulus of elasticity set-up.

are temporarily held together by two vertical supports during the installation of the set-up on the sample, two LVDTs (linear variable differential transformers), placed at 180° to each other, are recording the measured deformation (it was found that it was not necessary to complicate the set-up by using three LVDTs at 120°). An electronic device adds the two deformations recorded by each LVDT (Figure 15.26). This sum is used as the total deformation of the full length of the specimen (it should be remembered that the distance between the two yokes is only half the total length of the specimen).

Before loading the specimen, the two temporary supports are removed and the load and the relative displacement of the two rings are recorded simultaneously. The photographs in Figure 15.27 represent the different steps of the sample preparation and the recording of the loading–unloading cycles.

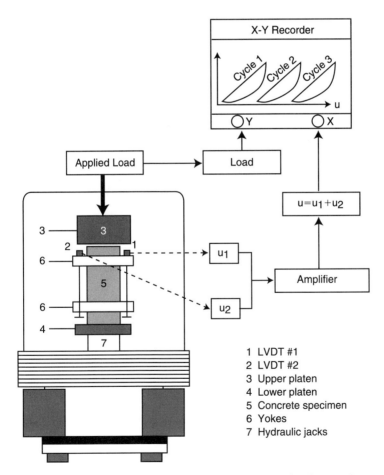

Fig. 15.26 Set-up for the recording of the loading and unloading cycles.

Fig. 15.27 Different steps in the measurement of the elastic modulus.

REFERENCES

AASHTO T-277 (1983) *Standard Method of Test for Rapid Determination of the Chloride Permeability of Concrete.*

Aïtcin, P.-C. and Riad, N. (1988) Curing temperature and very high-strength concrete. *Concrete International*, **10**(10), 69–72.

Aïtcin, P.-C., Miao, B., Cook, W.D. and Mitchell, D. (1994) Effects of size and curing on cylinder compressive strength of normal and high-strength concretes. *ACI Materials Journal*, **91**(4), July–August, 349–54.

ASTM C39 (1991) *Standard Test Method for Compressive Strength of Cylindrical Concrete Specimens, Annual Book of ASTM Standards*, Section 4, Construction, Volume 04.02, Concrete and Aggregates, ISBN 0-8031-1923-2, pp. 20–24.

ASTM C 157 (1993) *Standard Test Method for Length Change of Hardened Hydraulic Cement Mortar and Concrete, Annual Book of ASTM Standards*, Section 4, Construction, Vol. 04.02 Concrete and Aggregates, ISBN 0-8031-1923-2, pp. 103–8.

ASTM C 512 (1993) *Standard Test Method for Creep of Concrete in Compression, Annual Book of ASTM Standards*, Section 4, Construction, Vol. 04.02, Concrete and Aggregates, ISBN 0-8031-1923-2, pp. 278–281.

ASTM C 1202 (1993) *Test Method for Electrical Indication of Concrete's Ability to Resist Chloride Ion Penetration. Annual Book of ASTM Standards*, Section 4, Construction Vol. 04.02, Concrete and Aggregates, ISBN 0-8031-1923-2, pp. 627–32.

Bartlett, F.M. and MacGregor, J.G. (1994) *Assessment of Concrete Strength in Existing Structures*, Structural Engineering Report No. 198, University of Alberta, Department of Civil Engineering, May, 299 pp.

Bartlett, F.M. and MacGregor, J.G. (1996) Statistical analysis of the compressive strength of concrete in structure. *ACI Materials Journal*, **93**(2), March–April, 158–68.

Bickley, J.A., Ryell, J. and Read, P.H. (1990) *Preconstruction Testing for an 85 MPa Concrete Structure*, ACI Fall Convention, Philadelphia, November, personal communication.

Boulay, C. (1989) La boîte à sable pour bien écraser les bétons à hautes performances. *Bulletin de Laision des Laboratoires des Ponts et Chaussées*, November–December, No. 164, pp. 87–8.

Boulay, C. (1996) *Capping H.P.C. Cylinders with the Sand Box: New Developments*, Proceedings of the 4th International Symposium on Utilization of High-Strength/High-Performance Concrete, Paris, May, Vol. 2, pp. 197–202.

Boulay, C. and de Larrard F. (1993) A new capping system for testing HPC cylinders: the sand box. *Concrete International*, **15**(4), April, 63–6.

Burg, R.G. and Ost, B.W. (1992) *Engineering Properties of Commercially Available High-Strength Concretes*, RD 104.01T, Portland Cement Association, Skokie, IL, USA, 55 pp.

Carino, N.J., Guthrie, W.F. and Lagergen, E.S. (1994) *Effects of Testing Variable on the Measured Compressive Strength of High-Strength (90 MPa) Concrete*, NISTIR, 5405 National Institute of Standards and Technology, Gaithersburg, MD, 141 pp.

Carrasquillo, R.L., Nilson, A.H. and Slate, F.O. (1981) Properties of high strength concrete subject to short-term loads. *Journal of the American Concrete Institute*, **78**(3), May–June, 171–8.

Carrasquillo, P.M. and Carrasquillo, R.L. (1988a) Effect of using unbonded capping systems on the compressive strength of concrete cylinders. *ACI Materials Journal*, **85**(3), March–April, 144–7.

Carrasquillo, P.M. and Carrasquillo, R.L. (1988b) Evaluation of the use of current concrete practice in the production of high-strength concrete. *ACI Materials Journal*, **85**(1), January–February, 49–54.

Collins, M.P. and Mitchell, D. (1991) *Prestressed Structure*, Prentice Hall, Englewood Cliffs, New Jersey, 766 pp.

Collins, M.P. and Porasz, A. (1989) Shear design for high-strength concrete. *CEB Bulletin d'information*, No. 193, December, 77–83.

Collins, M.P., Mitchell, D. and MacGregor, J.G. (1993) Structural design considerations for high-strength concrete. *Concrete International*, **15**(5), May, 27–34.

Cook, J.E. (1989) 10 000 psi concrete. *Concrete International*, **11**(10), October, 67–75.

Date, C.G. and Schnormeier, R.H. (1984) Day-to-day comparison of 4- and 6-in diameter concrete cylinder strengths. *Concrete International: Design and Construction*, **6**(8), August, 24–6.

de Larrard, F., Torrenti, J.M. and Rossi, P. (1988) Le flambeme à deux échelle dans le rupture des bétons en compression. *Bulletin de Laision des Laboratoires des Ponts et Chaussées*, No. 154, pp. 51–5.

de Larrard, F., Belloc, A., Reenwez, S. and Boulay, C. (1994) Is the cube test suitable for high-performance concrete. *Materials and Structures*, **27**, 580–83.

Eckart, Von A., Ludwig, H.M. and Stark, J. (1995) Hydration of the four main Portland cement clinker phases. *ZKG International*, **48**(8), 443–52.

El-Dieb, A.S. and Hooton, R.D. (1995) Water-permeability measurement of high-performance concrete using a high-pressure triaxial cell. *Cement and Concrete Research*, **25**(6), 1199–1208.

Forstie, D.A. and Schnormeier, R. (1981) Development and use of 4 by 8 inch concrete cylinders in Arizona. *Concrete International: Design and Construction*, **3**(7), July, 42–5.

Freedman, S. (1971) *High-Strength Concrete*, IS176.01T, Portland Cement Association, Skokie, IL, USA, 19 pp.

Gagné, R., Aïtcin, P.-C. and Lamothe, P. (1993) *Chloride Ion Permeability of Different Concretes*, Proceedings of the 6th International Conference on Durability of Building Materials and Components, Omiya, Japan, E & FN Spon, London, ISBN 0-419-18670-0, pp. 1171–80.

Gonnerman, H.F. (1925) *Effect of Size and Shape of Test Specimen on Compressive Strength of Concrete*, Proceedings ASTM, Vol. 25, part 2, pp. 237–50.

Grygiel, J.S. and Amsler, D.E. (1977) *Capping Concrete Cylinders With Neoprene Pads*, Engineering Research and Development Bureau, New York State Department of Transportation, Research Report 46, April, pp. 1–6.

Haque, M.N. and Gopalan, M.K. (1991) Estimation of in situ strength of concrete. *Cement and Concrete Research*, **21**, 1603–10.

Hester, W.T. (1980) Field testing high-strength concretes: a critical review of the state-of-the-art. *Concrete International: Design and Construction*, **2**(12), December, 27–38.

Holland, T.C. (1987) Testing high-strength concrete. *Concrete Construction*, **32**(6), June, 534–6.

Howard, L.N. and Leatham, D.M. (1989) The production and delivery of high-strength concrete. *Concrete International: Design and Construction*, **11**(4), April, 26–30.

Jensen, O.M. and Hansen, P.F. (1995) A dilatometer for measuring autogenous deformation in hardening Portland cement paste. *Materials and Structures*, **28**(181), August–September, 406–9.

Johnson, C.D. and Mirza, S.A. (1995) Confined capping system for compressive strength testing of high-performance concrete cylinders. *Canadian Journal of Civil Engineering*, **22**(3), June, 617–20.

Jung, F. (1968) *Shrinkage Due to Hydration and its Relation to Concrete Strength*, RILEM International Conference on Shrinkage of Hydraulic Concretes, Madrid, Spain, March, 18 pp.

Kaan, P.H., Hanson, N.W. and Capell, H.T. (1978) *Stress Strain Characteristics of High-Strength Concrete*, RD051.01D, Portland Cement Association, Skokie, IL, USA, 10 pp.

Khayat, K.H., Bickley, J.A. and Hooton, R.D. (1995) High-strength concrete properties derived from compressive strength values. *Cement, Concrete, and Aggregates*, **17**(2), December, 126–33.

Lessard, M. (1990) *How to Test High-Performance Concrete*, Master's degree thesis (in French), Department of Civil Engineering, Université de Sherbrooke, Canada, 105 pp.

Lessard, M. and Aïtcin, P.C. (1992) Testing high-performance concrete, in *High-Performance Concrete – From Material to Structure* (ed. Y. Malier), E & FN Spon, London, ISBN 0419 176004, pp. 196–213.

References

Mak, S., Attard, M.M., Ho, D.W.S. and Darvall, P.L.P. (1990) *In-situ Strength of High-Strength Concrete*, Monash University Australia, Civil Engineering Report, ISBN 0 7326 0218 1, 90 pp.

Malhotra, V.M. (1976) Are 4 × 8 inch concrete cylinders as good as 6 × 12 inch cylinders for quality control of concrete? *Journal of the American Concrete Institute*, **73**(1), January, 33–6.

Matsushita, H., Kawai, T. and Mohri, K. (1987) Study on capping method for compressive strength test. *CAJ Review*, May, 226–9.

Miao, B., Aïtcin, P.-C., Cook, W.D. and Mitchell, D. (1993) Influence of concrete strength on in situ properties of large columns. *ACI Materials Journal*, **90**(3), May–June, 214–19.

Mirza, S.A. and Johnson, C.D. (1996) Compressive strength testing of HPC cylinders using confined caps. *Construction and Building Materials*, **10**(8), 589–95.

Mobasher, B. and Mitchell, T.H. (1988) *Laboratory Experience with the Rapid Chloride Permeability Test*, ACI SP-108, pp. 117–44.

Moreno, J. (1990) 225 W. Wacker Drive. *Concrete International: Design and Construction*, **12**(1), January, 35–9.

Nasser, K.W. and Al-Manaseer, A.A. (1987) It's time for a change from 6 × 12 to 3 × 6 in. cylinders. *ACI Materials Journal*, **84**(3), May, 213–16.

Nasser, K.W. and Kenyon, J.C. (1984) Why not 3 × 6 inch cylinders for testing concrete compressive strength? *ACI Materials Journal*, **81**(1), January, 47–53.

Neville, A.M. (1995) *Properties of Concrete*, Longman Scientific & Technical, 4th edn, pp. 581–94.

Neville, A.M. and Brooks, J.J. (1987) *Concrete Technology*, Longman Scientific & Technical, 1st edn, pp. 300–326.

Ozyildirim, C. (1985) Neoprene pads for capping concrete cylinders. *Cement, Concrete, and Aggregates*, **17**(1), Summer, 25–8.

Perraton, D., Aïtcin, P.-C. and Vézina, D. (1988) *Permeabilities of Silica Fume Concrete*, ACI SP-108, pp. 63–84.

Shah, S.P., Gokoz, U. and Ansari, F. (1981) An experimental technique for obtaining complete stress strain curves for high strength concrete. *Cement, Concrete, and Aggregates*, **3**(1), Summer, 21–7.

Sigvaldson, O.T. (1966) The influence of testing machine characteristics upon the cube and cylinder strength of concrete. *Magazine of Concrete Research*, **18**(57), December, 197–206.

Tazawa, E. and Miyazawa, S. (1993) Autogenous shrinkage of concrete and its importance in concrete technology, in *Proceedings of the 5th International RILEM Symposium on Creep and Shrinkage of Concrete*, Barcelona (ed. Z.P. Bazant and I. Carol), E & FN Spon, London, pp. 159–68.

Tazawa, E. and Miyazawa, S. (1995) Experimental study in the mechanism of autogenous shrinkage of concrete. *Cement and Concrete Research*, **25**(8), December, 1633–8.

Tazawa, E. and Miyazawa, S. (1996) *Influence of Autogenous Shrinkage on Cracking in High-Strength Concrete*, 4th International Symposium on Utilization of High-Strength/High-Performance Concrete, Paris, ISBN 2-85878-258-3, pp. 321–30.

Thaulow, S. (1962) Apparent strength of concrete as affected by height of test specimen and friction between the loading surfaces. *Bulletin R.I.L.E.M., Matériaux et Constructions – Recherches et Essais*, No. 17, December, 31–3.

Thorenfeldt, E., Tomaszewicz, A. and Jensen, J.J. (1987) Mechanical properties of high-strength concrete and application in design, in *Proceedings of the Symposium on Utilization of High-Strength Concrete*, Stavanger (ed. I. Holland et al.), Tapir, N-7034 Trondheim, NTH, Norway, ISBN 82-519-0797-7, pp. 149–59.

Tsutsumi, T., Yamamoto, A., Misra, S. and Motohashi, K. (1993) *Effect of Composition and Age on Chloride Permeability of Concrete*, Proceedings of the 6th International Conference on Durability of Building Materials and Components, Omiya, Japan, E & FN Spon, London, ISBN 0-419-18670-0, pp. 963–72.

Van Gysel, A. and Taerwe, L. (1996) Analytical formulation of the complete stress–strain curve for high-strength concrete. *Materials and Structures*, **29**, November, 529–33.

Vuturi, V.S., Lama, R.D. and Saluja, S.S. (1984) *Handbook on Mechanical Properties of Rocks*, Trans Tech Publications, Clausthall, Germany, Vol. 1, pp. 253–67.

Wee, T.H., Chin, M.S. and Mansur, M.A. (1996) Stress–strain relationship of high-strength concrete in compression. *Journal of Materials in Civil Engineering*, **8**(2), May, 70–76.

Whiting, D. (1981) *Rapid Determination of the Chloride Permeability of Concrete*, Final Report No. FHWA/RD-81/119, Federal Highway Administration, August, NTIS No. PB82140724.

Whiting, D. (1984) *In Situ Measurements of the Permeability of Concrete to Chloride Ions*, ACI SP-82, pp. 501–24.

Whiting, D. (1988) *Permeability of Selected Concretes*, ACI SP-108, pp. 195–212.

Wiegrink, K., Maribunte, S. and Shah, S.F. (1996) Shrinkage cracking of high-strength concrete. *ACI Materials Journal*, **93**(5), September–October, 409–15.

CHAPTER 16

Mechanical properties of high-performance concrete

16.1 INTRODUCTION

It is not the intention of this chapter to cover in full detail each of the mechanical properties of normal weight high-performance concretes, as an entire volume could be dedicated to such a subject. Moreover, there are so many types of high-performance concrete made from so many different materials that are used in so many applications and conditions that it would be presumptuous to try to cover all of the mechanical properties of these different mixes in one chapter. In this chapter, we will rather review the mechanical properties of high-performance concrete in a more general manner, keeping in mind what makes the mechanical properties of high-performance concrete different yet similar to those of usual concrete (Perenchio and Klieger, 1978; Swamy, 1986; Nilson, 1987; Olsen, Krenchel and Shah, 1987; Castillo and Durrani, 1990; Xie, Elwi and MacGregor, 1995).

It is wrong to believe that the mechanical properties of a high-performance concrete are simply those of a stronger concrete. It is also as wrong to consider that the mechanical properties of high-performance concrete can be deduced by extrapolating those of usual concretes as it would be wrong to consider that none of them are related. It is also wrong to apply blindly the relationships linking the mechanical properties of a usual concrete to its compressive strength that were developed through the years for usual concretes found in codes and textbooks. As will be seen in this chapter, it is better to examine the different mechanical properties case by case and to consider the influence of the fundamental mechanisms relating concrete microstructure to the macrostructural property under evaluation.

Certainly there are cases when high-performance concrete behaves simply as a stronger concrete, but there are also other cases when high-performance concrete behaves quite differently. These differences in the mechanical behaviour of high-performance concrete and usual concrete result from their different microstructures, so an external load applied on the concrete does not necessarily develop the same stress field within the concrete and the material does not respond in the same way to this stress field.

The high water/binder ratio of usual concretes is translated in microstructural terms by a porous microstructure, as presented in Chapter 5, specifically around the aggregate, where a more or less thick transition zone with a higher porosity can be observed. The higher the water/binder ratio the more porous the concrete microstructure and the thicker the transition zone can be. Therefore, in a usual concrete there is a limited degree of stress transfer between the hydrated cement paste and the aggregate, more particularly at the coarse aggregate level. In usual concretes the mechanical properties of the coarse aggregate do not greatly influence the mechanical properties of the usual concrete because it is the hydrated cement paste that constitutes the weakest link. Therefore, most of the mechanical properties of a usual concrete can be closely related to the strength of the hydrated cement paste or its water/binder ratio, and therefore to its compressive strength. This explains why it is easy to develop simple relationships between compressive strength and most other mechanical properties.

On the other hand, the microstructure of high-performance concrete is more compact, including the transition zone with the coarse aggregate, resulting in a thin or no transition zone at all. Therefore, the mechanical properties of the coarse aggregate influence some of the mechanical properties of high-performance concrete. Therefore, the sacrosanct water/binder ratio law is no longer true in the case of some high-performance concretes made with 'weak' coarse aggregates. For any coarse aggregate there is a critical value of the water/binder ratio below which any further decrease of the water/binder ratio does not result in a significant increase of the compressive strength. This critical value depends on the strength of the rock from which the coarse aggregate is made, but also on the maximum size of the coarse aggregate. This is because when crushing a particular rock the smallest fragments are usually stronger than the coarsest because they contain less defects. This phenomenon is sometimes referred as a 'size effect phenomenon'.

In broad terms, it can be said that usual concretes usually act as homogeneous and isotropic materials in which the weakest link is the hydrated cement paste and/or the transition zone. On the other hand, high-performance concretes essentially act like non-isotropic composite materials made of hydrated cement paste and aggregates that can have quite different mechanical properties. Evidently the properties of this composite material are influenced by the properties of each of its constituents as well as its water/binder ratio (Asselanis, Aïtcin and Mehta, 1989).

If it is easy from a mechanical point of view to make a clear distinction between the case of a usual concrete having a water/binder ratio of 0.60 and a high-performance concrete having a water/binder ratio of 0.30, it

Compressive strength

is becoming more difficult to make such a distinction when the water/binder ratio decreases below 0.50 or increases above 0.30. In fact, there is not a sharp discontinuity in the behaviour of concretes having an in-between water/binder ratio but rather a continuous evolution from one behaviour to the other. This is another reason why it is not safe to believe that the mechanical properties of high-performance concretes are only the properties of a stronger concrete. This will be seen when studying more closely the compressive strength, the modulus of rupture, the splitting strength, the tensile strength, the elastic modulus, the Poisson's ratio, the stress–strain curve, and the creep characteristics and fatigue resistance of high-performance concrete.

It is also not our intention to make an extensive review of all of the recent results published in the literature, because too often the precise conditions under which these results were obtained are not known with enough certainty, and therefore it is sometimes difficult to interpret these results correctly. This chapter will instead present results mainly obtained at the Université de Sherbrooke, focusing on general trends rather than on specific issues.

16.2 COMPRESSIVE STRENGTH

Obviously, the compressive strength of high-performance concrete is higher than that of usual concrete, and it is not as easy as many people believe to measure it properly when it goes over 60 MPa, as was discussed in Chapter 15.

As for usual concretes, the compressive strength of high-performance concrete increases as the water/binder ratio decreases. But unlike usual concrete, the water/binder 'law' is only valid until the 'crushing strength' of the coarse aggregate becomes the weakest link within the high-performance concrete. When coarse aggregates are no longer strong enough in comparison with the strength of the hydrated cement paste, the compressive strength of a high-performance concrete does not increase significantly as the water/binder ratio decreases. The only way to increase the compressive strength of such a high-performance concrete is therefore to use another type of coarse aggregate. In certain places, such a situation can result in a significant gap in the price of a given high-performance concrete and another having a 10 MPa higher strength, because the coarse aggregate might have to be hauled over a long distance. Designers should be aware of this price gap when selecting the compressive design strength for a particular project.

When the coarse aggregate is strong enough, it is still impossible to state a general relationship between the water/binder ratio and the high-performance concrete compressive strength that can be achieved, because of the multiple factors influencing the relationship between f'_c and the water/binder ratio. However, based on the personal experience of the author, the following broad guidelines can be used to predict the

Table 16.1 Compressive strength of high-performance concrete as a function of the water/binder ratio

W/B	Maximum compressive strength range (MPa)
0.40–0.35	50–75
0.35–0.30	75–100
0.30–0.25	100–125
0.25–0.20	< 125

maximum compressive strength (not the **design strength**) that can be achieved for different water/binder ratios, as seen in Table 16.1. In this table, it is supposed that the coarse aggregates are stronger than the resulting concrete. The proposed values in Table 16.1 could appear to pertain to too broad ranges, but considering the great number of material combinations and material properties used to make high-performance concrete, it is difficult to be more specific. Only trial batches can provide the actual values that can be achieved in a particular location.

There are also other issues related to compressive strength that are important and need to receive particular attention. Some of these are:

- the early compressive strength of high-performance concrete;
- the influence of the maximum temperature reached at early age on the compressive strength of the concrete;
- the long-term development of the compressive strength of high-performance concrete;
- the strength of cores compared to cast specimens.

16.2.1 Early-age compressive strength of high-performance concrete

For a contractor an ideal concrete should stay plastic as long as needed for it to be placed easily, but as soon as it is placed it should harden within a few hours without developing any excessive heat, shrinkage or creep and also without any need for curing! In spite of its numerous advantages and qualities, a high-performance concrete is far from being such an ideal concrete.

As in the case of usual concrete, setting and hardening are strongly influenced by the initial temperature of the concrete at its delivery point, the ambient temperature and the incorporated admixtures. A low ambient temperature can significantly delay concrete hardening, so in certain circumstances the use of insulating blankets should be considered. The hardening process of high-performance concrete is also greatly dependent on the amount of superplasticizer and set retarder (whenever used) that have been incorporated to decrease the water/binder ratio to the level needed to ensure the required compressive strength while maintaining the desired workability long enough to facilitate placement.

As the decrease of the water/binder ratio can be achieved in two different ways, by decreasing the water content through the use of a high dosage of superplasticizer or by increasing the binder content, and consequently the water content with less superplasticizer being needed to obtain the desired workability, it is possible to have some control of the early strength of a particular high-performance concrete when designing its composition (Rougeron and Aïtcin, 1994).

Without any heating it is possible to obtain a high early strength, between 20 and 30 MPa within 24 hours, with a water/binder ratio of between 0.30 and 0.35 at an ambient temperature of around 20 °C. But getting a high early strength before 12 hours is often difficult with high-performance concrete. Designers, field engineers and contractors should be aware that 2 or 3 additional hours of curing at 20 °C can make a great difference to the early compressive strength of high-performance concrete. It must be emphasized, as already pointed out, that the hydration reaction can be delayed significantly in high-performance concrete owing to the use of a high dosage of superplasticizer and sometimes a set retarder, but when hydration starts it develops very rapidly. For precast applications, it is therefore better to formulate a high-performance concrete with a high amount of binder rather than designing it with the lowest amount of water possible (Rougeron and Aïtcin, 1994).

When high-performance concrete mixes are properly designed, in terms of the amount of cement within the binder and mixing water content, it is possible to achieve 15 MPa within 12 hours, 20 MPa within 18 hours and 30 MPa within 24 hours. Using a Type III cement and a water/binder ratio of 0.22, a high-performance concrete having a 24 hour compressive strength of 75 MPa has been made at the Université de Sherbrooke, but the slump of such a concrete could not be maintained over 150 mm longer than 15 minutes.

Finally, it should be mentioned that the material used to make the forms can have a critical effect on the strength development of the in-place concrete, because ambient conditions can influence concrete hardening to a greater or lesser degree, depending on the thermal conductivity and the thickness of the concrete element (Khan, Cook and Mitchell, 1996). The use of metallic forms has to be carefully evaluated in order to avoid too great an influence of the ambient temperature on the hardening process or the creation of high thermal gradients. When concrete cools down, plywood forms that are less conductive protect high-performance concrete better from the early effect of ambient temperature and do not create such thermal gradients, as discussed in Chapter 14.

16.2.2 Effect of early temperature rise of high-performance concrete on compressive strength

Usually, the temperature of high-performance concrete increases significantly during the first 24 to 48 hours following its placement, as already shown in Chapter 14. Maximum temperatures of 65 to 70 °C have been recorded in some massive parts of high-performance concrete (Aïtcin, Laplante and Bédard, 1985; Cook *et al.*, 1992). Therefore it is reasonable to raise some questions about the actual compressive strength of the concrete that hardens within the structure under curing conditions so different from those used when testing standard specimens, as mentioned in Chapter 15. It is rather surprising that the importance of the maximum temperature reached by high-performance concrete is becoming an important issue because temperatures as high as those reached in high-performance concrete can be reached in massive structural elements when 30 to 40 MPa concretes are used, as shown by Cook *et al.* (1992). Most probably, as already mentioned in Chapter 14, this issue is drawing some attention because of the absolutely false idea that the maximum temperature reached by the concrete in a structural element is proportional to the amount of cement used in the mix. In fact – and this can never be emphasized enough – concrete temperature increase is a function of the amount of cement that is actually **hydrating**, and not the **total amount of cement** present in the mix. In very low water/binder ratio concretes, water scarcity becomes the factor that limits the amount of cement that is hydrating, and therefore the maximum temperature achieved, in spite of the fact that the mix contains a high amount of cement.

In usual concrete, early **external** heating of concrete results in an increase in its early strength but in a decrease in its 28 day strength. From the limited but well-documented experimental and field data available, it does not seem that this is always the case with high-performance concrete temperature.

In such cases, it has been found that the 28 day compressive strength measured on cores extracted from field structures shows similar compressive strength as the 28 day compressive strength of standard cured specimens.

16.2.3 Influence of air content on compressive strength

High-performance concrete always contains some entrapped air, from 0.5 to 2.5%, or it can be purposely air-entrained. The latter is necessary to enhance frost durability (i.e. to pass successfully the ASTM C666 Procedure A Standard Test for Resistance of Concrete to Rapid Freezing and Thawing that confirms its freeze–thaw resistance) or to improve its workability and finishability. The presence of 4 to 6% entrained air weakens the high-performance concrete. From the experience acquired in Québec with air-entrained, high-performance concrete used for the construction of several bridges, it has been found that in the range 4 to 6% air, a 1% difference in the air content of a high-performance concrete results in a 4 to 5% reduction in the compressive strength (Lessard, Baalbaki and Aïtcin, 1995), which is about the same rule of thumb as the one used for usual concrete. This point will be discussed in more detail in Chapter 18.

16.2.4 Long-term compressive strength

Some caution should be exercised on the effect of long-term water curing of high-performance concrete specimens. When performing standard compressive strength tests on water-cured high-performance concrete specimens beyond 28 days of age, it should be understood that some external water penetrates or diffuses into the high-performance concrete specimens through the external surfaces, so a more or less thick rim of concrete continues to hydrate all around a specimen. The additional hydration results in an increase in the compressive strength of the specimen in that area. On the other hand, in an actual structural element hydration has stopped due to a lack of water or too low a relative humidity within the pore system. Therefore, it is obvious that the 91 day or 1 year compressive strength of water-cured, high-performance concrete specimens has no direct relationship to that of the concrete in the structure and can lead to a very optimistic figure of the in-situ compressive strength. It is thus not surprising that the 1 year compressive strength of cores should not be too different from the compressive strength measured at 28 days, and that this core compressive strength is almost always significantly lower than that of standard specimens cured for 1 year in water.

The reference testing date for high-performance concrete has to be defined carefully. The author believes in using the 28 day sacrosanct testing age and does not advocate the use of a later age for mixes in which some supplementary cementitious materials have been used on the pretext that these materials (other than silica fume) have slow hardening rates and do not fully develop their potential at 28 days. Very often in high-performance concrete with a very low water/binder ratio, hydration stops within the concrete long before 28 days due to lack of water or when the partial pressure of water vapour within the pores has reached the 80% limit below which hydration is slowed down very significantly (Powers and Brownyard, 1948).

It has been found that some high-performance concrete laboratory specimens experienced a slight decrease in compressive strength after a long period of curing in air, particularly those containing silica fume (de Larrard and Aïtcin, 1993). This finding resulted in some controversy and in some engineers being reluctant to use high-performance concrete. This loss in compressive strength of air-cured specimens is also referred to in the literature as 'strength retrogression'. However, it is reassuring to find that cores extracted from a high-performance concrete column cast in 1984 have not shown such a strength retrogression over the years. Moreover, a careful microscopic examination of the concrete from this column and others has not shown any change in the microstructure of the high-performance concrete that could explain any retrogression (Aïtcin, Sarkar and Laplante, 1990). In some cases it has been shown rather that the observed small strength retrogressions were in fact due to the drying of a very thin layer of the skin of the high-performance concrete. This drying resulted in a significant tensile stress gradient, sufficient to affect the compressive strength values. The effects of this phenomenon were aggravated in the 100×200 mm specimens by the scale effect, owing to the small size of the specimens. Whatever the size of a structural element, drying shrinkage in high-performance concrete affects only a thin layer of concrete which has the same thickness, whatever the size of specimen; the thickness of this zone becomes absolutely negligible in the case of a column having a 1×1 m cross-sectional area, which is not the case in a standard specimen.

Another reason can be put forward to explain the so-called strength retrogression of high-performance concrete specimens: the degree of self-desiccation reached within the first 24 hours before they are placed under water or in a fog room. If the hardening of the high-performance concrete is delayed until 16 to 18 hours owing to the use of a high dosage of superplasticizer, with or without a small amount of a set retarder, very little autogenous shrinkage has a chance to develop during the 6 to 8 hours that precede the demoulding of the specimens and their immersion in lime-saturated water or before transferring them to a fog room. When immersed or placed in a fog room, some volumetric contraction will continue to develop in the presence of an external source of water, so no autogenous shrinkage should develop in the specimens.

On the contrary, if the hardening of high-performance concrete specimens started 4 to 6 hours after mixing, self-desiccation had time to develop very rapidly and intensely within the concrete specimens during the 18 to 20 hours during which their hardening occurred without any external supply of water. Consequently, a significant amount of autogeneous shrinkage had time to develop within these concrete specimens before they were demoulded at 24 hours. It can be supposed that these specimens, which were submitted to a strong autogenous shrinkage during their first 24 hours of curing at room temperature, will be more sensitive to air-drying later on because they possess a more important network of interconnected fissures that was developed during these first 24 hours of curing without any supply of external water.

The strength retrogression phenomenon that has been occasionally observed when high-performance concrete is exposed to severe air-drying conditions emphasizes the need for a proper initial water curing for high-performance concrete, and even perhaps in some critical cases the coating of some high-performance concrete structural elements with an impervious film or their painting in order to prevent any drying shrinkage, so that no tensile gradient can be created in the 'skin' of the high-performance concrete.

It can never be emphasized enough that proper early water curing (before 24 hours) is much more important in the case of high-performance concrete than for usual concretes, especially when most of the hydration reactions are taking place.

16.3 MODULUS OF RUPTURE AND SPLITTING TENSILE STRENGTH

As seen in the previous chapter, the direct measurement of the tensile strength of usual concrete is not easy because of the complicated set-up that must be used (though it is not impossible, as will be seen at the end of this section). Therefore, tensile strength is usually calculated using indirect measurements, such as the measurement of the modulus of rupture (MOR) (ASTM C78) and/or the splitting tensile strength (ASTM C496). Performing the MOR and splitting tensile strength measurements does not present any special difficulties in the case of high-performance concrete, so that the same set-ups and procedures used for usual concrete can also be used for high-performance concrete.

As the MOR and splitting strength measurements require the making of extra specimens, which are particularly heavy in the case of the MOR when it is measured on $150 \times 150 \times 500$ mm prisms weighing about 27 kg, and because designers like to reduce most mechanical properties of concrete to values correlated to compressive strength, a number of relationships linking the MOR and splitting strength to the compressive strength of high-performance concrete are proposed in different codes, as is the case for usual concrete.

In order to avoid using such big and heavy concrete specimens many people prefer to measure the splitting tensile strength of concrete because it requires the use of less than 15 kg of concrete for 150×300 mm specimens or 5 kg when 100×200 mm specimens are used. However, when observing the ruptured surface of the concrete after a MOR or splitting tensile strength measurement, it is easy to see that the rupture of the concrete was not due to the development of the same mechanism of failure when performing these two tests. In the case of the MOR measurement, very often the failure surface shows that failure occurs at the coarse aggregate/mortar interface which was the weakest link, while in the case of the measurement of the splitting strength, the crushing strength of the coarse aggregate is often mobilized when the failure occurs. It is therefore obvious that the two measured values are not related and not inevitably linked together.

Finding a relationship between the MOR and splitting strength for usual concrete is not so difficult, because the values of the MOR and splitting strength are quite low and do not vary very much, and because they are very much influenced by the tensile strength of the hydrated cement paste. However, this is no longer the case for high-performance concrete, for which the water/binder ratio and the compressive strength can vary over a wide range. Moreover, high-performance concrete is made using so many different cementitious compositions that MOR and splitting strengths can vary over a wide range for the same compressive strength value. Therefore, it must be emphasized that the relationships that have been suggested between compressive strength, MOR and splitting strength for usual concrete lose some of their predictive value when going from usual concrete to high-performance concrete. Some of these relationships are presented in the following:

- Comité Euro-International du béton CEB-FIP (1978)

  $f_{sp} = 0.273 f_c'^{2/3}$ (in MPa)

  where f_{sp} represents the splitting tensile strength and f_c' the compressive strength;
- Carrasquillo, Nilson and Slate (1981) suggested the following relationship for concretes with f_c' ranging from 21 to 83 MPa:

  $f_{sp} = 0.54 f_c'^{1/2}$ (in MPa);
- Raphael (1984) suggested the following relationship for concretes with f_c' below 57 MPa:

  $f_{sp} = 0.313 f_c'^{1/2}$ (in MPa);
- ACI Committee 363 on high-strength concrete (1984) suggested using the following equation for concretes with f_c' ranging from 21 to 83 MPa:

  $f_{sp} = 0.59 f_c'^{0.55}$ (in MPa);
- Ahmad and Shah (1985) suggested the following relationship for concretes with f_c' less than 84 MPa:

  $f_{sp} = 0.462 f_c'^{0.55}$ (in MPa);
- Burg and Ost (1992) suggest the following equation for moist-cured concrete with f_c' ranging from 85 to 130 MPa:

  $f_{sp} = 0.61 f_c'^{0.5}$ (in MPa).

It is not the intention here to recommend one over another, because the author is convinced that the best way to predict the value of the MOR and of the splitting strength of any high-performance concrete is to measure them directly.

Different relationships between f_r and f_c' have been presented in the literature, where f_r represents the MOR:

- Carrasquillo, Nilson and Slate (1981) suggested the following correlation between flexural strength and compressive strength:

  $f_r = 0.94 f_c'^{1/2}$ (in MPa);
- Burg and Ost (1992) suggested the following equation:

  $f_r = 1.03 f_c'^{1/2}$ (in MPa);

- Khayat, Bickley and Hooton (1995) suggested:

$f_r = 0.23 + 0.12 f'_c - 2.18 \times 10^{-4}(f'_c)^2$ (in MPa).

In this case, too, it is not the intention of the author to recommend the use of one particular formula rather than another; here too it is better to measure f_r directly.

16.4 MODULUS OF ELASTICITY

The knowledge of the modulus of elasticity of concrete is very important from a design point of view when the deformations of the different structural elements comprising the structure have to be calculated.

We have seen in Chapter 15 that it is not easy to measure directly the elastic modulus of concrete because it is necessary to measure simultaneously the applied load and the axial strain generated by this load. The separate measurement of a load and of a strain are easy in themselves, it is their simultaneous recording that makes the experimental set-up complicated. Not only does it involve the use of electronic devices to control the strain rate, it also involves the use of an X-Y recorder in order to draw the stress–strain curve necessary to measure the elastic modulus. When compared with compressive strength measurements carried out to characterize f'_c, it is definitely the most complicated and demanding test, except for the measurement of Poisson's ratio, where a lateral strain has to be measured in addition to the two previous ones. In order to avoid so complex a measurement, engineers and researchers have tried to find some short cuts to enable them to predict the elastic modulus of concrete using either a theoretical approach or an empirical one. In the latter case they usually express the modulus of elasticity as a function of the compressive strength.

16.4.1 Theoretical approach

When using the theoretical approach, the elastic modulus of concrete is calculated based on different models, more or less complicated, that are supposed to represent the elastic behaviour of concrete. The simplest models are two-phase models involving the aggregate and hydrated cement paste (Illston, Dinwoodie and Smith, 1987). In these models the constituents are supposed to carry either the same strain (in the Voigt model) or to develop the same stress (in the Reuss model), presented schematically in Figures 16.1 and 16.2.

If E_1 is the elastic modulus of the mortar, E_2 the elastic modulus of the coarse aggregate, g_1 the relative volume of the mortar and g_2 the relative volume of the coarse aggregate, then the Voigt model predicts the elastic modulus of the concrete to be:

$$E_c = E_1 g_1 + E_2 g_2 \quad \text{with } g_1 + g_2 = 1$$

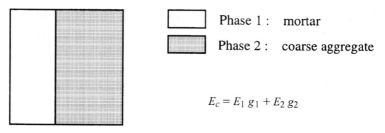

Fig. 16.1 The Voigt model.

$$E_c = E_1 g_1 + E_2 g_2$$

and the Reuss model predicts the elastic modulus of concrete to be equal to:

$$\frac{1}{E_c} = \frac{g_1}{E_1} + \frac{g_2}{E_2}$$

It can be demonstrated theoretically that these two models represent the upper and lower limits of the actual elastic modulus of a concrete in a two-phase model.

In the literature more complicated and complex models can be found, such as the one proposed by Hansen (1965):

$$\frac{E_c}{1-2v_c} = \left\{\left[g_1 \frac{E_1}{1-2v_1} + \left(\frac{1+v_1}{2(1-2v_1)} + g_2\right)\frac{E_2}{1-2v_2}\right]\right/$$
$$\left[\left(1 + \frac{1+v_1}{2(1-2v_1)}g_2\right)\frac{E_1}{1-2v_1} + \left(\frac{1+v_1}{2(1-2v_1)}g_1\right)\frac{E_2}{1-2v_2}\right]\right\}$$
$$\times \frac{E_1}{1-2v_1}$$

where v_c, v_1 and v_2 are, respectively, the Poisson's ratio of the concrete, of the mortar and of the aggregates; or the one proposed by de Larrard and Le Roy (1992):

$$E_c = \left[1 + 2g\frac{E_2^2 - E_1^2}{(g^* - g)E_2^2 + 2(2 - g^*)E_2 E_1 + (g^* + g)E_1^2}\right]E_1$$

where g^* represents the compactness of the dry-rodded aggregate.

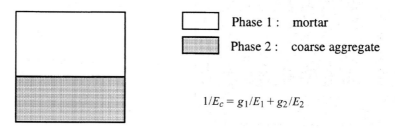

$$1/E_c = g_1/E_1 + g_2/E_2$$

Fig. 16.2 The Reuss model.

Modulus of elasticity

However, experience shows that even the more sophisticated models do not always yield an accurate predictive value because of the assumptions made when developing these models, such as:

- the applied load remains uniaxial and compressive throughout the model;
- the effect of lateral continuity between the different layers constituting the model is ignored;
- any local bond failure or crushing does not affect the deformation.

Even in the models where Poisson's ratio is introduced, which further complicates the calculations, the predictive value of the model is not significantly improved.

These models, however, are of some interest in that they show how the elastic modulus of concrete may vary when changing the elastic properties of one of its components or of two at the same time.

Recently Baalbaki (1997) proposed two models that seem to offer better predictive values than the previous ones. In his first model, the transition zone phase was considered and a parameter, a, depending on the ratio E_1/E_2, the nature of the aggregate and the use or not of silica fume, was employed. This model is schematically presented in Figure 16.3 and can be expressed as follows:

$$E'_c = ag_1 \times E_1 + \frac{(1 - ag_1)^2}{\frac{(1-a)}{E_1}g_1 + \frac{g_2}{E_2}}$$

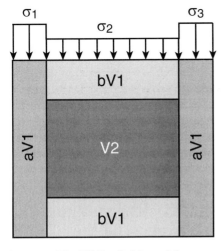

Fig. 16.3 The W. Baalbaki model.

bV1 hydrated cement paste in the transition zone

aV1 hydrated cement paste

V2 volume of the coarse aggregate

If $a = 1$, the value of E'_c is equivalent to that obtained with the Voigt model:

$$E'_c = g_1 E_1 + g_2 E_2$$

and if $a = 0$, the value of E'_c is equivalent to the one given by the Reuss model:

$$\frac{1}{E'_c} = \frac{g_1}{E_1} + \frac{g_2}{E_2}$$

Using data published by seven different authors, Baalbaki was able to show that the 124 experimental results could be predicted using his formula within a 5% margin of error. As shown in Figure 16.4, the correlation between the predicted values of the elastic modulus and the actual experimental values is excellent in spite of the fact that the specific gravity of the coarse aggregate used to obtain these 124 experimental values varied from 0.80 to 7.85, and that the elastic modulus of the coarse aggregate used to make these concretes varied from 5 GPa to 210 GPa.

When it is rather the elastic properties of the paste that are known instead of those of the mortar, Baalbaki proposes to use a second formula:

$$E'_c = (g^* - g)E_p + \frac{(1 + g - g^*)^2/g}{g/E_g + (1 - g^*)/E_p}$$

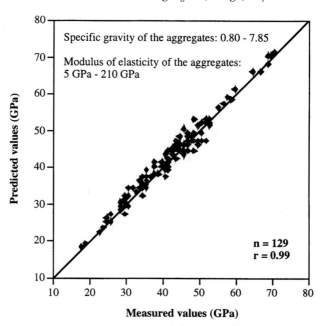

Fig. 16.4 Correlation between predicted and measured values of the elastic modulus using the W. Baalbaki model when the characteristics of the paste and the coarse aggregate are known.

Modulus of elasticity

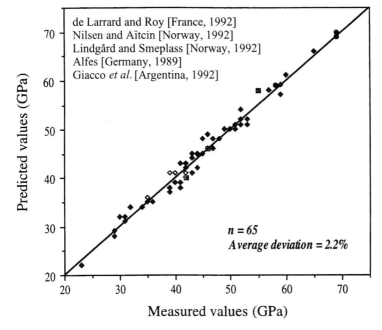

Fig. 16.5 Correlation between predicted and measured values of the elastic modulus using the W. Baalbaki model when the characteristics of the mortar and the coarse aggregate are known.

where E_p represents the modulus of elasticity of the paste, E_g that of the aggregates and g^* is a factor related to the amount of aggregate used in the mix, as proposed by Caquot (1937):

$$g^* = 1 - 0.47(d/D)^{0.2}$$

where d represents the maximum size of the fine aggregate and D the maximum size of the coarse aggregate.

This model was tested using 65 results already presented in the literature by five different authors. Figure 16.5 shows that the predicted values are within a 2% range when compared with the experimental values.

In spite of their more or less good predictive values, these models do not solve the fundamental problem of the complexity of the measurement of the elastic modulus of concrete, but rather double it, because it is necessary to know two moduli of elasticity instead of only one. However, such models again are useful in understanding the importance of various parameters on the elastic modulus of concrete.

16.4.2 Empirical approach

In the empirical approach, which is the most widely used by designers, the elastic modulus of concrete is linked to its compressive strength. This supposes in fact that the same main parameters influence at the same time both the compressive strength and the elastic modulus. Experience gained during many years of practice shows that this is almost true for usual concrete, because in fact in usual concretes it is the very porous hydrated cement paste that constitutes the weakest link of the concrete and therefore influences at the same time the compressive strength and the modulus of elasticity. Therefore, the compressive strength and elastic modulus values of concrete are essentially influenced by those of the hydrated cement paste. This is why all national codes present very simple relationships linking the elastic modulus to the compressive strength for usual concrete of the type $E'_c = \psi(f_c^{\prime 1/n})$. Unfortunately, due to chauvinism and nationalism or continentalism, engineers have not yet been able to end up with a universal relationship, but any way all these slightly different relationships give not so fundamentally different values in their range of validity, hopefully because concrete in itself is not a chauvinist material!

When high-performance concrete began to be used, it was tempting for engineers to try to extrapolate these very convenient and well-accepted relationships to it. But it was rapidly found that what was working well with usual concrete worked more or less well with high-performance concrete. Therefore, based on national experience, more diverging relationships were established for national codes in the case of high-performance concrete:

- The CEB-FIP (1990) proposed the following relationship:
  $E_{c\,28d} = 10(f'_c + 8)^{1/3}$ in (GPa);
- The Norwegian code (1992) proposes:
  $E_c = 9.5(f'_c)^{0.3} \left(\dfrac{\rho}{2400}\right)^{1.5}$ in (GPa) with ρ in kg/m³;
- Carrasquillo, Nilson and Slate (1981) and ACI Committee 363 suggest:
  $E_c = 3.32 \sqrt{f'_c} + 6.9$ in (GPa)
  or
  $E_c = (3.32 f_c^{\prime 0.5} + 6.9)\dfrac{\rho}{2346}$ in (GPa) with ρ in kg/m³;
- The European Code of Buildings CEB 1990 (1995) suggests:
  $E_c = 10(f'_c + 8)^{1/8}$ in (GPa);
- Gardner and Zhao (1991) propose:
  $E_c = 9(f'_c)^{1/3}$ for $f'_c > 27$ MPa.
- The Canadian code CAN A23.3-M90 Design of Concrete Structure for Buildings recommends the following equation:
  $E'_c = 5(f'_c)^{1/2}$ (in GPa).

Modulus of elasticity

In fact when looking at the major differences in the microstructure of usual concrete and high-performance concrete, it is seen that the very simple relationship between E'_c and f'_c that could be established for usual concrete, because there is very little stress transfer at the hydrated cement paste/aggregate interface due to the great porosity of the transition zone, could not work any more in high-performance concrete. In some high-performance concretes, coarse aggregate can be the weakest link of the concrete, or when it is stronger than the paste or the mortar, there is a much better stress transfer at the hydrated cement paste/aggregate interface. In fact when measuring the elastic modulus of high-performance concrete made of different types of coarse aggregate with quite different elastic moduli, it is found that elastic modulus values ranging from simple to double can be found in concretes having the same compressive strength, simply because they are made of aggregates having quite different elastic modulus and Poisson's ratio values (Nilsen and Aïtcin, 1992).

Based on results obtained from similar high-performance concretes made with crushed aggregates of different origin – limestone, sandstone and granite – the main mechanical properties and characteristics of which are given in Table 16.2, Baalbaki (1997) proposed a simple relationship:

$$E'_c = K_o + 0.2 f'_c \text{ (GPa)}$$

where K_o is a factor depending of the type of aggregate.

Based on his experimental results, he proposes that $K_o = 9.5$ GPa for the sandstone, $K_o = 19$ GPa for the granite and $K_o = 22$ GPa for the limestone. In order to fit experimental results within such a very simple relationship, it is seen that the K_o value has to vary over a wide range. In his thesis, Baalbaki (1997) was able to express K_o as a function of the elastic modulus of the coarse aggregate, so that the previous relationship can be expressed more generally as:

$$E'_c = -52 + 41.6 \log(E_a) + 0.2 f'_c$$

where E_a is the elastic modulus of the coarse aggregate.

Table 16.2 Mechanical properties and characteristics of the aggregates used by W. Baalbaki

	Limestone	Granite	Sandstone
Compressive strength, C_0 (MPa)	95	130	155
Modulus of elasticity, E_0 (GPa)	60	50	30
Poisson's ratio, v_0	0.14	0.13	0.07
Splitting tensile strength, T_0 (MPa)	7.5	12.0	7.0
Porosity (%)	2.9	3.0	6.4
Specific gravity	2.68	2.72	2.53
Absorption (%)	1.2	1.1	3.7

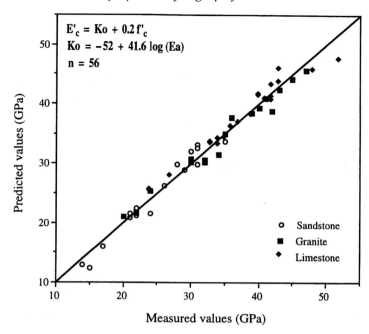

Fig. 16.6 Correlation between predicted and measured values of the elastic modulus using the W. Baalbaki formula, $E'_c = K_o + f'_c$.

Figure 16.6 shows that this formula worked well to predict the 56 sets of experimental values obtained using three coarse aggregates which had different mechanical characteristics and properties.

An abacus (Figure 16.7) predicting the elastic modulus of any type of concrete can be deduced from the knowledge of the elastic modulus of the coarse aggregate and the compressive strength of the concrete.

Baalbaki (1997) also found that the Poisson's ratio of the aggregate influences the elastic modulus of concrete and proposed the following empirical formula:

$$E'_c = 5.5(E_m)^{0.53}(E_a)^{0.22}(v_a)^{0.38}$$

where E_m is the elastic modulus of the mortar, E_a the elastic modulus of the aggregate and v_a the Poisson's ratio of the coarse aggregate. This formula is interesting in that it shows the relative influence of these different parameters on the value of the elastic modulus of a high-performance concrete.

16.4.3 Concluding remarks on elastic modulus evaluation

In spite of the merit of all these models, the author is suggesting that rather than relying on these theoretical, or empirical, or theoretico-empirical models to predict the elastic modulus of a high-performance concrete, it is better to measure it directly on actual specimens made

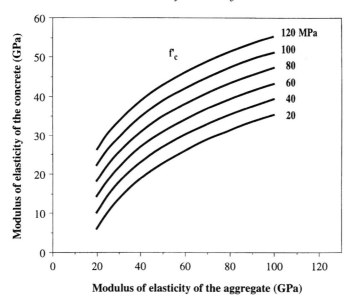

Fig. 16.7 Abacus predicting the value of the elastic modulus of a concrete according to the value of the elastic modulus of the coarse aggregate and concrete compressive strength.

under real field conditions, as already suggested by Khayat, Bickley and Hooton (1995), 'Rather than relying on a generic formula, it would be better for important projects to determine the modulus (of elasticity) directly for each high-strength concrete proposed for use. Even for a given aggregate, different moduli can result from changes in mixture proportions, so aggregate-specific and mixture-specific tests are desirable'.

Any concrete producer interested in developing a market for high-performance concrete should start by making three experimental batches having, for example, water/binder ratios of 0.35, 0.30 and 0.25 for each type of promising aggregate he can use (hard, clean and having a cubic shape). He should measure the compressive strength, the modulus of rupture, the elastic modulus and Poisson's ratio of these different concretes in order to be able to provide a data sheet to designers and public and parapublic agencies so that the appropriate values could be used in the different calculations used when designing the high-performance concrete structure.

Based on these experimental values, it will be possible to fit the parameters that will have to be used in the different models to get a good prediction of the actual mechanical properties of the high-performance concrete. Moreover, a price tag could be fixed to a certain level of compressive strength or a certain value of the modulus of elasticity.

It is worth while repeating that in the case of the elastic modulus of high-performance concrete, what has been gained in performance has been lost in simplicity.

16.5 POISSON'S RATIO

If it is not easy to measure the elastic modulus of concrete, what can be said about the measurement of Poisson's ratio, which involves the simultaneous measurement of the axial load, the axial strain and the transversal strain at the same time at a constant load rate! This is why limited data on the Poisson's ratio of concrete in general, and high-performance concrete in particular, are found in the literature. Ahmad and Shah (1985) reported values of between 0.18 and 0.24, while Kaplan (1959) reported values of between 0.23 and 0.32.

16.6 STRESS–STRAIN CURVES

A large number of equations representing the stress–strain curves of concrete have been proposed through the years (Desayi and Krishnan, 1964; Popovics, 1973; Wang, Shah and Naaman, 1978; Ahmad and Shah, 1985; Hatanaka, 1986; Taerwe and Vangyset, 1996; Wee, Chin and Mansur, 1996). The existence of such a long list of formulae traduces only the fact that the problem of finding a simple equation to fit experimental data is not easy because all the parameters that influence the shape of the stress–strain curve are related not only to the properties of concrete but also to experimental conditions. The ascending branch of the stress–strain curve is not always linear and depends on the quality of the matrix/aggregate interface, the rate of strain, the composition of the matrix and the nature of the aggregate; therefore a fairly good equation should incorporate the following parameters: the pic value of f'_c, the strain value at this peak value, the secant modulus, the tangent modulus and the strain at which the rupture criteria are defined.

As high-performance concrete is almost behaving as a real composite material, the author is convinced that the stress–strain equation is a field where concrete researchers and technologists should look closely at what has been found and done in rock mechanics. In fact, a high-performance concrete behaves as an artificial rock rather than a usual concrete. The great deal of experience available in the domain of rock mechanics is very valuable and can reduce the need to start from scratch.

In fact, from a stress–strain curve point of view, rocks can be broadly classified into three different categories according to the shape of the hysteresis curve obtained when performing a loading–unloading test, as proposed by Nishimatsu and Heroesewojo (1974) (Figure 16.8). These hysteresis shapes have also been found with high-performance concrete specimens (Aïtcin and Mehta, 1990; Baalbaki *et al.*, 1991).

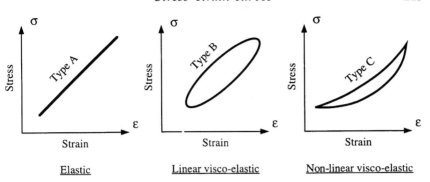

Fig. 16.8 Schematic representation of the response of a rock to its loading and unloading (hysteresis curve).

Houpert (1979) studied the influence of the characteristics of rocks on the shape of the stress–strain curve and found that in fact most of stress–strain curves are composed of parts that can be related to the three ideal ones presented in Figure 16.8. If we look closely at the general stress–strain curve that is represented in Figure 16.9, we see that it is composed of four main parts, identified by the letters A, B, C and D. From the origin to A the behaviour is non-linear visco-elastic corresponding to the closing of fissures, especially the ones perpendicular to the direction in which the axial load is applied. Between A and B the behaviour is linear elastic; strain is elastic and reversible. There are very few changes in the microstructure of the rock except perhaps under repeated loading and unloading cycles. The B–C part of the curve corresponds to a visco-elastic linear behaviour. Fissures are steadily developing but they remain stable. If the sample is unloaded, a permanent strain remains. The C–D part of the curve corresponds to the development of unstable fissures starting close to C, where the maximum stress has been reached. When point D is reached the rupture of the rock is well advanced and the remaining strength is essentially due to the friction between the fissured parts.

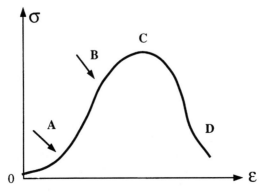

Fig. 16.9 Schematic stress–strain curve for a rock (Houpert, 1979).

This scenario also applies to high-performance concrete. Therefore designers should be aware that when they are applying the simplified equations that appear in national and international codes it can sometimes be a more or less valid equation to describe the behaviour of the high-performance concrete they are using.

The most widely used equations appearing in the codes are:

- CEB/FIB:

$$\varepsilon_0 = \frac{1}{1000}[2.0 + 0.005(f'_c - 50)]$$

$$\varepsilon_u = \frac{1}{1000}\left[2.5 + 2\left(1 - \frac{f'_c}{100}\right)\right]$$

- Norway:

$$\varepsilon_0 = \frac{1}{1000}[0.004 f'_c + 1.9]$$

$$\varepsilon_u = \frac{f'_c}{E_c} \times \frac{2.5 \, \varepsilon_0 \, E_c}{f'_c - 1.5}$$

The signification of ε_0, ε_u, f'_c and E_c can be found in Figure 16.10.

In this case too the author personally prefers to obtain the actual stress–strain curve experimentally rather than to rely on such equations.

In order to illustrate how different the shapes of stress–strain curves can be, Figure 16.11 has been taken from Baalbaki (1997). This figure

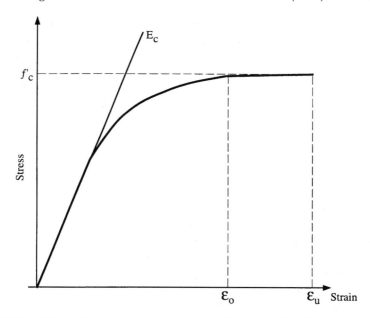

Fig. 16.10 Schematic stress–strain curve found in national codes.

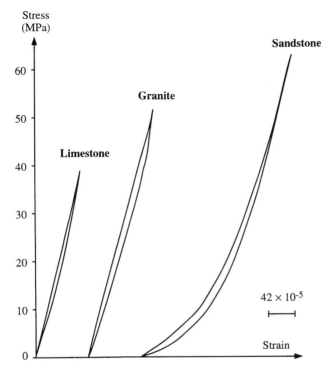

Fig. 16.11 Stress–strain curves for coarse aggregate studied by W. Baalbaki.

shows that the nature of the aggregate used to make concrete influences strongly the shape of the stress–strain curve of the concrete. This last point emphasizes once more the importance of selecting 'good' coarse aggregates when making high-performance concrete and how the properties of a particular high-performance concrete can be tailored to the designer's needs by selecting the appropriate coarse aggregate.

16.7 CREEP AND SHRINKAGE

In the previous chapter it was shown that currently it is not possible to characterize precisely the creep and shrinkage behaviour of high-performance concretes because most of the results that are available in the literature have been obtained using present testing standard procedures. It has been explained why these standard testing procedures, which give reliable results for usual concrete, are not valid when testing high-performance concrete. Using present standard testing procedures, the autogenous shrinkage developed during the first 24 hours is not taken into account.

Moreover, in Chapter 12 it was shown that a high-performance concrete does not necessarily present some autogenous shrinkage if it is properly water-cured just before the hydration starts.

While writing this book the author has undertaken a major fundamental study to characterize the shrinkage behaviour of high-performance concrete quantitatively using the definition he proposed with A. Neville and P. Acker (1997).

The same is true for creep measurement, but in this case it is hoped that the present comprehensive testing programme undertaken by W. Dilger from the University of Calgary, Canada, will bring extensive and reliable results in this field.

This does not mean that there are no long-term field results on creep and shrinkage. At least three field studies have been conducted by monitoring high-performance concrete columns in three high-rise buildings: one in Chicago (Russel and Corley, 1978; Russel and Larson, 1989; Russel, 1990) and two in Montréal (Laplante and Aïtcin, 1986; El Hindy et al., 1994; Miao and Aïtcin, 1996).

In the experiment reported by Laplante and Aïtcin (1986), a dummy column similar to an adjacent active column was cast by its side. These two columns were absolutely similar except that the dummy column was 50 mm shorter than the active one. Both columns were monitored in the same way with vibrating wires in order to follow the strains developed in two critical parts of the columns: at their centre and at their surface. Comparing the strains developed at the same location in the two columns, it has been possible to calculate the part of the shrinkage and of the creep of the strain obtained with the vibrating wires (Figure 16.12).

For example, during the first 4 days during which the forms were maintained in place, a 250 microstrain shrinkage was measured at the centre of both columns. This shrinkage definitely represents the autogenous shrinkage developed within the high-performance concrete, which did not receive any special curing. Moreover, between 4 days and 4 years, an additional autogenous shrinkage of only 50 microstrains was measured. These measurements show that autogenous shrinkage develops very rapidly at an early age in high-performance concrete, but also that it stops rapidly, very often owing to a lack of water to continue the hydration.

The forms were removed after 4 days, and two vibrating wires, one vertical and the other one horizontal, were placed on both columns to monitor the drying shrinkage. Figure 16.13 represents the recorded drying shrinkage over a period of one-and-a-half years. It can be seen that the drying shrinkage that was measured at the end of this period, 175 and 200 microstrains, is much lower than the one measured on usual concrete.

As the shrinkage measured at the centre of the column represents the autogenous shrinkage, which is almost isotropic, it can be concluded that

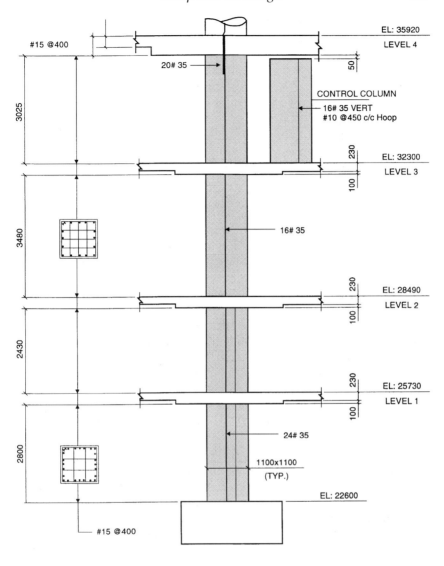

Fig. 16.12 Schematic representation of the experimental set-up to measure the long-term behaviour of the high-performance concrete cast in the Lavalin building (Laplante and Aïtcin, 1986).

the total shrinkage of the high-performance concrete at its surface was the sum of the autogenous shrinkage measured at the centre of the column and the drying shrinkage measure at its surface after 4 days.

The drying shrinkage increased very rapidly during the first 4 weeks up to a value of 125 microstrains; thereafter it increased only very slowly.

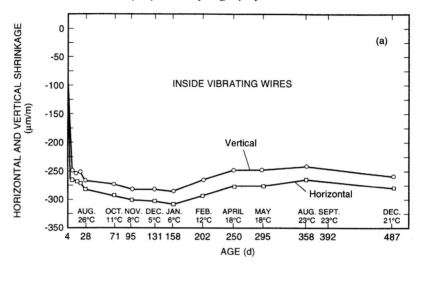

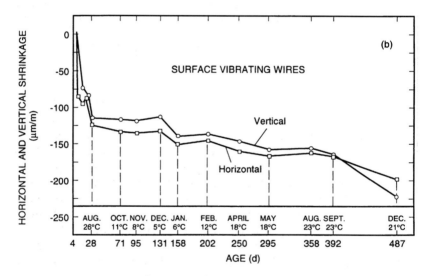

Fig. 16.13 Vertical and horizontal shrinkage measured with: (a) interior vibrating wires; (b) surface vibrating wires.

If we add this drying shrinkage to the autogenous shrinkage measured at the centre of the column, we end up with a total shrinkage of 425 to 450 microstrains, which is much lower than the 675 microstrain value measured on 100 × 100 × 375 mm prisms, as shown in Figure 16.14.

The measured creep at the centre and at the surface of the column was calculated by comparing the length change in the dummy and active column.

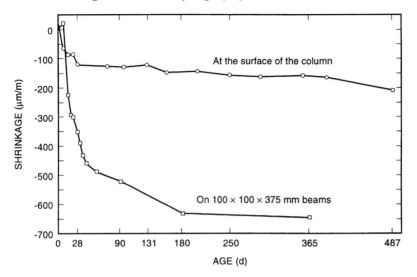

Fig. 16.14 Comparison of shrinkage of the concrete measured: (a) in the lab on 100 × 100 × 375 mm prisms; (b) with vibrating wires on the surface of the columns.

16.8 FATIGUE RESISTANCE OF HIGH-PERFORMANCE CONCRETE

16.8.1 Introduction

Usually concrete structures are designed to resist static or quasi-static loads. When concrete structures may be submitted to some dynamic loads, for design purposes these dynamic loads are transformed into static loads after being multiplied by an amplification factor; this is the case for structures such as bridges, pavements and offshore platforms which are subjected to cyclic loads generated by winds, trucks and cars, and waves. The frequency and amplitude of these cyclic loads usually vary with time. In some cases service failures have occurred due to fatigue in the form of larger deflections than expected or as an accumulation of fissures or even as a catastrophic failure of the whole structure.

As concrete is submitted to a high strength level in a high-performance concrete and as design methods become more and more demanding, it is very important to be able to predict the behaviour of high-performance concrete submitted to cyclic loading. This point has been recognized in the CEB-FIP (1990) code, which takes into account fatigue as a limit state factor when designing high-performance concrete structures having a design strength of up to 80 MPa. This compressive strength is measured on 150 × 300 mm specimens.

In spite of the importance of the fatigue behaviour of high-performance concrete, very few results are available in the literature on this subject, mostly because this type of research is very demanding in terms of equipment and time. Moreover, it gives scattered results that are not easy to interpret even when tests are done on several duplicates. Fatigue research is not an easy domain. This explains why the first studies on fatigue in high-performance concrete have been mostly oriented to see if the knowledge accumulated on the fatigue resistance of usual concretes can be transferred or extrapolated to high-performance concretes.

Another source of comparison is the fatigue behaviour of rocks. This has been studied for many years in rock mechanics. As has already been pointed out, some natural rocks are very similar to high-performance concrete in terms of compressive and tensile strength, elastic modulus and Poisson's ratio, so knowledge of the fatigue behaviour of these rocks should be very helpful, but it should not be forgotten that some of these rocks could strongly differ in terms of homogeneity, micro- and macro-structure from high-performance concrete, which should make any crude extrapolation quite dangerous, knowing the fundamental role of the microstructure and particularly its homogeneity in fatigue behaviour.

As fatigue studies are not usually as familiar to concrete engineers as to mechanical engineers, the author, who is not a particular expert in fatigue, felt it necessary to present some fatigue general background and definitions before presenting the present knowledge on the fatigue resistance of high-performance concrete.

16.8.2 Definitions

When materials are submitted to a great number of cyclic loadings two distinct types of behaviour are found: some materials, such as steel, will never fail under these cyclic loadings as long as the maximum stress in the cycle is lower than a specific stress called the fatigue limit. Most others, such as concrete, ultimately fail even though the peak load in the cycle is well within the elastic range. However, steel behaves in the same way as concrete when the peak stress during the cycle is higher than the fatigue limit. In such cases failure in fatigue is said to have occurred.

As fatigue failure occurs more or less rapidly, in terms of number of cycles, in order to define a safe working lifetime for practical purposes, an endurance limit has been arbitrarily defined. This is the stress required to cause failure for a specific value of the number of cycles N, usually 10^6 or 10^8 cycles, but this value may of course vary according to the circumstances. In mechanical engineering, this distinction between fatigue limit and endurance limit is essential. This is not, however, the case in concrete technology, since strictly speaking concrete does not have a fatigue limit. As the situation can be confusing, most of the literature erroneously deals with the fatigue strength of concrete and not the endurance strength of concrete.

(a) Wöhler diagrams

The results of fatigue studies are usually presented in terms of Wöhler diagrams, also known as S–N curves. These are semilogarithmic representations where the maximum loading, expressed as S_{max}, is plotted

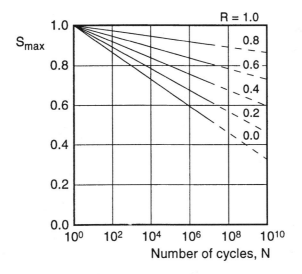

Fig. 16.15 S–N curves for constant R-values, where $R = S_{min}/S_{max}$ (RILEM, 1984).

algebraically on the y-axis while the number of cycles at which failure occurs is plotted logarithmically on the x-axis. Usually S_{max} is represented as a fraction of the static load at which ultimate failure occurs. Such S–N curves can be drawn for different values of the minimal loading or as a function of the ratio $R = S_{min}/S_{max}$, as in Figure 16.15.

(b) Goodman diagrams

Fatigue strength can also be represented by means of a Goodman diagram, with S_{min} and S_{max} plotted along the x- and y-axis, respectively. In this diagram S_{min} and S_{max} are expressed as a fraction of the static load at which ultimate failure occurs. In Figure 16.16 it is seen that if a usual concrete is submitted to a cyclic loading with $S_{min} = 0.4$ and $S_{max} = 0.75$, it will be able to sustain 10^6 cycles before fatigue failure occurs. From the fact that the log N lines are ascending to the right it seems that when S_{min} increases for a given life cycle (N = constant) S_{max} increases, or that N increases for a given S_{max}.

(c) Miner's rule

Most fatigue data have been obtained when subjecting concrete to a constant maximum and minimum stress value. To account for the effect of variable amplitude on the fatigue behaviour of concrete, Miner's hypothesis of linear accumulation of damage is widely used. Fatigue failure occurs if:

$$\sum \frac{n_i}{N_i} = 1$$

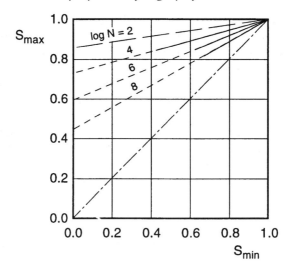

Fig. 16.16 Goodman diagram (RILEM, 1984).

where n_i is the number of cycles applied at a particular stress level and N_i is the number of cycles that will cause fatigue failure at the same stress level.

Miner's rule is still controversial: Neville (1981) and Shah (1984) in particular argued from experimental experience that the increase of damage in concrete with cyclic loading is highly non-linear and proposed a third power non-linear cumulative damage law.

As Miner's rule could give too optimistic values when designing offshore platforms, the '1' value of Miner's rule is replaced by the more pessimistic 0.2 value.

16.8.3 Fatigue resistance of concrete structures

Usually the fatigue resistance of concrete structures is subdivided into six main parts:

- the fatigue resistance of concrete in compression;
- the fatigue resistance of concrete in tension;
- the fatigue strength of concrete in flexure;
- the fatigue strength of the concrete/steel bond;
- the fatigue of structural elements made of reinforced concrete;
- the fatigue resistance of prestressed structural elements.

It would be impossible to cover so many subjects, and only some basic results covering the first two points will be presented here.

In their studies of some aspects of the fatigue resistance of high-performance concrete conducted at the Université de Sherbrooke and at the Laboratoire Central des Ponts et Chaussées in Paris, Do, Chaallal and Aïtcin (1993), Do *et al.* (1993) and Do (1994) compared the fatigue resistance and resulting deformations of high-performance concrete with that of usual concrete. The results suggest that the evolution of the deformations of high-performance concrete is very similar to the three-phase deformation pattern known for usual concrete: Phase I – initiation, Phase II – stabilization and Phase III – instability. However, the duration of each phase and the amplitude of the deformations depended on the type of high-performance concrete under study. The deformation of high-performance concrete containing silica fume was found not to evolve very much. However, the failure under fatigue could be brutal for high-performance concrete.

Usually, high-performance concretes lose less rigidity under fatigue testing than usual concrete: the higher the elastic modulus, the lower the rigidity loss. Under compressive cyclic loading with S_{max} higher than 0.75, high-performance concretes were reported to resist a higher number of cycles before failure compared with usual concrete. However, when S_{max} is smaller than 0.75 the reverse is true.

Under traction/compression cyclic loading, high-performance concretes seem to be less affected by tensile stresses than usual concrete. When testing reinforced high-performance columns Do (1994) found that:

- a fragile failure could happen when the percentage of transverse reinforcement is not high enough;
- under cyclic traction–compression loading, the cracking induced by tensile stresses rapidly annihilated the contribution of the concrete working in tension.

As far as the fatigue life cycle is concerned, it seems that the model proposed by Hsu (1981) is the most appropriate in the case of high-performance concrete and could be applied to the three modes of loading: compression, traction and traction/compression. It was also found that the S–N–P diagrams established from reliability computations do not take into account the dispersion of the results.

Those readers who have a strong interest in the fatigue resistance of high-performance concrete should refer to the following synthesis works: Bennett and Muir (1967), RILEM (1984), Lenschow (1987), Lambotte and Taerwe (1987), SINTEF (1989), CEB-FIP (1990) or the work from Petkovic *et al.* (1990) and Petkovic (1991).

16.9 CONCLUDING REMARKS

This reading of this chapter will leave some readers very disappointed because it shows that high-performance concrete's mechanical properties cannot be deduced from knowledge of f'_c. In a sense, each high-performance concrete is unique, this uniqueness coming from the cementitious system used, the water/binder ratio achieved and the characteristics of the aggregates, especially the coarse ones. As high-performance concretes act like true composite materials it is normal that the characteristics of the hydrated cement paste, of the transition zone (through the water/binder ratio) and of the aggregates influence their mechanical properties. High-performance concrete is not a simple material, and too simplistic an approach to it can lead to very poor results. The design work is complicated to some extent, but this is the price of designing an efficient structure with a high-performance material (Collins, Mitchell and MacGregor, 1993). It is hoped that with the tremendous volume of research that is presently being done world-wide on high-performance concrete, some families of high-performance concrete will be defined with appropriate and simple relationships in order to make the design process of high-performance concrete almost as simple as that of usual concrete.

REFERENCES

ACI Committee 363, (1984) State-of-the-art report on high-strength concrete. *ACI Journal, Proceedings* **81**(4), July–August, 364–411.

Ahmad, S.H. and Shah, S.P. (1985) Structural properties of high strength concrete and its implication for precast prestressed concrete. *PCI Journal*, **30**(6), November–December, 91–119.

Aïtcin, P.C. and Mehta, P.K. (1990) Effect of coarse-aggregate characteristics on mechanical properties of high-strength concrete. *ACI Materials Journal*, **87**(2), March–April, 103–7.

Aïtcin, P.-C., Laplante, P. and Bédard, C. (1985*) Development and Experimental Use of a 90 MPa (13 000 psi Field Concrete)*, ACI SP-87, pp. 51–67.

Aïtcin, P.-C., Neville, A.M. and Acker, P. (1997) The various types of shrinkage deformation in concrete. an integrated view. *Concrete International*, **19**(9), September, 35–41.

Aïtcin, P.-C., Sarkar, S.L. and Laplante, P. (1990) Long term characteristics of a very high strength field concrete. *Concrete International*, **12**(1), January, 40–44.

Alfes, C. (1989) High-strength silica-fume concretes of low deformability. *Concrete Precasting Plant and Technology*, Bauverlag, Wiesbaden, Germany, pp. 62–71.

Asselanis, J., Aïtcin, P.-C. and Mehta, P.K. (1989) Influences of curing conditions on the compressive strength and elastic modulus of very high-strength concrete. *Cement Concrete and Aggregate*, **II**, Summer, 80–83.

Baalbaki, W. (1997*) Analyse expérimentele et prévisionnelle du module d'élasticité des bétons*. PhD. Thesis, Université de Sherbrooke, Québec, Canada, 158 pp.

Baalbaki, W., Benmokrane, B., Chaallal, O. and Aïtcin, P.C. (1991) Influence of coarse aggregate on elastic properties of high-performance concrete. *ACI Materials Journal*, 88(5), September–October, 499–503.

Bennett, E.W., Muir, S.E. St J. (1967) Some fatigue tests of high-strength concrete in axial compression. *Magazine of Concrete Research (London)*, **19**(59), June, 113–17.

Burg, R.G. and Ost, B.W. (1992) *Engineering Properties of Commercially Available High-Strength Concretes*, RD 104-DIT, Portland Cement Association, 55 pp.

Caquot, A. (1937) Le rôle des matériaux dans le béton, *Mémoire de la Société des Ingénieurs Civil de France*, July–August, pp. 562–82.

Canadian Code (1990) CAN A23.3-M90.

Carrasquillo, R.L., Nilson, A.H. and Slate, F.D. (1981) Properties of high strength concrete subjected to short-term load. *ACI Journal*, **78**(3), 71–8.

Castillo, C. and Durrani, A.J. (1990) Effect of transient high-temperature on high-strength concrete. *ACI Materials Journal*, **87**(1), January–February, 47–53.

CEB Model Code (1978) *CEB FIP pour les Structures en Béton*, Bulletin d'information 124-125, pp. 14–16.

CEB Model Code (1995) *CEB-FIP pour les Structures en Béton*, Bulletin d'information No. 228, pp. 12–13.

CEB-FIP (1990) *State-of-the-Art Report: High Strength Concrete*, Bulletin d'information CEB No. 197, August, 212 pp.

CEB-FIP Model Code (1990) Bulletin d'information CEB No. 203, July, final draft, chapters 1–3.

Collins, M.P., Mitchell, D. and MacGregor, J.G. (1993) Structural design consideration for high-strength concrete. *Concrete International*, **15**(5), May, 27–34.

Cook, W.D., Miao, B., Aïtcin, P.C. and Mitchell, D. (1992) Thermal stresses in large high-strength concrete columns. *ACI Materials Journal*, **89**(1), January–February, 61–8.

de Larrard, F. and Aïtcin, P.-C. (1993) The strength retrogression of silica fume concrete. *ACI Materials Journal*, **90**(6), November–December, 581–5.

de Larrard, F. and Le Roy, R. (1992) Relation entre formulation et quelques propriétés mécaniques des BHP. *Materials and Structures*, **25**, 464–75.

Desayi, P. and Krishnan, S. (1964) Equation of the stress–strain curve of concrete. *ACI Journal*, **61**(3), March, 345–50.

Do, M.T. (1994) *Fatigue des Bétons à haute Performance*, PhD Thesis No. 788 (in French), Université de Sherbrooke, Québec, Canada, 190 pp.

Do, M.T., Chaallal, O. and Aïtcin, P.-C. (1993) Fatigue behavior of high-performance concrete. *Journal of Materials in Civil Engineering*, **5**(1), February, 96–111.

Do, M.T., Schaller, I., de Larrard, F. and Aïtcin, P.-C. (1993) Fatigue of plain and reinforced high-performance concrete, in *High-Strength Concrete* (ed. I. Holland and E. Sellevold), Norwegian Concrete Association, Oslo, ISBN 82-91341-00-1, pp. 146–54.

El Hindy, E., Miao, B., Chaallal, O. and Aïtcin, P.-C. (1994) Drying shrinkage of ready-mixed high-performance concrete. *ACI Materials Journal*, **91**(3), May–June, 300–305.

Gardner, N.J. and Zhao, J.W. (1991) Mechanical properties of concrete for calculating long-term deformations, in *Proceedings, Second Canadian Symposium on Cement and Concrete*, Vancouver, July (ed. S. Hindess), University of British Columbia Press, pp. 150–9.

Giacco, G., Rocco, C., Violini, D., Zappitelli, J. and Zerbino, R. (1992) High-strength concretes incorporating different coarse aggregate. *ACI Materials Journal*, **89**(3), May–June, 242–6.

Hansen, T.C. (1965) Influence of aggregate and voids on modulus of elasticity of concrete, cement mortar and cement paste. *American Concrete Institute Journal*, **62**(2), 193–216.

Hatanaka, S. (1986) Study on stress–strain relationship of concrete in plastic hinge range of reinforced concrete members, in *Proceedings of the Annual Meeting of AIJ, Tokai Branch*, Japan Concrete Institute, Tokyo, pp. 113–16.

Houpert, R. (1979) Le comportement à la rupture des roches, in *Proceedings of the International Conference on Rock Mechanics, Montreux, Switzerland*, Vol. 3, Balkema, Rotterdam, pp. 107–14.

Hsu, T.T.C. (1981) Fatigue of plain concrete. *ACI Journal*, **78**(4), July–August, 292–305.

Illston, J.M., Dinwoodie, J.M. and Smith, A.A. (1987) *Concrete, Timber and Metals*, Van Nostrand Reinhold, New York, ISBN 0-442-30145-6.

Kaplan, M.F. (1959) Ultrasonic pulse velocity, dynamic modulus of elasticity, Poisson ratio, and the strength of concrete made with thirteen different coarse aggregates. *RILEM Bulletin (Paris)*, New Series, No. 1, March, 58–73.

Khan, A.K., Cook, W. and Mitchell, D. (1996) Tensile strength of low, medium and high strength concretes at early ages. *ACI Materials Journal*, **93**(5), September–October, 487–93.

Khayat, K., Bickley, J. and Hooton, R.D. (1995) High-strength concrete properties derived from compressive strength values. *Cement Concrete and Aggregates*, **17**(2), December, 126–33.

Lambotte, H. and Taerwe, L. (1987) Fatigue of plain, high-strength concrete structures subjected to flexural tensile stress, in *Utilization of High-Strength Concrete*, Tapir, Trondheim (eds I. Holland et al.), ISBN 82-519-0797-7, pp. 331–42.

Laplante, P. and Aïtcin, P.-C. (1986) Field monitoring of creep and shrinkage on a 100 mpa (14 500 psi) concrete column in a 25-story building, in *Proceedings of the Fourth International Symposium on Creep and Shrinkage of Concrete: Mathematical Modeling*, Evaston, Ill., USA, pp. 777–86.

Lenschow, R. (1987) Fatigue of high strength concrete, in *Utilization of High Strength Concrete*, Tapir, Trondheim (eds I. Holland et al.), ISBN 82-519-0797-7, pp. 272–90.

Lessard, M., Baalbaki, M. and Aïtcin, P.-C. (1995) Mix design of air-entrained high-performance concrete, in *Concrete Under Severe Conditions. Environment and Loading*, Vol. 2, E and FN Spon, ISBN 0419-198601, pp. 1025–31.

Lindgård, J. and Smeplass, S. (1992) *High-Strength Concrete Containing Silica Fume – Impact of Aggregate Type on Compressive Strength*, ACI SP-132, Vol. 2, pp. 1061–75.

Miao, B. and Aïtcin, P.-C. (1996) *Five Years' Monitoring of the Behavior of HPC Structural Columns*. ACI Fall Convention, Montreal, ACI SP-167, pp. 193–210.

Neville, A.M. (1981) *Properties of Concrete*, Third Edition, Longman Scientific and Technical, pp. 338–44.

Nilsen, A.V. and Aïtcin, P.-C. (1992) Properties of high-strength concrete containing light–normal and heavyweight aggregate. *Cement, Concrete and Aggregates*, **14**(1), Summer, 8–12.

Nilson, A.M. (1987) *Properties and Performance of High-Strength Concrete*, IABSE Paris, Versailles Symposium, pp. 389–93.

Nishimatsu, Y. and Heroesewojo, R. (1974) Rheological properties of rocks under the pulsating loads, in *Proceedings of the Third International Congress on Rock Mechanics*, Denver, Vol. IIA, National Academy of Sciences, Washington, pp. 385–9.

Norwegian Code NS 3473 (1992) *Design of Concrete Structures*, Norwegian Council for Standardization, Oslo, Norway.

Olsen, N.H., Krenchel, H. and Shah, S.P. (1987) *Mechanical Properties of High-Strength Concrete*, IABSE Paris, Versailles Symposium, pp. 395–400.

Perenchio, W.F. and Klieger, P. (1978) *Some Physical Properties of High-Strength Concrete*, Portland Cement Research and Development, Bulletin No. 3, 6 pp.

Petkovic, G. (1991) *Properties of Concrete Related to Fatigue Damage with Emphasis on High-Strength Concrete*, PhD Thesis, Division of Concrete Structures, Norwegian Institute of Technology, Trondheim, 247 pp.

Petkovic, G., Lenschow, R., Stemland, H. and Rosseland, S. (1990) *Fatigue of High-Strength Concrete*, ACI SP-121, pp. 505–26.

Popovics, S. (1973) A numerical approach to complete stress–strain curve of concrete. *Cement and Concrete Research*, **3**(4), September, 583–99.

Powers, T.C. and Brownyard, T.L. (1948) *Studies of the Physical Properties of Hardened Portland Cement Paste*, Bulletin 22 of the Research Laboratories of the Portland Cement Association, March, 992 pp.

Raphael, J.M. (1984) Tensile strength of concrete. *ACI Materials Journal*, **18**(2), March–April, 158–65.

RILEM Committee-36 RDL (1984) Long term random dynamic loading of concrete structures. *Materials and Structures*, **17**(97), 1–28.

Rougeron, P. and Aïtcin, P.-C. (1994) Optimization of the composition of a high-performance concrete. *Cement Concrete and Aggregates*, **16**(2), 115–24.

Russel, H.G. (1990) *Shortening of High-Strength Concrete Members*, ACI SP 121, pp. 1–20.

Russel, H.G. and Corley, W.G. (1978) Time dependent behavior of columns in water tower place, in *Douglas McHenry International Symposium on Concrete and Concrete Structures*, ACI SP-55, pp. 347–73.

Russel, H.G. and Larson, S.C. (1989) Thirteen year of deformations in Water Tower Place. *ACI Structural Journal*, **86**(2), March–April, 182–91.

Shah, S.P. (1984) Predictions of cumulative damage for concrete and reinforced concrete. *Materials and Structures*, **17**(97), 65–8.

SINTEF (1989) *High-Strength Concrete; State of the Art*, SINTEF Report No. STF 65 A89 003, Trondheim, 139 pp.

Swamy, R.N. (1986) Properties of high strength concrete. *Cement, Concrete and Aggregates*, **8**(1), Summer, 33–41.

Taerwe, L. and Vangyset, A. (1996) Analytical formulation of the complete stress–strain curve for high-strength concrete. *Materials and Structures*, **29**, November, 529–33.

Wang, P.T., Shah, S.P. and Naaman, A.E. (1978) Stress strain curves of normal and lightweight concrete in compression. *ACI Journal*, **75**(11), November, 603–11.

Wee, T.H., Chin, M.S. and Mansur, H.A. (1996) Stress–strain relationship of high-strength concrete in compression. *Journal of Materials in Civil Engineering*, **8**(2), May, 70–6.

Xie, J., Elwi, A.E. and MacGregor, J.G. (1995) Mechanical properties of three high-strength concretes containing silica fume. *ACI Materials Journal*, **92**(2), March–April, 135–45.

CHAPTER 17

The durability of high-performance concrete

17.1 INTRODUCTION

The expression 'durability of concrete' is usually used to characterize, in general terms, the resistance of concrete to the attack of physical or chemical aggressive agents. The nature, intensity and mechanism implied in each of these different attacks can vary considerably, which is why the expression durability of concrete is sometimes perceived as too vague an expression. There does not exist any standardized method of measuring the durability of concrete in general. There are no units in which to evaluate the durability of concrete such as exist when the strength of a particular concrete has to be evaluated or for scaling resistance when concrete is exposed to slow freeze–thaw cycles in the presence of deicing salts. Some engineers still prefer to specify, immediately after the term durability, the type of attack that is involved. Therefore, we should really speak of the durabilities of concrete rather than the durability of concrete.

The aggressive agents that attack concrete can be classified schematically into two broad categories: external agents and internal agents. Among the external agents, chloride ions, carbon dioxide, sulfates, freeze–thaw cycles, bacteria and abrasives can be cited. Among the internal agents, chloride ions incorporated when certain accelerators are used and cement alkalis when potentially reactive aggregates are used can be cited. This chapter will deal only with the durability of high-performance concrete attacked by external agents, except for a very brief paragraph on alkali aggregate reactions. It is assumed that the materials used to make the high-performance concrete have been selected with care, according to the current state-of-the-art rules. For example, if the aggressive agents are sulfate ions contained in the soil, it is assumed that the high-performance concrete has been made using a Type V sulfate-resistant cement; if the external agent is rapid freezing and thawing cycles while the high-performance concrete is water saturated, then it is assumed that the high-performance concrete has been air-entrained.

Introduction

It is difficult to make statements about the durability of a 'new material', because durability relates to the long-term performance of a particular material, in a particular environment, under particular service conditions. Since durability cannot be assessed if this knowledge of its long-term behaviour is unreliable, it is very difficult in a more and more litigious human environment to specify a 'new material'.

What is particularly lacking in the case of high-performance concrete is well-documented field cases of its successful or unsuccessful use. High-performance concretes are so new on the market, and understandably not specifically specified and used in harsh environments, that we have very little experience in this area. Not enough errors have been made to date in the use of high-performance concrete, so its field track record is not sufficiently documented. It is the author's opinion that, at the present time, well-documented field case studies should be a priority in the area of high-performance concrete technology, not only for researchers, but also for users interested in developing safe use of high-performance concretes. A single conclusive field study is worth hundreds of laboratory tests, as representative of 'real life' as they may be.

Of course, there are different approaches to overcoming this 'chicken and egg' situation. First of all, the durability of the material can be assessed using the accelerated durability tests that have been standardized over the years, in which samples are usually exposed to concentrated aggressive solutions under extreme temperature, load or gradient conditions. It is implicitly assumed that these laboratory conditions represent field conditions fairly well and that only the destructive mechanism is accelerated. The second approach is to use the material in temporary or experimental full-scale structures and to monitor them through the years, so that the behaviour of the material can be well documented when these temporary structures are no longer in use or as they age. Such opportunities should be taken as often as possible because they represent full-scale applications without any major long-term financial risk, since either the structure will have to perform only on a short-term basis or the risk of failure is taken into account when building it. The third approach is to start using this new material in small projects where the financial risks are not very high, and from the experience and confidence gained in the suitability of the material, to use it in increasingly more elaborate projects.

A fourth approach consists of studying the causes of failure of similar or related materials in similar environments and under similar service loads, in order to understand the mechanisms of the failure process (Whiting, 1984). By comparing the fundamental properties of the failed material and that of the new material, it may be possible to find out whether the new material would have performed better than the failed one. However, in such a case it would be difficult to assess how much better it would have performed. A more global approach consists of trying to take the opportunity to use all these approaches at the same time to gain confidence in the assessment of the new material's durability. All of these approaches have been followed in the case of high-performance concrete and they will be briefly reviewed in this chapter. However, this review of the durability of some high-performance concretes under different environmental and service conditions will be preceded by some general considerations about the durability of usual concrete and some lessons from the past.

In order to tackle such a wide subject (because there are a great number of aggressive environments in which high-performance concrete has been, is or will be used), it has been decided to approach high-performance concrete durability by considering the types of aggression to which concrete can be subjected. High-performance concrete can be physically or mechanically attacked, such as when exposed to abrasion or freezing and thawing cycles; it can be chemically attacked directly by chloride ions (or chloride ions may attack the reinforcing steel the concrete should have protected), by sulfate ions, or by acidic or other types of aggressive chemicals. It can also be chemically attacked by a gas such as carbon dioxide and even by bacteria. Moreover, it can self-destruct in the presence of alkali reactive aggregates.

17.2 THE DURABILITY OF USUAL CONCRETES: A SUBJECT OF MAJOR CONCERN

The durability of usual concrete is becoming a subject of concern in most countries and most climates because too large a number of concrete structures display serious signs of advanced deterioration (Skalny, 1987). This situation is not the result of a lack of information on the durability of usual concrete (Neville, 1987; Mehta, 1991; Mehta, Schiessl and Raupach, 1992). Some researchers have even proposed models to try to predict the durability of usual concrete (Bentz and Garboczi, 1992; Dagher and Kulendran, 1992). But in spite of this abundance of information and data, this information is not reaching the right people. This state of affairs is due to the lack of communication that exists in the concrete field between the three solitudes: the materials people, the designers and the field engineers. These people often do not attend the same meetings or the same symposia, they may not read the same journals or the same papers, and generally they do not participate in the same committees. However, despite the abundant literature on concrete durability, it must be admitted that there are few technical or scientific papers which can be read by these three solitudes that deal with the crucial question: how does one make, place and cure a durable concrete that can be used to build, repair or rehabilitate concrete structures (Pomeroy, 1987; Philleo, 1989).

17.2.1 Durability: the key criterion to good design

One of the principal reasons for the deterioration of many concrete structures stems from the fact that, in the past and even now, too much importance has been given to concrete compressive strength when designing concrete structures and not enough to the environmental factors that the structure will have to face when performing its structural function (Ho and Cao, 1993). For example, from a structural point of view, it is easy to design an exterior parking garage with a 20 MPa concrete, but under very severe environmental conditions, the construction of such a parking garage would be absolutely catastrophic. In fact a 20 MPa concrete would never be able to protect efficiently the reinforcing steel against corrosion, owing to the deicing salts that would drip from cars that are always parked in the same place, however thick the concrete cover that is placed over the reinforcing steel.

The author is not in favour of a solution that consists of using a very porous and non-durable concrete in which expensive epoxy-coated reinforcing steel is incorporated (in fact the efficiency of these epoxy-coated rebars is not yet well accepted), and then using an expensive membrane or cathodic protection. Would it not be better to begin to use a high-performance concrete with a very low permeability? For many designers, such a concrete is too strong and brings 'unnecessary' MPa's. Unfortunately, the durability of a parking garage is not linked to the dead loads or live loads, but rather to its environment, and at the present time we do not know how to make a 20 MPa concrete that is durable enough against deicing salts. The durable concretes that we now make have compressive strengths of the order of 60 to 100 MPa.

When concrete is subjected to external chemical attack, there is only one way to reduce the intensity of this external aggression: to lower the porosity and the permeability of the concrete in order to reduce or to slow down as much as possible the penetration of the aggressive agents. Therefore, in order to offer the best resistance to external chemical attack (and even to physical attack, as will be seen later), it is necessary that the concrete be as compact and as impervious as possible. To achieve this, it is necessary that the concrete has a low water/cement ratio, or, as is more and more the case, a low water/binder ratio. In fact it is the water/binder ratio, and not the compressive strength, that has always been and will always be the key factor controlling the impermeability of a concrete and therefore its durability. If the water/binder ratio necessary to reach such a level of impermeability leads to a high compressive strength, it is the responsibility of the designer to use this additional strength more efficiently. In the future, high-performance concrete will be used not so much for its high compressive strength, but rather for its increased durability. The day the concrete community understands this fact, a new page will be turned in concrete technology.

The durability of concrete, which is also governed by the aggressiveness of the environment, has not yet received enough attention in most national codes. However, in recent years a new attitude can be perceived towards durability, and in particular national codes in Japan, in Australia, in Europe (Moore, 1993; Rostam and Schissl, 1993) and in Canada are reflecting this new trend.

The decrease of the water/binder ratio is a necessary condition to obtain durable concrete, but unfortunately it is not a sufficient condition. Other factors can affect concrete durability, in particular construction detailing. In spite of the fact that the engineer has no influence on environmental conditions, the engineer can, with appropriate design in certain details, avoid creating artificial catastrophic micro-climatic conditions (Norberg et al., 1993). In many cases, concrete beams have suffered because the particular construction details have resulted in the concentration of aggressive agents at a specific point in the structure, whereas had the same amount of aggressive agents been distributed uniformly over the structure, they would not have affected the durability of the concrete.

When a concrete structure is designed, it is essential to begin by defining in the best way possible the exact environmental conditions to which the concrete will be subjected during its service life. It is the role of the material specialist to adjust the composition and to select the right materials so that the concrete that will be used can resist these environmental conditions. Over the years, different types of cements and concretes have been developed for use in different environmental conditions, but they must be used in an appropriate manner.

When an owner decides to change the use of a concrete structure, it is not sufficient to check only whether the structure is able to support the new loads or the new overloads; it is absolutely essential to determine whether the concrete will be able to survive in its new environment. How many catastrophic changes in use could have been avoided if it had been verified that the concrete could not support such a change of environment?

17.2.2 The critical importance of placing and curing in concrete durability

As has already been mentioned, specifying a high-performance concrete with a low water/binder ratio is a necessary condition for obtaining a durable concrete, but it is not a sufficient condition. It is also necessary to be sure that this high-performance concrete is placed and cured in a correct manner. Even a properly specified and produced high-performance concrete will have only a mediocre durability if it is not placed and cured properly. As for any concrete, the durability of high-performance concrete is governed by the intrinsic quality of the material and the severity of the environment to which it is exposed, but it is also affected by the care that has been taken during its placing and curing.

The importance of the placing and curing of both usual concrete and high-performance concrete, as well as the difficulty of carrying out these operations properly, is one of the major weaknesses of concrete when compared with other construction materials. Excessive vibration generates internal bleeding even in high-performance concrete with a high slump and a very low water/binder ratio (lower than 0.30). This excessive vibration can create a network of interconnected capillaries that go through the entire volume of the concrete to its surface, and this network of capillaries constitutes an easy means of penetration for all aggressive agents. Weak transition zones can be found beneath coarse aggregates (with respect to the casting direction). In these weak transition zones, internal bleeding results in an accumulation of water just below the aggregates, as the aggregates trap the water during its rise towards the concrete surface. The same phenomenon can be observed around reinforcing steel and at the contact with the forms.

High-performance concrete has to be vibrated but not too much; it is sufficient only to consolidate it until it is in place. An average slump of 180 to 200 mm, and rarely greater than 230 mm, seems to be a good compromise.

Premature drying of the concrete surface can have a catastrophic effect on the durability of concrete in general, and high-performance concrete in particular, because, as in the case of excessive vibration, a network of capillaries will appear at the surface of the concrete which will constitute an easy pathway through which aggressive agents will penetrate the concrete. It is absolutely essential to protect fresh high-performance concrete surfaces against early drying, because high-performance concrete either doesn't bleed or bleeds very little.

Unfortunately, as is the case for usual concrete, the curing of high-performance concrete is too often neglected because good curing procedures often have a negative effect on the rapid execution of the work, or are not perceived as being essential for the durability of the structure. Contractors do not like to invest time and money in an operation that appears only to slow down the construction process. Yet such a simple thing as wetting the concrete surface in order to prevent it from drying, by using a good curing compound (with usual concrete), or better fogging or pounding, can make all the difference between a durable concrete structure and a structure that will deteriorate quite rapidly.

17.2.3 The importance of the concrete 'skin'

Concrete is still too often perceived as a homogenous and isotropic material. Some designers even go so far as to reduce it entirely to its 28 day compressive strength. It is only in a very few instances that concrete has been considered by designers as a heterogeneous material, for example, when the difference in bonding between steel and concrete is considered in terms of the position of the reinforcing steel relative to the direction of concreting, so that internal bleeding can be taken into account. While, at the metre or decimetre scale, concrete can be considered as a homogeneous material, at the centimetre scale, it is definitely heterogeneous. For a long time, concrete has been considered as a two-phase material (hydrated cement paste and aggregates) whose properties are mostly governed by the water/cement ratio of the paste. This simple model was universally accepted until it was realized that the contact zone between the aggregates and the hydrated cement paste constituted a special zone, the so-called transition zone, which had properties quite different from those of the bulk hydrated cement paste. This zone was found to be weaker than the bulk of the hydrated paste and to play a very important role with regard to the strength and durability of the concrete. It is weaker because when it hardens its actual water/binder ratio is higher than in the rest of the paste, i.e. higher than the value calculated by simply dividing the mass of water used during the mixing by the mass of cement introduced in the mix. This higher water/binder ratio in the vicinity of coarse aggregates is usually related to internal bleeding. Aggregates, which can be considered as rigid and non-absorptive inclusions in the fresh paste, trap water around them. When concrete is vibrated, coarse aggregates begin to vibrate and to compact the mortar around them, expelling some water from it. Not all this water bleeds up to the surface, because it is trapped by a denser mortar, and it creates a zone of high water/cement ratio around the coarse aggregate. Microstructural observation with a scanning electron microscope definitely shows this transition zone, where oriented portlandite crystals tend to grow, as well as needle-like ettringite crystals, making it quite porous and weak. This is why concrete is now sometimes considered to consist of three phases. The occurrence of such a weak zone around the aggregate can help to explain the relationships between the various mechanical properties of usual concrete, and also to explain the deterioration mechanisms that are observed in concrete exposed to aggressive agents.

There was some hope that high-performance concrete could be considered as a two-phase material when it was found that there was no longer a transition zone around the coarse aggregates in high-performance concretes. We have already seen in Chapter 16 that the disappearance of the transition zone makes high-performance concrete a true composite material with good stress transfer between the hydrated cement paste and the aggregate. However, when looking at concrete from a durability point of view, it has been found that the high slumps achieved when using superplasticizers create a new type of heterogeneous zone along the forms or at the top surface of the concrete. This zone has become known as the 'concrete skin' (Kreijger, 1987), 'outer skin' (Bentur and Jaegermann, 1991), 'concrete cover' (Halvorsen, 1993) or simply as 'covercrete'.

It is only quite recently, and still not very often, that the importance of the concrete skin (the outermost 5 to 10 mm) has been recognized from a durability point of view (Parrott, 1992), in spite of the fact that it has been well known for a long time that the concrete skin does not have exactly the same composition and microstructure as the interior of the concrete, owing to the so-called 'wall effect'. The packing of the aggregates is less dense adjacent to the forms, or around any solid inclusion in the concrete, so that in these regions the concrete is richer in hydrated cement paste. For a long time, the consequence of this wall effect has been ignored from a durability point of view because concretes were generally placed with low slumps of 20 to 100 mm, resulting in concrete that was very cohesive and not prone to too much segregation, and thus not very much affected by the wall effect. However, this is no longer the case, since many usual concretes and most high-performance concretes are cast with very high slumps using superplasticizers. The concrete skin is very rich in paste, owing to the high slump of the high-performance concrete. When a high-performance concrete is plastified so that its slump is maintained within the 150 to 200 mm range, there is little risk of segregation, because the mix is quite rich and quite thixotropic, but it is observed that the wall effect is greatly increased when the slump increases. The use of permeable forms seems to be an option very often used in Japan (Katayama and Kabayashi, 1991; Kumagai, Arioka and Tanabe, 1991; Sugawara et al., 1993; Tsukinaga, Shoya and Sugawara, 1993) to improve the durability and aesthetics of concrete skin.

Moreover, when different supplementary materials are used to produce high-performance concretes, and of course any usual concretes, it is possible to see some segregation of the fine particles of the cementitious part, particularly when the fine particles of the different supplementary cementitious materials have different specific gravities. For instance, during the construction of a bridge, there was an occurrence of dark spots in the lower parts of the girders. The explanation was quite simple: the cement used to build the bridge was a blended silica fume cement, and the silica fume used in this blended cement contained a certain amount of very fine carbon black particles. The dark spots were located just at the points where very powerful vibrators were used to place the concrete in the lower part of the girders, which were densely reinforced. The excessive vibration that was applied at these particular places created a kind of segregation within the concrete which resulted in a concentration of the carbon black particles in the liquid part of the concrete, just at the location of the vibrators.

The author is convinced that in future there will be more and more problems with the concrete skin because contractors will only be interested in placing almost fluid concretes that will not receive proper curing, and because the cements of tomorrow will contain more and more supplementary cementitious materials or other kinds of filler with different specific gravities, which become a problem in a very fluid concrete. It is no longer realistic to imagine that concrete contractors will resume the placing of concretes having a 100 mm slump in countries with very high salaries for construction workers. However, it will be necessary to be quite prudent each time a 200 mm slump is specified. For instance, self-placing high-performance concretes, that can be placed practically without any vibration even in structural elements that are densely reinforced, have been developed to avoid segregation. To achieve this, it is only necessary to modify the ratio between the coarse and the fine aggregate, to add an appropriate amount of superplasticizer, and to use a colloidal agent and a coarse aggregate having a smaller maximum size.

17.2.4 Why are some old concretes more durable than some modern ones?

A simplistic answer to this question would be that all the ancient concretes that were not durable disappeared long ago and the only ones that have survived are the ones that were well made, well placed and well cured, or they are those that were not exposed to very aggressive environmental conditions. This is one explanation, but there is another.

One cause of the repeated failure of some concretes in terms of durability can be related to the evolution of cement and concrete technology over the past 50 years, an evolution that seems to be completely ignored by those who design concrete structures. For a long time, a strong concrete was synonymous with a durable concrete because in order to obtain a high compressive strength at 28 days, it was necessary to use a high amount of Portland cement in order to be able to lower the water/binder ratio to within the 0.40 to 0.45 range. In the past, the cement particles would not all be deflocculated because water reducers were not too efficient or were not used at all. The only practical way to reduce the water/binder ratio was to increase the cement content of the concrete.

Moreover, 40 to 50 years ago Portland cements were not ground as finely as they are presently, so that their hydration proceeded slowly. This meant that it was necessary to use more cement than today in order to obtain a given 28 day compressive strength. At 28 days, these concretes had not yet developed all of their potential strength because only a part of the cement was hydrated at that time. It is thus not suprising to see that the compressive strengths that are measured on some old concretes, that were designed for relatively low compressive strengths, are now quite high.

It should also be mentioned that during the past two decades, under pressure from contractors, cement companies have developed Portland cements that harden faster in order to accelerate form removal. Portland cement can harden faster if it is ground finer and if its phase composition is changed. Present-day cements, for example, have higher C_3S and C_3A contents (the two most reactive phases of concrete) than the cements of 40 years ago. These phases, which hydrate more rapidly, allow for the achievement of a given compressive strength more quickly, but with the consequence that the compressive strength of the concrete does not increase much after 28 days, because there are fewer cement particles that are unhydrated, so there is less potential for the compressive strength to increase.

This technological evolution in the manufacture of Portland cement has created major changes in the composition of modern concretes, because the structural specifications of concrete are still related only to the 28 day concrete compressive strength. According to Wischers (1984), from 1945 to 1947 in England it was necessary to use a 0.47 water/binder ratio in a concrete containing 300 kg of cement per cubic metre of concrete in order to obtain a compressive strength of 30 MPa at 28 days. Between 1975 and 1980, it was possible to get the same strength with a water/binder ratio of 0.72 with only 250 kg of cement per cubic metre of concrete. While from the point of view of the structural designer these two concretes are the same in terms of compressive strength, in terms of microstructure and permeability they are quite different, and their durability will be different when they are exposed to aggressive environments.

As has been stated earlier, for many years it was impossible to reduce the water/binder ratio of concrete below the critical value of 0.40, even when using the most efficient water reducers. It was therefore impossible to deflocculate sufficiently all of the cement particles. At the present time, this situation has changed completely with the use of superplasticizers. Superplasticizers have very powerful deflocculating properties that can be used to great advantage to lower considerably the water/binder ratio of modern concretes. The synthetic polymers that are the basis of these modern superplasticizers are so efficient in deflocculating cement grains that it is now possible to make workable concretes, even flowable concretes, with a water/binder ratio as low as 0.25, and even in some cases with a water/binder ratio as low as 0.20.

Consequently, when hydration stops, these concretes still contain a good proportion of unhydrated cement particles. These unhydrated cement particles may eventually play a very important role, because they constitute, in a certain way, a reserve of cement within the concrete, such as is still found in ancient concretes that performed quite well. If for any reason, structural or environmental, cracks develop in the concrete, these as yet unhydrated cement particles can hydrate as soon as water penetrates the concrete. These unhydrated cement particles thus offer a 'self-healing' potential for high-performance concrete.

17.3 WHY HIGH-PERFORMANCE CONCRETES ARE MORE DURABLE THAN USUAL CONCRETES

Figure 17.1 represents schematically two cement pastes with different water/binder ratios (0.65 and 0.25) in both the fresh and hardened states. It can be seen that, when hardened, each of these cement pastes has a different microstructure. In this schematic representation, the ratio between the surface representing water and the surface representing cement was equal to the water/cement ratio by a mass that is indicated. It can be seen that in a cement paste with a water/binder ratio of 0.65, which should give a concrete with a compressive strength of about 25 MPa, the cement particles are quite far from each other compared with their respective positions in the paste having a water/binder ratio of 0.25, which should give a concrete of about 100 MPa compressive strength. In the latter case the hydrated product fills most of the intergranular space, leading to a very rapid strength increase. Moreover, in this case it is not necessary to develop very much hydration product to obtain a structure that is compact and quite strong. However, in the case of the paste with a water/binder ratio of 0.65, it is necessary that the external hydration products develop crystals over a large distance before reaching the hydration products that have been developed from an adjacent grain.

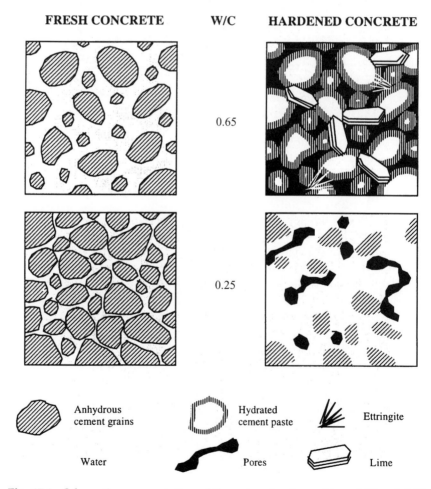

Fig. 17.1 Schematic representation of the microstructure of two 0.65 and 0.25 water/binder ratio fresh and hardened cement pastes.

Why high-performance concretes are more durable 469

Figure 17.2 shows two micrographs of the microstructures of two concretes obtained in the field as seen with an electron microscope. One has a 28 day compressive strength of about 20 MPa, and the other of about 100 MPa. The first concrete had a water/binder ratio of approximately 0.65, while the high-performance concrete had a water/binder ratio of 0.25. The microstructure of the hydrated paste of the 20 MPa concrete is wide open and it is possible to see pores surrounded by crystals of ettringite and large hexagonal crystals of portlandite – $Ca(OH)_2$ – and smaller needle-like hydrated calcium silicate crystals. On the other hand, it is quite difficult to perceive any crystal-like material within the microstructure of the high-performance concrete with a

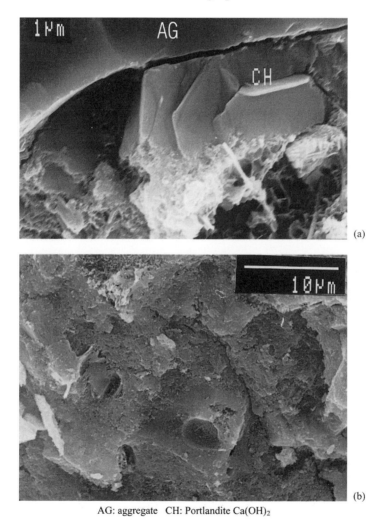

AG: aggregate   CH: Portlandite $Ca(OH)_2$

Fig. 17.2 Comparison of the microstructure of (a) a high water/cement ratio and (b) a low water/binder ratio concrete.

water/binder ratio of 0.25. The hydrated cement paste looks more like a quasi-amorphous, very compact paste. It is also possible to see the usual transition zone existing between the cement paste and the aggregate, which is very porous in the case of usual concrete.

This difference in the microstructure of high-performance concrete has two very important consequences for its compressive strength and permeability. The compressive strength of a high-performance concrete increases in a very spectacular manner and can reach 100 MPa, instead of 20 MPa for the higher water/binder ratio concrete. Moreover, the permeability of high-performance concrete is considerably lower than that of usual concrete. High-performance concrete is so impervious to water that it is almost impossible to measure its water permeability (Torrent and Jornet, 1991). But if the chloride ion permeability is measured according to ASTM C 1202-91, *Standard Test Method for Electrical Indication of Concrete's Ability to Resist Chloride Ion Penetration*, for a 0.45 water/binder ratio concrete, the chloride ion permeability would be found to be around 3000 to 5000 coulombs, but only between 100 and 500 coulombs for a high-performance concrete containing silica fume. This very low chloride ion permeability of high-performance concrete indicates that there still exists an interconnected continuous network of very fine and tortuous capillaries, but also that these capillaries are so fine that water cannot flow through them. In Figure 15.24, the kind of relationship that exists between the water/binder ratio and chloride ion permeability of different concretes is shown.

The penetration of chloride ions into high-performance concrete is so low that it is almost impossible to corrode unprotected reinforcing steel, even when accelerated corrosion tests are performed. It is therefore evident that the best way to protect reinforcing steel against corrosion is to begin to cover it with a very compact and impervious high-performance concrete having a sufficient thickness, and to prevent this high-performance concrete from cracking by curing it appropriately, rather than using epoxy-coated reinforcing steel with a very porous concrete open to all kinds of chemical aggression.

It has also been possible to demonstrate that high-performance concretes are less sensitive to carbonation and to external chemical attack, because their network of interconnected pores is much less developed than in usual concretes, as will be seen in sections 17.11 and 17.12.

17.4 DURABILITY AT A MICROSCOPIC LEVEL

Concrete is, and always will be, a porous material, either because too much water has been used during mixing in usual concretes, or in high-performance concretes owing to autogenous shrinkage when water hydrates cement particles. It is the total volume of this porosity, and its degree of interconnection, that will make a concrete more or less impervious to its environment, i.e. more or less durable.

What makes high-performance concrete more durable than usual concrete, at a microscopic level, is its reduced porosity because it is made with a reduced amount of water. As discussed earlier, this does not mean that all the water used during mixing is totally consumed by hydration, because as the microstructure of the concrete becomes more compact and less pervious, some water can remain trapped in certain locations and be unable to reach anhydrous cement particles located elsewhere.

Definitely concrete durability is closely linked to hydrated cement paste permeability, but we have seen in Chapter 15 that it is not always easy to characterize and measure the permeability of high-performance concrete. Recently, the theory of leaching (percolation), well known to rock mechanics scientists, has been applied by Bentz and Garboczi (1992) to the permeability of cement pastes. In such an approach, leaching theory is applied to mass transfer in a badly connected fissured network that is characterized by a critical diameter d_c, which represents the minimum diameter of geometrically continuous pores. It has been proposed that this critical diameter can be deduced from mercury porosimetry curves. This new approach seems to be promising.

It has been shown in Chapter 15 that air permeability measurements have also been explored to try to evaluate high-performance concrete durability, but researchers are facing problems in interpreting their results, owing to the difficulty of drying high-performance concrete samples and the modification of the microstructure of samples associated with this drying.

It is hoped that these experimental difficulties can be solved, partially or totally, so that simple experimental measures can be used to create models to predict the durability of a particular concrete in a particular environment.

Before ending this section, it should be emphasized that researchers should focus their studies more on the microstructure of the concrete skin, which is definitely different from that of the mass of the concrete, and is in the front line of aggression.

17.5 DURABILITY AT A MACROSCOPIC LEVEL

Concrete durability depends not only on the microstructural characteristics of the hydrated cement paste, of the transition zone and of the concrete skin, but also on the development of macrocracks which can originate from drying shrinkage, thermal gradients, autogenous shrinkage, thermal shrinkage and from overloading. Therefore it is very important to control all these aspects at which a high-performance concrete can develop some cracks, because these cracks are particularly damaging when they extend down to the reinforcing steel, which is then no longer protected by the covercrete.

As already pointed out in Chapter 12, proper curing of a high-performance concrete is crucial in order to avoid the development of cracks. The curing of high-performance concrete has to begin much earlier than the curing of usual concretes, not after 24 hours but rather just after the placing and finishing. If a sufficient amount of water is provided to a high-performance concrete just after its placing, plastic shrinkage can be completely prevented and autogenous shrinkage can be reduced drastically, at least in the covercrete, which is the most important and sensitive part of concrete elements to ensure a good durability. Generally speaking, thermal gradients, thermal shrinkage and drying shrinkage are easier to control than autogenous shrinkage.

It can never be emphasized enough that proper curing practices are essential to ensure the durability of a structural element. It is a pity to see some high-performance concrete structures built with a concrete that is very durable between its cracks!

17.6 ABRASION RESISTANCE

17.6.1 Introduction

Abrasion resistance can be a critical design factor in certain circumstances, or in certain parts of concrete structures: highway pavements, braking and accelerating zones at toll booths on highways, approaches to urban highway tunnels (Aïtcin and Khayat, 1992), and some parts of hydraulic structures such as spillways, stilling basins, and bridge piles subjected to the action of water containing solid particles in suspension. More recently, ice abrasion has become a major concern for designers of Arctic offshore platforms and bridges in Canada (Carino, 1983; Tadros *et al.*, 1996). The importance of abrasion resistance is such that some codes and specifications now require a minimum level of concrete strength for abrasion resistance under various traffic conditions (Guirguis, 1992). The recent prevalent use on highways of abrasive material instead of deicing salt under winter conditions will make the use of high-performance concrete more attractive because it will add pressure to pavement designers to specify concrete with a high resistance to abrasion.

Generally speaking, scant attention has been paid to concrete abrasion resistance; however, it is well established from tests and field experience that the abrasion resistance of usual concrete is a direct function of its compressive strength, as well as coarse aggregate volume and hardness (Liu, 1981; Neville, 1981; Ozturan and Kocataskin, 1987; Dhir, Hewlett and Chan, 1991; Mehta, 1993a). High-quality cement paste and abrasion resistant aggregates are essential to produce an abrasion resistant concrete (ACI 201.2R-92, 1993). Moreover, abrasion resistance is closely related to the surface properties of the concrete, which are in turn closely related to the finishing and curing of the surface (Fentress, 1973; Kettle and Sadezzadeh, 1987; ACI 201.2R-92, 1993). Since abrasion occurs at the surface, it is critical, when possible, to maximize the strength of the surface using high-quality aggregates, either by the dry-shake method or as part of a high-strength topping. In some extreme cases metallic aggregates are used to provide additional service life (ACI 201.2R-92, 1993).

The use of high-performance concrete can be expected to increase significantly the service life of concrete structures exposed to severe abrasion, because this type of concrete is strong and is usually made of high-quality aggregates. The question remains: how large is the benefit?

Testing abrasion resistance is complicated because there are several types of abrasive action, and no single test method has been found to be adequate for all conditions. It is therefore proposed, first, to present some laboratory and pilot project results on the abrasion resistance of high-performance concrete. The field applications cover pavements and hydraulic structures, including those subject to ice abrasion, but as they are quite recent it is difficult at the present time to give precise forecasts of the extended service life that can be expected from the use of high-performance concrete in such circumstances. So far, however, these experiences are promising.

17.6.2 Factors affecting the abrasion resistance of high-performance concrete

Two fundamental studies on the abrasion resistance of high-performance concrete have clearly shown that this type of concrete has a greater abrasion resistance than usual concrete.

Gjørv, Baerland and Ronning (1987, 1990) studied the abrasion resistance of different high-performance concretes using an accelerated load facility for full-scale testing of abrasion resistance of highway pavements exposed to heavy traffic with studded tyres. They compared the abrasion resistance of high-performance concrete with that of usual concrete, and with the top quality asphalt mixture used in Norway for highway pavements.

In this study, concretes having a compressive strength of up to 150 MPa, made with different types of coarse aggregate, were tested. Manufactured sand was also used to replace natural sand. The tests were carried out under dry and wet conditions (Figure 17.3).

Their main findings were as follows:

- by increasing the compressive strength from 50 to 100 MPa, the abrasion of the concrete was reduced by roughly 50%;
- at 150 MPa, the abrasion resistance of concrete is almost equal to that of massive granite;
- the service life of a 150 MPa concrete is 10 times longer than that of a top quality asphalt pavement under abrasion by studded tyres;

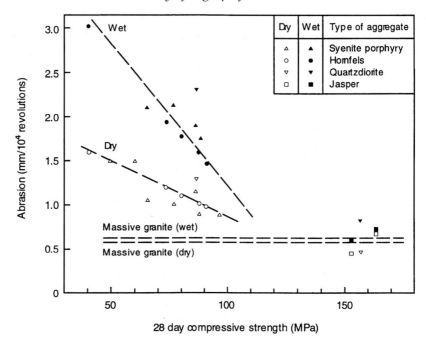

Fig. 17.3 Relationship between abrasion resistance and concrete strength (Gjørv, Baerland and Ronning, 1987).

- the abrasion resistance is decreased under wet conditions as compared with dry conditions, but the higher the compressive strength, the less the difference;
- the quality of aggregate was found to play a critical role, especially in concretes with a higher compressive strength;
- the quality of the sand was also found to influence the abrasion resistance of concrete. The partial substitution of a sand manufactured from a hard rock for natural sand was found to decrease compressive strength but to increase abrasion resistance.

In another study, Laplante, Aïtcin and Vézina (1991) explored to what extent abrasion resistance can be increased by decreasing the water/binder ratio and by the use of silica fume and of different types of coarse aggregate. The water/binder ratio varied from 0.48 down to 0.27, corresponding to compressive strengths between 30 and 90 MPa. Three types of aggregate were studied: a coarse grained metamorphic limestone, a quite hard finely grained dolomitic limestone, and a finely grained granite (Figure 17.4). It was found that the quality of the coarse aggregate is the most important factor affecting concrete abrasion resistance under the procedure of the ASTM C 779 standard (1993), with the water/binder ratio ranking second. Silica fume was found to increase the abrasion resistance of concrete, but was less significant than either the coarse aggregate quality or the water/binder ratio. The abrasion resistance is strongly influenced by the relative abrasion resistance of the coarse aggregate and of the mortar. In the case of the dolomitic limestone, which displayed similar abrasion resistance to the mortar under the selected abrasion testing method, this behaviour can result in an unsafe

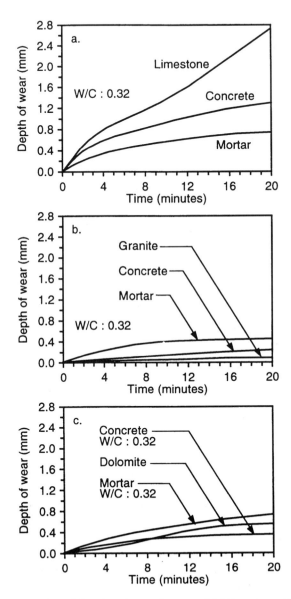

Fig. 17.4 Comparison of the abrasion resistance of (a) limestone, (b) granite and (c) dolomitic limestone with the respective mortars and concretes (Laplante, Aïtcin and Vézina, 1991). Reproduced by permission of ASCE.

476 The durability of high-performance concrete

skidding and slipping performance under wet conditions, owing to the fairly uniform and polished surface wear.

At a very low W/B ratio, high-performance concrete containing granite coarse aggregate can be nearly as abrasion resistant as the high-quality rock alone.

17.6.3 Pavement applications

Helland (1990) reports that high-performance concrete with a compressive strength of 80 to 100 MPa has been used in Norway in different projects: pavements, tunnels and bridge decks. In most of these field projects, it was possible to reproduce the very high abrasion resistance already found in laboratory mixes by Gjørv, Baerland and Ronning (1987). Owing to the success of these first field uses, several other projects have been completed in Norway, as seen in Table 17.1 (Aïtcin, 1992). However, finishing the pavement surface within the specified tolerances was not easy, and in many cases the pavement surface had to be partially ground with a grinding machine to achieve the desired profile and smoothness. The same situation has been faced in Québec.

Helland (1990) also mentions that the use of fibre-reinforced high-performance concrete on a bridge deck was not very easy, owing to the extreme harshness of the concrete, which also necessitated grinding of the surface to provide driving comfort. However, based on such a limited experience and service life, it has not yet been possible to develop a model for an accurate calculation of the possible reduction in the thickness of the pavement slab in service.

In Canada, an experimental high-performance concrete slab was built in 1993 at the Lafarge cement plant entrance, near Montreal, Province of Québec, and is under monitoring to evaluate its long-term performance.

Table 17.1 Use of high-performance concrete in pavement applications in Norway (Aïtcin, 1992)

Project	Length (km)	Compressive strength (MPa)[a]	Year
Smestad tunnel, Oslo	2 × 0.4	60–70	1983
E18, Klinestad-Tassebekk	5.6	75	1986
E6, Klett	0.2	90	1987
E18, Porsgrunn tunnel	0.88	75	1989
E69, Aalesund	2.1	80	1989
E6, Grillstadhaugen tunnel	1.1	75	1989
E6, Jessheim-Mogreina	9.0	75	1989
E18, Gulli-Holmene	6.6	75	1989–90
E6, Kroppan bridge	0.9	75	1990
E18, Holmene-Tassebekk	7.6	75	1991
Helgeland bridge	1.06	65	1991

[a] Measured on 100 mm cubes.

17.6.4 Abrasion–erosion in hydraulic structures

One thousand five hundred cubic metres of high-performance concrete was used to repair the Kinzua Dam stilling basin (Holland et al., 1986), which was severely damaged by the abrasion–erosion action of the Allegheny River in western Pennsylvania. A high-strength concrete made with silica fume and limestone aggregate was found to be suitable; it had a 28 day strength of 86 MPa. Diver inspection of the concrete in service indicates that the high-performance concrete is performing as intended (Holland, personal communication, 1994).

17.6.5 Ice abrasion

Inspection of concrete lighthouses in the Baltic Sea has shown that in some severely damaged lighthouses, the reinforcement has been torn out (Janson, 1986). With the development of Arctic activities, particularly in the field of oil exploration and exploitation, ice abrasion of concrete is becoming a very important design factor (Hoff, 1988). In spite of the fact that ice abrasion is more complex than simply the abrading of the concrete, there is no doubt that high-performance concrete will be more resistant to the abrading action of the ice floes. Different field studies are in progress in monitored sites to determine the extent of this improvement (Tadros et al., 1996). In parallel, different accelerated testing methods are being developed in order to predict the service life of a particular concrete exposed to ice abrasion.

17.7 FREEZING AND THAWING RESISTANCE

Although high-performance concretes were first developed in the early 1970s, it is surprising that such an important issue, from a durability point of view, is still a matter of controversy among the research community. In fact, in the early days of their development, the freezing and thawing resistance of low water/binder ratio concretes did not receive too much attention, because these concretes were in fact only used for their high strength in indoor applications. It was only when the use of low water/binder ratio concretes began to be explored for field applications, where repeated cycles of freezing and thawing were to be experienced during the service life of the concrete, that this question became very important. Quickly, the $64,000 question surfaced: is it necessary to entrain air in high-performance concrete to make it resistant to repeated freezing and thawing cycles? It is obvious that the addition of some air within a high-performance concrete conflicts with the search for a very high compressive strength. As already mentioned, a variation of 1% air content within the 4 to 8% range results in an opposite variation of 5% in compressive strength.

17.7.1 Freezing and thawing durability of usual concrete

It is well known that the best way to make usual concretes resistant to repeated freezing and thawing cycles is to entrain a network of closely spaced air bubbles. In fact, it is well established that it is not the total volume of entrained air that protects concrete against repeated freezing and thawing cycles, but rather the uniform distribution of this air as a multitude of very small bubbles (Powers, 1949). (As an example, it is obvious that a 60 l big hole, representing 6% of the volume of a 1 m concrete cube, will not protect it against repeated freezing and thawing cycles.) The space distribution of the air-void system is characterized by the so-called spacing factor, which represents very approximately the average half-distance between two adjacent air bubbles, i.e. the average maximum distance that water will have to flow to reach the closest air bubble. This spacing factor is measured using the ASTM C 457-90 (1993) *Standard Test Method for Microscopic Determination of Parameters of the Air-Void System in Hardened Concrete*.

Therefore, the measurement of the total air content in a fresh concrete as an assessment of its resistance to freezing and thawing can be questioned. In fact, however, the acceptance of the fresh concrete based on this measurement is good practice because when the manufacture, delivery and placing of concrete is well controlled, the spacing factor depends essentially on the total amount of entrained air. If the total amount of entrained air is constant, or varies slightly, so does the spacing factor.

In North America, the freeze–thaw durability of concrete is generally established using Procedure A (freezing and thawing in water) of the ASTM C 666 (1993) *Standard Test Method for Resistance of Concrete to Rapid Freezing and Thawing*. The nominal freezing and thawing cycle consists of cycling the temperature of the specimens between -17.8 and $+4.4\,°C$ in not less than 2 or more than 5 hours, not less than 25% of the time being used for thawing. The specimens used for this method should be not less than 76 mm, or more than 127 mm in width, depth or diameter, and not less than 279 mm or more than 406 mm in length. Cores or prisms cut from hardened concrete can be tested. This test is performed for 300 cycles, or until the relative dynamic modulus of elasticity reaches 60% of the initial modulus, whichever occurs first, though other limits may be specified.

Unless some other age is specified, the specimens should be removed from curing and freezing and thawing tests started when they are 14 days old. The specimens should be tested for fundamental transverse frequency at intervals not exceeding 36 cycles of exposure to the freezing and thawing cycles. However, in many cases, the measurement is taken once a week, every 50 cycles.

As this test is quite long, it takes about 8 weeks to be completed: 2 weeks for initial curing, plus at least 6 weeks of testing, if the testing is done at a rate of 50 cycles per week. This is not very practical for the

Freezing and thawing resistance

rapid acceptance of any concrete, and therefore from a practical point of view it has been necessary to find more rapid criteria to assess whether a given concrete is freeze–thaw resistant.

For usual concretes, a good correlation has been found between the results of ASTM C 666 and the value of the spacing factor. Research done at the Université Laval of Québec City has clearly established that for a usual concrete having a water/cement ratio within the 0.40 to 0.45 range, the critical spacing factor that makes it resistant to the 300 freezing and thawing cycles is around 400 μm. From a practical point of view, it is generally accepted that 200 μm represents a good design value for the spacing factor, although this value can be considered as too conservative. This low value of the spacing factor is not a critical issue, because in usual concrete it is relatively easy to produce a concrete having a satisfactory air-void system, and to preserve the spacing factor of this system until the placement of the concrete in the formwork (except when it is pumped, which may lead to higher spacing factors). This is why in Canada the A23.1 CSA standard, which states that, for usual concretes, the average spacing factor should not be greater than 230 μm, with no individual values greater than 260 μm, was accepted without any protest by the concrete industry.

Incidentally, it is very often forgotten that this spacing factor limiting value is necessary, but not sufficient, to make concrete resistant to freezing and thawing. To avoid surface scaling, usual concrete must also have a water/cement ratio no larger than 0.45 when the class of exposure is the most severe: frequent cycles of freezing and thawing in saturated conditions in the presence of chlorides.

17.7.2 Freezing and thawing durability of high-performance concrete

As previously mentioned, when the use of low water/binder concretes for outdoor field applications began to be considered, high-performance concretes were tested for freeze–thaw durability, which initiated a considerable controversy between researchers, engineers and specifiers (Hammer and Sellevold, 1990; Gagné, Pigeon and Aïtcin, 1991a,b; Pigeon et al., 1991; Marchand et al., 1993, 1996; Nili, Kamada and Katsura, 1993; Jacobsen, Marchand and Hornain, 1995; Jacobsen et al., 1995; Gagné, Boisvert and Pigeon, 1996; Pigeon, 1996; Jacobsen, Soether and Sellevold, 1997). The as yet unanswered questions are:

- what is the most appropriate testing method for assessing freeze–thaw durability?
- is the ASTM C 666 standard testing method appropriate for testing the freeze–thaw durability of high-performance concrete?
- is there any need for entrained air to make a high-performance concrete freeze–thaw resistant?

- if entrained air is necessary to protect high-performance concrete against repeated cycles of freezing and thawing, do the 200 μm or 230 μm limits still apply?
- how many field cycles does one ASTM C 666 cycle represent?
- how many cycles should a concrete pass before being declared freeze–thaw resistant?
- what is the influence of the freezing rate (which can vary by a factor of 2 in ASTM C 666) on the freeze–thaw durability?

The controversy related to the freeze–thaw durability of high-performance concrete, and more precisely the one on the critical spacing factor, has very important implications for the future use of high-performance concrete because it is not as easy as for usual concrete to entrain air in low water/binder mixtures. It is not as easy as for usual concrete to get a spacing factor lower than 200 μm, and it is extremely difficult to maintain such a low value of the spacing factor after pumping a high-performance concrete (Lessard, Baalbaki and Aïtcin, 1995; Pleau et al., 1995). The difficulty of maintaining an average spacing factor lower than 230 μm has led the Québec Ministry of Transportation in Canada to forbid the placing of high-performance concrete with a pump on all the high-performance concrete bridges that were built before 1996. Only some very limited field trials have been made where high-performance concrete was allowed to be pumped. Still, today, the Québec Ministry of Transportation specifies that high-performance concrete must be placed with buckets, though it is seriously considering relaxing this specification.

There are researchers who consider that, owing to the very low water/cement or water/binder ratio of high-performance concretes, it is not necessary that they contain entrained air to make them freeze–thaw resistant. Others consider that entrained air is absolutely necessary.

Considering the wide range of compositions and properties of high-performance concretes, there is no single answer to this controversy. Theoretical considerations, as well as available experimental data using the ASTM C 666 test method show that, for each binder, there seems to exist a critical water/binder ratio below which air entrainment is no longer required. For most Portland cements (with or without silica fume) this critical water/binder ratio lies between 0.25 and 0.30. It is thus sound engineering practice to use entrained air when the water/binder ratio is higher than 0.25. For high-performance concrete made with significant amounts of supplementary cementitious materials such as fly ash and slag, the limited available freeze–thaw data suggest caution, so such mixes should be tested according to ASTM C 666. But are ASTM C 666 and similar freezing and thawing cycle tests, which were developed to test the freeze–thaw durability to rapid cycling of usual concrete, well adapted to assess that of high-performance concrete? High-performance concrete usually self-desiccates somewhat, even when it is water-cured during the first 24 hours, and the result of the freeze–thaw test will therefore depend in good part on the absorption of water during the curing period, which will determine the degree of saturation at the beginning of the cycling. For certain high-performance concretes, capillary porosity at 24 hours might be so small and mostly disconnected that the water will not be able to penetrate and fully saturate the specimens to be tested. In the field, up to now, water curing is not usually strictly enforced, and thus only under special conditions will high-performance concrete freeze at close to saturation. Considering also the very low freezing rates generally observed under natural exposure conditions, the critical spacing factor value to assess freeze–thaw durability could be relaxed, in many cases, to the values found in the laboratory studies.

In order to support this approach, reference will be made to the work of Philleo (1986), who was asked by the Transportation Research Board to write a state-of-the-art report on this subject. At that time Philleo compiled 56 papers, some more fundamental than others, some very practical and descriptive only; but only very few, of course, were related to 'an examination of field structures because there are not enough documented durability studies of sufficient structures of sufficient age available for a statistically valid sample'. Philleo's main conclusion was that the resistance to freezing is directly dependent on the concrete capacity for, and its probability of containing, freezable water, a point that had been raised by Powers as early as 1955 (Powers, 1955). Philleo also expressed some reservations about the validity of the ASTM C 666 standard, which he found too severe to be of practical value because this test 'exposes specimens to freezing at an intermediate level of maturity with no opportunity for drying before the test and exposes them to a very rapid freezing cycle ... Although the test is excellent for assessing the frost resistance of mature specimens, for most typical exposures frost resistance might better be assessed by altering the age-at-test and specimen-conditioning requirements in C 666, or by replacing it with a critical dilatation test such as ASTM C 671'. As far as this conclusion is concerned, it must be admitted, 10 years later, that the same old controversy still haunts the research community about the relevance of ASTM C 666 standard as an appropriate means of evaluating the freezing and thawing resistance of concrete.

Philleo discussed how self-desiccation can create conditions favourable to increasing the freezing and thawing resistance of high-performance concrete, beyond pure strength considerations. For example, he noted that if an 83 MPa concrete 'is as strong as it is ever going to be, it may well have no capacity for freezable water', while an 83 MPa concrete that is 'on its way to 103 MPa concrete ... certainly has room for freezable water'.

Finally, he stated that 'high-strength concrete containing freezable water, like usual concrete containing freezable water, must contain entrained air, possibly with a different bubble spacing. If high-strength concrete is to be durable without entrained air, it must contain no freezable water' and 'the combinations of exposure and water/cement ratios not requiring intentionally entrained air should be delineated'.

The only research work known to the author that is related to the amount of freezable water present in high-performance concrete is that of Hammer and Sellevold (1990). Using low-temperature calorimetry, down to $-50\,°C$, they measured the formation of ice after a drying/resaturation treatment in eight non-air-entrained concretes having water/cement or water/binder ratios between 0.25 and 0.40. Four of these concretes contained silica fume, while the other four did not. The cement used in this experiment was a special high-strength cement designed for use in marine structures. The main difference between this cement and an ordinary Portland cement is a lower C_3S/C_2S ratio and a low alkali content. In the case of this high-strength cement, it was shown that the amount of freezable water was very low, below 10% of the total evaporable water, but also that these concretes performed very poorly when subjected to the ASTM C 666 test. A thermal incompatibility between aggregate and binder was suggested as a possible cause of this behaviour. On the other hand, using a more conventional Type III cement, Gagné, Pigeon and Aïtcin (1991a) found that a non-air-entrained high-performance concrete, having a water/binder ratio of 0.30, was freeze–thaw resistant just 24 hours after its manufacture, owing to the rapid self-desiccation that developed in this concrete, which resulted in the elimination of freezable water and the creation of a well-distributed, very fine porosity due to chemical contraction. This result was the opposite of other experimental results in which a conventional Type I cement was used at the same or an even lower water/binder ratio. Pigeon et al. (1991) concluded that 'for the Type III cement that was used (with or without silica fume) the limiting value of the water/binder ratio below which air entrainment is not necessary for adequate frost protection is most probably higher than 0.30, even after only one day of curing. For the Type I cement that was used (with or without silica fume) the results indicate that the limiting value is of the order of 0.25 ... It is thus clear that the cement has a large influence on the frost resistance of high-performance concretes'. Stark and Ludwig (1993) found the same.

It would be dangerous to extrapolate these results to blended cements. Unpublished results obtained at the Université de Sherbrooke have shown that non-air-entrained high-performance concretes having a 91 day compressive strength of 120 MPa were unable to pass the ASTM C 666 test when subjected to freezing and thawing cycles after only 14 days of curing. This behaviour is not surprising, considering Bob Philleo's remark about the influence of the percentage of the final strength reached when concrete prisms are subjected to freezing and thawing cycles.

Based on these results, Pigeon (1996) suggests the following guidelines: 'air entrainment is not needed for good frost resistance when the water/binder ratio is lower than 0.25 but it should be used when the water/binder ratio is equal to or higher than 0.30. For intermediate values, it is not possible to formulate a precise recommendation and tests should be carried out to determine the required air void parameters. These guidelines are of course only valid if good quality and low porosity aggregates are used'.

17.7.3 How many freeze–thaw cycles must a concrete sustain successfully before being said to be freeze–thaw resistant?

Usually, ASTM C 666 standard considers that if a usual concrete is able to sustain 300 freeze–thaw cycles with a durability factor greater than 60%, it is freeze–thaw resistant. In the case of some recent high-performance concrete specifications, the author has seen the number of cycles increased to 500, which is definitely tougher to meet, but the rationale behind this increase in the number of cycles has never been scientifically justified. It would be naive to think that any concrete with a spacing factor lower than 200 μm will resist indefinitely the rapid freeze–thaw cycles of Procedure A of ASTM C 666. Mother Nature teaches us that in the mountains even the strongest rocks end their life as grains of sand in the sea after being fragmented by freezing and thawing; it is only necessary to be patient to see this happen!

Figure 17.5 shows schematically some results obtained at the Université Laval and at the Université de Sherbrooke. Spacing factors are

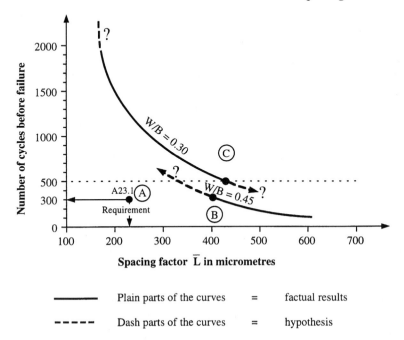

Fig. 17.5 Number of freeze–thaw cycles to reach failure as a function of the spacing factor for two concretes having a 0.45 and 0.30 water/binder ratio.

reported on the x-axis in micrometres and the number of cycles before failure are reported on the y-axis. The two curves corresponding to water/binder ratios of 0.45 and 0.30 shown in this figure are composed of a solid part and a dashed part. The solid parts correspond to factual results, while the dashed parts represent extrapolated results. Three points, A, B and C, have also been identified on this figure. Point A represents the CSA A23.1 requirement for freeze–thaw durability (which is in fact a requirement for deicing salt scaling resistance, as will be seen later). Point B is the critical value of the spacing factor found when exposing a concrete with a water/binder ratio of 0.45 to 300 cycles of Procedure A of ASTM C 666. The approximate 400 μm value shown in the figure is only indicative of the range in which this critical value is situated. (The number of cycles to reach failure as the spacing factor is decreased has not been measured, because in universities where the freeze–thaw durability of concrete is tested, the students who are on the waiting list do not give one a chance to explore this side of the curve!) Point C is the critical spacing factor for a high-performance concrete having a water/binder ratio of 0.30, similar to the concrete used to build the Confederation bridge in Canada. It is seen that the curve obtained for the water/binder ratio of 0.30 becomes very steep for spacing factors lower than 200 μm, indicating that the number of freeze–thaw cycles before failure increases rapidly.

17.7.4 Personal views

The author is strongly convinced that all high-performance concretes should contain a small amount of entrained air to improve their workability, placing and finishing. He also believes that this entrained air will efficiently protect most field concretes from freezing and thawing damage, under most field conditions. The author suggests:

- a total air content of 3.5 to 4.5%, which implies about 2 to 3% of actual entrained air on top of the 1.5 to 2.0% entrapped air that is present in any case in high-performance concrete;
- a spacing factor lower than 400 μm.

In order to validate this proposal, in 1993 the author was permitted by the City of Sherbrooke to carry out a field experiment during the reconstruction of the two entrances of one of the McDonald's restaurants in Sherbrooke (Lessard et al., 1994). One entrance was rebuilt with a non-air-entrained concrete (1.8% entrapped air) that did not pass the ASTM C 666 standard (spacing factor of 520 μm), and the other with a mix having the same composition except that it contained 6.8% air. The spacing factor of this air-entrained high-performance concrete was 120 μm, and it successfully withstood the 300 cycles of freezing and thawing of ASTM C 666.

The first part of the experiment was very convincing. The placing crew did not like the placing and finishing of the non-air-entrained concrete and much preferred that of the air-entrained high-performance concrete, though they still found it somewhat more difficult than the usual concrete they were used to placing. As far as the second part of the experiment is concerned, i.e. the comparison of the freeze–thaw durability of the two concretes, after four winters no difference in the behaviour of these two concretes can be noticed. During these four winters, the two high-performance concretes have been exposed to about 50 freeze–thaw cycles in the presence of deicing salts. Thus as far as the validity of the ASTM C 666 standard for assessing the durability of these two concretes under Sherbrooke field conditions is concerned, it will be necessary to wait until 1998, when two more years will have elapsed.

17.8 SCALING RESISTANCE

It is well known from both laboratory tests and field studies that air-entrainment is a necessary (but not sufficient) condition for good resistance to deicer salt scaling for usual concrete, as long as the spacing factor is of the order of 200 μm. This explains in particular why the critical spacing factor limit for freeze–thaw durability for usual concrete has been placed so low; with such a low critical spacing factor value both freeze–thaw durability and scaling resistance are obtained. One of the interesting aspects of the frost durability of high-performance concrete is that, regardless of the controversy concerning its resistance to rapid freezing and thawing cycles, the resistance of non-air-entrained high-performance concrete to scaling due to freezing in the presence of deicing salts has generally been found to be very good. This is clearly due to the very low porosity of this type of concrete, which reduces the rate of chloride penetration, and increases the time required for the saturation of the surface layer.

Being a surface phenomenon, finishing and curing take on a special importance, since it is possible to weaken considerably the microstructure of the skin of an otherwise adequately air-entrained concrete. A low water/binder ratio is equally important to reduce on the one hand salt penetration, and on the other full saturation of the surface layer. There is very little laboratory test data, and almost no long-term field data, on high-performance concrete resistance to deicing salt scaling. However, when high-performance concrete is tested according to the ASTM C 672 or Swedish SS 137244 standard for 50 or 56 cycles of freezing and thawing in the presence of a calcium chloride solution, it is clear that high-performance concrete shows a very good resistance to scaling (Peterson, 1986; Foy, Pigeon and Banthia, 1988; Hammer and Sellevold, 1990; Gagné, Pigeon and Aïtcin, 1991b; Zhang, Bouzoubaâ and Malhotra, 1997).

For resistance to deicing salt scaling, the type of cement seems to be less important if the C_3A content is kept low, below 6%. Hammer and Sellevold (1990) and Gagné, Pigeon and Aïtcin (1990, 1991b) found that the use of a Type III cement could bring a very rapid resistance to deicer salt scaling. Gagné, Pigeon and Aïtcin also found that at a water to cement ratio of 0.30 in concrete specimens, a curing period as short as 24 hours could provide resistance to 150 cycles.

From the laboratory data available, it seems that it is easier to produce a high-performance concrete resistant to deicing salts, by lowering the water/binder ratio, than to produce a freeze–thaw-resistant high-performance concrete. It is hoped that these laboratory results will rapidly be confirmed by field test data. Finally, it should be pointed out that, to the author's best knowledge, there are almost no data on the scaling resistance of high-performance concretes containing significant amounts of supplementary cementitious materials.

17.9 RESISTANCE TO CHLORIDE ION PENETRATION

Chloride ion penetration is probably the most devastating phenomenon for structures made of usual concrete. When chloride ions penetrate into the interstitial solution, they can react with unhydrated C_3A to form monochloroaluminates ($3CaO \cdot Al_2O_3 \cdot CaCl_2 \cdot 10H_2O$), which can modify the microstructure of concrete favourably or, more devastatingly, they can reach reinforcing steel rebars and corrode them quite rapidly. This corrosion usually begins with the development of a network of microcracks, which facilitate further the penetration of additional chloride ions, and ends up by the spalling of the covercrete owing to the expansive forces resulting from the formation of rust. This spalling of the covercrete exposes a new concrete surface to the action of chloride ions, and so on.

It has been accepted (as a first approximation) that the diffusion of chloride ions in a steady-state flow within concrete is governed by Fick's law. Fick's law states that the flux of ions is proportional to the concentration gradient:

$$J = - D_x \frac{\delta C}{\delta x}$$

where J corresponds to the flux and is measured in mol/cm^2/s; D_x is the coefficient of diffusion measured with a cell diffusion in which two solutions containing different amounts of ions are placed on either side of a 50 mm-thick disk of pure paste, expressed in cm^2/s; and C represents the chloride ion concentration in mol/cm^3. D_x is the coefficient that characterizes the ease with which the ions will diffuse through the porous concrete. This approach is beginning to be seriously contested (Maage, Helland and Carlsen, 1993; Delagrave, Marchand and Pigeon, 1996). However, whatever the theoretical approach followed, it has been found that the water/binder ratio is the major parameter that influences chloride ion penetration within concrete (Hansson *et al.*, 1985; Fukute et al., 1996). The explanation of this slowdown of chloride ion penetration in low water/binder ratio pastes is linked to the refinement of the capillary porosity (Hansson et al., 1985). In low water/cement ratio pastes, chloride ions are obliged to diffuse through a network of pores that is more tortuous and disconnected. Moreover, it seems that less chloroaluminates are formed within pastes with a low water/binder ratio in spite of the fact that they contain more unhydrated cement particles. Finally, different studies have shown that quite often supplementary cementitious materials tend to reduce significantly chloride ion mobility within concrete (Zhang and Gjørv, 1991; Maage, Helland and Carlsen, 1993; Fukute et al., 1996). The lower values of the coefficient of diffusion found in these cases have been attributed to refinement in the pore system. A reduction in the electronegativity of the internal pore surface has also been cited.

Recently Gagné, Aïtcin and Lamothe (1993) discussed the importance of the pH of the solution on the penetration of chloride ions. They found that this penetration increased drastically as the pH decreased.

17.10 CORROSION OF REINFORCING STEEL

The corrosion of reinforcing steel has been and still is the major deterioration mechanism for reinforced concrete structures. Corrosion of reinforcing steel occurs whenever the covercrete that was supposed to protect the reinforcing rebars against oxidation or rusting does not fulfil its role for various reasons: too high a water/binder ratio, poor curing practice or the absence of curing, misplacing of reinforcing rebars too close to the forms, ingress of chloride ions (Neville, 1995), bacteria, carbonation (Hansson and Sorensen, 1990), etc. The corrosion mechanism of steel in concrete is well known. The loss of passivation when the concrete pH decreases is the basic phenomenon leading to oxidation or rusting of the steel. Steel oxidation or rusting is accompanied by an increase in volume which first generates microcracks, or increases the number of pre-existing microcracks present in the covercrete owing to inadequate curing. These first microcracks make the penetration of the aggressive agent easier, so that steel corrosion becomes easier and finally leads to the spalling of the covercrete. When such a level of degradation is reached, not only are the reinforcing rebars exposed to direct corrosion, but also new concrete at a greater depth from the surface is exposed directly to the chloride ions that are the primary source of corrosion, so that the second level of reinforcing steel will face the same situation.

For too long a time, and still now, too many people have believed that corrosion of the reinforcing steel is an unavoidable phenomenon, inherent in reinforced concrete, which makes them buy any well-marketed alternative solution that is supposed to solve the problem. Various alternatives to replace ordinary steel reinforcing rebars by 'corrosion-free' rebars have been proposed; different anticorrosion admixtures are regularly advertised, with some companies even promoting the protection of the entire structure through an expensive cathodic protection.

In fact, to solve the corrosion problem of ordinary steel rebars from a materials point of view, two approaches can be followed:

1. Very pervious concretes continue to be specified, for which it is absolutely mandatory to specify so-called 'corrosion-free' rebars or even a sophisticated cathodic protection for the whole structure. In the author's opinion a concrete having a specified compressive strength lower than 30 MPa is a pervious concrete, whatever the environment in which it will be used. Moreover, it is well known and documented that such a concrete does not offer a proper protection from carbonation. When adopting such an easy but costly solution it is forgotten that the two major causes of steel corrosion are too high a water/binder ratio and poor curing practice. It is also forgotten that it is easy, and not at all expensive, to fight these two causes without the necessity of implementing such very costly and not-so-foolproof solutions. It has been repeated throughout this book that any concrete with a water/binder ratio greater than 0.50 is a concrete having a very open microstructure offering large avenues for the penetration of aggressive agents.
2. Impervious concretes are specified and well cured, and it is no longer necessary to specify so-called corrosion-free rebars; ordinary steel will be protected for the entire service life of the concrete structure. In the author's opinion a concrete having a specified water/binder ratio in the 0.30 to 0.35 range is impervious enough to provide proper corrosion protection to ordinary steel if there is an adequate thickness of covercrete and an adequate curing. The thickness of this covercrete has to be adjusted to the severity of the environment.

Of course, such a first step is necessary to solve the steel corrosion problem, but it is not sufficient: this impervious concrete must be well placed and cured in order to protect reinforcing steel from corrosion. When placing and curing are done properly there is no need to look at the use of corrosion-free rebars, there is no need for anticorrosion admixtures, and there is no need for cathodic protection. A low water/binder ratio, and adequate placing and curing of the concrete, are the 'price' that has to be paid to protect reinforced concrete structures against the corrosion of reinforcing steel. When looked at on the basis of service life and societal cost, this is not too high a price that has to be paid!

Most unfortunately this second solution has never been promoted with enough strength and conviction and brought to the attention of designers. With the recent breakthrough in concrete technology, this solution is not only feasible but also, in the author's opinion, the most practical, the most economical and the safest to solve the corrosion problem of reinforcing steel if it is properly understood and applied.

Before documenting this new approach let us see the hidden drawbacks of the so-called 'foolproof' solutions to corrosion problems that are presently in use:

- stainless steel rebars;
- galvanized rebars;
- epoxy-coated rebars;
- glass fibre-reinforced rebars.

17.10.1 Use of stainless steel rebars

Stainless steel is a steel that in a normal environment (that of a kitchen, for example) will not rust. However, all metallurgical engineers know that this kind of steel is still subject to corrosion in certain environments. Usually, stainless steel has a higher elastic modulus and a higher tensile strength than plain steel. However, its bond strength to concrete seems to be somewhat lower than that of ordinary steel.

Ping Gu *et al.* (1996) have shown recently that stainless steel rebars present an excellent behaviour in chloride-contaminated concretes, but the cost of stainless steel reinforcing rebars is very high. This very high price does not mean that stainless steel rebars should not be specified, particularly in some extreme cases, but when such an extreme solution has to be used all other solutions should have been studied carefully and eliminated in the particular aggressive environment to which the concrete will be exposed.

17.10.2 Use of galvanized rebars

It is well known that a thin coating of zinc can drastically improve the corrosion resistance of steel. This protective effect of zinc is easily explained by electrochemical considerations. In fact, it is the zinc that is oxidized instead of the steel. However, it must be understood that the steel is protected against corrosion only as long as there is still some zinc to be oxidized. When the zinc coating has been consumed, the steel is no longer protected and can then begin its normal corrosion process. Therefore the thickness of the zinc coating and the manufacturing process used to apply it (the cheap cold process, or the more expensive hot process) influence the life of the protection.

Galvanized rebars are often specified in architectural panels by architects who greatly dislike seeing their panels stained by rust within a few years, although galvanized rebars cost much more than ordinary steel rebars, depending on the thickness of their coating and the galvanizing process used.

Moreover, the use of galvanized rebars presents some practical drawbacks. First of all, if the protective coating is not thick enough, it can be cracked when rebars are curved, for example to form stirrups. A minute proportion of the steel is then exposed to corrosion and a galvanic cell is formed which can result in the sectioning of the steel rebar in a very short period of time. Second, galvanized steel should not be permitted to come into contact with plain steel, because a galvanic cell is immediately created, which results in the rapid corrosion of the plain steel.

Third, galvanized steel cannot be welded, because during welding the zinc coating evaporates (zinc vapour is particularly noxious and special care must be taken by the welder). The protective coatings that are sold to restore corrosion protection to the welding are more or less efficient owing to the practical difficulty of covering the complete welded area.

Fourth, when concrete is cast with galvanized rebars two chemical reactions can develop:

1. The zinc may react with the free lime contained in the cement and the lime liberated by the early hydration of the C_3S, to produce hydrogen gas. These hydrogen bubbles can stick to the reinforcing bar and decrease significantly the bond between concrete and the rebars.
2. The zinc hydroxide that is formed during the reaction between zinc and free lime is a strong retarding agent of C_3S hydration. Zinc reacts with C_3S to form a very impervious coating that delays further hydration of C_3S. In fact, zinc oxide is sometimes used as a retarding agent. This retarding effect is very localized in concrete to just around the galvanized rebars, the bulk of the concrete being unaffected, so it hardens normally. From a practical point of view this localized retarding effect can be detrimental, especially in the case of precast panels that are removed from their forms and handled very rapidly and very often quite roughly; therefore a poor bond is developed between the galvanized rebars and the concrete, which results in premature cracking of the concrete panel because tensile stresses are not carried by the reinforcing steel.

17.10.3 Use of epoxy-coated rebars

An epoxy coating can be used to isolate steel completely against any aggressive environment. This coating is usually some micrometres thick. Epoxy is a very strong polymeric material perfectly impervious to gas, to chloride and sulfate ions, to acids, etc., so epoxy-coated rebars have been aggressively promoted, with some success, as the solution to steel corrosion. However, there are some drawbacks when using epoxy-coated rebars that must not be ignored by those who would like to specify this kind of reinforcement.

First of all, the protection provided by the epoxy coating is perfect only as long as the epoxy coating is perfect. However, the protection provided by the epoxy coating is severely impaired if for any reason this coating spalls off or is scratched through (McHattie, Perez and Kehr, 1996). Unfortunately, spalling can occur for several reasons: epoxy is a brittle material that does not usually glue well on to steel; rough hauling on a construction site can make the coating spall from place to place. In addition, epoxy-coated rebars are not always used in their manufactured state: they may be bent and may be forced into constructed spaces, and these operations can create discontinuities in the epoxy coating. In order to restore the continuity of the epoxy coating all of the exposed surface of the steel must be absolutely covered by a new epoxy coating.

As in the case of the zinc coating, wherever the coating is missing becomes a preferred zone of corrosion, where the rebar can be corroded in a very short period because the corrosion process is concentrated at only a few places instead of being distributed all over the rebar surface.

Another major drawback is that hydrated cement paste does not bond chemically to epoxy-coated rebars as does plain steel. This drawback is serious, and anchor lengths have to be increased. In order to improve the stress transfer at the epoxy–steel interface, fine sand grains can be projected on to the epoxy surface before it has started to harden, to provide a rough external surface that will develop improved mechanical bonding.

Finally, the cost of epoxy-coated rebars is much higher than the price of ordinary steel rebars, and it is necessary to use more rebars to get the same level of stress transfer.

So, in the author's opinion, it is not apparent that the use of epoxy-coated rebars represents an economical and totally foolproof solution to steel corrosion. To reach a reasonable level of efficiency, great care must be taken to be sure that when the concrete is cast within the forms there is not a single place likely to be exposed to aggressive agents where the steel is not protected.

17.10.4 Use of glass fibre-reinforced rebars

Quite recently glass fibre-reinforced rebars have been introduced. These rebars, also called composite rebars, are made according to a pultrusion process. Usually, glass strands are glued together with a polyester resin. Glass fibre-reinforced rebars are available with different surface finishes: they can have a plain glossy finish, they can have a sand finish like the epoxy-coated rebars, or they can have an indentation pattern like that of ordinary steel rebars. One advantage of this type of rebar is that it does not conduct electricity, and therefore it can be used to reinforce concrete exposed to eddy currents or to induced Foucault's currents, as might be found in an aluminium processing plant.

This type of rebar is absolutely corrosion-free. The glass fibre reactivity with concrete does not represent a serious problem because the fibres are usually made of an alkali-resistant glass and because each fibre strand is embedded in polyester resin.

Apart from their reduced bond with concrete, glass fibre-reinforced rebars suffer from a second weakness: they are not ductile. Failure occurs instantaneously without any plastic strain, unlike steel. This can be a serious disadvantage for structural elements where ductility after peak value is a necessity to bring safety to the structure. Moreover, glass fibre-reinforced rebars cannot be bent and special attachments have to be used to make the complicated shapes that are sometimes necessary.

Finally, glass fibre-reinforced rebars cost more than ordinary steel rebars on a length basis. However, the simultaneous use of glass fibre-reinforced rebars in the part of a concrete structure most exposed to corrosion, with ordinary steel in the less exposed parts, can constitute an interesting alternative solution to the corrosion of reinforcing steel.

When all the advantages and disadvantages of these various alternatives to ordinary steel are weighed up, the second alternative to solve the steel corrosion problem, i.e. the use of high-performance concrete, has to be looked at seriously.

17.10.5 Effectiveness of the improvement of 'covercrete' quality

It has already been noted that the lower the water/binder ratio, the more impermeable the concrete. So, in the author's opinion, the first step to improving the resistance to corrosion of ordinary steel is to decrease the water/cement or water/binder ratio, and then to see whether it is still necessary to provide further corrosion protection. The efficiency of silica fume to protect steel reinforcement has also been advocated (Wolssiefer, 1991), as well as the use of permeable sheets and surface coatings (Sugawara et al., 1993; Tsukinaga, Shoya and Sugawara, 1993).

The efficiency of low water/cement or low water/binder concrete in protecting ordinary steel from corrosion has been qualitatively demonstrated by Gagné, Aïtcin and Lamothe (1993), using the following simple experimental set-up. This special set-up used for accelerated corrosion tests is based on the usual 'lollipop' test (Yuan and Chen, 1980; Ravindrarajah and Ong, 1987). The shape of the specimens was changed in order to expose simultaneously three groups of 12 bars embedded in three different types of concrete (W/B = 0.30, 0.43, 0.60). Three thicknesses of concrete cover were used for each concrete (20 mm, 40 mm and 60 mm). The three concrete slabs were arranged to form three faces of a square tank containing a 5% NaCl solution (Figure 17.6). A constant voltage of 5 V was maintained between each bar and the wire mesh. Each bar had its own switch to allow individual measurement of the current and to isolate it from the others if necessary. The evolution of the current as a function of time was recorded until the current became too high, indicating that the expansion of corrosion products had induced a crack through the concrete cover.

(a) *Time to initiate cracking*

Accelerated corrosion tests consist mainly of measuring, over a period of 200 days, the current between the reinforcing bar and a wire mesh placed

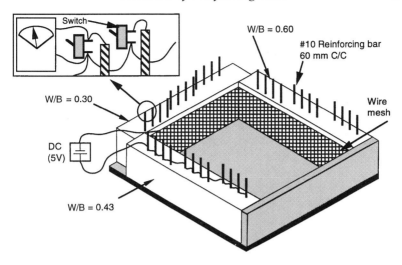

Fig. 17.6 Schematic view of the set-up for the accelerated corrosion test (Gagné, Aïtcin and Lamothe, 1993).

in a 5% NaCl solution near one face of the concrete slab. Typically, the current intensity slowly decreases to a minimum value, usually followed by a rapid increase. This sudden increase in the current can be explained by the opening of a crack produced by the expansion of corrosion products around the bar. The time required to obtain this sudden current increase is therefore described as the time required to initiate the first crack. The typical evolution of the current as a function of time and concrete cover for concrete with a water/binder ratio of 0.30 is shown in Figure 17.7. Each of these curves gives the average current measured on four contiguous bars. The time corresponding to the initiation of the first crack is also indicated.

The time required to initiate cracking, as a function of the water/binder ratio and concrete cover, is presented in Table 17.2. For a given water/binder ratio, a thicker concrete cover delays the initiation of the first crack. For a water/binder ratio of 0.30 and a concrete cover of 60 mm, no sudden increase of current was observed, indicating that no cracking occurred during the first 200 days. Low water/binder ratio concrete offers much greater protection against reinforcing-steel corrosion because of the lower and finer capillary porosity and lower conductivity which reduces the ionic exchange in the cement paste.

The chloride permeability of the concretes used for this rapid corrosion test were also measured on samples made from three separate batches. Table 17.2 gives the total charge and the initial current measured after 6 hours. It can be seen that increasing the water/binder ratio always yields an increase in the initial current and the total charge after 6 hours. Increasing the water/binder ratio from 0.30 to 0.43 results in an increase in the initial current and total charge by factors in the 1.7 to 2.0 range.

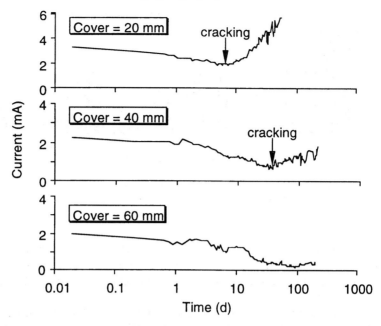

Fig. 17.7 Evolution of current as a function of time and thickness of cover for a water/binder ratio of 0.30 (Gagné, Aïtcin and Lamothe, 1993).

The same phenomenon occurs when the water/binder ratio is increased from 0.43 to 0.60, but the factor is about 1.4 (1.2 to 1.5). These results emphasize again the close relationship between the results of the accelerated corrosion test and the rapid chloride permeability test.

(b) Relationship between time to initiate cracking and initial current

Chloride permeability results have shown that the initial current can be used to obtain a good assessment of the total charge passed after 6 hours. Because rapid corrosion testing also involves measuring a current

Table 17.2 Time to initiate cracking, initial current and rapid chloride permeability

W/B	Accelerated corrosion test Concrete cover (mm)						Rapid chloride permeability	
	20		40		60		Total charge after 6 h (coulomb)	Initial current (mA)
	Initial current (mA)	Time to initiate cracking	Initial current (mA)	Time to initiate cracking	Initial current (mA)	Time to initiate cracking		
0.30	3.93	9 d	3.15	50 d	2.28	> 200 d	2100	70
0.43	8.88	1.5 d	5.50	40 d	4.03	80 d	4000	125
0.60	11.56	0.5 h	7.63	30 d	5.44	45 d	5200	155

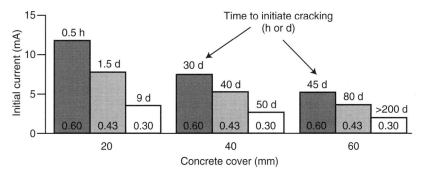

Fig. 17.8 Initial current and time to initiate cracking as a function of the water/binder ratio and thickness of covercrete (Gagné, Aïtcin and Lamothe, 1993).

passing through a concrete sample, it is of interest to see whether the initial current measured during the rapid chloride ion permeability test can also be used to assess the corrosion protection offered by different types of concrete.

Accelerated corrosion tests clearly show that a low initial current signals better protection against corrosion (Figure 17.8). For a given concrete cover, the reduction in water/binder ratio from 0.60 to 0.30 always yields a drop in the initial current and a significant increase in the time required to initiate cracking. Test results also demonstrate that, for a given concrete, better protection is obtained by increasing the concrete cover from 20 to 60 mm, which is again associated with a lower initial current.

From these results, it seems clear that even if the exact relationship between the initial current in the rapid chloride ion permeability test and the time to initiate corrosion were not fully understood, both parameters (water/binder ratio and thickness of cover) that reduce the initial intensity of current should lead to better protection against reinforcing-steel corrosion.

17.10.6 Concluding remarks

The corrosion resistance of reinforcing steel is closely associated with the quality of what is more and more frequently referred to as 'covercrete'. The critical importance of this very important part of the concrete has not always received the attention it should have. It is only recently that some researchers have begun to pay attention to it, but it is still not a universally accepted viewpoint. There are still many people in the construction industry who continue to think that to solve the reinforcing steel corrosion problem it is better to continue to use an inferior concrete and to focus instead on the use of corrosion-free rebars costing much more than ordinary steel rebars. The author does not share this attitude, in spite of the fact that he does not condemn the use of corrosion-free rebars in some very specific applications. Ordinary steel can be well protected against corrosion if it is covered by an appropriate thickness of very impervious covercrete. The thickness of the concrete and its impermeability should be adapted to the harshness of the environment to which concrete will be exposed.

The author also strongly believes that it would be a great mistake for engineers to decide that because the quality of the covercrete can be improved, savings can be achieved by decreasing the thickness of the concrete cover over the rebars. These engineers should be reminded that those that place or check the steel reinforcement are not always aware of the importance of the thickness of the covercrete in protecting reinforcing steel against corrosion.

17.11 RESISTANCE TO VARIOUS FORMS OF CHEMICAL ATTACK

Very few papers concerning the resistance of high-performance concrete to the attack of different chemicals have been published (Bentur and Ben-Bassat, 1993; Ravindrarajah and Mercer, 1993; Pigeon, 1996). The basic mechanism of chemical attack is well understood. The different hydration products are in equilibrium with the surrounding pore solution until the ingress of external ions modifies the solution equilibrium. A new equilibrium is then reached through the dissolution of certain hydration products as well as the formation of new products. In high-performance concrete, it has been clearly shown that the controlling factor for the kinetics of the process is the diffusivity of the paste; the lower the water/binder ratio, the lower the diffusivity. It must be pointed out that even pure water constitutes an aggressive environment for any concrete. Pure water will not only dissolve the portlandite crystals formed during cement hydration, but also the $C-S-H$ through a leaching process.

The improved resistance of high-performance concretes to chemical attack in comparison with that of usual concrete is essentially due to their much lower permeability. However, high-performance concrete cannot create a completely insurmountable barrier to those chemicals that attack usual concrete; it can only slow down the process of the chemical attack and increase significantly, in many cases, the durability of the concrete in that particular environment.

It must be remembered also that chemicals penetrate within any concrete element, not only through the capillary network, but also, and more easily, through the cracks that are present in the concrete, whatever their origin: thermal shrinkage, thermal gradients, plastic shrinkage, autogenous shrinkage, drying shrinkage, loads, etc. Therefore, any means of decreasing or eliminating concrete cracking should be considered seriously if a high-performance concrete has to face an aggressive environment.

17.12 RESISTANCE TO CARBONATION

High-performance concrete offers much better protection from carbonation than usual concrete owing to the compactness of its microstructure in the concrete skin area, as shown by C. Lévy (1992). Lévy compared the depth of penetration and speed of carbonation of the 60 MPa concrete used to build the Joigny bridge (Malier, Brazillier and Roi, 1991) with that of a 40 MPa concrete, neither of which contained any silica fume. In order to make this comparison, Lévy used an accelerated test for which 36 days in a special enclosure theoretically corresponded to 300 years of natural carbonation. He concluded that, 'in fact, even after 72 days of accelerated carbonation, there was no progress made by carbonation in the high-performance concrete, whereas the ordinary concrete was carbonated over several millimetres as from eighteen days'.

17.13 RESISTANCE TO SEA WATER

High-performance concrete has been used extensively to construct offshore platforms in the North Sea for over 20 years, and with great success. Here, too, it is the great compactness of high-performance concrete that is responsible for this improved durability. High-performance lightweight concretes are also performing very well in sea water.

This improved durability is of great interest when the exposure to sea water is combined with freeze–thaw cycles or to drying in the tidal and splash zones. Although we do not have a long track record of the behaviour of high-performance concrete in such a harsh environment, the first field results and the experimental results obtained at the Treat Island site (T.W. Bremner, personal communication, 1996) are very encouraging and will undoubtedly demonstrate the much greater superiority of high-performance concrete in such harsh environments.

17.14 ALKALI–AGGREGATE REACTION AND HIGH-PERFORMANCE CONCRETE

This section has been written in close co-operation with my colleague Marc-André Bérubé from the Université Laval who is a well-known AAR specialist (AAR stands for alkali–aggregates reaction).

17.14.1 Introduction

It has not been possible to find in the literature a set of experimental data or field data clearly demonstrating that AAR cannot develop in a high-performance concrete, in spite of the fact that some people are convinced of it. C. Desfossés has written an unpublished paper on this subject with such a conclusion (personal communication, 1996), but his demonstration was based on rather speculative considerations and lacked any experimental demonstration.

The same kind of reasoning will be adopted in this section and some observations made on usual concrete will be extrapolated to the case of high-performance concrete.

17.14.2 Essential conditions to see an AAR developing within a concrete

It is necessary to find simultaneously the three following conditions within a concrete if an AAR is to develop. First, the aggregate must be potentially reactive; second, the concentration of alkalis must be high enough within the interstitial solution of the concrete; and third, the concrete must be humid.

(a) Moisture condition and AAR

It is well known that for usual concrete the ambient external moisture conditions have a great influence on concrete expansion due to the AAR. At the Université Laval, Bérubé has shown that it is sufficient to have a 65% relative humidity at 38 °C to develop significant expansions greater than 0.04% per year when the aggregates are particularly reactive. In this particular case, the aggregates were a Spratt limestone and a Beauceville tuff. In the case of less reactive aggregates, such as Potsdam sandstone, the necessary relative humidity was 85%. It must be emphasized that these concretes had a water/cement ratio of 0.50, which is far from the water/cement ratio used when making high-performance concrete. It is most probable that ambient moisture conditions would not be so important in the case of sealed concretes or of high-performance concrete having a lower water/binder ratio, which is not as absorptive.

(b) Cement content, water/binder ratio and AAR

High-performance concrete usually has a high cement content and a low water/binder ratio, so it contains more alkalis per cubic metre and these alkalis have less residual water to dissolve; consequently this situation results in an interstitial solution that is richer in alkalis and has a high pH owing to the presence of a large amount of portlandite.

Experimental results obtained at the Université Laval by M. Landry (personal communication, 1994) tend to show that higher expansions were obtained for the lowest water/cement ratio that was tested (0.40 in this particular case). This conclusion has been reproduced several times for water/cement ratios of between 0.45 and 0.65. Owing to its higher alkali content and higher pH, the interstitial solution is more aggressive with the aggregates and as the porosity of a concrete decreases with its water/cement ratio there is less space to accommodate some expansion of the aggregates. These two phenomena are not counteracted by the lower permeability and the greater mechanical strength of the concrete.

17.14.3 Superplasticizer and AAR

High-performance concretes not only have a high cement content, they are also made using a high superplasticizer dosage. As most naphthalene and melamine superplasticizers are neutralized with soda (see Chapter 6), the use of a high dosage of superplasticizer increases the availability of alkali ions within the interstitial solution and therefore could result in an increase in the expected expansions, as found by Wang and Gillott (1989) and Buil and Zelwer (1989). However, a rapid calculation shows that the amount of alkali brought by the superplasticizer is very low compared with the amount brought by the cement, though it has been advanced that the alkali ions of the superplasticizer could be more concentrated near the surface of reactive aggregate. This explains why some researchers and agencies are specifically specifying a calcium naphthalene superplasticizer when they have the least doubt or when they don't know the potential reactivity of the aggregates used in the high-performance concrete.

17.14.4 AAR prevention

In order to fight, or at least to minimize, an AAR, it is necessary to act on one of the three conditions necessary for the development of an AAR. First, the aggregate specifications clearly indicate that high-performance concrete must be made with a potentially non-reactive aggregate. This specification is clearly put in force by the Ministry of Transportation of Québec and the City of Montreal, Quèbec, Canada. Second, a specification can limit to a maximum of 3 kg/m^3 the total mass of alkalis in the concrete, as required by the CSA A23.1 standard. In this calculation all the alkalis brought not only by the cement but also by the aggregates, the admixtures, the supplementary cementitious materials and even the water have to be added. Third, it should be recommended that the concrete surface be sealed using a good silane or siloxane, so that the concrete is allowed to dry progressively.

It has also been suggested adding some supplementary cementitious materials (silica fume, fly ash, slag) that are commonly used to fight efficiently or to reduce, when properly dosed, AAR expansions. In the case of silica fume, although some recent laboratory experiments tend to show that the protection provided by the silica fume does not last, it is well accepted that a replacement dosage of 7.0 to 8.5% is not high enough to fight AAR efficiently (M.A. Berubé, personal communication). However, it should be mentioned that in Iceland silica fumes have been used very efficiently for 20 years to fight AAR, though at the same time non-reactive basalt aggregates have started to be used instead of rhyolites.

17.14.5 Extrapolation of the results obtained on usual concrete to high-performance concrete

The major distinctions between a usual concrete and a high-performance concrete are the higher cement content, the lower water/cement ratio, the use of a superplasticizer and very often the use of silica fume. The first three characteristics tend to increase the development of an AAR, while the last one should reduce the risk. It is possible that the risk of seeing an AAR developing in a high-performance concrete could be lower than expected or even zero, because most of the mixing water is used during the hydration reaction so that there is practically no interstitial solution left within a high-performance concrete. Moreover, owing to its very low permeability and the poor connectivity of the capillary and pore system, there is a very low risk that a high-performance concrete will become saturated. Moreover, as a high-performance concrete has a greater tensile strength, it should better resist the expansive pressure developed by the gel. Therefore in order to confirm or dismiss these speculative conclusions it would be appropriate to undertake a research programme in the field of AAR in high-performance concrete.

17.15 RESISTANCE TO FIRE

One important feature of usual concrete as a construction material is its fire resistance, making concrete structures the safest in cases of conflagration. Very often concrete is used to protect structural steel against fire. In fact when exposed to a very high temperature the structural capacity of concrete is decreased (Neville, 1995), although not to a point that results in a collapse of the structure. Very often experts are asked to examine how much the structural capacity of concrete elements have been affected to determine if the concrete structure can be saved and what kind of additional support or retrofit it needs to recover its initial structural integrity fully (Felicetti *et al.*, 1996; Jensen, Opheim and Aune, 1996).

During a fire the porous nature of usual concrete, the amount of free water and the amount of more tightly bound water it contains contribute to the excellent behaviour of concrete in case of a fire.

17.15.1 Is high-performance concrete a fire-resistant material?

When high-performance concrete started to be used in the construction of high-rise buildings in Chicago, it was taken for granted that it possessed a good fire resistance, which was almost true because in the early

1970s the high-strength concretes that were used to build these high-rise buildings were only slightly stronger concretes than the ones for which there was no question about fire resistance.

However, when the water/binder ratio of high-performance concretes started to decrease drastically to about 0.30 or even lower, their microstructure became so dense that it was difficult to consider high-performance concrete as a porous material containing some free water, and it was necessary to consider seriously its resistance (Noumowe et al., 1996). This marked the rise of a controversy: is high-performance concrete as safe as usual concrete when considering its fire resistance (Sanjayan and Stocks, 1993; Chan, Peng and Chan, 1996; Khoylou and England, 1996)?

It is not easy to perform convincing fire tests on concrete, because the fire resistance has to be considered at two levels: the level of the material itself and the level of the structural element in which concrete is used in conjunction with steel.

From a material point of view, high-performance concrete is definitely not as porous as a usual concrete, as has been shown throughout this book. It contains practically no free water and when subjected to a very rapid increase in temperature it has a tendency to spall. However, steel reinforcing bars can stop the development of this spalling and help the structural element to keep enough residual strength to maintain the integrity of the whole construction.

For many years I have been joking with my students, saying that the best test to evaluate the actual resistance to fire of high-performance concrete was to wait until a fire happened in a high-rise building with columns built with a high-performance concrete. Now, when writing this section, we have had our first major fire in a structure built with high-performance concrete, but it did not happen at a positive elevation (with reference to the ground floor) but rather at a negative elevation in the Channel Tunnel linking France and England under the English Channel.

The Channel Tunnel fire has clearly demonstrated that the high-performance concrete used to make the tunnel segment was unable to resist the very high temperature reached at the centre of the fire. However, owing to the intensity of the fire, no other material could have resisted the very high temperature reached, except highly refractory materials. When moving away from the central zone of the fire, it is possible to distinguish different zones where the high-performance concrete has been exposed to lower and lower temperatures, and of course has been less and less affected by the exposure to very high temperature. As the author writes this paragraph it is too early to draw precise conclusions about the actual resistance of high-performance concrete, but it is certain that this fire will be very helpful in analysing the actual fire resistance of high-performance concrete.

Several attempts have already been proposed to improve the fire resistance of the material itself. One particularly clever proposal is to add some plastic fibres within the high-performance concrete so that these fibres melt in case of an elevation of the temperature of the concrete, creating channels through which the water vapour developed within the high-performance concrete could be released (Diederich *et al.*, 1993; Breitenbücker, 1996; Jensen and Aarup, 1996). The release of the built-up water vapour just behind the concrete surface exposed to the fire should significantly reduce the spalling tendency of high-performance concrete. In fact, the melting of the plastic fibres makes high-performance concrete a porous material like usual concrete, the good fire resistance of which is unquestionable.

While the author is writing this section, a major fire test programme on loaded high-performance concrete columns is under progress in Canada as a major research project of Concrete Canada, the Network of Centres of Excellence on High-Performance Concrete. This research programme involves the McGill University Centre (Professor Denis Mitchell) and the National Fire Test Laboratory of the National Research Council of Canada (Dr T.T. Lie). It is hoped that this major research programme will cast some light on the actual fire resistance of high-performance concrete columns.

17.15.2 The fire in the Channel Tunnel

(a) The circumstances

In spite of the fact that the *Train à Grande Vitesse* (TGV) operating in the Channel Tunnel has very sophisticated security systems, it was not foreseen that two incidents in which two contradictory orders were given to the engine man simultaneously could ever occur. As he was travelling within the tunnel, the engine man received one order asking him to speed up the train because there was a fire (150 trucks burn every year on the roads in France and as many in England), and a second one asking him to stop the train because a hydraulic jack used to support the access ramps had become detached. If it became blocked in the rails, this jack could destabilize the train and throw it off the tracks. When faced with these two contradictory orders, the engine man decided to stop the train. The consequence was that the Channel Tunnel faced its first fire.

Fortunately, this fire occurred in a part of the Channel Tunnel that cuts through a particularly impervious layer of white chalk. When the tunnellers crossed this layer of chalk, it was necessary to spray water continuously on the debris in order to protect the workers from the dust. But if the fire had occurred some kilometres further along, in a highly fractured zone, it could have been catastrophic because the tunnel could have been flooded and lost for ever. It would have been impossible to dry out the tunnel again.

(b) The damage

The fire occurred in a zone built by French contractors with high-performance concrete inverts. The fire lasted 9 hours, the maximum temperature reached was most probably above 1000 °C. The most damaged zone is about 30 m long. At the centre of this zone the concrete lining has completely disappeared over a length of several metres, and it was possible to see the white chalk behind the reinforcing bars of the inverts.

When leaving this zone where the concrete has disappeared totally, the remaining concrete surface looks like the dome of the Pantheon: disintegrated concrete is trapped behind the first reinforcing bars, but the centre is more deeply disintegrated. In the next zone, undestroyed concrete arches appear at the level of each joint where there was no reinforcement. This zone constitutes the last transition zone with the unattacked concrete.

The concrete sidewalk on which the TGV travels, which is made of usual concrete, has not suffered from the fire because it was never exposed to it. As soon as the alarm was given, a very strong flow of fresh air was activated from the floor in order to keep the smoke away from the head of the train, where the coaches holding the truck drivers were.

The high-performance concrete has been damaged in the shape of small chips, 30 mm thick and 50 to 75 mm long. The section where the fire took place has been cleaned and shotcreted (wet mix process).

17.16 CONCLUSIONS

Concrete durability, which was correlated with concrete strength for so long, can no longer be associated with it, owing to the technological progress in cement and concrete technology that has taken place in recent decades. Concrete durability is still associated with the concrete water/binder ratio, as it has always been, because the water/binder ratio represents the real concrete parameter that reflects its compactness and permeability to aggressive agents.

By using superplasticizers, it is now possible to make concretes with very low water/binder ratios, so they are as impervious as the most durable rocks. Moreover, as in high-performance concretes, as there is not enough water available to hydrate all the cement grains fully, there is a fair reserve of unhydrated cement particles. These unhydrated cement particles play a very important role. For example, if for any reason the environmental conditions are harsher than initially anticipated, or the concrete gets cracked, then they will hydrate as soon as any water penetrates the concrete. This means that the unhydrated cement has a potential for self-healing.

When engineers realize that concrete must no longer be specified in terms of compressive strength but rather in terms of its water/binder ratio, they will be able to solve the durability problem that has plagued concrete for so long. They will, however, have to make sure that this potentially durable concrete is placed and cured correctly.

At the turn of the 21st century, there is no excuse whatsoever for making concrete that is not durable. Low water/cement concrete can be easily made with or without entrained air. As for the extra MPa's offered by these durable concretes, designers will have to learn how to make the best use of them in their designs.

It must be emphasized that the concrete skin always plays a critical role in concrete durability and that all things that will result in the improvement of the concrete skin have to be implemented in order to improve the life cycle of concrete structures.

REFERENCES

ACI Manual of Concrete Practice, ACI Standard 201.2R-92 (1993) *Guide to Durable Concrete*, pp. 201.2R-12, 201.2R-16.

Aïtcin, P.-C. (1992) *The Use of High-Performance Concrete in Road Construction*, 27th Annual Congress of the AQTR, Sherbrooke, Québec, Canada, April, 12 pp.

Aïtcin, P.-C. and Khayat, K. (1992) *L'Utilisation des Bétons à haute Performance en Construction routière*, 27th Annual Congress of the AQTR Sherbrooke, Québec, Canada, April, 19 pp.

ASTM C 457-90 (1993) *Standard Test Method for Microscopic Determination of Parameters of the Air-Void System in Hardened Concrete* (1993) Annual Book of ASTM Standards, Section 4, Construction, Vol. 04.02, Concrete and Aggregates, ISBN 0-8031-1923-2, pp. 234–46.

ASTM C 666 (1993) *Standard Test Method for Resistance of Concrete to Rapid Freezing and Thawing*, Annual Book of ASTM Standards, Section 4, Construction, Vol. 04.02, Concrete and Aggregates, ISBN 0-8031-1923-2, pp. 326–31.

ASTM C 672 (1993) *Standard and Test Method for Scaling Resistance of Concrete Surface Exposed to Deicing Chemicals*, Annual Book of ASTM Standards, Section 4, Construction, Vol. 04.02, Concrete and Aggregates, ISBN 0-8031-1923-2, pp. 345–7.

ASTM C 779 (1993) *Standard Test Method for Abrasion Resistance of Horizontal Concrete Surfaces*, Annual Book of ASTM Standards, Section 4, Construction, Vol. 04.02, Concrete and Aggregates, ISBN 0-8031-1923-2, pp. 372–6.

ASTM C 1202-91 (1993) *Standard Test Method for Electrical Indication of Concrete's Ability to Resist Chloride Ion Penetration*, Annual Book of ASTM Standards, Section 4, Construction, Vol. 04.02, Concrete and Aggregates, ISBN 0-8031-1923-2, pp. 627–35.

Bentur, A. and Ben-Bassat, M. (1993) Durability of high-performance concrete in highly concentrated magnesium solution, in *Proceedings of the 6th International Conference on Durability of Building Materials and Components*, Omiya, Japan, Vol. 2, E & FN Spon, London, ISBN 0 419 18690 5, pp. 1021–30.

Bentur, A. and Jaegermann, G. (1991) Effect of curing and composition of the outer skin of concrete. *Materials in Civil Engineering*, **3**(4), November, 252–62.

Bentz, D.P. and Garboczi, E.J. (1992) Modelling the leaching of calcium hydroxide from cement paste: effects on pore space percolation and diffusivity. *Materials and Structures*, **25**, 523–33.

Breitenbücker, R. (1996) High strength concrete C 105 with increased fire resistance due to polypropylene fibers, in *Utilization of High Strength/High Performance Concrete*, Paris, ISBN 2 85978 2583, pp. 571–8.
Buil, M. and Zelwer, A. (1989) *Extraction de la Phase Liquide des Ciments Durcis*, Rapport du Laboratoire Central des Ponts et Chaussées, Paris, France, 15 pp.
Carino, N. (ed.) (1983) *Proceeding of the International Workshop on the Performance of Offshore Concrete Structures in the Arctic Environment*, National Bureau of Standards NBSIR 83-271, 67 pp.
Chan, S.Y.N., Peng, G.-F. and Chan, J.K.W. (1996) Comparison between high strength concrete and normal strength concrete subjected to high temperature. *Materials and Structures*, **29**, December, 616–19.
Dagher, H.J. and Kulendran, S. (1992) Finite element modeling of corrosion damage in concrete structures. *ACI Structural Journal*, **89**(6), 699–707.
Delagrave, A., Marchand, J. and Pigeon, M. (1996) Durability of high performance cement pastes in contact with chloride solutions, in *Utilization of High Strength/High Performance Concrete*, Paris, ISBN 2 85978 2583, pp. 479–88.
Dhir, R.K., Hewlett, P.C. and Chan, Y.N. (1991) Near surface characteristics of concrete: abrasion resistance. *Materials and Structures*, **24**(140), 122–8.
Diederich, U., Spitzner, J., Sandvik, M., Kepp, B. and Gillen, M. (1993) The behavior of high-strength lightweight aggregate concrete at elevated temperatures, in *High Strength Concrete*, ISBN 82 91341-00-1, pp. 1046–53.
Felicetti, R., Gambavora, P.G., Rosati, G.P., Corsi, F. and Gianuzzi, G. (1996) Residual strength of HSC structural elements damaged by hydrocarbon fire or impact loading, in *Utilization of High Strength/High Performance Concrete*, Paris, ISBN 2 85978 2583, pp. 579–88.
Fentress, B. (1973) Slab construction practices compared by wear tests. *Journal of the American Concrete Institute*, **70**(7), 486–91.
Foy, C., Pigeon, M. and Banthia, N. (1988) Freeze–thaw durability and deicer salt scaling resistance of a 0.25 water–cement ratio concrete. *Cement and Concrete Research*, **18**(4), 604–14.
Fukute, T., Hamada, H., Mashimo, M. and Watanabe, Y. (1996) Chloride permeability of high strength concrete containing various mineral admixtures, in *Utilization of High Strength/High Performance Concrete*, Paris, ISBN 2 85978 2583, pp. 489–98.
Gagné, R., Aïtcin, P.-C. and Lamothe, P. (1993) Chloride-ion permeability of different concretes, in *Proceedings of the 6th International Conference on Durability of Building Materials and Components*, Omiya, Japan, Vol. 2, E & FN Spon, London, ISBN 0 419 18690 5, pp. 1171–80.
Gagné, R., Boisvert, A. and Pigeon, M. (1996) Effect of superplasticizer dosage on mechanical properties, permeability and freeze–thaw durability of high-strength concretes with and without silica fume. *ACI Materials Journal*, **93**(2), 111–20.
Gagné, R., Pigeon, M. and Aïtcin, P.-C. (1990) *Deicer Salt Scaling Resistance of High-Performance Concrete*, ACI SP-122, pp. 29–37.
Gagné, R., Pigeon, M. and Aïtcin, P.-C. (1991a) *The Frost Durability of High-Performance Concrete*, Second Canada/Japan Workshop on Low Temperature Effects on Concrete, Ottawa, pp. 75–87.
Gagné, R., Pigeon, M. and Aïtcin, P.-C. (1991b) *Deicer Salt Scaling Resistance of High Strength Concretes Made with Different Cements*, ACI SP-126, pp. 185–99.
Gjørv, O.E., Baerland, T. and Ronning, H.R. (1987) High strength concrete for highway pavements and bridge decks, in *Proceedings of the Symposium on Utilization of High Strength Concrete*, Stavanger (ed. I. Holland *et al.*), Tapir, N-7034 Trondheim NTH, Norway, ISBN 82-519-0797-7, pp. 111–22.

Gjørv, O.E., Baerland, T. and Ronning, H.R. (1990) Abrasion resistance of high-strength concrete pavements. *Concrete International*, 12(1), January, 45–8.

Guirguis, S. (1992) *High-Strength Concrete*, published jointly by the Cement and Concrete Association of Australia and the National Ready Mixed Concrete Association of Australia, ISBN 09471 132511, pp. 1–20.

Halvorsen, G.T. (1993) Concrete cover. *Concrete Construction*, **38**(6), June, 427–9.

Hammer, T.A. and Sellevold, E.J. (1990) *Frost Resistance of High-Strength Concrete*, ACI SP-121, pp. 457–87.

Hansson, C.M. and Sorensen, B. (1990) The threshold value of chloride concentration for the corrosion of reinforcement in concrete, in *Corrosion of Steel in Concrete* (ed. N.S. Berbe), ASTM STP-1065, pp. 3–16.

Hansson, C.M., Strunge, H., Markussen, J.B. and Frolund, T. (1985) The effect of the cement type on the diffusion of chloride. *Nordic Concrete Research*, No. 4, pp. 70–80.

Helland, S. (1990) High strength concrete used in highway pavements, in *Proceedings, 2nd International Symposium on Utilization of High Strength Concrete* (ed. W. Hester), ACI SP-121, pp. 757–66.

Ho, D.W.S. and Cao, H.T. (1993) Concrete durability – strength or performance criteria, in *Proceedings of the 6th International Conference on Durability of Building Materials and Components*, Omiya, Japan, Vol. 2, E & FN Spon, London, ISBN 0 419 18690 5, pp. 856–64.

Hoff, G.C. (1988) *Resistance of Concrete to Ice Abrasion – A Review*, ACI SP-109, pp. 427–55.

Holland, T.C., Krysa, A., Luther, M.D. and Liu, T.C. (1986) *Use of Silica-Fume Concrete to Repair Abrasion. Erosion Damage in the Kinzua Dam Stilling Basin*, ACI SP-91, Vol. 2, pp. 841–63.

Jacobsen, S., Marchand, J. and Hornain, H. (1995) SEM observations of the microstructure of frost deteriorated and self-healed concretes. *Cement and Concrete Research*, **25**(8), 1781–90.

Jacobsen, S., Soether, D.H. and Sellevold, E.J. (1997) Frost testing of high strength concrete: frost/salt scaling at different cooling rates. *Materials and Structures*, **30**, January–February, 33–42.

Jacobsen, S., Gran, H.G., Sellevold, E. and Bakke, J.A. (1995) High strength concrete – freeze–thaw testing and cracking. *Cement and Concrete Research*, **25**(8), 1775–80.

Janson, J.-E. (1986) Swedish investigations of ice–structure interaction in the Baltic, in *Proceedings, International Workshop on Concrete for Offshore Structures*, St John's Newfoundland, Canada, 10–11 September, 5 pp (available from CANMET, Ottawa, Canada).

Jensen, B.C. and Aarup, B. (1996) Fire resistance of fibre reinforced silica fume based concrete, in *Utilization of High Strength/High Performance Concrete*, Paris, ISBN 2 85978 2583, pp. 551–60.

Jensen, J.J., Opheim, E. and Aune, R.B. (1996) Residual strength of HSC structural elements damaged by hydrocarbon fire on impact loading, in *Utilization of High Strength/High Performance Concrete*, Paris, ISBN 2 85978 2583, pp. 589–98.

Katayama, K. and Kabayashi, S. (1991) Study on concrete surface microcraks when using permeable forms. *Transactions of JSCE*, **156**(433), 161–77.

Kettle, R. and Sadezzadeh, M. (1987) *The Influence of Construction Procedures and Abrasion Resistance*, ACI SP-100, pp. 1385–1410.

Khoylou, N. and England, G.L. (1996) *The Effect of Elevated Temperatures on the Moisture Migration and Spalling Behaviour of High Strength and Normal Concretes*, ACI SP-167, pp. 263–8.

Kreijger, P.C. (1987) The 'skins' of concrete – research needs. *Magazine of Concrete Research*, **3**(140), September, 122–3.

Kumagai, T., Arioka, M. and Tanabe, D. (1991) Experimental study on an improved zone in concrete by a permeable form. *Transactions of JSCE*, **15**(433), 215–32.

Laplante, P., Aïtcin, P.-C. and Vézina, D. (1991) Abrasion resistance of concrete. *Journal of Materials in Civil Engineering*, **3**(1), Febuary, 19–28.

Lessard, M., Baalbaki, M. and Aïtcin, P.-C. (1995) *Effect of Pumping on Air Characteristics of Conventional Concrete*, Transportation Research Board Record 1532, ISBN 0 3909 05904 6, pp. 9–14.

Lessard, M., Dallaire, E., Blouin, D. and Aïtcin, P.-C. (1994) High-performance concrete speeds reconstruction for McDonald's. *Concrete International*, **16**(4), September, 47–50.

Lévy, C. (1992) Accelerated carbonation: comparison between the Joigny Bridge high performance concrete and an ordinary concrete, in *High Performance Concrete – From Material to Structure* (ed. Y. Malier), E & FN Spon, London, pp. 305–11.

Liu, T.C. (1981) Abrasion resistance of concrete. *Journal of the American Concrete Institute*, **78**(5), pp. 341–50.

Maage, M., Helland, S. and Carlsen, J.E. (1993) Chloride penetration in high-performance concrete exposed to marine environment, *in High-Strength Concrete, 1993*, ISBN 82-91341-00-1, pp. 838–46.

Malier, Y., Brazillier, D. and Roi, S. (1991) The bridge of Joigny, a high-performance concrete experimental bridge, *Concrete International*, **13**(5), 40–42.

Marchand, J., Gagné, R., Pigeon, M., Jacobsen, S. and Sellevold, E.J. (1993) The frost durability of high-performance concrete, in *Concrete Under Severe Conditions: Environment and Loading*, E & FN Spon, London, ISBN 0 419 19850 4, pp. 273–88.

Marchand, J., Gagné, R., Jacobsen, S., Pigeon, M. and Sellevold, E.J. (1996) La résistance au gel des bétons à haute performance. *Canadian Journal of Civil Engineering*, **23**(5), 1070–80.

McHattie, J.S., Perez, J.L. and Kehr, J.A. (1996) Factors affecting cathodic disbondment of epoxy coatings for steel reinforcing bars. *Cement and Concrete Composite*, **18**(2), 93–103.

Mehta, P.K. (1991) *Durability of Concrete – Fifty Years of Progress?* ACI SP-125, pp. 1–3.

Mehta, P.K. (1993a) *Concrete Microstructure, Properties and Materials*, 2nd edn, McGraw-Hill, New York, ISBN 0 07 041344 4, 548 pp.

Mehta, P.K. (1993b) *Concrete Technology at the Crossroads – Problems and Opportunities*, ACI SP-144, pp. 1–30.

Mehta, P.K., Schiessl, P. and Raupach, M. (1992) *Performance and Durability of Concrete Systems*, 9th International Congress on the Chemistry of Cement, New Delhi, India, Vol. 1, pp. 559–71.

Moore, J.F.A. (1993) Harmonised European design guidance for durability of concrete, in *Proceedings of the 6th International Conference on Durability of Building Materials and Components*, Omiya, Japan, Vol. 2, E & FN Spon, London, ISBN 0 419 18690 5, pp. 1413–20.

Neville, A.M. (1981) *Properties of Concrete*, 3rd edn, Longman, London, p. 511.

Neville, A.M. (1987) *Why we Have Concrete Durability Problems*, ACI SP-100, pp. 21–30.

Neville, A.M. (1995) Chloride attack of reinforced concrete: an overview, *Materials and Structures*, **28**(176), March, 63–70.

Nili, M., Kamada, E. and Katsura, O. (1993) An evaluation study on frost resistance of high strength concrete, in *Proceedings of the 6th International Conference on Durability of Building Materials and Components*, Omiya, Japan, Vol. 1, E & FN Spon, London, ISBN 0 419 18680 8, pp. 223–30.

Norberg, P., Sjöström, C., Kucera, V. and Rendahl, B. (1993) Microenvironment measurements and materials degradation at the Royal Palace in Stockholm, in *Proceedings of the 6th International Conference on Durability of Building Materials and Components*, Omiya, Japan, Vol. 1, E & FN Spon, London, ISBN 0 419 18680 8, pp. 589–97.

Noumowe, A.N., Clastres, P., Delvicki, G. and Costaz, J.-L. (1996) Thermal stresses and water vapour pressure of high-performance concrete at high temperature, in *Utilization of High Strength/High Performance Concrete*, Paris, ISBN 2 85978 2583, pp. 561–70.

Ozturan, T. and Kocataskin, F. (1987) Abrasion resistance of concrete as a two-phase composite material. *International Journal of Cement Composites and Lightweight Construction*, **9**(3), 169–76.

Parrott, L.J. (1992) Water absorption in cover concrete. *Materials and Structures*, **25**(149), June, 284–92.

Peterson, P.E. (1986) The influence of silica fume on the salt frost resistance of concrete, in *Proceedings of the International Seminar on Some Aspects of Admixtures and Industrial By-Products on the Durability of Concrete*, Götenburg, Sweden, 10 pp.

Philleo, R.E. (1986) *Freezing and Thawing Resistance of High Strength Concrete*, National Cooperative Highway Research Programme Synthesis of Highway Practice 129, Transportation Research Board, National Research Council, Washington, DC, 31 pp.

Philleo, R.E. (1989) Working to make efficient, safe and durable concrete. *Concrete International*, **II**(9), 29–31.

Pigeon, M. (1996) The durability of HS/HPC, Third General Report on Durability, in *4th International Symposium on the Utilization of High Strength/High Performance Concrete*, Paris, Vol. 1, ISBN 2-85978-257-5, pp. 39–45.

Pigeon, M., Gagné, R., Banthia, N. and Aïtcin, P.-C. (1991) Freezing and thawing tests of high-strength concretes. *Cement and Concrete Research*, **21**, 844–52.

Ping Gu, Elliott, S., Beaudoin, J.J. and Arsenault, B. (1996) Corrosion resistance of stainless steel in chloride contaminated concrete. *Cement and Concrete Research*, **26**(8), 1151–6.

Pleau, R., Pigeon, M., Lamontagne, A. and Aïtcin, P.-C. (1995) *Influence of Pumping on the Characteristics of the Air-Void System of High-performance Concrete*, presented at the 74th Annual Meeting of the Transportation Research Board, Washington, DC, January, 19 pp.

Pomeroy, D. (1987) *Concrete Durability from Basic Research to Practical Reality* (ed. J.M. Scanlon), ACI SP-100, Detroit, pp. 111–30.

Powers, T.C. (1949) The air requirement of frost resistant concrete, in *Proceedings of the Highway Research Board*, Vol. 29, pp. 184–211.

Powers, T.C. (1955) Basic considerations pertaining to freezing and thawing tests. *Proceedings, ASTM*, **55**, 403–10.

Ravindrarajah, R. and Ong, K. (1987) *Corrosion of Steel in Concrete in Relation to Bar Diameter and Cover Thickness*, ACI-SP 100, pp. 1667–77.

Ravindrarajah, R.S. and Mercer, C.M. (1993) Sulphuric acid attack on high strength concrete, in *Proceedings of the 6th International Conference on Durability of Building Materials and Components*, Omiya, Japan, Vol. 1, E & FN Spon, London, ISBN 0 419 18680 8, pp. 326–34.

Rostam, S. and Schissl, P. (1993) Next-generation design concepts for durability and performance of concrete structures, in *Proceedings of the 6th International Conference on Durability of Building Materials and Components*, Omiya, Japan, Vol. 2, E & FN Spon, London, ISBN 0 419 18690 5, pp. 1403–12.

Sanjayan, G. and Stocks, L.J. (1993) Spalling of high-strength silica fume concrete in fire. *ACI Materials Journal*, **90**(2), March–April, 170–73.

Skalny, J.P. (1987) *Concrete Durability: a Multibillion-Dollar Opportunity*, Nat'l Mat. Solv. Board Comm. Eng. and Tech. Systems, National Research Council, National Academy Press, NMAB-437, Washington, DC.

Stark, J. and Ludwig, H.-M. (1993) The influence of the type of cement on the freeze–thaw/freeze-deicing salt resistance of concrete in *Concrete Under Severe Conditions: Environment and Loading* (eds K. Sakai, N. Banthia and O.E. Gjørv), E & FN Spon, London, ISBN 0 419 19850 4, Vol. 1, pp. 245–54.

Sugawara, T., Saeki, N., Shoya, M. and Tsukinaga, Y. (1993) Frost resistance of concretes with permeable sheets and surface coating, in *Proceedings of the 6th International Conference on Durability of Building Materials and Components*, Omiya Japan, Vol. 1, E & FN Spon, London, ISBN 0 419 18680 8, pp. 497–506.

Tadros, G., Combault, J., Bilderbeek, D.W. and Fotinos, G. (1996) *The Design and Construction of the Northumberland Strait Crossing Fixed Link in Canada*, 15th Congress of IABSE, Copenhagen, Denmark, 16–20 June, 24 pp.

Torrent, R.J. and Jornet, A. (1991) *The Quality of the Concrete of Low-Medium and High-Strength Concretes*, ACI SP-126, pp. 1147–61.

Tsukinaga, Y., Shoya, M. and Sugawara, T. (1993) Air void character and pull-off strength in the near surface of concrete using permeable sheets, in *Proceedings of the 6th DBMC International Conference*, Omiya, Japan, Vol. 1, E & FN Spon, London, ISBN 0 419 18680 8, pp. 507–16.

Wang, H. and Gillott, J.E. (1989) The effect of superplasticizer on alkali–silica reactivity, in *Proceedings of the 8th International Conference on AAR*, Kyoto, Japan (eds K. Okada, S. Nishibayashi and M. Kawamura), Society of Materials Science of Japan, Kyoto, pp. 187–92.

Wischers, G. (1984) *The Impact of the Quality of Concrete Construction on the Cement Market*, Report to Holderbank Group, No. DIR 84/8448/4.

Wolssiefer, J.T. (1991) *Silica Fume Concrete: A Solution to Steel Reinforcement Corrosion in Concrete*, ACI SP-126, pp. 527–58.

Yuan, R.L. and Chen, W.F. (1980) *Behavior of Sulfur-Infiltrated Concrete in Sodium Chloride Solutions*, ACI SP-65, pp. 292–307.

Whiting, D. (1984) *In Situ Measurements of the Permeability of Concrete to Chloride Ions*, ACI SP-82, pp. 501–24.

Zhang, M.H. and Gjørv, O.E. (1991) Effect of silica fume on pore structure and chloride diffusivity of low porosity cement paste interface. *Cement and Concrete Research*, **21**(6), November, 1006–14.

Zhang, M.H., Bouzoubaâ, N. and Malhotra, V.M. (1997) *Resistance of Silica fume Concrete to Deicing Salt Scaling. A Review*, ACI 1997 International Conference on High-Performance Concrete: Design and Materials and Recent Advances in Concrete Technology, December, Kuala Lumpur, 30 pp.

CHAPTER 18

Special high-performance concretes

18.1 INTRODUCTION

Although the development of high-performance concrete is quite recent, there are already high-performance concretes that have gained the qualification of 'special'. As is the case in usual concrete, these special high-performance concretes depart in some way from the standard composition of more usual high-performance concretes. For example, a special high-performance concrete could have been made with a heavyweight aggregate to increase its unit mass or, alternatively, with a lightweight aggregate to decrease it, or fibres could have been added to increase its toughness, its impact resistance or its ductility. High-performance concrete can be confined in a thin, highly resistant (in tension) envelope in order to increase further its compressive strength and ductility. The placing method could be different from the usual placing method, as in the case of roller-compacted high-performance concrete or self-levelling high-performance concrete. It is certain that there are and there will be other types of special high-performance concrete with which the author is not familiar, but all these special high-performance concretes have one important thing in common: they have a very low water/binder ratio, a high compacity and consequently a great durability.

In this book a distinction is made between these special high-performance concretes and the ultra high-strength composite materials made of Portland cement that are discussed in the next chapter. These 'concretes' are actually too different from 'usual' or 'special' high-performance concrete to be treated in this chapter.

The first special high-performance concretes that will be presented will be air-entrained high-performance concretes. In spite of the fact that several references have been made to this type of high-performance concrete in the different chapters of this book, it was decided that it would be a good thing to regroup most of the disseminated information about air-entrained high-performance concrete in a comprehensive section devoted only to this subject.

Air-entrained concretes have been developed through the years to improve the freeze–thaw durability and the scaling resistance to deicing salt of usual concretes, and in a northern country with such severe climatic conditions as Canada it is not necessary any more to promote the use of entrained air in usual concrete (Lessard, Gendreau and Gagné, 1993; Mitchell *et al.*, 1993; Hoff and Elimov, 1995; Blais *et al.*, 1996; Tadros *et al.*, 1996). It was quite natural that there was no questioning about the necessity of designing high-performance concrete structures exposed to freezing and thawing with air-entrained high-performance concrete (Hoff and Elimov, 1995; Tadros *et al.*, 1996), in spite of the fact that the presence of entrained air would result in a lower compressive strength.

18.2 AIR-ENTRAINED HIGH-PERFORMANCE CONCRETE

18.2.1 Introduction

In the preceding chapter it was shown that as long as the freeze–thaw durability of high-performance concrete is evaluated according to Procedure A of ASTM C 666 standard (freezing and thawing in water), high-performance concrete having a water/binder ratio over 0.25 (corresponding to a 28 day compressive strength of between 75 and 125 MPa) must in the present state of the art contain entrained air to pass the requested 300 freezing and thawing cycles successfully (Gagné, Pigeon and Aïtcin, 1991).

It is not so easy to make a non-air-entrained high-performance concrete with a very low water/binder ratio, so adding an adequate air bubble system to make the concrete pass the ASTM C 666 standard test method represents another challenge and difficulty. However, field experience has shown that it is possible to make air-entrained high-performance concretes containing a stable air bubble system that protects the concrete from repeated freezing and thawing cycles. As is the case for non-air-entrained concrete, the materials used to make this concrete have to be carefully selected.

It is obvious that introducing air bubbles in the mortar of a high-performance concrete results in a decrease of its compressive strength. During the construction of different bridges in Québec Province in Canada, it has been found that each per cent of entrained air decreased on average the 28 day compressive strength by 5%. This relation is supported by the results obtained when measuring the compressive strength of several hundred concrete loads that were used during the construction of three bridges built with air-entrained high-performance concrete (Lessard, Gendreau and Gagné, 1993; Aïtcin and Lessard, 1994; Blais *et al.*, 1996).

It is interesting to note that this shows that the old 5% rule of thumb so familiar to field engineers for usual air-entrained concrete still works.

18.2.2 Design of an air-entrained high-performance concrete mix

This relationship between the percentage of air and compressive strength can be used to optimize the composition of an air-entrained high-performance concrete having a given 28 day specified compressive strength using the four-step method already discussed in section 9.7. The first two steps consist of designing a stronger non-air-entrained concrete and the third one of introducing the right air bubble system to develop the right air content and spacing factor. The last step is to verify the freeze–thaw durability of the developed concrete.

In order to define the average strength of the non-air-entrained concrete with the object of clearing the second step some hypotheses have to be made:

- The first one is related to the air content that will be needed to acquire the right spacing factor. From the experience gained at Sherbrooke and Laval universities, it has been found that when an efficient air-entraining agent is used an air content of 4 to 6% usually gives a spacing factor of around 200 μm, which will provide an excellent freeze–thaw resistance.
- The second one is related to the amount of entrapped air present in a non-air-entrained concrete. Experience shows that usually non-air-entrained high-performance concretes entrap between 1.5 and 2.5%. Rarely has the reported amount of entrapped air been found outside this range, so that a value of 2% can be taken as a fair average value for entrapped air in a high-performance concrete.
- The last hypothesis is related to the quality control level at which the concrete will be produced. The ACI 363 Committee suggests the grading of quality control based on the value of the coefficient of variation of the concrete production, as shown in Table 18.1.
- Finally, it is obvious that the strength acceptance criteria for this concrete have to be known.

(a) Sample calculation

An air-entrained concrete that will contain 6% of entrained air, for which the average of three consecutive compressive strength results must be higher than 60 MPa 99 times out of 100, and which has a coefficient of variation, V, of 10%, has to be made in a concrete plant. What is the average compressive strength of the non-air-entrained concrete that has to be optimized if the previous mix design method is used to find the

Table 18.1 Rating of quality control according to the value of the coefficient of variation given by the ACI 363 Committee on high-strength concrete

	Excellent	Very good	Good	Fair	Poor
Coefficient of variation (%)	0–6	6–8	8–10	10–13	> 13

mix composition of this air-entrained high-performance concrete? Using the equation presented in Figure 18.1 we have:

$$f'_{na} = 60\left[1 + \frac{5}{100}(6-2)\right] = 72 \text{ MPa} \tag{18.1}$$

$$f'_{cr} = 72 + \frac{2.33}{1.732} \times \sigma \tag{18.2}$$

As $\sigma = V \times f'_{cr}$:

$$\sigma = 0.1 \times f'_{cr} \tag{18.3}$$

Inserting this value of σ in (18.2):

$$f'_{cr} = 72 + 1.345 \times 0.1 \times f'_{cr} \tag{18.4}$$

which gives

$$f'_{cr} = \frac{72}{0.8655} = 83.2 \text{ MPa}$$

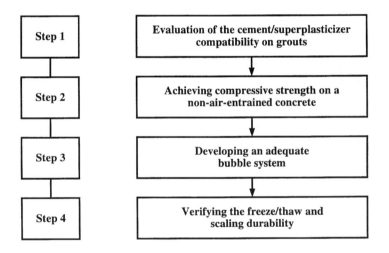

Fig. 18.1 Proposed mix design method for air-entrained high-performance concrete (after Lessard, Gendreau and Gagné, 1993).

So the first goal when optimizing the composition of the air-entrained high-performance concrete is to develop a non-air-entrained high-performance concrete having an average 28 day compressive strength of 83.2 MPa. When optimizing this non-air-entrained concrete, cement/superplasticizer compatibility can be checked, the water/cement ratio can be optimized, and the performance of the different coarse aggregates available can be compared, etc. In fact, this is the regular procedure already developed to optimize the composition of a non-air-entrained high-performance concrete which has already been discussed in Chapter 8.

The next step consists of introducing an air-entraining admixture into the mix in order to develop a stable and adequate air bubble system within the concrete, and adjusting the amount of superplasticizer, taking into account the improved workability of the mix.

It has been recognized that it is not always easy to entrain air in a high-performance concrete and that usually a higher amount of air-entraining admixture is necessary to entrain an adequate bubble system (Gagné, Pigeon and Aïtcin, 1990). Moreover, recently it has also been advocated that some types of superplasticizer do not perform as well as others in entraining air in high-performance concrete. Okkenhaug and Gjørv (1992) found that with the air-entraining admixture, with the naphthalene and lignosulfonate superplasticizers they were using the air bubble system was destabilized, while when they were using a melamine superplasticizer the air bubble system was stable. It could be dangerous to generalize this conclusion to all air-entraining admixtures and all commercial superplasticizers. There are at least two air-entraining admixtures available in North America that are perfectly compatible with two efficient pure naphthalene superplasticizers found in the eastern part of North America. One of these two air-entraining admixtures was used during the construction of Portneuf bridge in Québec Province in October 1992 and resulted in an air bubble system perfectly within the specifications (Lessard, Gendreau and Gagné, 1993). The average spacing factor of this air-entrained high-performance concrete obtained on the 26 loads of concrete that were tested was 180 μm, well below the 230 μm average value specified in the A23.1 CSA Standard for freeze–thaw durable concrete, with no individual value being higher than 260 μm. The average air content necessary to reach such a bubble spacing factor was 6.2% with a standard deviation of 0.7%, as shown in Figure 9.6. Only two values of the air content were quite low, but in only one case was the spacing factor just over the 260 μm maximum limit.

If specifications are asking for a particular spacing factor instead of an average air content, it is suggested making three trial concrete batches containing enough air-entraining agent to generate about 4, 6 and 8% of entrained air and checking the spacing factor obtained in each case as well as the compressive strength with the actual materials and mixing equipment. As air-entrained high-performance concrete hardens quite fast, the concrete surfaces necessary to measure the spacing factor can be polished two or three days following the production of the concrete, so the correlation between the amount of air-entraining admixture, the air content and the bubble spacing can be known quite rapidly. From the three sets of values, it is easy to interpolate the right amount of air-entraining admixture necessary to obtain the desired spacing factor and the necessary amount of entrained air to get such a spacing factor. A fourth trial batch can be made to fine-tune the final amount of entrained air to be sure to obtain the desired spacing factor.

When doing these three (or four) trial batches, it is suggested studying the stability of the bubble system for 1 hour, or more specifically during the time forecast to deliver and place the air-entrained high-performance concrete. It is strongly suggested that concrete sampling be done at the end of this period to simulate better the effects of hauling conditions on the stability of the air bubble system. Of course, during this experiment it would be interesting to check the air content just after batching in order to know how much air can be lost, if any, during this period of time and how the spacing factor is altered if it is. Knowing this value in advance, it will be easy to anticipate if any particular load leaving the mixing plant could meet the field specifications. This experiment will also be very helpful in establishing the correlation between the amount of air measured in the fresh concrete and the corresponding spacing factor in the hardened concrete, because the only applicable acceptance field measurement is the amount of entrained air in the fresh concrete. Moreover, it will also be possible to establish the relationship $f'_c = \psi(a)$ for this concrete. It is obvious that in order to reproduce the same spacing factor in a given concrete one must start to reproduce the same amount of air in the fresh concrete.

The last step could be the verification of the freezing and thawing durability of this air-entrained concrete according to the standard test method specified. In that respect, it must be remembered that if it is the ASTM C 666 test procedure acceptance criteria that are used, it will take 8 to 10 weeks before the 300 freezing and thawing cycles are completed.

18.2.3 Improvement of the rheology of high-performance concretes with entrained air

From the experience gained with air-entrained high-performance concrete, it has been generally found that the introduction of a small amount of entrained air improves the workability of a high-performance concrete significantly and can result in a slight decrease in the superplasticizer dosage. Air-entrained high-performance mixes are less sticky, and have a better workability, so it is easier to place them and to finish flat surfaces (Lessard, Gendreau and Gagné, 1993; Hoff and Elimov, 1995).

It is the opinion of the author that in the future a small amount of entrained air will be incorporated in some unexposed high-performance concretes in order to improve their workability, as has been done deliberately during the construction of the submerged part of the Hibernia offshore platform that will never be exposed to any freezing and thawing (Hoff and Elimov, 1995). The extra amount of cement and superplasticizer needed to compensate for the lost MPa's due to the introduction of air bubbles was easily and profitably compensated for by the saving in the placing time and the time spent in correcting the aesthetics of the surface with a mix too difficult to place.

18.2.4 Concluding remarks

It is possible to make consistent air-entrained high-performance concretes if the materials used to make them are carefully selected. A four-step mix design method is suggested, consisting of developing in the first two steps a stronger non-air-entrained concrete and optimizing its composition, then introducing the air-entraining admixture and adjusting the final dosage of superplasticizer if necessary.

It is the opinion of the author that in the future it will be interesting to entrain a small amount of air in harsh mixes in the 50 to 70 MPa strength range in order to improve their workability. The cost of the extra amount of cement and superplasticizer that would be added to the mix to keep the same compressive strength could be advantageously overcome by the savings realized when placing the concrete.

18.3 LIGHTWEIGHT HIGH-PERFORMANCE CONCRETE

18.3.1 Introduction

Knowing the importance of the strength of the coarse aggregate when looking for high strength, it could be surprising to use a lightweight aggregate to make a high-performance concrete. Lightweight aggregates are porous, not very strong and can be crushed quite easily. However, decreasing the unit mass of a 50 to 60 MPa concrete by around 2000 kg/m^3 can represent an economical advantage; the decrease of the desired mass of a particular structure can result in significant savings in the overall cost (Hoff, 1990; Novokshchenov and Whitcomb, 1990).

It has been shown that it is possible to make lightweight high-performance concrete with a compressive strength higher than 50 to 60 MPa (Malhotra, 1987, 1990), but it seems that in the present state of the art 100 MPa represents the upper boundary. A compressive strength slightly higher than 100 MPa on 100 mm cubes for a 1865 kg/m^3 unit mass of the fresh concrete has been achieved by Zhang and Gjørv (1990) and a

97.7 MPa 91 day compressive strength on 100×200 mm cylinders has been reported by Nilsen and Aïtcin (1992) for a lightweight concrete having a fresh unit mass of 2085 kg/m^3.

The 1850 to 2000 kg/m^3 range for minimum unit mass should not be considered as a lower limit for high-performance concrete since Berra and Ferrara (1990) have shown that it is possible to make a lightweight high-strength concrete having a compressive strength of 60 MPa on 150 mm cubes and a unit mass of 1700 kg/m^3. In that particular case the fine aggregate was a lightweight sand.

However, all the authors are very careful to mention that these impressive results obtained on strength and unit mass can only be achieved with appropriate lightweight aggregates. Not all lightweight aggregates can be used with the same degree of success when making high-performance concrete (Wasserman and Bentur, 1996). Some lightweight aggregates are not strong enough to retard as much as possible concrete failure during a compressive strength test.

The absorptive characteristics of a particular lightweight aggregate are very important (Zhang and Gjørv, 1991a). In some cases the use of a too absorptive lightweight aggregate can be problematic from a rheological point of view in spite of the fact that relatively high compressive strength and low unit mass can be achieved. Lightweight aggregate absorption has a direct effect on the control of the slump and of the workability of lightweight concrete. Finally, if the absorption of a lightweight aggregate is not constant from one delivery to another, or if these variations cannot stay within a reasonable range, it can be difficult to produce industrially a consistent high-performance lightweight concrete.

Another difficulty when making a lightweight high-performance concrete is to select the state in which it will be used: dry, pre-soaked or any intermediate state. Some researchers advocate the use of totally dry lightweight aggregate, taking into account the amount of water it will absorb during mixing and transportation, but Novokshchenov and Whitcomb (1990) consider that this can be done exceptionally with lightweight aggregate that has a low absorption. Other researchers prefer to soak aggregates in water so that they won't absorb any water from the mix (Malhotra, 1990; Hoff and Elimov, 1995); this absorbed water can be considered as a water supply for further hydration in very low water/binder ratio mixes that will decrease significantly autogenous shrinkage. From a practical point of view, coarse and lightweight aggregates are constantly sprinkled with water prior to mixing (Hoff and Elimov, 1995). Other researchers suggest starting with dry lightweight aggregates and pre-wetting them for about 10 minutes by mixing them with an appropriate amount of water corresponding to the absorption of the specific aggregate; this approach is recommended by Zhang and Gjørv (1990) and Berra and Ferrara (1990) but it could be difficult to implement in a concrete plant owing to too great a slowdown of the production of the plant in terms of m^3/h.

18.3.2 Fine aggregate

High-performance lightweight aggregate has been made using either a natural sand (Novokshchenov and Whitcomb, 1990; Nilsen and Aïtcin, 1992; Hoff and Elimov, 1995) or a lightweight sand (Berra and Ferrara, 1990; Malhotra, 1990; Novokshchenov and Whitcomb, 1990; Zhang and Gjørv, 1990) or a mixture of lightweight sand and natural sand (Zhang and Gjørv, 1990). Of course, the use of a natural or lightweight sand has a direct impact on the unit mass and compressive strength, and also on the workability of the mix; the use of lightweight sand gives harsher mixes that are significantly less workable than those where some of the lightweight sand has been partially or totally replaced by a natural sand. However, when using a natural sand it is difficult to lower the unit mass of the fresh concrete much below 2000 kg/m^3, while when using a lightweight sand the unit mass of the fresh concrete can be decreased to around 1850 kg/m^3.

18.3.3 Cementitious systems

Different combinations of cementitious systems have been used when making lightweight high-performance concrete but most of the time silica fume has been introduced into the mix at a rate of 7 to 10% of the total mass of the cementitious system. The use of silica fume is recommended to increase the final compressive strength of the mix and also to lower the risk of segregation by making the mortar paste somewhat more viscous (Novokshchenov and Whitcomb, 1990). In some mixes 20 to 30% of the cement was replaced by a fly ash (Berra and Ferrara, 1990; Malhotra, 1990), essentially for economic reasons. The rationale of using silica fume and fly ash as cement replacement materials is to combine the high and rapid pozzolanicity of silica fume and the slower one of the fly ash. Type I or Type III cements have been used (Zhang and Gjørv, 1990) and Malhotra (1990) found that the use of high early strength cement yielded concrete with higher compressive strength both at early and later ages. Finally, a rapid hardening blast-furnace slag cement has also been used by Berra and Ferrara (1990).

The cement dosage rates varied from 400 kg/m^3 up to 600 kg/m^3, but generally speaking the authors agree that there is a maximum limit to the amount of cementitious material that can be used to increase the compressive strength of a lightweight concrete. When such a limit is reached the increase of the strength of the hydrated cement paste has no more effect on the compressive strength of the lightweight concrete; it is rather the strength of the aggregate that is becoming the controlling factor of concrete strength. This maximum limit depends directly on the type of aggregate that is used, so it is difficult to give any general rule. This limit has to be found through trial batches.

The development of a pozzolanic reaction between lightweight aggregates and the lime liberated by the hydration of Portland cement is often

Lightweight high-performance concrete

Fig. 18.2 Pantheon dome. Courtesy of M. Collins.

mentioned as an advantage of lightweight aggregate. The Romans already knew this advantage when they built the Pantheon dome about 1850 years ago (Figure 18.2).

Seven concretes with decreasing unit mass were used to build the dome, the top part of which was built with a lightweight pozzolanic concrete in which the coarse aggregate was very porous and light volcanic ash (Bremner and Holm, 1995).

18.3.4 Mechanical properties

(a) Compressive strength

It will be difficult in this section to present data that could be used universally; the mechanical properties of lightweight concretes are so dependent on the type of aggregate used and the achieved unit mass (Slate, Nilson and Martinez, 1986; Zhang and Gjørv, 1991b). As previously mentioned, in some cases it would be difficult to produce a concrete having a unit mass lower than 2000 kg/m^3 and a compressive strength of 50 MPa, while in another case with a 'performant' lightweight aggregate it would be possible to produce a high-performance concrete having a compressive strength of around 100 MPa and a fresh unit mass of 1865 kg/m^3, as reported by Zhang and Gjørv (1990). Obviously for a given aggregate the higher the unit mass, the higher the compressive strength (Zhang and Gjørv, 1990).

(b) Modulus of rupture, splitting strength and direct tensile strength

Berra and Ferrara (1990) measured the direct tensile strength of different lightweight concretes and found values of 1.9 to 2.4 MPa. Malhotra (1990) found that the flexural strength of the mixes studied varied from 6.0 to 8.7 MPa, while the splitting strength varied from 3.5 to 5.2 MPa. Novokshchenov and Whitcomb (1990) obtained 7.6 and 6.7 MPa for the same characteristics for their lightweight high-strength concrete.

(c) Elastic modulus

The elastic modulus of high-performance lightweight concrete is obviously much lower than that of a normal weight high-performance concrete with the same compressive strength owing to the lower stiffness of lightweight aggregate. Values reported in the literature range from 17 GPa for specimens cured at 20 °C and 95% RH (Berra and Ferrara, 1990) up to 30 GPa (Novokshchenov and Whitcomb, 1990; Nilsen and Aïtcin, 1992). However, the values reported by Novokshchenov and Whitcomb (1990) decreased to 13 to 18 GPa for specimens cured at 20 °C and 50% RH.

(d) Bond strength

The bond strength between a 20 mm smooth reinforcing bar vertically embedded within a 200 × 200 × 200 mm cube specimen has been found to vary between 1.9 and 2.3 MPa (Berra and Ferrara, 1990). These values were obtained on specimens cured at 20 °C and 95% RH as well as on specimens cured at the same temperature but at only 50% RH.

(e) Shrinkage and creep

Berra and Ferrara (1990) found that high-performance lightweight concretes have a lower shrinkage rate owing to the presence of water in the aggregates, but they also found that the value at infinite time was higher. These authors found also that the specific creep of their concrete was twice that of a normal weight concrete. Malhotra (1990) found that the 1 year shrinkage of lightweight mixes that incorporated supplementary cementitious material was considerably lower (400×10^{-6} to 517×10^{-6})

than those containing Portland cement only (580×10^{-6} to 630×10^{-6}). He also found the same trend for specific creep values; he gave values of 640 and 685×10^{-6} for the mixes made with Portland cement, while for the mixes containing fly ash he found values ranging from 460 to 510×10^{-6}. Malhotra (1990) explained this difference as being due to the amount of fly ash particles that had not reacted and which behaved like aggregate, providing increased resistance to shrinkage and creep. Nilsen and Aïtcin (1992) also found a lower shrinkage for the two lightweight concretes they studied when compared with the shrinkage of a reference concrete having the same water/binder ratio but made with a normal weight aggregate. Their two lightweight concretes were made using a natural sand as fine aggregate. One of the lightweight aggregates, the best performing in terms of strength, showed a very low shrinkage of only 70×10^{-6} after a 28 day curing period, while the other showed a 260×10^{-6} shrinkage.

(f) Post-peak behaviour

Wang, Shah and Naaman (1978) and Berra and Ferrara (1990) found that the stress–strain curves were almost linear up to the peak load but that the descending branches were quite steep. They found that the ductility factor values decreased with increasing strength, but that at the same time lightweight concrete was less ductile than normal weight concrete.

(g) Fatigue resistance

According to Hoff (1990) many researchers have found that lightweight concretes are better suited to fatigue resistance than normal weight concretes. Having a modulus of elasticity closer to that of mortar, stress concentration at the aggregate/mortar interface is reduced, resulting in a more uniform stress distribution. A recent study on the fatigue resistance of 45 to 65 MPa lightweight concretes has shown that the endurance limit defined as the fatigue flexural stress at which a beam could resist 2 million non-reversed fatigue loading cycles resulted in improvements of 10 to 16% over normal weight concrete.

A limited number of fatigue tests have been performed in Norway on cores drilled from high-strength lightweight concrete, but it was found that there were no significant deviations in the fatigue life of the lightweight aggregate concrete when compared with that of high-strength normal weight concrete. However, trends favouring lightweight aggregate concrete at the highest stress level were found.

(h) Thermal characteristics

Berra and Ferrara (1990) found that the thermal properties (thermal expansion coefficient, thermal conductivity and thermal diffusivity) of the lightweight concrete they measured were equal to half that of normal weight concrete of similar composition. These authors concluded that these better thermal properties coupled with the lower elastic modulus should make lightweight high-performance concrete more resistant to thermal stresses than normal weight concrete. For their part, Novokshchenov and Whitcomb (1990) found that the coefficient of thermal conductivity increased drastically with the strength of the lightweight concrete.

18.3.5 Uses of high-performance lightweight concrete

Up to now high-performance lightweight concretes have been used in offshore construction essentially for two reasons, first because of the greater buoyancy they provide to the platform both in the dry dock and during the towing operation, and second for their high specific strength, i.e. the ratio between strength and unit mass (Kepp and Roland, 1987; LaFraugh and Wiss, 1987; Seabrook and Wilson, 1988; Wilson and Malhotra, 1988; Hoff, 1989a, b; Hoff, 1992; Hoff and Elimov, 1995). This second advantage also makes the use of high-performance lightweight concrete attractive for the construction of long-span box girder-type bridges. Zhang and Gjørv (1991c) and Kefenc, Zhang and Gjørv (1994) have studied the permeability and diffusivity of chlorides of high-strength lightweight concrete. A few major structures have already been built using lightweight high-performance concrete. One of the first major ones was Glomar Beaufort Sea I, a mobile offshore platform built in a dry dock in Japan, then towed through the Pacific Ocean and Bering Strait to be installed in the Beaufort Sea (LaFraugh and Wiss, 1987). The main problem for this offshore platform was that it had to be towed through Point Barrow Strait in the Beaufort Sea (north of Alaska), where the water depth is limited. It would not have been possible to tow a platform built of normal weight concrete through this particular place, owing to the lack of water depth. Some data related to the aggregates and concrete used to build this offshore platform are available (LaFraugh and Wiss, 1987).

The Hibernia offshore platform was built using a so-called semi-lightweight concrete in which half the volume of the coarse aggregate was a lightweight concrete (Figure 18.3).

In Norway, in 1991, six bridges were under construction using lightweight concrete (Aïtcin, 1992), as shown in Table 18.2. In spite of the fact that Norway is not producing any lightweight aggregate, so these lightweight aggregate had to be imported from Denmark or Germany, a lightweight design was found to be the more economical: the saving made on the dead mass of these long-span box girder bridges compensated for the extra cost of the imported lightweight aggregates (Figure 18.4).

18.3.6 About the unit mass of lightweight concrete

Most of the time confusion exists about the unit mass of lightweight concrete because people are not thinking about the same unit mass.

Lightweight high-performance concrete

Fig. 18.3 Hibernia offshore platform under construction.

According to ASTM standards there are presently two ways to measure the unit mass of a lightweight concrete: ASTM C 138 Test for Unit Weight, Yield and Air Content (Gravimetric) of Concrete and ASTM C 567 Standard Test Method for Unit Weight of Structural Lightweight Concrete, which indicate how to calculate the fresh unit weight, the calculated oven-dry weight, the calculated equilibrium unit weight, the observed oven-dry weight and the approximate dry weight.

Measuring the unit mass of the freshly mixed lightweight concrete is very easy, but it is not this value that is of interest to the designer because the lightweight concrete will dry to some extent after its placing. The dry unit mass of the lightweight concrete is also a well-defined value that can be easily measured, but in this case too it is pretty certain that the

Table 18.2 Some Norwegian bridges built with lightweight high-performance concrete

Name of bridge	Span (metres)	Compressive strength* (MPa)
Sundhormøya (box girder)	110 + 150 + 110	55
Boknasundet (box girder)	97.5 + 190 + 97.5	60
Eidsvold Sundbru (cable stay)	8 × 40	55
Bergsoysundet (floating bridge)	–	55
Salhus	–	55
Stovseth (box girder)	100 + 220 + 100	55

*100 mm cube strength.

Fig. 18.4 Norwegian bridge built with a lightweight aggregate.

lightweight concrete in a structure will not be in such a dry state under any circumstances.

In order to avoid any confusion during the construction of a particular structure involving the use of a lightweight high-performance concrete, it is very important to define which of these two methods, or any other method, will be used to control the unit mass of the lightweight concrete. When, for any reason, it is decided that the unit mass of the lightweight concrete will be measured differently it is very important to define precisely the procedure that will have to be used. In any case, it will be very useful to establish a correlation between this value and the values obtained with the two previous ASTM standard test methods, which are particularly well suited to quality control.

In one particular case in which the author was involved, it was the unit mass of cores taken from the structure 28 days after the placing of the concrete that was selected as the control value of the unit mass of the lightweight concrete. In that particular case there was no other way than casting big blocks of lightweight concrete and drilling cores out of them to establish a correlation between the unit mass of the fresh concrete according to ASTM C 138, or the different values calculated when using ASTM C 567, and the unit mass of the cores that interested the designer.

18.3.7 About the absorption of lightweight aggregates

A second point of confusion when dealing with lightweight aggregates is the value of their absorption. Usually lightweight aggregates have a much higher absorption than normal weight aggregates and the problem is how properly to measure it.

Sometimes the 10 minute absorption value is measured for mix design purposes when making lightweight concrete because it is admitted and found reasonable from experience that dry lightweight aggregates never become fully saturated when they are used in a lightweight concrete. But if for any reason another value has to be measured and used in calculations it should be clearly defined and the way to measure it also precisely defined. For example, LaFraugh and Wiss (1987) explained how the lightweight aggregate was mixed with 4% extra water to accommodate absorption during mixing and delivery.

18.3.8 About the water content of lightweight aggregates when making concrete

Another point of controversy when using lightweight aggregate to make high-performance concrete is to decide in which state of dryness the lightweight aggregate should be used: completely dry (LaFraugh and Wiss, 1987), completely soaked (Hoff and Elimov, 1995) or in an intermediate state (Zhang and Gjørv, 1990). If the first two cases are well defined and non-controversial, any time an intermediate state is suggested it is very important to define the water content of the lightweight aggregates in that intermediate state and the way to obtain it.

Obviously when lightweight aggregates are used in a dry state they absorb some water from the mix which affects concrete slump and workability, which is not an ideal situation when making concrete. It has been found that in such a case mixing the dry aggregate with an appropriate amount of water for 10 minutes could avoid any slump or workability loss of the high-performance lightweight concrete; however, such a process slows down the mixing process and is unacceptable from an industrial point of view.

When lightweight aggregates are pre-soaked for 24 hours or more they are almost fully saturated when they are used to make the lightweight concrete and they will not absorb any more water from the mix, so there should not be any slump and workability loss problems. But in such a case the unit mass of the fresh and hardened concrete will be significantly higher, so the advantage of using a lightweight aggregate will not be optimized.

In the case of high-performance concrete, the author recommends whenever it is possible using pre-soaked lightweight aggregate in order first to avoid any slump and workability loss and second to have some water available within the high-performance concrete when self-desiccation develops with the mass of the high-performance concrete in order to reduce significantly autogenous shrinkage. If the unit mass of such a lightweight concrete is really critical, it can be adjusted by entraining a certain amount of air within the concrete. In such a case it should be remembered that a 1% increase of the air content in a lightweight concrete will result in about a 20 kg/m^3 decrease in the fresh unit mass of this concrete and a 5% strength reduction.

18.3.9 Concluding remarks

It is reasonable to think that in the future there will be more civil engineering structures in which lightweight high-performance concrete will be used, because it will be found that the reduced unit mass could become a critical and/or economic advantage. Moreover, the reduced autogenous shrinkage that this type of concrete develops will also be interesting. It must be borne in mind that the production cost of this type of high-performance concrete will be higher, that it will be more difficult to produce it and that it will never be as strong and as stiff as a normal weight high-performance concrete having the same water/cement ratio.

18.4 HEAVYWEIGHT HIGH-PERFORMANCE CONCRETE

Heavyweight concretes are used only in very specific applications when increasing the unit mass of concrete can be considered an added value which can offset the higher price of heavyweight aggregates. Normal strength heavyweight concretes are mainly used when building huge counterweights or shields for radioactive γ-rays. Usually they are made using iron-rich aggregates such as ilmenite, haematite or even steel pieces as coarse aggregate and finely crushed ilmenite or haematite as fine aggregate. Their unit mass ranges from 3000 to 6000 kg/m^3, depending on the specific gravity of the aggregates that are used.

In fact heavyweight concrete is not so special a concrete except that it is made of aggregates having a higher specific gravity than usual ones. Heavyweight high-performance concrete can be made without any special problems. For example, using ilmenite as coarse and fine aggregate, Nilsen and Aïtcin (1992) made two heavyweight high-performance concretes having water/binder ratios of 0.30 and 0.31. The first mix was made using a natural sand as a fine aggregate so that its fresh unit mass was 3340 kg/m^3; the second mix was made using crushed ilmenite as a fine aggregate and its unit mass was 3805 kg/m^3. Both mixes reached almost the same 28 day compressive strength which was around 80 MPa, but the all-ilmenite mix had a systematically slightly lower strength than the one where the fine aggregate was the natural sand. For example, at 91 days the all-ilmenite concrete had a strength of 94.1 MPa, while the one made with natural sand had a strength of 98.5 MPa. A reference concrete having the same water/binder ratio at the same age had a strength of 117.8 MPa. So it seems that using heavyweight aggregates results in a decrease in the compressive strength that could have been achieved if a normal weight aggregate had been used.

This is not the case for the elastic modulus. Nilsen and Aïtcin (1992) found that the elastic modulus of heavyweight high-performance concrete was significantly higher than that of its normal weight counterpart. They reported an elastic modulus as high as 60 GPa at 28 days for the all-ilmenite high-performance concrete and 52 GPa for the heavyweight mix where they used natural sand as a fine aggregate. The normal weight high-performance reference concrete with the same water/binder ratio had a significantly lower elastic modulus of 40 GPa. At 91 days Nilsen and Aïtcin reported values as high as 65 and 59 GPa for their two heavyweight concretes compared to 43 GPa for their normal weight reference concrete.

Based on the results obtained with high-performance concretes of different unit mass, the two following equations linking elastic modulus, unit mass and compressive strength for any kind of concrete are proposed (Nilsen and Aïtcin, 1992):

$$E_c = 0.31 \, \rho^{1.19} f_c'^{0.35} \text{ (GPa)}$$

where ρ is expressed in kg/m^3 and f_c' in MPa, or more specifically for heavyweight concrete:

$$E_c = 0.00845 \, \rho^{0.80} f_c'^{0.29} \text{ (GPa)}$$

using the same units.

Nilsen and Aïtcin (1992) also found that the shrinkage of heavyweight high-performance concrete was much lower than that of its normal weight counterpart. At 128 days the all-ilmenite heavyweight concrete had shrunk by half as much as the normal weight one (140 × 10^{-6} versus 320 × 10^{-6} after 28 days of wet curing). This difference in the behaviour of these two concretes, which had exactly the same water/binder ratio, can be explained by the greater stiffness of the aggregate and of the resulting concrete. The heavyweight concrete made with a natural sand had of course an intermediate shrinkage of about 200 × 10^{-6}.

In conclusion it can be said that heavyweight aggregates can be used as well as normal weight aggregates to make high-performance concrete. In spite of the fact that such heavyweight concretes seem to have a slightly lower compressive strength, they have a much higher elastic modulus, owing to the higher stiffness of the heavyweight aggregate. The drying shrinkage of heavyweight high-performance concrete is much lower than that of a normal weight concrete of the same water/binder ratio.

18.5 FIBRE-REINFORCED HIGH-PERFORMANCE CONCRETE

Different types of fibres are currently used to improve the toughness and the impact resistance of usual concretes; consequently it is not surprising that all kind of fibres have already been added to high-performance matrices in order to try to achieve the same objectives (Wafa and Ashour, 1992).

In the previous chapter we discussed very briefly the merit of adding plastic fibres within a high-performance concrete in order to improve its fire resistance. In this section we will rather concentrate on the use of steel fibres in high-performance concrete.

Some projects and field tests involving the use of fibre-reinforced high-performance concrete have recently been made by the Université de Sherbrooke concrete research group. One particular project dealt with the replacement of the asphalt overlay on an orthotropic bridge deck by a 50 mm layer of a fibre-reinforced high-performance concrete. The results of this experimental project have not yet been published, but this project has shown that it is possible to design, produce and deliver a very strong fibre-reinforced high-performance concrete.

It is the opinion of the author that the physical characteristics of all the fibres that are presently available on the market are not perhaps the best ones for high-performance concrete applications. It is also the opinion of the author that the use of plastic fibres presents little interest in high-performance matrices, owing to their intrinsic properties, except perhaps when the impact capacity of the high-performance matrix has to be increased. These types of fibres have too low an elastic modulus to improve significantly concrete toughness, the property that is essentially looked for when adding fibres within a concrete at the present time.

There is definitely a need for steel (or carbon) fibres in high-performance concrete whenever its intrinsic brittleness represents a limitation for its use. For example, steel fibres can be used in areas with high seismic risk or in areas where the shear strength of high-performance concrete has to be increased in such a way that the area would become so congested with shear reinforcements that it would be almost impossible to place any concrete in between the steel bars, or when the cost of placing so many stir-ups would be very expensive.

The main problem found by material researchers is that the steel fibres that are presently available on the market have been conceived and developed for usual concrete, i.e. for low-performance matrices where the bond between the fibre and the hydrated cement paste is not very strong. Under such conditions when some tensile stress is developed within the fibre-reinforced concrete, elementary fibres start more or less rapidly to slip within the matrix when the tensile stress developed within the fibre is greater than the bond stress developed between the fibre and the matrix. When slipping within its print the fibre develops a friction resistance that imparts toughness to the fibre-reinforced concrete matrix. As the length, the aspect ratio and the geometry of present fibres have been optimized for working conditions that are far from the ones that exist in a high-performance matrix, the fibres do not behave at their best in high-performance concrete. In high-performance matrices the bond between the fibres and the matrix is so strong that, quite often, the resisting bond strength developed at the fibre/matrix interface is greater than the ultimate stress that can be developed within the fibres; therefore the fibres break before any pull-out develops. Consequently a fibre-reinforced high-performance matrix behaves like a slightly stronger but brittle matrix (Chanvillard and Aïtcin, 1996).

There are two ways to restore some toughness to fibre-reinforced high-performance concrete. The first one is to keep the same aspect ratio, same length and same geometry of the present fibres and increase the ultimate tensile stress of the steel used in high-performance matrices to such a level that the fibre will not break before it debonds from the matrix.

The second one is to develop shorter fibres with a smaller diameter in order to decrease the bond strength so that the stress developed within the fibre due to its bond to the matrix is lower than the ultimate tensile stress of the steel.

As the reinforcing of the matrix has to be as uniform as possible, it is the author's opinion that the most promising solution seems to be the second one, because when reducing the diameter of the fibres, the number of fibres per unit volume of concrete increases and there are many more fibres per unit volume in the second case. Unfortunately, very thin and short fibres are rather rare on the market, but not totally absent. Using 15×0.6 mm fibres, H.H. Bache (personal communication, 1995) from Aalborg Cement has been able to cast a high-performance mortar having a water/cement ratio of 0.18 and containing 6.5% fibres (500 kg/m^3) made with a sand with coarser particles with a maximum diameter of 4 mm. The panel in which this fibre-reinforced mortar was cast contained 20% longitudinal steel and 7% vertical steel.

In the next chapter we will also see the importance of the aspect ratio of the steel fibre when designing a reactive powder concrete.

Usually it is found that the addition of steel fibres does not improve high-performance concrete compressive strength. It is usually found that the modulus of rupture is slightly increased, but surprisingly not as much as could be expected; it is rather the post-peak behaviour of the stress–strain curves (for the modulus of rupture) that is improved (Figure 18.5). Instead of having a brittle failure, as in the case of a non-fibre-reinforced concrete, the fibre-reinforced high-performance concrete shows a pseudo-ductile behaviour. The shape of the post-peak curve is a direct function of the fibre dosage. For very high fibre dosage, more than 1.5% per volume (i.e. higher than 120 kg/m^3), an increase in the ultimate load can be observed, which is not the case with the usual industrial dosages that range from 40 to 80 kg/m^3 in usual concretes.

The main limitation in increasing the dosage of steel fibres in a high-performance concrete is the same as in a usual concrete, it is the significant decrease in the workability of the mix. In that sense, shorter fibres would help to increase the fibre dosage.

The second main limitation in increasing the dosage of steel fibre in a high-performance concrete is the price of the fibres. At a dosage rate of between 40 and 80 kg/m^3, the high-performance concrete price is

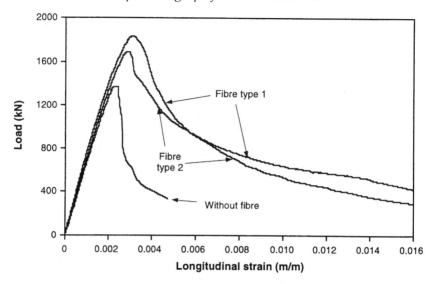

Fig. 18.5 Load–strain curves for fibre-reinforced high-performance concrete.

increased by 50 to 100%; therefore it is absolutely necessary that the investment in the fibres has a significant payback in the improvement of the property that is looked for.

The use of steel fibres is gaining popularity in shotcrete applications but the author has no experience in that field. Readers interested in such an application are advised to read the numerous papers written on this subject by his colleague R.M. Morgan of Concrete Canada, the Canadian Network of Centres of Excellence on High-Performance Concrete.

18.6 CONFINED HIGH-PERFORMANCE CONCRETE

The idea of confining concrete in a steel envelope is not new in itself, since papers related to confined concrete date back as far as the early part of the century (Sewel, 1902; Burr, 1912). It was found that the confinement of concrete in steel tubes resulted in a significant increase in the strength of the composite material when compared with the strength of each of the components. But at that time the compressive strength of the concrete used by the industry was around 20 MPa, so the contribution of concrete in the composite material was not significant enough. Concrete was, rather, essentially used to protect the structural steel from corrosion and fire.

However, the confinement of concrete in thin steel tubes recently regained interest with the development of stronger concrete or even high-performance concrete. This time it is not a low-strength concrete that is confined but rather a high-performance concrete with a compressive strength approaching or in excess of 100 MPa (Lahlou, 1994) or even reactive powder concrete having a compressive strength in excess of 200 MPa, as will be shown in the next chapter.

During the past 30 years some researchers such as Furlong (1967), Gardner and Jacobson (1967), Sen (1969), Bertero and Moustafa (1970), Knowles and Park (1970), Bode (1976), Takahashi and Yoshida (1982), and Tsukagoshi, Kurose and Orito (1990) resumed some research work, this time taking into account the confinement of normal strength concrete in a steel tube. But it was only very recently that the use of high-performance concrete in steel tubes attracted some interest from the research community (Thorenfeldt and Tomaszewicz, 1989; Cederwall, Engstrom and Graners, 1990; Lahlou, Aïtcin and Chaallal, 1991, 1992). It should be pointed out that the concept of confining high-performance concrete in steel tubes has already been exploited by the designers of the Two Union Square high-rise building in Seattle (see section 4.7.6) and in Melbourne, Australia (Webb, 1993).

The effect of a triaxial confinement on rock samples has been studied since the pioneering work of Von Karman (1911). He studied in particular the role of confining pressure on Carrara marble; later on Paterson (1958, 1978), Vuturi, Lama and Saluja (1974) and more recently Fredrich, Evans and Wong (1989) published similar results but for different rocks. All these authors found that increasing the confining pressure has three important effects in the stress–strain curves of rock samples:

1. As the confining pressure increases, the ultimate compressive strength increases.
2. The strain reached before macroscopic failure increases very markedly when the confining pressure increases, exceeding what has been called the threshold pressure. A value of 3 to 5% of strain at failure is often taken as defining the brittle–ductile transition at which the rock specimen behaves as a ductile material.
3. The slope of the stress–strain curve in its elastic part stays the same, showing that the confinement has no effect on the elastic modulus of the confined solid.

Recently, at the 1993 Transportation Research Board Meeting in Washington, Hammons and Neeley (1993) presented some results on the triaxial characterization of high-strength Portland cement concrete (105 MPa). They found that high-strength Portland cement concretes are capable of large plastic strains and flow under states of high confinement.

All the recent studies on the confinement of a high-performance concrete in a steel tube already mentioned have shown that:

- the use of a thin steel tube to confine, for example, an 85 MPa concrete results in multiplying by a factor ranging from 1.6 to 2.2 the loading capacity of a column submitted to an eccentric load, the effect of this confinement was also compared with a reinforced column made of a 45 MPa concrete which was confined with a spiral;

- the confinement is particularly interesting whenever the increase of the ductility of a square tube column is looked for;
- the bond between the steel tube and the concrete and the way in which the load is applied to the composite material is an important factor.

In the case of high-performance concrete it seems that the confinement of the concrete in a steel tube is more effective than the confinement with spirals and stir-ups. This improved behaviour is due to several factors:

- the confinement of a high-performance concrete by spirals or rectangular stir-ups is less efficient than the confinement of normal strength concrete (Martinez, Nilson and Slate, 1984; Bjerkeli, Tomaszewicz and Jenzen, 1990);
- in order to get a similar increase in relative strength, high-performance concrete needs a higher pressure of confinement than a usual concrete. Ahmad and Shah (1982) explain this result by saying that for a given relative value of the axial stress the radial strain in a high-performance concrete is smaller than in a usual concrete;
- a sufficient ductility was obtained by Mugumura and Wanatabe (1990) when they used a high-strength steel with a 115 MPa concrete. Using a special transverse and longitudinal reinforcement layout, Bjerkeli, Tomaszewicz and Jenzen (1990) and Cusson, Paultre and Aïtcin (1992) were also able to improve ductility;
- the post-peak behaviour of columns made of high-performance concrete is characterized by the spalling of the concrete cover, which results in a drop of the sustained load before the lateral confinement of the reinforcing steel becomes efficient (Martinez, Nilson and Slate, 1984; Yong, Nour and Nawy, 1988; Cusson, Paultre and Aïtcin, 1992). This behaviour is not observed for columns without any cover (Martinez, Nilson and Slate, 1984; Abdel-Fattah and Ahmad, 1989);
- on the contrary to normal strength columns, rupture occurs by the opening of a shear crack within the concrete core (when $f'_c > 80$ MPa) before the buckling of the vertical reinforcing cage develops (Yong, Nour and Nawy, 1988).

From a practical and manufacturing point of view, confining high-performance concrete is very interesting; it is only necessary to fill a steel tube with concrete. Placing can be done without any vibration by pushing the concrete inside the tube with a pump, the tube being the form. There is no need to place any reinforcing steel and there is no need to cure the concrete. The confinement can, if necessary, be combined with internal and external post-tensioning, so that very light and rigid segmental structural elements can be designed perfectly suited for the construction of bridges, offshore structures and long-span structures. Moreover, external post-tensioning can be used for assembling.

Figure 18.6 presents some experimental results obtained by Lahlou, Aïtcin and Chaallal (1992). Three different concretes having compressive

Confined high-performance concrete

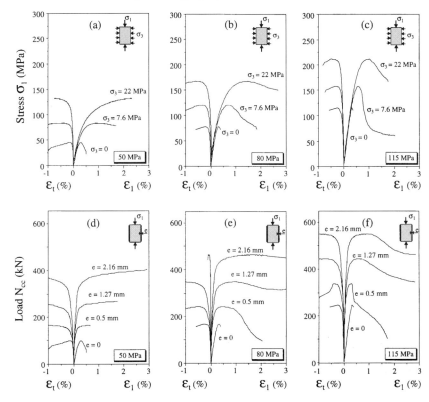

Fig. 18.6 Load–strain curves for different confined concrete samples: (a)–(c) hydrostatic pressure; (d)–(f) steel tube.

strengths of 50, 80 and 115 MPa were studied. Figures 18.6(a), (b) and (c) represent the results obtained under three lateral confining pressures in a triaxial cell with σ_3 respectively equal to 0, 7.6 and 22 MPa. Figures 18.6(d), (e) and (f) represent the results obtained with three steel tubes having a thickness of 0.5, 1.27 and 2.16 mm which correspond to a percentage of Grade 450 steel of 3.8, 9.7 and 17% as compared with the concrete section.

In this series of curves, a significant increase in the bearing capacity and in the ductility of the composite is observed whatever the compressive strength of the concrete. However, it is also observed that as the compressive strength of the concrete increases, the effect of the confinement is less important.

In the use of a passive confinement in a steel tube the actual stress in the concrete can be calculated from strain gauge measurements. Figure 18.7 shows the different types of relation that can be found between the maximum axial stress in the concrete and the confining pressure due to the steel tube.

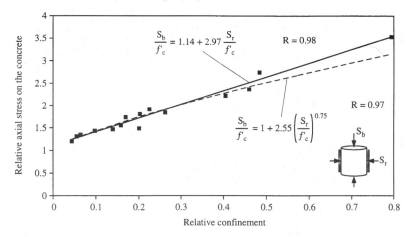

Fig. 18.7 Strength criteria for the concrete core.

However, before the analysis and modelling of compressive segments made of confined high-performance concrete can be developed further it will be necessary to study the following aspects:

- the triaxial behaviour of the concrete core;
- the behaviour of the thin envelope under biaxial loading;
- the effect of the interaction between the envelope and the concrete core;
- the behaviour of the concrete that could at any time affect the behaviour of the composite;
- the buckling behaviour of the composite according to how the loads are applied and the aspect ratio of the confined concrete element.

18.7 ROLLER-COMPACTED HIGH-PERFORMANCE CONCRETE

A very new and interesting development that happened in the Province of Québec in Canada is the use of roller-compacted high-performance concrete to build very large industrial slabs. In a pulp and paper plant near Sherbrooke, an 87 000 m² industrial slab equivalent to the surface of 16 adjacent football fields has been built within one and half months using standard asphalt paving plant and equipment (Figure 18.8). This large slab is composed of two 150 mm-thick layers of a high-performance concrete having a compressive strength of 60 MPa at 28 days (Figure 18.9). This slab does not contain any steel reinforcement and has no joints. The roller-compacted concrete placed was found to be freeze–thaw resistant and presented a very strong abrasion resistance to the steel belts of some pieces of equipment used to handle the wood logs.

The roller-compacted high-performance concrete was prepared in a conventional asphalt plant (Figure 18.10), placed with an asphalt paver (Figures 18.11 and 18.12) and compacted as any usual asphalt layer

Roller-compacted high-performance concrete 535

Fig. 18.8 General aspect of the 87 000 m² concrete slab equivalent to 16 adjacent football fields.

Fig. 18.9 Aspect of two 150 mm layers after their compaction. Note that the longitudinal joints of the two layers are slightly apart from each other.

Fig. 18.10 Preparation of the roller-compacted high-performance concrete mix in an asphalt plant.

Fig. 18.11 Placing the roller-compacted high-performance concrete with an asphalt paver.

Fig. 18.12 The surface of the roller-compacted high-performance concrete just after the placing by the asphalt paver.

Fig. 18.13 Compacting the concrete surface and checking the unit mass with a nucleodensimeter.

(Figure 18.13). The unit mass of the roller-compacted high-performance concrete was checked using a standard nucleodensimeter (Figure 18.13). The aspect of the concrete surface was slightly open, as shown in Figure 18.14, but its general aspect was very satisfying (Figure 18.15).

Fig. 18.14 Aspect of the concrete surface after the compaction by the vibrating rollers.

Fig. 18.15 General aspect of the finished surface. the white colour results from the use of a curing membrane.

Similar slabs have been built in a slag processing operation, in an industrial scrap iron yard and in a chemical plant. The main advantages advocated by the owners are the durability of the roller-compacted high-performance slab, its high wear resistance, the rapidity of the construction and the rapidity at which the industrial slab can be in operation.

Roller-compacted high-performance mix composition has to be optimized taking into account the composition of the binder system used, the available aggregates, the mixing equipment and the laying equipment.

As for any roller-compacted concrete, the consistency of the mix has to be adjusted using the Vebe test (ASTM C 1170-91). A 60 ± 10 s Vebe time is usually looked for. Table 18.3 provides the mix composition of a high-performance roller-compacted concrete and Table 18.4 its properties.

Table 18.3 Composition of a typical roller-compacted high-performance concrete

Silica fume blended cement (kg/m^3)	285
Water (l/m^3)	90
Aggregate (kg/m^3)	
Coarse	1305
Fine	770
Admixtures	
AEA (ml/100 kg cement)	200
Water reducer (ml/m^3)	600–900

Table 18.4 Properties of fresh and hardened roller-compacted high-performance concrete

Vebe time (s)	60 ± 10
Unit mass (kg/m^3)	2450
Compressive strength (MPa)	
1 d	30
7 d	50
28 d	60
91 d	65
Flexural strength (MPa)	
7 d	7–8

The quality control of a roller-compacted concrete does not imply the development of any new technique; the usual quality control techniques used when making and placing roller-compacted concrete, usual concrete and high-performance concrete can be used. This is a very promising field of application for high-performance concrete.

18.8 CONCLUDING REMARKS

As in the case of usual concrete, some high-performance concrete characteristics can be modified to fulfil a special technological need. Air-entrained, lightweight, heavyweight, fibre-reinforced, confined and roller-compacted high-performance concrete have been already used. It is certain that in the future other types of special high-performance concrete will be developed to fulfil particular technological needs; this adaptability of concrete to satisfy different uses is one of its main advantages. It is the imagination of engineers and the economics of special high-performance concrete that will determine the extent to which it is used in the future.

REFERENCES

Abdel-Fattah, H. and Ahmad, S.H. (1989) Behavior of hoop-confined high-strength concrete under axial and shear loads. *ACI Structural Journal*, **86**(6), 652–9.

Ahmad, S.H. and Shah, S.P. (1982) Stress–strain curves for concrete. *Journal of the Structural Division, ASCE*, **108**(ST4), 728–42.

Aïtcin, P.-C. (1992) *High-Performance Concrete Review of World Wide Activities*, CANMET/ACI International Symposium on Advances in Concrete Technology, September, Toronto, Ontario, Canada, 29 pp.

Aïtcin, P.-C. and Lessard, M. (1994) Canadian experience with air-entrained high-performance concrete. *Concrete International*, **16**(10), October, 35–8.

References

ASTM C 138 (1993) *Standard Test Method for Unit Weight, Yield and Air Content (Gravimetric) of Concrete*, in *1993 Annual Book of ASTM Standards*, Sec. 4, Construction, Vol. 04.02, Concrete and Aggregates, ISBN 0-8031-1923-3, pp. 83–5.

ASTM C 567 (1993) *Standard Test Method for Unit Weight of Structural Lightweight Concrete*, in *1993 Annual Book of ASTM Standards*, Sec. 4, Construction, Vol. 04.02, Concrete and Aggregates, ISBN 0-8031-1923-3, pp. 290–92.

ASTM C 666 (1993*) Standard Test Method for Resistance of Concrete to Rapid Freezing and Thawing*, in *1993 Annual Book of ASTM Standards*, Sec. 4, Construction, Vol. 04.02, Concrete and Aggregates, ISBN 0-8031-1923-3, pp. 326–31.

ASTM C 1170 (1991) *Standard Test Methods for Determining Consistency and Density of Roller-Compacted Concrete Using a Vibrating Table*, pp. 618–21.

Berra, M. and Ferrara, G. (1990) *Normalweight and Total-Lightweight High-Strength Concretes: A Comparative Experimental Study*, ACI SP-121, pp. 701–33.

Bertero, V.V. and Moustafa, S.E. (1970) Steel encased expansive cement concrete column. *Journal of the Structural Division, ASCE*, **96**(ST11), November, 2267–82.

Bjerkeli, L., Tomaszewicz, A. and Jensen, J.J. (1990) *Deformation Properties and Ductility of High Strength Concrete*, ACI SP-121, May, pp. 215–38.

Blais, F.A., Dallaire, E., Lessard, M. and Aïtcin, P.-C. (1996) *The Reconstruction of the Bridge Deck of the Jacques-Cartier Bridge in Sherbrooke (Quebec) using High-Performance Concrete*, 1st Structural Specialty Conference Canadian Society for Civil Engineering, Edmonton, June, ISBN 0-921303-60-0, pp. 501–7.

Bode, H. (1976) Columns of steel tabular sections filled with concrete: design and applications. *Acier-Stahl-Steel*, **11**(12), 388–93.

Bremner, T.W. and Holm, T.A. (1995) *High-Performance Lightweight Concrete – A Review*, ACI SP-154, pp. 1–19.

Burr, W.H. (1912) Composite columns of concrete and steel. *Proc. Inst. Civ. Eng.*, **188**, 114–26.

Cederwall, K., Engstrom, B. and Graners, M. (1990) *High Strength Concrete used in Composite Columns*, 2nd International Symposium on Utilization of High Strength Concrete, Berkeley, CA, May.

Chanvillard, G. and Aïtcin, P.-C. (1996) Pull-out behavior of corrugated steel fibers: qualitative and statistical analysis. *Advances in Cement Based Materials*, **4**(1), July, 28–41.

Cusson, D., Paultre, P. and Aïtcin, P.-C. (1992) *Le Confinement des Colonnes en Béton à Haute Performance par des Etriers Rectangulaires*, CSCE Annual Conference, Québec, May, Vol. IV, pp. 21–30.

Fredrich, J.T., Evans, B. and Wong, T.-F. (1989) Micromechanics of the brittle to plastic transition in Canada marble. *Journal of Geophysical Research*, **94**(84), 4129–45.

Furlong, R.W. (1967) Strength of steel-encased concrete beam columns. *Journal of the Structural Division, ASCE*, **93**(ST1), January, 267–81.

Gagné, R., Pigeon, M. and Aïtcin, P.-C. (1990) *Deicer Salt Scaling Resistance of High Performance Concrete*, Paul Klieger Symposium on Performance of Concrete, San Diego, California, ACI SP-122, pp. 29–44.

Gagné R., Pigeon, M. and Aïtcin, P.-C. (1991) *The Frost Durability of High-Performance Concrete*, Proceedings of the 2nd Canada/Japan Workshop on Low Temperature Effects on Concrete, Ottawa, August, pp. 75–87.

Gardner, N.J. and Jacobson, E.R. (1967) Structural behaviour of concrete filled steel tubes. *ACI Journal*, **64**(7), July, 404–13.

Hammons, M.I. and Neeley, B.D. (1993) *Triaxial Characterization of High-Strength Portland Cement Concrete*, 72nd Transportation Research Board Annual Meeting, Paper No. 930126, Washington, January, 22 pp.

Hoff, G.C. (1989a) *High Strength, Lightweight Concrete for the Arctic*, Proceedings, Symposium on International Experience with Durability of Concrete in Marine Environments, University of California at Berkeley, January, pp. 9–19.

Hoff, G.C. (1989b) *Evaluation of Ice Abrasion of High-Strength Lightweight Concretes for Arctic Applications*, Proceedings of the 8th International Conference on Offshore Mechanics and Arctic Engineering, Vol. 3, The Hague, The Netherlands, March, pp. 583–90.

Hoff, G.C. (1990) *High-Strength Lightweight Aggregate Concrete – Current Status and Future Needs*, ACI SP-121, pp. 619–44.

Hoff, G.C. (1992) *High Strength Lightweight Aggregate Concrete for Arctic Applications, Parts I, II and III*, ACI SP-136, Structural Lightweight Aggregate Concrete Performance, pp. 1–246.

Hoff, G.C. and Elimov, R. (1995) *Concrete Production for the Hibernia Platform*, Supplementary papers, 2nd CANMET/ACI International Symposium on Advances in Concrete Technology, Las Vegas, Nevada, 11–14 June, pp. 717–39.

Kefenc, T., Zhang, M.-H. and Gjørv, O.E. (1994) Diffusivity of chlorides from seawater into high-strength lightweight concrete. *ACI Materials Journal*, **91**(4), 447–52.

Kepp, B. and Roland, B. (1987) *High-Strength LWA-Concrete for Offshore Structures – Ready for Action*, Utilization of High Strength Concrete, Stavanger, Norway, June, ISBN 82-519-0797-7, pp. 679–88.

Knowles, R.B. and Park, R. (1970) Axial load design for concrete filled steel tubes. *Journal of the Structural Division, ASCE*, **96**(ST10), October, 2125–53.

LaFraugh, R.W. and Wiss, J.E. (1987) *Design and Placement of High Strength Lightweight and Normalweight Concrete for Glomar Beaufort Sea 1*, Utilization of High Strength Concrete, Stavanger Norway, June, ISBN 82-519-0797-7, pp. 497–508.

Lahlou, K. (1994) *Comportement des Colonnes courtes en Béton à Hautes Performances confiné dans des Tubes circulaires en Acier Soumises à des Efforts de Compression*, PhD Thesis, Université de Sherbrooke, Québec, Canada, No. 82.

Lahlou, K., Aïtcin, P.-C. and Chaallal, O. (1991) *Étude préliminaire sur le Confinement de Bétons de différentes Résistances en Compression dans des Enveloppes extérieures*, CSCE Annual Conference, Vancouver, BC, Canada, Vol. Supp., Part I, 13pp.

Lahlou, K., Aïtcin, P.-C. and Chaallal, O. (1992) Behaviour of high strength concrete under confined stress. *Cement and Concrete Composites*, **14**(3), 185–93.

Lessard, M., Gendreau, M. and Gagné, R. (1993) *Statistical Analysis of the Production of a 75 MPa Air-Entrained Concrete*, 3rd International Symposium on High Performance Concrete, Lillehammer, Norway, June, ISBN 82-91 341-00-1, pp. 793–800.

Malhotra, V.M. (1987) *CANMET Investigations in the Development of High-Strength Lightweight Concrete*, Utilization of High Strength Concrete, Stavanger, Norway, June, ISBN 82-519-0797-7, pp. 15–25.

Malhotra, V.M. (1990) *Properties of High-Strength Lightweight Concrete Incorporating Fly Ash and Silica Fume*, ACI SP-121, pp. 645–66.

Martinez, S., Nilson, A.H. and Slate, F.O. (1984) Spinally reinforced high-strength concrete columns. *ACI Journal*, **81**(5), September–October, 431–42.

Mitchell, D., Zaki, A., Pigeon, M. and Coulombe, L.-G. (1993) *Experimental Use of High-Performance Concrete in Bridges in Québec*, 1993 CSCE/CPCA Structural Concrete Conference, May, pp. 63–75.

Mugumura, H. and Wanatabe, F. (1990) *Ductility Improvement of High Strength Concrete Column with Lateral Confinement*, ACI SP-121, p. 47.

Nilsen, A.U. and Aïtcin, P.-C. (1992) Properties of high-strength concrete containing light-, normal- and heavy-weight aggregate. *Cement Concrete and Aggregate*, **14**(1), Summer, 8–12.

Novokshchenov, V. and Whitcomb, W. (1990) *How to Obtain High-Strength Concrete Using Low Density Aggregate*, ACI SP-121, pp. 683–700.

Okkenhaug, K. and Gjørv, O.E. (1992) Effect of delayed addition of air entraining admixtures to concrete. *Concrete International*, **14**(10), October, 37–41.

Paterson, M.S. (1958) Experimental deformation and faulting in Wonbeyan marble. *Geological Society of America Bulletin*, **69**, 465–76.

Paterson, M.S. (1978) *Experimental Rock Deformation – The Brittle Field*, Springer-Verlag, pp. 162–7.

Seabrook, P.T. and Wilson, H.S. (1988) High strength lightweight concrete for use in off-shore structures: utilization of fly ash and silica fume. *The International Journal of Cement Composites and Lightweight Concretes*, **10**(3), 183–92.

Sen, M.K. (1969) *Triaxial Effects in Concrete Filled Tubular Columns*, PhD Thesis, University of London.

Slate, F.O., Nilson, A.H. and Martinez, S. (1986) Mechanical properties of high-strength lightweight concrete. *ACI Journal*, **83**(4), 606–13.

Sewel, J.S. (1902) Columns for buildings. *Engineering News*, **48**(17).

Tadros, G., Combault, J., Bilderbeek, D.W. and Fotinos, G. (1996) *The Design and Construction of the Northumberland Strait Crossing Fixed Linked in Canada*, IABSE Congress, Copenhagen, Denmark, 16–20 June, 24 pp.

Takahashi, M. and Yoshida, F. (1982) *Elasto-Plastic Behaviors of Expansive-Concrete Filled Steel-Pipe Members under Cyclic Axial Loadings*, the 25th Japan Congress on Materials Research – Non-Metallic Materials, pp. 255–60.

Thorenfeldt, E. and Tomaszewicz, A. (1989) Thin steel mantle as lateral confinement in reinforced concrete columns, test results. *Nordic Concrete Research*, Oslo, No. 8, December, pp. 199–208.

Tsukagoshi, H., Kurose, Y. and Orito, Y. (1990) *Unbonded Composite Steel Tube Concrete Columns*, IABSE Symposium on Mixed Structures Including New Materials, Brussels, pp. 499–500.

von Karman, T. (1911) Festigheitsversuche underallseitgein Druck. *Z. Ver. Deutsch. Ing.* **55**, 1749.

Vuturi, V.S., Lama, R.D. and Saluja, S.S. (1974) *Handbook on Mechanical Properties of Rocks*, Trans. Tech. Publications, pp. 188–97.

Wafa, F.F. and Ashour, S.A. (1992) Mechanical properties of high-strength fiber reinforced concrete. *ACI Materials Journal*, **89**(5), September–October, 449–55.

Wang, P.T., Shah, S.P. and Naaman, A.E. (1978) Stress–strain curves of normal and lightweight concrete in compression. *ACI Journal*, **75**(11), 603–11.

Wasserman, R. and Bentur, A. (1996) Interfacial interactions in lightweight aggregate concretes and their influence on the concrete strength. *Cement and Concrete Composite*, **18**(1), 67–76.

Webb, J. (1993) High-strength concrete: economics, design and ductility. *Concrete International*, **15**(1), January, 27–32.

Wilson, H.S. and Malhotra, V.M. (1988) Development of high-strength lightweight concrete for structural applications. *The International Journal of Cement Composites and Lightweight Concrete*, **10**(2), 79–90.

Yong, Y.-K., Nour, M.G. and Nawy, E.G. (1988) Behavior of laterally confined high-strength concrete under axial loads. *Journal of Structural Engineering, ASCE*, **114**(2), 332–51.

Zhang, M.-H. and Gjørv, O.E. (1990) *Development of High-Strength Lightweight Concrete*, ACI SP-121, pp. 667–81.

Zhang, M.-H. and Gjørv, O.E. (1991a) Characteristics of lightweight aggregates for high-strength concrete. *ACI Materials Journal*, **88**(2), 150–58.

Zhang, M.-H. and Gjørv, O.E. (1991b) Mechanical properties of high-strength lightweight concrete. *ACI Materials Journal*, **88**(3), 240–47.

Zhang, M.-H. and Gjørv, O.E. (1991c) Permeability of high-strength lightweight concrete. *ACI Materials Journal*, **88**(4), 463–9.

CHAPTER 19

Ultra high-strength cement-based materials

19.1 INTRODUCTION

High-performance concretes are not the strongest materials that can be made with Portland cement. For many years, efforts have been devoted to pushing cement-based materials to their ultimate strength limit. The scientific literature regularly reports the fruits of such efforts. Most of these 'cold ceramics', as they are termed by Pierre Richard of the Bouygues Company, get their very high strength from an extremely low porosity that is achieved using different techniques. However, only on a few rare occasions have these laboratory compositions found commercial application, because most of the time they are too expensive to be implemented on an industrial scale, their processing is too complicated, or their overall characteristics are not as attractive as their increased compressive strength. This limited success does not imply, however, that all attempts to increase the compressive strength of cement-based materials beyond what is currently achieved in high-performance concrete, are doomed to commercial failure.

In spite of the fact that the materials that will be very briefly described in the following paragraphs are not what we usually call concrete, they are cement-based materials that display all of the strength potential that cement can provide. Most of them use Portland cement as part of the cementitious system, except for one which uses aluminous cement.

In order to comprehend further the scientific principles that govern the mechanical strength of cement-based materials, the fundamental principles governing the behaviour of brittle materials can be applied to concrete, as shown in Chapter 5. Such materials include ceramics, which present several microstructural similarities to concrete (Mehta and Aïtcin, 1990).

It is well known that the tensile strength of a brittle material can be related to its porosity by an exponential law of the type $S = S_0 e^{-bP}$, where S represents the tensile strength of the material with a porosity P, S_0 represents its intrinsic tensile strength when it has no porosity, and b is a factor that depends on the dimensions and the shapes of the pores. Studies that have attempted to establish relationships between strength and microstructure in ceramic materials have demonstrated that in addition to porosity, the dimensions of the material grains (in the microstructural sense) and the non-homogeneity are important factors that also control tensile strength.

The compressive strength of brittle materials is far greater than the tensile strength. This is because a material subjected to a tensile stress fails following the rapid propagation of a few microcracks, whereas a certain number of tensile cracks need to be joined together in order to cause failure of the material when subjected to a compressive stress. Therefore, there must be a considerably greater amount of energy developed in order to initiate and propagate the network of cracks necessary to cause the failure in compression of a brittle material.

By applying Griffith's failure criterion in tension and the laws of mechanics of continuous media, the compressive strength of a homogeneous ceramic material can be estimated to be eight times greater than its tensile strength. Even if a theoretical approach has not yet been developed to enable the derivation of the compressive strength of a material from its microstructural characteristics, it is still possible to use empirical relations to estimate its compressive strength. Such relationships can be of the type $C = C_0 (1 - P)^m$, where C represents the compressive strength of a brittle material with a porosity P, C_0 represents its intrinsic compressive strength when the porosity is nil, and m is a factor that depends on the bond strength of the solid, the shape and dimensions of the cracks, the dimensions of the grains and the presence of impurities. Moreover, it is well known that compressive strength decreases as the pore size increases, and increases as the dimension of the material grains in the microstructural sense decreases.

Rice, cited by Illston, Dinwoodie and Smith (1979), made an interesting observation on the behaviour of ceramics: while the majority of material properties (other than mechanical properties) are related more to mean values rather than extreme values, mechanical properties are affected more by extreme values. In other words, the failure of a material is a function of the strength of the weakest link, which is why, in addition to the number, dimension and form of the pores and their spatial distribution, their concentration at a precise point may constitute a preferential source of damage where premature initiation of material failure occurs.

For a long time, researchers and engineers did not know how to master the porosity of Portland cement-based materials because of technological limitations occurring during the application of their cement-based materials, which were essentially a total absence of workability during mixing when using a minimum amount of water. This lack of workability of mixtures of cement and water is, as has already been discussed, a consequence of the presence of numerous non-saturated electrical charges which are found on the surface of cement grains after

Introduction

their grinding. These electrical charges favour the flocculation of cement particles when in contact with water. But as seen earlier, this flocculation can be easily and economically counteracted nowadays using superplasticizers.

However, the use of a superplasticizer alone is not sufficient to reach a compressive strength of several hundred MPa's, but it is only after knowing how simply and economically to make cement-based materials containing a minimum amount of water that it is possible to address the next technological barrier. This technological barrier is to find a way to decrease further the porosity within the compacted granular materials by optimizing the grain size distribution of the granular skeleton.

For quite some time, the laws governing the packing of spherical particles have been common knowledge; however, the granular materials commonly used in concrete are far from being spherical, and it must be admitted that traditionally very little care was devoted to their optimal packing. Furthermore, these granular materials have a particle size distribution that does not necessarily correspond to that which is necessary for spheres to achieve a dense packing. The problem of granular packing was partially overcome when silica fume, made up of spherical particles, began to be incorporated into cementitious materials. In order to compensate for the lack of optimization of the grain size distribution of powders in the finest part of their distribution, and the lack of optimization of the shape of each of these elementary particles, high amounts of silica fume were incorporated. It is important to note that the ratio of the mean diameter of minuscule spheres of silica fume to that of cement particles is approximately 1:100, which is far from the optimum ratio dictated by the laws of spherical packing to achieve optimum compaction.

Moreover, in the search for a material of minimal porosity, the use of Portland cement presents a major inconvenience, because during Portland cement hydration the volume of hydration products is smaller than the sum of the volumes of the cement and water that combine. Portland cement-based materials increase their internal porosity during hardening. This phenomenon, known as the Le Chatelier contraction, or chemical contraction, has been well known since the early studies of cement chemistry. Such volumetric contraction of the hydrating cement paste measures roughly 10% and is inevitable because it is inherent to the development of the chemical reactions that cause the hardening of Portland cement-based materials. Therefore, an additional intrinsic porosity of the order of 10% is created within Portland cement-based materials from a microstructural point of view.

Nevertheless, in view of the fact that cement hydration is not an instantaneous reaction but develops over a period of several hours, it is possible to eliminate mechanically part of the chemical contraction by applying a pressure during the first stage of hardening as long as the cement paste remains in a plastic state and has not yet developed an internal cohesion sufficient to oppose the compressive effect of this external force. As early as 1925, the famous French concrete engineer Freyssinet was pressing the low-strength concretes produced in his time in order to squeeze out most of the entrapped air and some water and to cancel partially some of the early chemical contraction in order to increase their compressive strength.

Therefore, to help solve the second technological problem found in this quest for ultra high-strength materials, it is now possible to contemplate taking advantage of proven technologies that are becoming economical and that enable us to optimize the grain size distribution of the granular skeletons, to enhance the shapes of the granular particles, and partially to eliminate the effects of the chemical contraction of Portland cement paste during its hydration. Therefore, it is possible to produce a Portland cement-based material with a minimal porosity and a cost–performance ratio that is fully competitive in certain structural applications.

Because these Portland cement-based materials draw their 'strength' from ionic forces, they are brittle materials, unlike ductile metallic materials. Therefore, for structural applications, this type of material can pose a problem for engineers concerned about ductility and safety. This inconvenience can be circumvented by incorporating steel fibres into the cement matrix in order to render it more ductile, or by confining these materials in thin metallic tubes (Lahlou, Aïtcin and Chaallal, 1991, 1992). Thus a particularly effective composite material is created, partially combining the ductility of the steel into which the concrete is cast and confined, or that from the steel fibres, and the rigidity and the lightness of concrete (in comparison to steel). Therefore, it is possible nowadays to fabricate Portland cement-based materials that are at the same time very strong and that exhibit sufficient ductility to be safely used in high-performance structural applications.

The results of careful thinking to exploit these technological breakthroughs and/or some fundamental knowledge about low-porosity materials has led to the development over the past 25 years of a number of Portland cement-based materials that present remarkable mechanical properties that were quite unimaginable before. In 1972–73 Brunauer published several papers on hardened Portland cement paste of low porosity having a compressive strength of 240 MPa. Another example that is particularly noteworthy is the DSP (for Densified with Small Particles) perfected and patented by H.H. Bache of the Aalborg Cement Company of Denmark which can attain a compressive strength of 150 to 200 MPa. Similarly, the MDF (Macro Defect Free) material developed by Birchall *et al.* can develop a flexural strength that can reach 200 MPa. Another example is the pressed cement paste of D.M. Roy (1972) which can attain a compressive strength of 650 MPa. Finally, the DASH 47 material, a commercial product marketed in the United States by CEMCOM, is another example of an ultra high-strength material; part of the aggregate in the DASH 47 is made up of stainless steel powder.

Reactive powdered concrete (RPC), developed by P. Richard, is the latest ultra high-strength Portland cement-based material that has been developed. It can achieve a compressive strength of up to 800 MPa, though this is not the ultimate strength that is likely to be achieved.

All these materials have one thing in common: their water/cement ratio is much lower than that of high-performance concrete. It lies in the 0.10 to 0.20 range in order to achieve the densest packing possible in the hardened product. They are only different in the way in which this increased packing is achieved.

It is not the intention here to review in detail all of these new materials, but rather to show that they all form part of a continuum if they are examined only in terms of their water/binder ratio. Moreover, the description of these materials will show that when the basic principles of cement hydration and particle packing are applied wisely, the compressive strength achieved in high-performance concrete is only a fraction of what can be achieved in the present state of Portland cement technology.

Respecting an historical order, the following sections will present the early first attempt to increase the compressive strength of Portland cement-based materials by Brunauer *et al.*, and will then focus on the principles on which DSP, MDF and RPC are based.

19.2 BRUNAUER *et al.* TECHNIQUE

Brunauer published several papers on the subject and took several patents to cover the technology. Essentially, the Brunauer *et al.* technique consists of:

- grinding very finely an ordinary Portland cement clinker (600 to 900 m^2/kg) in a special grinding machine in the presence of a very efficient grinding aid;
- adding 0.5% potassium carbonate to control C_3A hydration, since the C_3A hydration is not controlled by gypsum, the dissolution of which can be blocked by lignosulfonate;
- adding 1% lignosulfonate in order to make a paste having a W/C ratio of 0.20. In 1972–73, Brunauer was most probably not aware of the more powerful dispersing properties of polynaphthalene and polymelamine sulfonate-based superplasticizers. This technology has not yet been applied on a wide industrial basis.

19.3 DSP

When he developed DSP materials, H.H. Bache took advantage of the combined action of silica fume and superplasticizer in order to lower the porosity of his DSP compositions. He also noted that the coarse aggregate was becoming the weakest link in his compositions. Therefore, in

Table 19.1 Influence of maximum size and type of aggregate on the compressive strength of DSP

ϕ_{max} (mm)	Type of aggregate	f'_c (MPa)	Dynamic elastic modulus (GPa)
16	granite	125	68
16	diabase	168	65
10	calcined bauxite	218	109
4	calcined bauxite	268	108

order to increase their compressive strength, he used very strong aggregates: granite, diabase and calcined bauxite. Moreover, he noted also that the smaller the size of the coarse aggregate, the stronger the DSP, as shown in Table 19.1

Different DSP compositions are marketed under the trade name of DENSIT® by Aalborg Cement for different industrial markets, such as industrial wear-resistant toppings, industrial floors, the chemical fertilizer industry, etc.

19.4 MDF

MDF compositions were developed by Birchall et al. in the early 1980s. Birchall obtained an impressive flexural strength of 200 MPa with such a material. Presently, most of the research work on MDF is being carried out by Young and his colleagues as part of the NSF-ACBM centre research programme. This research work has clearly established that the name MDF cement initially selected by Birchall and his collaborators is not appropriate, since the polymer used during the processing of the MDF material plays an active role, and furthermore MDF systems can be processed without a polymer. Young (1995) suggests instead calling this new type of material 'organo-cement-composite'. Therefore, the abbreviation OCC should be used rather than MDF to properly define this new family of materials. However, this abbreviation is not yet in use, and the materials are still known as MDF materials.

Initially Birchall's idea was based on the analysis of the Griffith criterion, which states that the tensile strength of a brittle elastic material having a critical crack of length l is:

$$\sigma = \sqrt{\frac{E\gamma}{\pi l}}$$

where σ is the tensile strength of the material, E the elastic modulus and γ the specific surface energy of rupture. Therefore in order to increase σ it is only necessary to decrease the length of the critical crack, l.

The mixing of MDF is carried out using a Banbury mixer (a mixer currently used in the plastic and rubber industries) in the presence of a polyvinyl alcohol (PVA) polymer which acts first as a powerful dispersing and then as a reactive binder, developing some partial cross-linking during the process with aluminate ions. Young (1995) says, 'The final product is a rubbery dough which can be calendered into a sheet or extruded to desired shapes. When the cross linking becomes too dense, the dough starts to degrade as the polymer becomes too stiff and macro-defects are reintroduced. Hence a processing window can be defined beyond which macro-defects are introduced into the material on further processing'. Young (1995) has shown that aluminous cement works better than Portland cement when making OCC (MDF) materials. His exploration is that, 'The resulting composite is thus an example of a ceramic body with interpenetrating matrix phases'.

At the present time, the main problem limiting the use of OCC (MDF) composites is that they degrade on immersion in water. First, water is adsorbed by PVA, from which it diffuses to the cement grain where further hydration occurs. Young writes: 'Strength drops rapidly as the polymer swells and softens, and later the interphase region is gradually replaced by a conventional hydrated phase'. Research is presently in progress to solve this drawback of (OCC) MDF composites. An organo-titanate has been found to be promising in that respect.

19.5 RPC

The impressive mechanical performance of RPCs is obtained by:

- rendering the RPC much more homogeneous than usual or high-performance concretes by limiting the maximum size of particles used in its composition to 300 μm. This is going one step further than in DSP compositions in down-scaling the size of the coarse aggregate. It will be shown below that this down-scaling of the size of the coarse aggregate is of the utmost importance in explaining the mechanical behaviour of RPC;
- increasing the density of the granular mixture by using optimal proportions of particles with mean diameters well spaced over the entire granulometric scale;
- applying pressure on the material over the course of hardening to compensate partially for the chemical shrinkage of the cement;
- enhancing the microstructure through an appropriate thermal treatment which transforms the C—S—H into tobermorite and then xonotlite;
- improving ductility through the incorporation of selected steel microfibres, the size and diameter of which are very important.

By integrating all of these individually well-known technologies into one easily implemented manufacturing process, it is possible to manufacture a new type of 'concrete' which responds to two quite unusual objectives. These objectives are high flexural strength and ductility, as well as high compressive strength. During a test completed in October 1994 at the Université de Sherbrooke, it was possible to demonstrate that it was feasible to manufacture, from commercial products found in the Sherbrooke region, a fibrous RPC that had an unconfined compressive strength of 200 MPa, and a compressive strength of 350 MPa when confined in a 3 mm-thick stainless steel tube. Such an RPC was manufactured in a ready-mix truck. A second RPC was then fabricated in a counter-current horizontal mixer in Béton Bolduc precast concrete plant located near Sherbrooke.

As previously mentioned, one area in which RPC behaves quite differently from other ultra high-strength materials that have been developed is a consequence of what can be called a scale effect. In fact, it is easy to demonstrate that the very short and thin steel fibres that are incorporated in RPC, don't act in the same way as the longer and coarser fibres that are usually incorporated in usual or high-performance concretes. If a comparison is made between the size of the fibres ($L = 12$ mm, $\phi = 0.5$ mm) and the maximum size of the coarser powder grain found in an RPC, which is of the order of 300 μm, and the maximum size of the coarse aggregate of a usual or high-performance concrete, which is around 20 mm, it is found that the small and thin fibres used in RPC would correspond to a 10 mm steel reinforcing bar 1 m long in usual or high-performance concrete. In a usual or high-performance concrete such a piece of steel would act not as a fibre but rather as a reinforcing bar and provides a ductility to the concrete, which has a completely different nature than when we reinforce usual or high-performance concrete with one of the commonly used steel fibres.

Another domain in which RPC is quite different from other ultra high-strength materials is that in certain types of application, particularly when it is confined in steel tubes, it can be pressed during its hardening in order to increase its compactness mechanically, by eliminating most of the entrapped air, by squeezing out some of the mixing water and by eliminating a significant part of the chemical porosity that is created during Portland cement hydration.

The performance of RPC that has a very high compressive strength can be readily exploited on site using the post-tensioning technique, which is now widely used in the construction industry. Therefore, it is necessary to design reticulated structures in which the composite RPC elements work only in compression, using steel to take care of tensile stresses. This is necessary to succeed in implementing structural concrete elements made with RPC while using a construction technique similar to those used in steel construction. RPC has already been used in the construction of an experimental cyclopedestrian walkway in Sherbrooke.

SELECTED REFERENCES

GENERAL INTEREST

Buil, M. (1984) Quelques techniques récentes d'obtentin de pâtes de ciment à très haute résistance. Étude bibliographique. *Bulletin de Liaison des Laboratoires des Ponts et Chaussées*, No. 134, November–December, pp. 67–74.

Illston, J.M., Dinwoodie, J.H. and Smith, A.A. (1979) *Concrete, Timber and Metals. The Nature and Behaviour of Structural Materials*, Van Nostrand Reinhold, New York, pp. 421–2, 465.

Lahlou, K., Aïtcin, P.-C. and Chaallal, O. (1991) *Preliminary Study on Confinement of Concrete of Various Compressive Strengths in External Envelopes* (in French), Proceedings, 10th Conference of Canadian Society of Civil Engineering, Supplementary Volume, Vancouver, 13 pp.

Lahlou, K., Aïtcin, P.-C. and Chaallal, O. (1992) Behavior of high strength concrete under confined stresses. *Cement and Concrete Composite*, **14**(3), 185–94.

Mehta, P.K. and Aïtcin, P.-C. (1990) *Microstructural Basis of Selection of Materials and Mix Proportions for High-Strength Concrete*, ACI SP-121, pp. 265–6.

Rossi, P., Renuez, S. and Belloc, A. (1995) Les bétons fibrés à ultra-hautes performances. *Bulletin de Liaison des Laboratoires des Ponts et Chaussées*, No. 196, March, pp. 61–6.

Roy, D.M. (1987) New strong cement materials: chemically bonded ceramics. *Science*, **235**, February, 651–8.

Roy, D.M., Gouda, G.R. and Bobrousky (1972) Very high strength cement pastes prepared by hot pressing and other high pressure techniques. *Cement and Concrete Research*, **2**(3), May, 349–65.

Skalny, J. and Odler, I. (1976) *Use of Admixtures in Production of Low-Porosity Pastes and Concretes*, TRB Transportation Research Record 564, Washington, DC, pp. 27–8.

Young, J.F. (1994) *Engineering Microstructures for Advanced Cement-based Materials*, Proceedings of the Conference in Tribute to Micheline Moranville Regourd, Sherbrooke, published by Concrete Canada, Faculty of Applied Sciences, Université de Sherbrooke, Sherbrooke, Québec, Canada, 20 pp.

Young, J.F. (1993) *High Performance Cement-Based Materials*, Chapter 9, Teaching the materials science, engineering and field aspects of concrete, Part I, NSF-ACBM Center, pp. 253–77.

BRUNAUER TECHNIQUE

Brunauer, S., Yundenfreund, M., Odler, I. and Skalny, J. (1973a) Hardened Portland cement pastes of low porosity, VI mechanism of the hydration process. *Cement and Concrete Research*, **3**(2), March, 129–47.

Brunauer, S., Skalny, J., Odler, I. and Yundenfreund, M. (1973b) Hardened Portland cement pastes of low porosity, VII further remarks about early hydration – composition and surface area of tobermorite gel, summary. *Cement and Concrete Research*, **3**(3), May, 279–93.

Odler, I., Yundenfreund, M., Skalny, J. and Brunauer, S. (1972a) Hardened Portland cement pastes of low porosity, III degree of hydration – expansion of paste – total porosity. *Cement and Concrete Research*, **2**(4), July, 463–80.

Odler, I., Hagymassy, J. Jr, Bodor, E.E., Yundenfreund, M. and Brunauer, S. (1972b) Hardened Portland cement pastes of low porosity, IV surface area and pore structure. *Cement and Concrete Research*, **2**(5), September, 577–89.

Yundenfreund, M., Odler, I. and Brunauer, S. (1972) Hardened Portland cement pastes of low porosity, I materials and experimental methods. *Cement and Concrete Research*, **2**(3), May, 313–29.

Yundenfreund, M., Skalny, J., Mikhail, R.S. and Brunauer, S. (1972a) Hardened Portland cement pastes of low porosity, II exploratory studies – dimensional changes. *Cement and Concrete Research*, **2**(3), May, 331–48.

Yudenfreund, M., Hanna, K.M., Skalny, J., Odler, I. and Brunauer, S. (1972b) Hardened Portland cement pastes of low porosity, V compressive strength. *Cement and Concrete Research*, **2**(6), November, 731–43.

DSP

Bache, H.H. (1981) *Densified Cement/Ultra-Fine Particle-Based Materials*, presented at the 2nd International Conference on Superplasticizers in Concrete, Ottawa, 10–12 June, published by Aalborg Cement, PO Box 165, DK-9100 Aalborg, Denmark, 12 pp.

Bache, H.H. (1987) *High-Strength Development through 25 Years*, CBL Reprint No. 17, Published by Aalborg Portland, PO Box 165, DK-9100, Aalborg, Denmark, 28 pp.

Bache, H.H. (1989a) *Fracture Mechanics as the Basis for Design of Brittle Matrix Composites*, Proceedings of the 5th SMS Scandinavian Symposium on Materials Science DSM, Copenhagen, pp. 135–141.

Bache, H.H. (1989b) Fracture mechanics in integrated design of new, ultra-strong materials and structures, in *Fracture Mechanics of Concrete Structures. From Theory to Applications*. Report of Technical Committee 90. FMA Fracture Mechanics to Concrete. Applications. RILEM (ed. L. Elfgren), Chapman & Hall, London, pp. 382–98.

Bache, H.H. *Densified Cement Ultra-Fine Particle-Based Materials*, CBL NR40.

Bache H.H. Design for ductility, in *Concrete Technology: New Trends, Industrial Applications*, E & FN Spon, London, ISBN 049 20150 5, pp. 113–25.

MDF

Birchall, J.D. (1983) Cement in the context of new materials for an energy-expensive future. *Philosophical Transaction of the Royal Society of London*, **A310**, 31–42.

Birchall, J.D. and Kelly, A. (1983) New inorganic materials. *Scientific American*, **248**(5), May, 104–15.

Birchall, J.D., Howard, A.J. and Kendall, K. (1981) Flexural strength and porosity of cements. *Nature*, **289**(5796), 29 January, 388–90.

Birchall, J.D., Howard, A.J. and Kendall, K. (1982) A cement spring. *Journal of Materials Science Letters*, pp. 125–6.

Birchall, J.D., Howard, A.J., Kendall, K. and Raistrick, J.H. (1983) *Cement Composition and Product*, US Patent 4 410 366, 18 October.

Young, J.F. (1993) *High-Performance Cement-Based Materials*, Chapter 9, teaching the materials science, engineering and field aspects of concrete, Part I, NSF-ACBM Center, pp. 253–77.

Young, J.F. (1995) Engineering advanced cement-based materials for new applications, in *Concrete Technology: New Trends, Industrial Applications*, E & FN Spon, London, ISBN 049 20150 5, pp. 103–12.

RPC

Behloul, M. (1995) Définition d'une loi de comportement du BPR. *Les Annales de l'ITBTP*, No. 532, Série béton 320, March–April pp. 122–7.

Selected references

Bonneau, O., Poulin, C., Dugat, J., Richard, P. and Aïtcin, P.-C. (1996) Reactive powder concrete from theory to practice. *Concrete International*, **18**(4), April, 47–9.

Bonneau, O., Lachemi, M., Dallaire, É., Dugat, J. and Aïtcin, P.-C. (1997) Mechanical properties and durability of two industrial reactive powder concretes. *ACI Materials Journal*, **94**(4), 286–90.

Cheyrezy, M., Maret, V. and Frouin, L. (1995) Analyse de la microstructure du béton de poudres réactives. *Les Annales de l'ITBTP*, No. 532, Série béton 320, March–April, pp. 103–11.

Dugat, J., Roux, N. and Bernier, G. (1995) Étude expérimentale de la déformation sous contrainte et du comportement à la rupture du béton de poudres réactives. *Les Annales de l'ITBTP*, No. 532, Série béton 320, March–April, pp. 112–21.

Dugat, J., Roux, N. and Bernier, G. (1996) Mechanical properties of reactive powder concretes. *Materials and Structures*, **29**(188), May, 233–40.

Feylessoufi, A., Villiéras, F., Michot, L.J., DeDonato, P., Cases, J.M. and Richard, P. (1996) Water environment and nanostructural network in a reactive powder concrete. *Cement and Concrete Composite*, **18**(1), February, 23–9.

Richard, P. and Cheyrezy, M.H. (1994a) *Reactive Powder Concrete with High Ductility and 200–800 MPa Compressive Strength*, ACI SP-144, pp. 507–18.

Richard, P. and Cheyrezy, M. (1994b) *Les Bétons de Poudres réactives*, Proceedings of the Conference in Tribute to Micheline Moranville Regourd, Sherbrooke, published by Concrete Canada, Faculty of Applied Sciences, Université de Sherbrooke, Sherbrooke, Québec, Canada, 28 pp.

Richard, P. and Cheyrezy, M.H. (1995) Les bétons de poudres réactives. *Les Annales de l'ITBTP*, No. 532, Série béton 320, March–April, pp. 85–102.

Roux, N. and Barranco, M. (1995) Mise en oeuvre des bétons de poudres réactives (BPR) dans le BTP. *Les Annales de l'ITBTP*, No. 532, Série béton 320, March–April, pp. 128–32.

Roux, N., Andrade, C. and Sanjuan, M.A. (1995) Étude expérimentale sur la durabilité des bétons de poudres réactives. *Les Annales de l'ITBTP*, No. 532, Série béton 320, March–April, pp. 133–41.

CHAPTER 20

A look ahead

Regularly, eminent researchers or engineers are asked to present their views on the future of concrete, on concrete in the year 2000 or on future trends (Tassios, 1987; Walther, 1987; Kukko, 1993; Mather, 1993, 1995; Mehta, 1993; Richard, 1996). It is not my intention to review all of these texts or to try to synthesize them. I would rather express my own perception of the future trends that we should see developing in the years to come in the area of high-performance concrete. I am ready to accept the credit for my correct predictions as well as the consequences of the wrong ones.

20.1 CONCRETE: THE MOST WIDELY USED CONSTRUCTION MATERIAL

According to CEMBUREAU, more than 1 billion tonnes of cement were produced every year between 1990 and 1995 (Table 20.1). If it is assumed that on average 250 kg of cement are necessary to produce a cubic metre of concrete, it can be estimated that more than 4.4 billion cubic metres of concrete were produced in each of these years. As one cubic metre of concrete weighs around 2.5 tonnes, these 4.4 billion cubic metres of concrete weighed about 11 billion tonnes, i.e. almost 2 tonnes of concrete per person per year. Only fresh water was more widely used during the same period of time (Aïtcin, 1995).

However, concrete is not used uniformly in all countries (Table 20.2). Generally speaking, concrete is used more in industrialized countries than in developing ones, though when scrutinizing the CEMBUREAU statistics, it may be observed that the consumption of cement is rather stagnant in industrialized countries, but is increasing in developing

Table 20.1 World-wide production of cement, according to CEMBUREAU

	Millions of tonnes				
	1990	1991	1992	1993	1994
Total production world-wide	1141.5	1167.3	1239.5	1296.8	1388.4
As a percentage of 1990 figure	Reference year	+ 2%	+ 9%	+ 14%	+ 22%

Concrete: the most widely used construction material

Table 20.2 Production of cement in different countries, according to CEMBUREAU

	Cement production percentage		
	Millions of tonnes per year		Variation as a percentage of 1990
	1990	1994	
China	208	405	+ 95%
Japan	84.4	91.6	+ 8.5
USA	68.6	75.9	+ 10.5%
India	47.3	61.5	+ 30%
Italy	40.8	33.0	− 19%
South Korea	33.3	51.7	+ 55%
Spain	28.1	25.2	− 9%
France	26.4	20.0	− 24%
Brazil	25.8	25.2	− 0.7%
Turkey	24.4	30.1	+ 23%
Mexico	23.8	27.6	+ 16%
Taiwan	19.4	23.4	+ 21%
Thailand	18.8	31.1	+ 65%

countries (Tables 20.2 and 20.3). Therefore, on a world-wide basis, the future of the concrete industry is well secured, at least for the first half of the next century, owing to the predictable development of the most highly populated developing countries.

This great consumption of concrete by human beings is due to some of the inherent qualities of concrete as a construction material: concrete remains a cheap construction material incorporating around 85% of local materials (sand, coarse aggregate and water), it has a very good

Table 20.3 Consumption of cement per inhabitant per country, according to CEMBUREAU

	Consumption per inhabitant		
	kg/year		Variation as a percentage of 1990
	1990	1994	
Luxembourg	1150	1185	+ 0.3%
Switzerland	831	678	− 18%
Greece	751	686	− 9%
Italy	748	609	− 19%
Spain	704	612	− 13%
Portugal	698	768	+ 10%
Japan	680	643	− 5%
France	448	347	− 23%
USA	322	329	+ 0.2%
China	184	331	+ 80%

resistance to water and fire, it is not attacked by insects (except by some bacteria in highly polluted urban environments), it does not rot, it is usually durable under a great variety of environments, and its production and use do not require a highly sophisticated technology.

More particularly, the use of high-performance concrete will increase not only in industrialized countries but world-wide, owing to the rapid development of communications. The manufacture of high-performance concrete is no longer a well-kept secret by some concrete producers. Not only is the technology of high-performance concrete well known, but its science is also well understood in most places. Considerable knowledge has been acquired on the selection of the materials that are required in order to make economical and durable high-performance concrete, and producers can often take advantage of some local economic advantages to produce an economical high-performance concrete meeting all the required specifications. This relative flexibility in the fabrication of high-performance concrete will be a key factor in the success of the development of its use in the years to come.

20.2 SHORT-TERM TRENDS FOR HIGH-PERFORMANCE CONCRETE

Based on trends that can already be observed, it can be predicted that high-performance concretes made with lower and lower water/binder ratios will find a use. If at the present time the 0.30 to 0.35 water/binder ratio range seems to be the most widely used, high-performance concretes with water/binder ratios in the 0.25 to 0.30 range have been used for the construction of some impressive structures, such as the Two Union Square building in Seattle and the Confederation bridge which links Prince Edward Island and New Brunswick in Canada, and will be used more and more in the near future. Tomorrow's high-performance concretes will contain less and less Portland cement but more and more supplementary cementitious materials or even fillers. As pointed out earlier in this book, the substitution of a certain amount of Portland cement by a supplementary cementitious material is not only advantageous from an economical point of view, but is also usually very advantageous from a rheological and sometimes from a heat development point of view. Replacing a certain amount of cement by a less rheologically reactive material results in a better control of the rheology of the fresh high-performance concrete and its easier placing. Of course, such high-performance concretes are not as strong as Portland cement concretes at 24 hours, but as their water/binder ratios can be decreased somewhat because they contain less highly reactive Portland cement, they can develop 24 hour compressive strengths in the range of 15 to 30 MPa, which is high enough for most structural applications. What has been lost in terms of the early age binding properties of Portland cement can be compensated for by an additional decrease of the water/binder ratio. However, it has been shown recently that limestone filler can be used to increase the very early strength at 12 to 15 hours when making high-performance concrete (Kessal et al., 1995; Nehdi, Mindess and Aïtcin, 1995). Finally, the decrease in the amount of cement used to make a high-performance concrete and its replacement by a less reactive cementitious material results most of the time in the use of less superplasticizer to get a given workability, so that a significant decrease in the cost of a cubic metre of concrete can be achieved.

Slag sources are not as numerous as fly ash ones, and therefore it is not very wreckless to predict that in general world-wide high-performance concrete will evolve to containing more and more fly ash, just like the first high-strength concretes made by the pioneers who were trying to lower the water/binder ratio of their concrete by the few means they had at their disposal. But the major difference between this first generation of high-performance concrete and the next one will be the replacement of lignosulfonate-based water reducers by more efficient modern superplasticizers.

Portland cement and superplasticizers in the years to come will become more efficient and compatible in the domain of the low water/binder ratios because cement and admixture companies will realize that in spite of its relatively low present share of the concrete market, high-performance concrete use will develop rapidly in a market in which the competition is not uniquely based on price but also on quality. Very often the ability to produce and deliver the 10 or 20% of high-performance concrete required to build a structure will give as a premium the delivery of the 80 to 90% of the usual concrete necessary to complete the structure, making the economical production of high-performance concrete particularly advantageous.

It is easy to produce almost anywhere a cement or a superplasticizer meeting the present acceptance standards for usual concretes, but this is not the case for a Portland cement and a superplasticizer that have to be used together in a low water/binder ratio mixture. Portland cement and superplasticizer manufacturing and quality control have to be tightened in order to produce a Portland cement and a superplasticizer that will make possible the delivery of a high-performance concrete with predictable and constant rheological and strength properties under changing ambient temperatures.

The technology of manufacturing superplasticizers has improved greatly in recent years and it is not presumptuous to say that the superplasticizers that are available today are much more effective than the 'first generation' superplasticizers. By no means, however, can it be stated that there are no further improvements to be achieved in that domain. Superplasticizer manufacturers are in a strong position to succeed in their search to improve the efficiency of their products, as their market is growing rapidly and a great deal of research is presently being done using sophisticated modern analytical techniques to try to explain and understand the fundamental action of superplasticizers on cement particles.

Different types of superplasticizer will be developed to better fit the different rheological requirements of different types of high-performance concrete. Ranges of formulation will be developed to allow for the chemical differences in cement composition which can become important when these cements are used at low water/binder ratios. The increase of superplasticizer efficiency will result in a significant decrease in the dosage required to obtain a certain level of workability or compressive strength or durability, and this is likely to result in further significant decreases in the unit cost of high-performance concrete, and of course, in increases in its competitiveness.

Certainly these will not be the only innovations that will develop in the near future in high-performance concrete. It is always risky to try to predict technological developments, because sometimes chance or an unpredictable environment or political developments can play a great role in a new technological achievement. It is always much easier for scientists and researchers to explain these kinds of development after the event rather than before. As has very often been the case with most engineering technologies, concrete technology will probably precede concrete science for many years to come.

20.3 THE DURABILITY MARKET RATHER THAN ONLY THE HIGH-STRENGTH MARKET

Certainly, by the end of the first part of the next century, usual concretes will still be used and, in spite of all of its qualities, high-performance concrete will never capture the entire concrete market. There are many applications for which it is not necessary to use a low water/binder ratio concrete and for which the use of a usual concrete is entirely satisfactory. Moreover, there are very few applications for which a high-strength concrete is absolutely necessary: perhaps 5 to 10% of the present concrete market, or even less. In fact, the market in which high-performance concrete applications will develop in the near future is the durability market: any concrete structure that will have to face severe environmental conditions will be made with a high-performance concrete in order to increase its life cycle.

It can be estimated that such a market can represent, depending on the country's location, from 25 to 35% of the total concrete market. However, if because of environmental considerations more stress is placed on a better use of our natural resources, which include fuel, sand, coarse aggregate and a decrease of CO_2 emission, the market share of high-performance concrete will increase more than that. It is easy to demonstrate that to obtain 1 MPa in a structural element requires the use of less concrete and cement and the use of less aggregates when using a high-performance concrete rather than a usual concrete. Also, using less cement means the emission of less CO_2. It must be remembered that the use of 1 tonne of cement entails roughly the release of 1 tonne of CO_2 into the atmosphere.

In order to reinforce this personal vision, it must be noted that more and more agencies are realizing that it is no longer possible to build an infrastructure with high water/binder ratio concretes with a very short life cycle, owing to the increasing severity of most urban environments and the skyrocketing direct and social costs of maintenance and repair works. No country in the world is rich enough to pursue its development using cheap concrete with a high water/binder ratio. When an infrastructure is built on the built–own–operate and transfer mode, it is amazing how contractors become conscious of the quality of concrete.

20.4 LONG-TERM TRENDS FOR HIGH-PERFORMANCE CONCRETE

High-performance concretes will always remain low water/binder concretes incorporating superplasticizers, but their technology of fabrication and placing will change. In the future high-performance concrete will contain less cement, less water, more supplementary cementitious materials and more admixtures. Portland cement composition will evolve towards a more belitic composition; it will contain more C_2S and less C_3S and probably more C_4AF than C_3A. C_3A will no longer be necessary to boost concrete's early strength.

In parallel, high-performance concrete will have to be made a more robust material, i.e. a material less sensitive to variations in the quality of the cement and the superplasticizer or of the ambient temperature. Heat development, shrinkage and creep will be decreased.

The increased use of high-performance concrete and its intelligent monitoring will allow us to refine present design methods and the prediction of the life cycle of high-performance concrete structures, but as a consequence, what will be gained in performance will be lost in simplicity and robustness. Concrete technology will have to evolve from a low-tech to a high-tech level.

Not only the material itself but also its placing must evolve rapidly. According to Pierre Richard (1996), the use of 1 litre of superplasticizer corresponds to the cost of 20 minutes of a helper in France, and seven-eighths of the time needed to fill a cofferdam with concrete is devoted to the erection of forms and the placing of the steel reinforcement, which is too labour-intensive. Too much labour is still needed to build and place concrete in structural elements. In particular, in Japan it is said that concrete is afflicted with the 4 K syndrome, which can be translated as the 4 D syndrome in English: concrete is **d**ifficult, **d**angerous, **d**irty and desperately undervalued. Different placing techniques have already been developed to facilitate the use of high-performance concrete, such as self-placing concretes (also called high-performance concretes in some Japanese publications), roller-compacted concretes, etc.

On the design side, a better use of the extra MPa's provided by high-performance concrete will have to be made. Prestressing and post-tensioning of high-performance concrete will become more and more widely used. Whenever the thickness of a structural element is dictated by minimum cover considerations, the use of external post-tensioning will have to be considered and provisions will have to be made for the possibility of changing any external post-tensioning cables, as in the Joigny bridge (Chapter 4).

20.5   HIGH-PERFORMANCE CONCRETE COMPETITION

Already, 'high-performance' concrete is no longer the strongest and most high-performing concrete on the market, since the recent development of reactive powder concretes. In spite of the fact that this new type of concrete is much more expensive on a unit price basis than high-performance concrete, there are already some industrial applications for which its use is economical because of the savings realized by eliminating the need for reinforcement and the lightness of the final structure. Moreover, if at the present time there is a gap between high-performance concrete and reactive powder concrete owing to the early development stage of this new type of concrete, it is certain that within the next few years a continuum will fill this gap and that concretes that are partially reactive powder concrete and partially high-performance concrete will find some applications for which reactive powder concrete performance could not be fully utilized and for which the present high-performance concretes still fall somewhat short in terms of performance. As there is no discontinuity between usual concrete and high-performance concrete, there is no reason for a discontinuity between high-performance concrete and reactive powder concrete. It is not being too optimistic to predict that a market for concrete in the near future will develop for high-performance concretes having strengths between 1 and 1000 MPa.

20.6   RESEARCH NEEDED

In spite of the fact that a tremendous research effort is presently underway in the domain of high-performance concrete, as could be observed at the last International Symposium on High-Performance Concrete held in Paris in May 1996, there are still some areas in which the final answers are not yet known and for which additional research and development will have to be undertaken.

From a materials point of view, which is the author's research field of interest, progress has to be made in the following fields:

- Portland cement–superplasticizer compatibility;
- durability of high-performance concrete, particularly with respect to freezing and thawing and deicing salt attacks;
- need of entrained air for freeze–thaw durability;
- decrease of heat development;
- decrease of autogenous shrinkage;
- increase of toughness;
- resistance to fire.

Great progress in the understanding of high-performance concrete and of concrete in general has followed from a better knowledge of concrete microstructure. It seems that the next major progress will follow from a better knowledge of concrete **nanostructure**. This is the challenge of the years to come, but one thing is already certain: as a construction material, concrete has not yet said its last word. Many new discoveries and developments will occur in the concrete field, which will be a consequence of a research effort linking science and technology, because in spite of its complexity, concrete is a material that still obeys the laws of physics and chemistry.

REFERENCES

Aïtcin, P.-C. (1995) *Concrete: The Most Widely Used Construction Material*, Adam Neville Symposium on Concrete Technology, Las Vegas, USA, June (ed. V.H. Malhotra), pp. 257–66.

Kessal, M., Nkinamubanzi, P.-C., Tagnit-Hamou, A. and Aïtcin, P.-C. (1996) Improving initial strength of a concrete made with type 20 M cement. *Cement, Concrete, and Aggregates*, **18**(1), June, 49–54.

Kukko, H. (ed.) (1993) *Concrete Technology in the Future*, VTT Symposium 138, RILEM Workshop, ISBN 951-38-4089-1, 125 pp.

Mather, B. (1993) Concrete – year 2000, revisited, in *Concrete Technology – Past, Present, and Future*, Proceedings of V. Mohan Malhotra Symposium (ed. P.K. Mehta), ACI SP-144, pp. 31–9.

Mather, B. (1995) Concrete – year 2000, revisited in 1995. *Adam Neville Symposium on Concrete Technology* (ed. V.M. Malhotra), Las Vegas, USA, June, pp. 1–9.

Mehta, P.K. (1993) Concrete technology at the crossroads. Problems and opportunities, in *Concrete Technology – Past, Present and Future*, Proceedings of V. Mohan Malhotra Symposium, (ed. P.K. Mehta), ACI SP-144, pp. 1–30.

Nehdi, M., Mindess, S. and Aïtcin, P.-C. (1995) Use of ground limestone in concrete: a new look. *Building Research Journal*, **43**(4), 245–61.

Richard, P. (1996) *The Future of HS/FIPC*, 4th International Symposium on Utilization of High Strength/High-Performance Concrete, Paris, Vol. 1, ISBN 2-859 78-257-5, pp. 101–6.

Tassios, T.P. (1987) *Concrete Structures for the Year 2000*, Versailles, Paris 1987 IABSE Symposium, IABSE, Zurich, ISBN 3-85748-053-1, pp. 639–46.

Walther, R. (1987) *Concrete Structures for the Year 2000*, Versailles, Paris 1987 IABSE Symposium, IABSE, Zurich, ISBN 3-85748-053-1, pp. 631–8.

Afterword

Pierre-Claude Aïtcin has shared with us his expertise and passion for the development of concrete with higher strength and enhanced quality. His expertise comes from years of work and innovation. In fact, many of the advances in the field of high-performance concrete originate with him.

His passion is evident in his approach. His drive to improve the quality of concrete structures has led him to become involved in civil engineering projects ranging from a modest to large scale. And this has been critical in giving added value to his teaching.

One might wonder why it takes so long for such an advanced material to be universally accepted. Because of its qualities – durability, resistance to natural and industrial attack, ease of placement and enhanced appearance – it should be used more often.

In my opinion, the main obstacle to the greater use of high-performance concrete is simply inertia. Some point out that high-performance concrete costs more when it comes out of the mixer truck, ignoring that this difference disappears when the total costs of the structure are considered. The significant reduction in volume required, faster placement and the elimination of surface defects that need correction easily compensate for unit price differences, as indicated by the gradually increasing use of high-performance concrete.

Others claim that high-performance concrete is brittle. As with cost, this statement is both true and false. It is true when applied to laboratory testing, but high-performance concrete, owing to its higher homogeneity, produces reinforced or post-tensioned concrete structures that are much more ductile than those made with usual concrete.

The only real criticism that can be levelled at high-performance concrete is that its tensile strength remains inherently low, only slightly better than that of its usual counterpart. But even with this, high-performance concrete has made an invaluable contribution in this regard by providing the impetus for scientists across the world, especially in Canada, the United States and France, to take up the challenge of exploring this aspect. Their research will surely lead to higher tensile strength in high-performance concrete. It is easy to imagine what a boon this will be to the art of building, and it is just around the corner.

Pierre-Claude Aïtcin stands out as a pioneer in this scientific journey and a key individual in its progress. This book stands as proof of that. He has and will continue to have my friendship and gratitude.

Pierre Richard
27 March, 1997

Suggested reading

INTERNATIONAL CONFERENCES ON HIGH-STRENGTH/HIGH-PERFORMANCE CONCRETE

Utilization of High Strength Concrete (1987) (eds I. Holland, D. Holland, B. Jakobsen and R. Lenschow), published by TAPIR, N-7034 Trondheim, NTH, Norway, ISBN 82-519-0797-7.

High Strength Concrete (1990) 2nd International Symposium (ed. Weston T. Hester), ACI SP-121.

High Strength Concrete (1993) Vols 1 and 2, Proceedings of the Lillehammer Symposium, (ed. I. Holand and E. Sellevold), published by the Norwegian Concrete Association, PO Box 2312 Solli, No. 201, Oslo, Norway, ISBN 82-91341-00-1.

Utilization of High Strength/High Performance Concrete (1996) 4th International Symposium, Paris (ed. F. de Larrard and R. Lacroix), published by Les Presses de l'École Nationale des Ponts et Chaussées, ISBN 2-85978-257-(1-3-5).

STATE-OF-THE-ART REPORTS

Literature Review of High Strength Concrete Properties (1988) by L.J. Parrott. Review carried out by C and CA Services, January, 87 pp.

High Performance Concretes (1991) by P. Zia, M.L. Leming and S.M. Ahmad. A State-of-the-Art Report by the Strategic Highway Research Programme, National Research Council, 2101 Constitution Avenue, N.W., Washington, DC, 20418, USA.

High Strength Concrete (1992) Compilation 17, American Concrete Institute.

State-of-the Art Report on High Strength Concrete (1992) Report by Committee 363, ACI 363R-92.

ANNOTATED BIBLIOGRAPHY

High Performance Concretes: An Annotated Bibliography (1989–1994) by the Federal Highway Administration, US Department of Transportation. WEB Site, Turner Fairbank Highway Research Center Home page http://www.thhrc.gov.

BOOKS

High Performance Concrete: From Material to Structure (1992) (ed. Y. Malier), published by E & FN Spon, 11 New Fetter Lane, London EC4P 4EE, ISBN 0-419-17600-4.

High Performance Concrete: Properties and Applications (1994) by S.P. Shah and S.H. Ahmad, McGraw-Hill, ISBN 0-07-056974-6.
Production of High Strength Concrete (1986) by M.B. Peterman and R.L. Carrasquillo, Noyes Publications, Park Ridge, New Jersey 07656, USA, ISBN 0-8155-1057-8.
High-Strength Concrete (1985) (ed. H. Russel) ACI SP-87.
High Performance Concrete in Severe Environments (1993) (ed. P. Zia), ACI SP-140.

GUIDES

High Strength Concrete – Phase 3 – SP1 Design Guide (1994) Report 1.2 Design Guide, STF70 A94044, published by SINTEF Structures and Concretes, N-7034 Trondheim, Norway.
High Performance Concrete Recommended Extensions to the Model Code 90, Research Needs (1995) CEB Bulletin d'information 228, July.

MISCELLANEOUS

High-Performance Construction Materials and Systems. An Essential Programme for America and its Infrastructure (1993), CERF Technical Report 93-50, ISBN 0-87262-938-2.
High Strength Concrete (1987) Seminar Course Manual SCM-15(87), ACI.
High-Strength Concrete (1992) Cement and Concrete Association of Australia, National Ready Mixed Concrete Association of Australia. Published by Cement and Concrete Association of Australia, Tel.: (02) 923–1244, Fax: (02) 923–1925.
High-Strength Concrete in Chicago High-Rise Buildings (1977) Task Force Report, Chicago Committee on High-Rise Buildings, Report No. 5, February, 63 pp.
Engineering Properties of Commercially Available High-Strength Concretes (1992) by R.G. Burg and B.W. Ost, published by the Portland Cement Association, RD 104.01T.
In-Situ Strength of High Strength Concrete (1990) Report No 4/90, by S.L. Mak, M.H. Attard, D.W.S. Ho and P.L.P. Darwell, ISBN 07326 02181, Monash University, Clayton, Victoria 3168, Australia.
Conference in Tribute to Micheline Moranville Regourd (1994) Concrete Canada, Department of Civil Engineering, Faculty of Applied Sciences, Université de Sherbrooke, Sherbrooke, Québec, Canada.
Effects of Testing Variables on the Measured Compressive Strength of High-Strength (90 MPa) Concrete (1994) by N.J. Carino, W.F. Guthrie, E.S. Lagargen and E.S. Lagergen, NISTIR 5405, National Institute of Standards and Technology, Gaithersburg, MD, USA, 141 pp.
VTT Symposium 138, RILEM Workshop Concrete Technology in the Future (1996) ISBN 951-38-4089-1.

IN FRENCH

Les bétons à hautes performances, caractérisation, durabilité, applications (1992) Édité par Y. Malier, Publié par les Presses de l'École Nationale des Ponts et Chaussées, ISBN 2-85978-187-0.
Synthèse bibliographique sur les bétons à très hautes performances (1992) par F. de Larrard, École Nationale des Ponts et Chaussées, Département de génie civil et transport, 28 Rue des Saints-Pères, Paris 75007.

Extension du domaine d'application des règlements de calcul BAEL/BPEL aux bétons à 80 Mpa (1996) Special XIX, Laboratoire Central des Ponts et Chaussées, ISBN 1269-1496.

Bétons à hautes performances (BHP) SIA D 068 (1991) publié par SIA Avenue Domini 8, CP 1471 Lausanne 1001.

Propriétés mécaniques des bétons durcissants: analyse comparée des bétons classiques et à très hautes performances (1993) par P. Laplante, publié par le Laboratoire Central des Ponts et Chaussées, OA 13.

Structures en béton à hautes performances. Fissuration. Étanchéité. Durabilité SIA D 0702 (1995) publié par SIA Avenue Domini 8, CP 1471 Lausanne 1001, Switzerland.

ADDRESSES OF SOME NATIONAL AND INTERNATIONAL ORGANIZATIONS REFERRED TO IN THIS BOOK

American Ceramic Society, PO Box 6136, Westerville, OH 43086–6136, USA, Tel.: 614/890-4700, Fax.: 614/899–6109, Internet: http://www.acers.org.

American Concrete Institute, PO Box 9094, Farmington Hills, MI 48333, USA, Tel.: 810–848–3700, Fax: 810–848–3701, World Wide Web: http://www.aci-int.inter.net.

American Society for Testing and Materials, 100 Barr Harbor Dr., W. Conshahocken, PA, 19428–2959, USA, Tel.: 610/832–9500, Fax: 610/832–9555, ASTM Web Site: http://www.astm.org.

Canadian Standard Association (CSA), 178 Rexdale Street, Toronto, Ontario, M9W 1R3, Canada.

CANMET (a/s Mohan Malhotra), 405 Rochester Street, Ottawa, Ontario, K1A 0G1, Tel.: (613) 996–5449, Fax: (613) 992–9389.

CEB Comité Euro-International du Béton (CEB), BP 88, CH-1015 Lausanne, Switzerland.

IABSE/AIPC/IVBH, ETH Hönggerberg CH-8093 Zurich, Tel.: 411–633–2647, Fax: 411–371–2131.

Laboratoire Central des Ponts et Chaussées, 58 boulevard Lefebvre, F-75732 Paris Cedex 15, Tel.: 1–40–43–50–00, Fax: 1–40–43–54–98, Internet http://www.lcpc.inrest.fr/.

Portland Cement Association, 5420 Old Orchard Road, Skokie, IL 60077-1083, USA, Tel.: 708–966–6200, Fax: 708–966–9781.

RILEM The International Union of Testing and Research Laboratories for Materials and Structures. Pavillon des Jardins, 61 Av. du Président Wilson, 94235, Cachan Cedex, France, Tel: 1–47–40–23–97, Fax: 1–47–40–01–13, Web Server: http://www.ens.cachan.fr/rilem/.

Strategic Highway Research Programme, National Research Council, 2101 Constitution Avenue, N.W., Washington, DC 20418, Tel.: 202–334–3774.

Transportation Research Board, Business Office, National Research Council, 2101 Constitution Avenue, N.W., Washington, DC 20418, Tel.: 202–334–3214.

Author index

Aarup, B. 502
Abdel-Fattah, H. 532
Abdul-Maula, S 130
Absi-Halabi, H. 126, 170
Acker, P. 62, 312, 325, 350, 370, 413, 446
Addis, B.J. 257
Ahmad, S.H. 432, 442, 532
Aïtcin, P.-C. 26, 27, 28, 29, 58, 64, 87, 88, 95, 131, 133, 136, 140, 141, 143, 149, 151, 154, 162, 170, 173, 175, 178, 185, 195, 197, 198, 201, 202, 203, 204, 205, 206, 207, 208, 209, 210, 211, 215, 225, 234, 236, 257, 266, 267, 270, 271, 274, 275, 276, 277, 278, 279, 280, 282, 287, 299, 300, 301, 302, 303, 304, 306, 311, 312, 330, 342, 344, 345, 346, 349, 350, 353, 354, 355, 356, 357, 358, 359, 360, 361, 362, 363, 364, 365, 367, 368, 369, 370, 376, 381, 383, 396, 398, 413, 414, 424, 427, 428, 429, 430, 437, 439, 442, 446, 447, 452, 472, 474, 475, 476, 479, 480, 482, 484, 485, 487, 492, 493, 494, 495, 511, 514, 517, 518, 520, 521, 522, 526, 527, 529, 531, 532, 545, 548, 554, 555, 556, 559
Albinger, J. 26
Alègre, R. 113
Alexander, M.G. 257, 322
Alfes, C. 437
Ali, M.A. 126, 170
Allard, M.M. 290
Allum, D. 295
Al-Manaseer, A.A. 393
Alou, F. 29
Amsler, D.E. 384
Anderson, F.D. 297
Andersen, P.J. 130, 133
Andrade, C. 555
Ansari, F. 399
Aoyama, H. 29
Arioka, M. 466
Arsenault, B. 489
Ashour, S.A. 527
Asselanis, J. 424
Attard, M.M. 398
Aune, R.B. 500
Autefage, F. 154, 195

Baalbaki, M. 29, 136, 170, 184, 185, 197, 234, 271, 274, 275, 276, 277, 282, 301, 302, 345, 346, 368, 369, 428, 480
Baalbaki, W. 95, 435, 439, 440, 442, 444
Bache, H.H. 27, 93, 97, 146, 529, 554
Baerland, T. 473, 474, 476
Bakke, J.A. 479
Baldini, G. 111
Ballivy, G. 95
Balogh, A. 329
Banthia, N. 479, 482, 485
Baroghel-Bouny, V. 325, 354
Baron, J. 98, 320
Barranco, M. 555
Bartlett, F.M. 398
Barton, R.B. 2
Baudot, J. 47
Baussant, J.-B. 136
Beaudoin, J.J. 123, 489
Bédard, C. 27, 133, 345, 349, 356, 357, 428
Behloul, M. 555
Bélanger, P.R. 366
Belloc, A. 202, 383, 390, 553
Ben-Bassat, M. 496
Benmokrane, B. 95, 442
Bennett, E.W. 453
Bentur, A. 96, 146, 465, 496, 517
Bentz, D.P. 460, 471
Bernhardt, C.J. 28
Bernier, G. 555
Berra, M. 517, 518, 520, 521
Berry, E.E. 140
Bertero, V.V. 531
Bérubé, M.-A. 497, 499
Bessho, S. 29
Bhatty, J.I. 190
Bickley, J.A. 50, 51, 52, 140, 197, 269, 274, 281, 288, 290, 292, 293, 301, 307, 345, 346, 359, 367, 373, 397, 433, 441
Bilderbeek, D.W. 76, 472, 477, 511
Birchall, J.D. 93, 553, 554
Bissonnette, B. 323
Bjerkeli, L. 532
Black, B. 123
Blais, F.A. 266, 267, 270, 271, 274, 275, 276, 277, 278, 279, 280, 291, 299, 300, 304, 363, 367, 368, 369, 511
Blick, R.L. 4, 24, 25, 123, 257

Blouin, D. 274, 304, 306, 342, 360, 361, 484
Bobrowsky, A. 92, 553
Bode, H. 531
Bodor, E.E. 553
Bois, A.P. 356, 360
Boisvert, A. 479
Bolomey, J. 96
Bonneau, O. 555
Bonzel, J. 123
Bosc, F. 178, 180, 271, 341
Bossanyi, F. 128
Boulay, C. 66, 271, 341, 383, 384, 385, 390
Bouzoubaâ, N. 64, 356, 485
Bradley, G. 126
Bramforth, P.B. 199, 368
Branco, F.A. 366
Brazilier, D. 59, 497
Breitenbücker, R. 502
Bremner, T.W. 497, 519
Brooks, J.J. 378
Brownyard, T.L. 429
Brunauer, S. 553
Buil, M. 97, 113, 132, 320, 499, 553
Burg, R.G. 349, 398, 432
Burnett, I. 29
Burr, W.H. 530
Bye, G.C. 102

Cadoret, G. 47, 53, 55, 56, 115, 346
Cail, K. 193
Canitrot, B. 257, 261
Cánovas, M.F. 257
Cao, H.T. 461
Capell, H.T. 398
Caquot, A. 96, 437
Carette, C.G. 140
Carin, V. 133
Carino, N.J. 290, 373, 472
Carles-Gibergues, A. 154, 195
Carlsen, J.E. 486, 487
Carrasquillo, P.M. 384, 393
Carrasquillo, R.L. 257, 384, 393, 432, 438
Cases, J.M. 555
Casey, K. 267, 271
Castillo, C. 423
Catherine, C. 178, 180
Causse, G. 53
Cechner, R. 144, 192
Cederwall, K. 531
Chaallal, O. 95, 330, 358, 442, 446, 452, 531, 532, 548, 554
Chan, J.K.W. 501
Chan, S.Y.N. 501
Chan, Y.N. 472
Chanvillard, G. 529
Charif, H. 29
Chatterji, V.S. 121
Chen, W.F. 492
Chern, J.C. 29
Cheyrezy, M.H. 554

Chibnowski, S. 130
Chin, M.S. 399, 442
Clastres, P. 501
Clifton, J.R. 290
Coche, D.L. 122
Collepardi, M. 111
Collins, M.P. 399, 454
Combault, J. 76, 472, 477, 511
Conradi, M. 111
Cook, J.E. 271, 297, 393, 394
Cook, W.D. 303, 349, 350, 353, 354, 356, 357, 359, 360, 370, 398, 427, 428
Corley, W.G. 42, 446
Corsi, F. 500
Costa, V.B. 130
Costaz, J.-L. 501
Coulombe, L.-G. 65, 511
Cubaynes, J.-F. 322
Cunningham, J.C. 132
Cusson, D. 532

Dagher, H.J. 460
Daimon, M. 170
Dallaire, É. 266, 267, 270, 271, 274, 275, 276, 277, 278, 279, 280, 299, 300, 304, 306, 330, 342, 360, 361, 363, 367, 368, 369, 484, 511, 555
Dalziel, J.A. 198
Darvall, P. le P. 290, 398
Date, C.G. 393
Day, K.W. 215, 295
DeDonato, P. 555
Deflorenne, F. 178, 180, 271, 341
Delagrave, A. 486
de Larrard, F. 29, 62, 66, 97, 98, 132, 178, 180, 192, 202, 215, 257, 261, 271, 341, 350, 383, 384, 387, 390, 429, 434, 437, 452
Delvicki, G. 501
Desayi, P. 442
Desfossés, C. 498
Detrez, M. 132
Detwiler, G. 29
Detwiler, R.J. 97
Dhir, R.K. 472
Diamond, S. 130
Diederich, U. 502
Dinwoodie, J.M. 87, 94, 433, 546, 554
Do, M.T. 452, 453
Dodson, V.H. 113, 136
Domone, L.J. 257
Dubouchet, A. 364
Dugat, J. 555
Durekovic, A. 145, 147
Durrani, A.J. 423
Dury, B.L. 132

Eckart, V.A. 115, 407
Eguchi, H. 136, 165, 166, 184
El-Dieb, A.S. 412
El Hindy, É. 330, 446

Author index

Elliott, S. 489
Elimov, R. 71, 74, 269, 270, 274, 299, 304, 307, 345, 511, 515, 516, 517, 518, 522, 525
Elwi, A.E. 423
England, G.L. 501
Engstrom, B. 531
Erlacher, A. 325
Evans, B. 531
Ezeldin, A. 95

Faury, J. 96
Feldman, R. 28
Felicetti, R. 500
Fentress, B. 472
Féret, R. 8, 27, 91, 257
Fernon, V. 132
Ferrara, G. 517, 518, 520, 521
Feylessouffi, A. 555
Foissy, A. 130
Foot, K. 285
Forstie, D.A. 393
Fotinos, G. 76, 472, 477, 511
Foy, C. 485
François, R. 322
Fredrich, J.T. 531
Freedman, S. 4, 22, 25, 392
Frolund, T. 486, 487
Frouin, L. 555
Fukute, T. 486, 487
Fuller, W.B. 96
Fundel, A. 343
Furlong, R.W. 531
Fuyaka, Y. 166, 184

Gagné, R. 299, 413, 479, 482, 485, 487, 492, 493, 494, 495, 511, 513, 514, 515
Gaidis, J.M. 133, 145
Gambavora, P.G. 500
Ganju, T.N. 215
Garboczi, E.J. 460, 471
Gardner, N.J. 438, 531
Garvey, M.J. 130
Garvin, S. 130
Gaumy, A. 66, 69
Gawsewitch, J. 325, 354
Gebauer, J. 110
Gendreau, M. 184, 234, 274, 276, 282, 299, 346, 369, 511, 513, 514, 515
Gerwick, B.C. Jr 343
Giacco, G. 437
Gianuzzi, G. 500
Gillen, M. 502
Gillet, G. 257, 261
Gillott, J.E. 499
Gjørv, O.E. 193, 271, 340, 341, 473, 474, 476, 487, 514, 516, 517, 518, 519, 520, 522, 525
Godfrey, K.A. Jr 28, 56
Godin, J. 325, 354
Golaszewski, J. 340, 341

Goldman, A. 96, 146
Gonnerman, H.F. 393
Gopalan, M.K. 290, 359, 398
Gordon, J.E. 37
Gorse, J.F. 178, 192
Gouda, G.R. 92, 553
Gokoz, U. 399
Gran, H.G. 479
Graners, M. 531
Gray, M. 133
Gregory, T. 132
Groves, G.W. 145
Grygiel, J.S. 384
Grzeszczyk, S. 110, 121
Guirguis, S. 29, 472
Guthrie, W.F. 373
Gutteridge, W.A. 198
Guttiérez, P.A. 257

Haddad, G. 345
Hagymassy, J. Jr 553
Halle, R. 133
Halvorsen, G.T. 465
Hamada, H. 486, 487
Hammer, T.A. 479, 482, 485
Hammons, M.I. 531
Hanehara, S. 134, 135, 139, 190
Hanna, É. 133
Hanna, K.N. 553
Hansen, P.F. 326, 404, 405
Hansen, T.C. 434
Hanson, N.W. 398
Hansson, C.M. 486, 487
Haque, M.N. 290, 359, 398
Hassaballah, A. 98
Hatanaka, S. 442
Hattori, K. 26, 123
Haug, A.K. 42, 257, 276, 299, 301, 304, 307
Hayden, T. 113, 136
Helland, S. 476, 486, 487
Heroesewojo, R. 442
Hers, T.R. 122
Hester, H.T. 343, 381
Hewlett, P.C. 123, 133, 472
Hinrichs, N. 151
Hiraishi, H. 29
Ho, D.W.S. 290, 359, 398, 461
Hoff, G.C. 29, 71, 74, 269, 270, 272, 274, 299, 304, 307, 345, 477, 511, 515, 516, 517, 518, 521, 522, 525
Holand, I. 29
Holland, T.C. 192, 381, 477
Holm, T.A. 519
Hooton, R.D. 140, 359, 397, 412, 433, 441
Hornain, H. 92, 110, 113, 355, 368, 479
Houpert, R. 443
Hover, K. 301
Howard, A.J. 554
Howard, N.L. 58, 266, 267, 299, 393
Howarth, J.M. 126

Hsu, T.T.C. 453
Hu, C. 261, 271, 341
Hua, C. 325
Hughes, B.P. 257
Hughes, Y. 285
Humbert, P. 364
Huynh, H.T. 175, 342
Hu Qingchang 29
Hwang, C.L. 29

Illston, J.M. 87, 94, 433, 546, 554
Imoto, Y. 106
Isabelle, H. 197, 368

Jaccoud, J.-P. 29
Jacobsen, S. 479
Jacobson, E.R. 531
Jaegermann, G. 465
Janson, J.E. 477
Jensen, B.C. 502
Jensen, J.J. 398, 500, 532
Jensen, O.M. 326, 404, 405
Johnson, C.D. 385, 387, 388
Johnson, R.B. 37
Jolicoeur, C. 128, 131, 133, 134, 135, 175, 178, 342, 344, 368
Jornet, A. 470
Jung, F. 404

Kaar, P.H. 398
Kabayashi, S. 466
Kamada, E. 479
Kantro, D.L. 176
Kaplan, M.F. 442
Kasai, T. 312
Kawai, T. 384
Kasmatka, S.H. 2
Katayama, K. 466
Katsura, O. 479
Keck, R. 267, 271
Kepp, B. 502, 522
Kefenc, T. 522
Kehr, J.A. 490
Kelly, A. 93, 554
Kendall, K. 554
Kenyon, J.C. 393
Kessal, M. 198, 204, 559
Kettle, R. 472
Khalifé, M. 133
Khan, A.K. 427
Khayat, K.H. 140, 145, 343, 359, 397, 433, 441, 472
Khorami, J. 173
Khoylou, N. 501
Klieger, P. 423
Knab, L.I. 290
Knowles, R.B. 531
Kocataskin, F. 472
König, G. 29
Korman, R. 57, 267
Kreijger, P.C. 121, 122, 465

Krenchel, H. 423
Krishnan, S. 442
Kristmann, M. 110
Krysa, A. 477
Kucera, V. 462
Kucharska, L. 121
Kulendran, S. 460
Kumagai, T. 466
Kurose, Y. 531

Lachemi, M. 64, 349, 350, 356, 360, 361, 362, 364, 365, 367, 368, 555
LaFraugh, R.W. 522, 525
Lagergen, E.S. 373
Lahalih, S.H. 126, 170
Lahlou, K. 530, 531, 532, 548, 554
Lama, R.D. 383, 531
Lambotte, H. 453
Lamontagne, A. 480
Lamothe, P. 413, 487, 492, 493, 494, 495
Landry, M. 498
Laplante, P. 27, 133, 349, 356, 357, 428, 430, 446, 447, 474, 475
Larson, S.C. 42, 446
Leatham, D.M. 58, 266, 267, 299, 393
Le Chatelier, H. 315
Leistikow, E. 199
Lenschow, R. 453
LeRoy, R. 62, 350, 434, 437
Leshchinshy, A.M. 290
Lessard, M. 64, 184, 234, 257, 266, 267, 270, 271, 274, 275, 276, 277, 278, 279, 280, 282, 287, 299, 300, 301, 302, 304, 306, 330, 342, 345, 346, 349, 350, 356, 360, 361, 362, 363, 364, 367, 368, 369, 376, 377, 381, 383, 385, 387, 388, 389, 392, 393, 397, 428, 480, 484, 511, 513, 514, 515
Lévy, C. 28, 133, 497
L'Hermite, R.G. 320
Lindgård, J. 437
Liu, T.C. 472, 477
Ludwig, H.-M. 115, 407, 482
Luke, K. 133
Luther, M.D. 477

Maage, M. 486, 487
McCarter, W.J. 130
MacGregor, J.G. 175, 178, 342, 344, 398, 399, 423, 454
McHattie, J.S. 490
Mack, W.W. 199
Mailvaganam, N.P. 123, 126
Mak, S.L. 290, 398
Maki, I. 106
Malhotra, M. 28, 123, 140, 387, 393, 485, 516, 517, 518, 520, 521, 522
Malier, Y. 4, 29, 39, 59, 497
Manai, K. 343
Mansur, M.A. 399, 442
Marchand, J. 479, 486

Author index

Maret, V. 555
Markussen, J.B. 486, 487
Martinez, S. 519, 532
Mashimo, M. 486, 487
Massazza, F. 130
Matsushita, H. 384
Mehta, P.K. 87, 88, 89, 90, 95, 97, 140, 195, 202, 257, 342, 424, 442, 460, 472, 545, 554
Mendes, P.A. 366
Mercer, C.M. 496
Meyer, A. 26, 123
Miao, B. 303, 330, 349, 350, 353, 354, 356, 357, 358, 359, 360, 370, 398, 428, 446
Michot, L.J. 555
Mikhail, R.S. 553
Miller, F.M. 136
Mindess, S. 111, 115, 198, 559
Mirambell, E. 366
Mirza, S.A. 385, 387, 388
Mirza, W.H. 328
Misra, S. 413
Mitchell, D. 303, 349, 350, 353, 354, 356, 357, 359, 360, 370, 398, 399, 427, 428, 454, 511
Mitchell, T.H. 413
Mobasher, B. 413
Mohri, K. 384
Moksnes, J. 42
Mollah, M.Y. 122
Monachon, P. 66, 69
Monteiro, P.J.M. 89, 90, 342
Moore, J.F.A. 461
Moranville, M. *see* Moranville-Regourd, M.
Moranville-Regourd, M. 27, 92, 108, 111, 113, 133, 147, 165, 355, 368
Moreno, J. 22, 26, 27, 306, 386, 393
Mortureux, B. 92, 113
Motohashi, K. 413
Moustafa, S.E. 531
Muan, A. 102, 103, 105
Mugumura, H. 532
Muir, S.E. St J. 453
Murato, T. 29
Myazawa, S. 312, 326, 328, 404, 405, 406

Naaman, A.E. 442, 521
Nasser, K.W. 393
Nawa, T. 136, 165, 166, 184
Nawy, E.G. 532
Neeley, B.D. 531
Nehdi, M. 198, 559
Neville, A.M. 1, 2, 94, 195, 215, 311, 312, 342, 354, 367, 375, 378, 381, 382, 388, 389, 446, 452, 460, 472, 487, 500
Nielsen, L.F. 89
Nili, M. 479
Nilsen, A.U. 437, 439, 517, 518, 520, 521, 526, 527
Nilsen, T. 140
Nilson, A.H. 393, 423, 432, 438, 519, 532

Nishimatsu, Y. 442
Nkinamubanzi, P.-C. 133, 134, 135, 198, 204, 368, 559
Nonat, A. 115
Norberg, P. 462
Noumowe, A.N. 501
Nour, M.G. 532
Novokshchenov, V. 295, 516, 517, 518, 520, 522
Noworyta, G. 144, 166, 193

Odler, I. 111, 130, 151, 167, 553
Ohsato, H. 106
Okamura, T. 132, 139
Okkenhaug, K. 514
Okkubo, M. 165
Ollivier, J.-P. 98
Olsen, N.H. 423
Ong, K. 492
Onofrei, M. 133
Opheim, E. 500
Orito, Y. 531
Osborn, E.F. 102, 103
Ost, B.W. 349, 398, 432
Ouellet, C. 65
Ozturan, T. 472
Ozyildirim, C. 384

Page, K.M. 301
Paillère, A.M. 113, 132
Palmer, G. 105
Palta, P. 122
Park, R. 531
Parrott, L.J. 413, 465
Paterson, M.S. 531
Paulini, P. 130
Paultre, P. 532
Pauri, M. 111
Peng, G.-F. 501
Penttala, U.E. 190
Perenchio, W.F. 4, 22, 162, 423
Perez, J.L. 490
Perraton, D. 133, 358, 368, 413, 414
Perreault, F. 128
Persson, B. 115
Peterman, M.B. 257
Petersen, C.F. 4, 24, 25, 123, 257
Peterson, P.E. 485
Petkovic, G. 453
Petrie, E.M. 130
Philleo, R.E. 460, 481
Philips, B. 102, 105
Pierre, A. 130
Pierre, P. 323
Pigeon, M. 184, 234, 274, 276, 282, 301, 323, 346, 369, 479, 480, 482, 485, 486, 496, 511, 514
Ping Gu 489
Piotte, M. 127, 128, 129, 133, 134, 135
Pistilli, M.F. 144, 192
Pleau, R. 480

Plumat, M.   345
Pomeroy, D.   460
Pons, G.   322
Popovics, J.S.   215
Popovics, S.   215, 442
Porasz, A.   399
Potter, R.J.   29
Poulin, C.   555
Powers, T.C.   5, 321, 324, 429, 478, 481
Puch, C.   178, 192
Punkki, J.   340, 341

Quinn, P.J.   50

Rad, F.N.   37
Radiquet, B.   47
Raistrick, J.H.   554
Ralston, M.   57, 267
Ramachandran, R.   28, 123
Ramezanianpour, A.A.   140
Ranc, R.   28, 113, 133
Randall, V.   285
Raphael, J.M.   432
Rau, G.   144, 192
Raupach, M.   460
Ravindrarajah, R.S.   492, 496
Read, P.H.   373
Renwez, S.   383, 390, 553
Regourd, M.   *see Moranville-Regourd, M.*
Rendahl, B.   462
Riad, N.   396
Richard, P.   46, 47, 53, 56, 346, 554, 555, 556, 561
Richardson, I.G.   145
Ringot, E.   322
Rixom, M.R.   123, 126
Rocco, C.   437
Roi, S.   59, 497
Roland, B.   522
Ronneberg, H.   25, 26, 42, 200, 270, 276, 299, 300, 301, 304, 307, 346
Ronning, H.R.   473, 474, 476
Rosati, G.P.   500
Rosenberg, A.M.   145
Rosseland, S.   453
Rossi, P.   387, 553
Rostam, S.   461
Rossington, D.R.   123
Rougeron, P.   175, 204, 205, 206, 207, 208, 209, 210, 211, 236, 427
Roussel, J.P.   49
Roux, N.   555
Roy, D.M.   92, 133, 170, 553
Russell, H.G.   42, 446
Ryell, J.   50, 51, 52, 140, 197, 269, 274, 281, 301, 307, 346, 367, 373

Sadezzadeh, M.   472
Saeki, N.   466, 492
Saluja, S.S.   383, 531
Samman, T.A.   328
Sanjayan, G.   501
Sanjuan, M.A.   555
Sarkar, S.L.   28, 133, 197, 355, 368, 430
Sandvik, M.   25, 26, 42, 200, 257, 270, 275, 276, 299, 300, 301, 304, 307, 346, 502
Sawaki, D.   134, 135, 139, 170, 190
Schaller, I.   62, 350
Schiessl, P.   460, 461
Schaller, I.   452
Schlaich, J.   35
Schnormeier, R.   393
Seabrook. P.T.   522
Sedran, T.   215, 257, 261, 271, 341
Sellevold, E.J.   140, 145, 479, 482, 485
Sen, M.K.   531
Sewel, J.S.   530
Shah, S.P.   399, 423, 432, 442, 452, 521, 532
Shilstone, J.M.   2
Shilua, Z.   123
Shirlaw, M.R.   366
Shirosaka, T.   139
Shoya, M.   466, 492
Sicard, V.   322
Siebel, E.   123
Sigvaldson, O.T.   399
Simard, M.-A.   133, 134, 135, 368
Simons, B.P.   59, 285
Singh, A.C.   170
Singh, N.B.   170
Sivasundaram, V.   140
Sjöström, C.   462
Skaggs, C.D.   343
Skalny, J.P.   460, 553
Slate, F.O.   432, 438, 519, 532
Smeplass, S.   437
Smith, A.A.   87, 94, 433, 546, 554
Smith, G.J.   37
Soether, D.H.   479
Sone, T.   170
Sonebi, M.   343
Sorensen, B.   487
Soutsos, M.N.   257
Spitzner, J.   502
Stark, J.   115, 407, 482
State, F.O.   393
Stemland, H.   453
Stocks, L.J.   501
Struble, L.J.   130
Strunge, H.   486, 487
Sudret, J.P.   62, 350
Sugawara, T.   466, 492
Sung-Woo Shin   29
Swamy, R.N.   195, 423

Tadros, G.   76, 472, 477, 511
Tadros, T.A.   130
Taerwe, L.   399, 442, 453
Tagnit-Hamou, A.   136, 185, 198, 204, 342, 559
Takahashi, M.   531
Tanabe, D.   466

Author index

Tang, F.J. 136
Taniska, T. 106
Tassios, T.P. 556
Tazawa, E.I. 312, 326, 328, 404, 405, 406
Thao, P.X. 47, 49
Thaulow, S. 381
Thomas, H. 193
Thompson, S.E. 96
Thorenfeldt, E. 398, 531
Tkalcic-Cibici, B. 145
Tomita, R. 329
Tomaszewicz, A. 398, 531, 532
Torrent, R.J. 470
Torrenti, J.M. 387
Tsai, T.H. 29
Tsukagoshi, H. 531
Tsukinaga, Y. 466, 492
Tsutsumi, T. 413
Tucker, G.R. 26, 123, 171

Uchida, S. 132, 139
Uchikawa, H. 132, 134, 135, 139, 170, 190

Vaquier, A. 154, 195
Van Damme, H. 115, 315, 316
Van Gysel, A. 399, 442
Vempati, R.K. 122
Vernet, C. 115, 116, 117, 118, 119, 131, 132, 144, 166, 193
Vézina, D. 413, 414, 474, 475
Vichot, A. 136
Villiéras, F. 555
Violini, D. 437
Virlogeux, M. 53, 66
Volant, M. 133
Von Karman, T. 531
Vuturi, V.S. 383, 531

Wafa, F.F. 328, 527
Walther, R. 556
Wang, H. 499

Wang, P.T. 442, 521
Wasserman, R. 517
Watanabe, F. 532
Watanabe, Y. 486, 487
Webb, J. 531
Wee, T.H. 399, 442
Weinland, L.A. 123
Welsh, G.B. 257
Wenzel, T.H. 98
Whitcomb, W. 516, 517, 518, 520, 522
Whiting, D. 175, 413, 414, 415, 459
Wilson, H.S. 522
Winter, M.E. 4, 24, 25, 123, 257
Winterstein, R. 144, 192
Wischers, G. 467
Wiss, J.E. 522, 525
Witier, P. 132
Wittmann, F.H. 312, 320
Wolssiefer, J.T. 492
Wong, T.S. 531
Woodhead, H.R. 70, 269, 274

Xie, J. 423

Yahia, A. 343
Yamamoto, A. 413
Yong, Y.-K. 532
Yoshida, F. 531
Young, J.F. 93, 111, 115, 123, 133, 551, 553, 554
Yuan, R.L. 492
Yundenfreund, M. 553

Zaki, A. 511
Zappitelli, J. 437
Zelwer, A. 499
Zerbino, R. 437
Zhang, M.H. 167, 485, 487, 516, 517, 518, 519, 520, 522, 525
Zhao, J.W. 438
Zhu Jinquam 29

Subject index

Abrasives 458
Abrasion resistance 18, 37, 76, 472
  ice abrasion 76
  influence
    of aggregates 472, 473, 474
    of compressive strength 472, 473
    of fine aggregate 473, 474
    of finishing and curing 472
    of the nature of the aggregate 474
    of silica fume 474
Absolute volume method 216, 217
Absorption of aggregates 223
  effect
    on workability 517
    on slump 517
    on slump loss 223
    lightweight aggregates 517, 524, 525
Accelerated test
  durability 459
Adiabatic curing 314
Adjustment of the slump on the site 59, 300
Advantage of using HPC 35, 36, 37, 38, 39, 40, 559
Aggregates 199, 200, 201, 202, 203
  absorption 223
  content 262
  control of the shape 268
  effect
    on shrinkage 372
    on slump 341
  grain size distribution 268
  non reactive 80
  performance 9
  preparation 58
  selection of the MSA 258
  storage 266
Aggressive agent
  carbon dioxide 458
  external 458
  internal 458
Air content
  air entrained HPC 63, 64, 76, 304, 343
  drop after pumping 301
  effect on compressive strength 428
  entrapped air 277, 344
  estimation 217, 220, 237, 258
  influence of lignosulfonate 122
  non air entrained HPC 343

Air entrained HPC 11, 19, 62, 72, 76, 276, 277, 287, 290, 344, 345, 480, 481, 482, 483, 484, 511
  effect
    on placing and finishing 20, 93
    on pumping 345
    on rheology 20
    on strength loss 277, 345
    on workability 20
  entrapped air 91, 277
  influence of pumping 12
  mix design 512
Air entraining admixture 62
  compatibility with the superplasticizer 515
  dosage rate 231, 277, 278
  effect on the spacing factor 515
Alite 101
  crystals 108
Alkalies
  in clinker 110, 111, 112, 136
  effect on C_3A polymorphism 108, 110, 165
  solubility 9, 113
  sulfates 103
Alkali-aggregate reaction 19, 497, 498, 499, 500
  influence
    of cement content 498
    of early temperature rise 14, 19
    of moisture 498
    of superplasticizer type 499
    of water/binder ratio 498
  non reactive aggregates 72, 80
  prevention 499
Alkali sulfates 103, 110, 111
  in clinker 112
  solubility 113
α position
  in naphthalene superplasticizer 126, 127
Aluminate phase
  in clinker 118, 119
Aphithalite 112
Ambient temperature 349, 350, 360
  effect
    on compressive strength 64
    on temperature increase 64
    on thermal gradients 64

Subject index

Anhydrite 112, 136
Anticorrosion admixture 488
Arcanite 112
ASTM C666 Standard Test 478
Ash content 103
Assurance manual on quality 59
Autogenous shrinkage 5, 12, 312, 322, 329, 330, 402, 403, 404, 405, 406
  development 13
    under water 406
  difference with drying shrinkage 321
  effect
    on durability 472
    on freeze-thaw durability 18
  how to reduce it 323
  importance in HPC 324, 325
  influence
    of cement content 325
    of cement composition 406, 407
    of curing 325
    of pre-soaked lightweight aggregate 517
    of retarders 13
    of water/binder ratio 326
    of water curing 320, 321
  mechanism 326
  prediction 330
  rate of development 326
Average
  air content 63, 280, 292
  compressive strength 45, 63, 74, 277, 292
  molecular weight of a superplasticizer 174
  slump 45, 55, 63, 280, 286, 464
  spacing factor 278, 290, 292
  temperature 63, 69, 280
    of fresh concrete 274, 275

Bacteria 458
BANBURY mixer 551
Batch composition 252
  batching sheet 253
  calculations 254
  mix calculation 252
Batching sheet 253, 255
BAY ADELAIDE CENTER 289
Belite 101, 106
  nest 106, 109
Belitic cement 561
β position
  in naphthalene superplasticizer 126, 127
Binary phase diagram 103
Binder
  selection of 261
Blaine fineness 549
  of cement 167
  effect on flow time 186
  of fly ash 196
  of silica fume 194

Bleeding
  effect on finishing 304
  of grout 56
  of HPC surfaces 464
  internal bleeding 94, 463, 465
  rate 328
Blended cement 2, 3
Blended silica fume cement 64, 69, 72, 80
Bond
  bond strength of lightweight HPC 520
  chemical bonding of aggregates 94
  influence of the cleanliness of the surface of aggregate 201
Branching
  in polynaphtalene superplasticizer 127
Bridge
  use of lightweight aggregate 522, 523
BRUNAUER technique 549, 553
Bubbling HPC 344

C_3A 9, 103, 104, 106, 107, 108, 110, 111, 112, 130, 131, 133, 136, 314
  effect
    on autogenous shrinkage 406, 407
    on cement/superplasticizer interaction 130, 133
    on rheology 166
    on scaling resistance 485
  interest of a SAL treatment 168
  monoclinic 108
  orthorhombic 108
  polymorphin 120, 165, 166, 167, 168
  reactivity 111
  solubility rate 136
C_4AF 103, 104, 106
  effect on cement/superplasticizer interaction 133
Calcined bauxite 27, 550
Calcium langbeinite 136
Calcium sulfate 101, 102, 111, 112, 116, 131, 136
  different forms 136
  effect on cement/superplasticizer interaction 133, 136
  phase composition 113
  role of 111
  solubility 112, 113, 136
    of the different forms 9
  synthetic 112
Cantilever construction method 50, 53, 69
Capping compound 295
  confined 385
  effect on compressive strength 375
  ends preparation 383, 385, 387
  strength 15
Carbonation 17, 470, 497
Cardboard mould 392
Cast in plast joint 56
Cathodic protection 488
Cement
  -admixture interaction 115

578 Subject index

consumption 557
content 44, 218, 258
  effect on maximum temperature 349, 350
evolution of C_3S content 467
fineness
  effect
    on cement behaviour 9
    on compressive strength 22
influence
  of cement fineness 9
  of the cement type on rheology 9, 24
production 556, 557
quality 44
Cement dosage
  in lighweight HPC 518
Cement/superplasticizer combination
compatibility 58, 175, 268
effect
  on strength 9
  on rheology 9
interaction 130, 131
  influence
    of calcium sulfate 131, 132, 136
    of cement fineness 133
    of gypsum 131
    of hemihydrate 131
Cementitious system
  selection of 190, 191
Central mixer 269
Champagne effect 344
Chemical attacks 496
  influence
    of cracks 496
    of permeability 496
    of porosity 496
Chemical
  bonding 94
  composition
    of clinker 106
    of fly ash 154
    of portland cement 102
    of silica fume 143
    of slag 149
  contraction *see* volumetric
    contraction 90, 319
  soundness 113
Chloride ion 458
  penetration 486
  permeability 470, 494
    effect on compressive strength 360
  scale 414, 415
Chloroaluminates 487
CHUNNEL fire 501, 502
  circumstances 502
  damage 503
  repair 503
Classes of HPC 163, 164
Clinker 107, 108, 109, 110, 111, 112
  chemical composition 106
  grinding 112

manufacturing 103, 104
microstructure 105, 106
phases 105
  aluminate 119
  silicate 119
Closure joint
  elimination 50
Coarse aggregate
content 219, 220, 236, 258
effect on
  compressive strength 424, 425
  modulus of elasticity 439
influence
  of the crusher on the shape of 200
  of the shape on rheology 200
  of the type on compressive strength 202
maximum size 10, 202, 203
mechanical bond 58
selection 42, 199, 200, 202
  of the maximum size 217, 258
Coefficient of variation 51, 66, 74, 296, 297
Cold weather concreting 198, 274, 276, 287, 361
use of hot water 369
Colloidal admixtures
  effect on segregation 272
Compressive strength 56
of confined HPC 530, 531, 533
of core 45, 288, 289
design strength 232
determination 261
early compressive strength 426
of fibre-reinforced HPC 529
of heavyweight HPC 526
influence
  of age 94
  of aggregate 95, 199, 201, 202
  of the air content 428
  of ambient temperature 427
  of the amount of cement 428
  of cleanliness 9
  of the coarse aggregate 16, 200, 201, 202
  of crumbly particles 9
  of curing 359
  of early temperature rise 427
  of entrained air. 513. 514. 515
  of flaws 89
  of gravel smoothness 9
  of the hydrated cement paste 8
  of impurities 89
  of inhomogeneities 89
  of lightweight aggregate 20
  of the nature of the lightweight aggregate 517
  of porosity 88, 89, 94
  of sand fineness 200
  of self desiccation 430
  of shape and size of pores 89

of soft particles 9
of the transition zone 8
of the water/binder ratio 424
of the water/cement ratio 256
of water curing 429
of lightweight HPC 519, 520
long term 16, 429
for offshore platforms 44, 45
reduction due to air 428
required 232
required average strength 232, 257
specified 232
standard deviation 233
testing 374, 375
  core strength versus specimen strength 16, 397
  influence
    of capacity of the testing machine 375
    of capping compound 375
    of curing 395
    of eccentricity 388, 389, 390
    of ends preparation 383, 386
    of specimen
      diameter 392
      mould 391
      shape 374, 375, 390
      size 374, 375, 376, 386, 387, 392, 393, 394
    of spherical head diameter 377, 378, 379, 380
    of testing
      age 395, 396
      machine 375
        capacity 375, 376
        rigidity 381
        procedures 382
    of the verticality of the mould axis 391
Concrete
  composition 45
  confined 56
  consumption 556
  skin see Covercrete 312, 464, 465, 471
Condensation reaction
  naphthalene superplasticizer 127
Cone crusher
  effect on the shape of the particles 200
CONFEDERATION Bridge 76
Confined
  capping compound 385, 387, 388
  concrete 20, 56
  RPC 552
Confined HPC 530, 531, 532, 533, 534
  economical advantages 532
  effect
    on compressive strength 530, 531, 533
    on ductility 531, 533
    on stress-strain curve 531, 533
  placing 532

Congested area 43, 49, 72
Conical failures 86, 381, 382
Construction
  cost 64
  schedule 36, 53, 56
Control
  of aggregate moisture 58
  of HPC 11
  of the properties of the fresh concrete
    at the batching plant 338, 339
    on the job site 338
    statistical treatment 339
Cooling HPC 274, 366
  cost of 274, 275, 276
  different ways 366
  effect
    on thermal gradient 353
    on workability 276
  influence of the level of substitution 368
  nitrogen cooling 274, 275
  rate 353
  use
    of a low heat of hydration cement 369
    of a retarder 367, 368
    of SCM 368
  with crushed ice 62, 274, 275, 367
  with liquid nitrogen 51, 274, 275, 367
Core strength 45, 62, 358, 359, 428, 429
  versus specimen 359, 397, 398
Corrosion
  influence of initial current on initial cracking 495
  of reinforcing steel 487, 488
Cost of HPC
  cost saving 23
  influence
    of fly ash 10, 195
    of limestone filler 10
    of silica fume 10, 11, 194
    of slag 10, 196
    of superplasticizer 11
Cost reduction see Cost savings
Cost savings 52, 58
  comparison with UC 84
  influence of the use of SCM 191
Covercrete see also concrete skin 487, 492
Creep 37, 38, 445
  long term results 446
  measurement 407
    development of different deformations 407
    early loading 408
    influence
      of early loading
        for a HPC 409, 447
        for a UC 408
      of the water/binder ratio 408
    proposed curing 410
    sample curing 407

580　Subject index

reduced 55
Creep and shrinkage
　of lightweight HPC 520, 521
　of heavyweight HPC 527
Critical diameter 471
Cross linking
　effect on superplasticizer 128
　in MDF 551
Crumbly particles
　effect on compressive strength 9
Crushed ice 367
　effect
　　on cooling 62
　　on the temperature of fresh concrete 275
Crushed stone *see* Coarse aggregate
Crushing strength
　of aggregates 425
C_2S 27, 103, 104, 106
　effect on cement/superplasticizer interaction 130
C_3S 103, 104, 106, 167
　effect
　　on cement/superplasticizer interaction 130, 133
　　on maximum temperature 350
　evolution of C_3S content 467
　reactivity 9
　size of C_3S crystals 106, 108
C-S-H 88, 93, 117, 120, 316
Cube facing 383, 390
Curing 28, 290, 311, 327, 328, 464, 474
　compound 63
　duration 311, 312, 323
　effect
　　on autogenous shrinkage 323
　　on compressive strength 395
　　on durability 335
　　on total shrinkage 328
　how to cure specimens 396
　importance 312
　membrane 13, 325, 326, 327, 329
　　effect
　　　on autogenous shrinkage 327
　　　on HPC 327
　　　on self desiccation 327
　　　on usual concrete 327
　new procedure 397
　　for creep measurement 410
　　for shrinkage measurement 405, 406
　reason for 312
　when to stop 323
Curling 334

DASH 47 548
Dead load
　reduction of 37, 38, 61
Degree of sulfurization of the clinker 166
De LARRARD method 259, 260, 261
Delayed addition
　of superplasticizer 190, 270

Delivery of HPC 285, 299
Design strength 53, 56, 61, 66, 76, 232, 278, 426
Diabase 200, 202
Diffusion of water 120
Diorite 200
Disposable mould 392
Dolomitic limestone 42, 200
　effect on abrasion resistance 474, 475
Dosage
　air entraining 231
　superplasticizer 226, 227
　water reducer 231
Drop-in spans 76
Drum mixer capacity 299
Dry-batch plant 51, 62, 269, 281
　production of HPC 11
Dry-dock construction 42, 45
Drying shrinkage 94, 304, 313, 322, 430
　difference with autogenous shrinkage 321, 326
　elimination of 323
　influence of the W/B ratio 326
　measurement on UC 401, 402
　mechanism 326
　rate of development 326
Dry-rodded unit weight 217
DSP 548, 549, 550, 554
Ductility
　influence
　　of confinement 549
　　of fibres 549
Durability 37, 459
　a major concern 460
　definition 459
　influence
　　of aggressiveness of the environment 461
　　of autogenous shrinkage 472
　　of compactness 461
　　of compressive strength 461
　　of permeability 461
　　of placing and curing 463, 464
　　of the water/binder ratio 461
　market 560
　of old concrete 466, 467

Early compressive strength 426
　influence
　　of the superplasticizer dosage 426, 427
　　of a set retarder 426
Early strength 3, 49, 55, 66, 286
　how to increase it 346
　influence
　　of alkalies 346
　　of fly ash 98
　　of limestone filler 98
　　of the slag dosage 197
　　of SMC 97, 98
Eccentricity

Subject index 581

effect on compressive strength 388, 389, 390
Ecological advantage 560
Economical advantage
  of using HPC 7
  of using SMC 10, 11, 97, 194, 195, 196
Elastic modulus 37, 56, 433
  BAALBAKI formula 439, 440
  BAALBAKI models 435, 436
  of confined concrete 56
  of confined HPC 531
  correlation with f'_c 436, 437, 438, 439
  De LARRARD and ROY model 434
  empirical approach 437
  as a function of f'_c 16
  HANSEN model 434
  of heavyweight HPC 527
  influence
    of coarse aggregate 439
    of lightweight aggregate 20
  of lightweight HPC 520
  measurement 412, 415
    set up 416, 417, 418
  models 17
  relationships with dynamic modulus 16
  REUSS model 433, 434, 436
  theoritical approach 433
  VOIGT model 433, 434, 436
Electrostatic repulsion 134, 135
Ends (specimen)
  capping compound 383
  effect on compressive strength 386
  grinding 59, 66, 383
  neoprene pads 384
  preparation 59, 383
  of rock specimens 383
  sand box 383
Entrained air *see also* Air-entrained HPC
  effect
    on compressive strength 511, 515
    on finishing 306
    on freeze-thaw durability 511
    on slump 342
    on the spacing factor 515
    on unit mass 513, 514, 515, 526
    on workability 484, 485, 515, 516
  finding the right admixture dosage 515
  in lightweight HPC 526
  stability with time 515
Entrainment of air
  influence of lignosulfonate 22
Entrapped air 91, 277
  in lightweight concrete 512
Environmental advantages of HPC 8
Epoxy coated rebars 461, 470, 490, 491
Ettringite 111, 112, 355
  formation 170, 314, 465, 469
  role of the interstitial phase 116
Expansion joints 50
External post tensionning 49, 59, 61

External vibration 69
  effect on micro-concrete 27
Extra cost
  of crushed ice 367
  of nitrogen cooling 367
  of HPC production 265

Factorial design 203, 204, 205, 206
Failure surface 85, 86, 95
False set 112, 131, 132, 136
FARRIS
  model 260
  theory 11
Fatigue failure 450
  Goodman diagrams 451
  Wöhler diagrams 450, 451
Fatigue resistance 448, 449, 450
  of concrete structure 452, 453
  Goodman diagrams 450, 451
  of lightweight HPC 521
  Miner's rule 451
  Wöhler diagrams 450, 451
FÉRET's law 8, 91
Fibre dosage 529
  effect
    on compressive strength 529
    on modulus of rupture 529
    on thoughness 528, 529,
    on unit price 529
    on workability 529
Fibre reinforced HPC 20, 476, 527, 528
  effect of the plastic fibres
    on fire resistance 528
    on mechanical properties 528, 529
  influence
    of the aspect ratio 529
    of the dosage 529
    of steel ultimate strength 529
  polypropylene 20
  thin overlay 528
FICK's law 486
Field testing of HPC
  influence
    of early curing 59
    of the mould 59
    of the size of the specimen 59
  tests 287, 290
Filler effect of silica fume 145
Filtration of a superplasticizer 129
Fine aggregate *see* Sand
Fineness modulus 217
  of the sand 10, 199
Finishability
  influence of entrained air 20
Finishing 304, 305
  influence
    of bleeding 304
    of entrained air 304
Finite element calculation methods 369
  calculation
    of maximum temperature 352

of thermal gradients 352
Fire resistance 19
  influence of polypropylene fibres 19
Flash set 131, 132, 136
Flaws
  effect on compressive strength 89
Flocculation 117, 121, 124, 125, 547
Flow table 69
Fluidification of concrete 123
Fluidifiers *see* Superplasticizers 8, 26
Fluvioglacial gravel 58, 201
Fluvial gravels 201
  cleanliness 201
  effect on the strength 201, 202
  smoothness of the surface 201, 202
Fly ash 2, 23, 140, 152, 153, 154, 156
  characteristics 42
  chemical composition 154
  different types 153, 154
  dosage 195
  effect
    on the cost of HPC 10
    on the rheology 10
    on the slump loss 25
  grain size distribution 152
  hydration 356
  selection of 195
  ternary composition 195
  use in HPC 41, 42
Fog misting 324
Forms
  effect
    on temperature increase 363
    on thermal gradients 363
  insulated 69, 369
Formwork removal 25, 29,55
4K syndrome 561
Fracture surface 256
Freezable water 18
  amount of 481, 482
Freeze-thaw
  cycles 458
    correlation with the spacing factor 479
    duration 478
    number 478, 483
    specifications 479
  durability *see* Freezing and thawing resistance 80
  resistance *see* Freezing and thawing resistance 17
Freezing and thawing resistance 477
  guideline 482
  of HPC 479, 511
  influence
    of air entrained concrete 18
    of non-air entrained concrete 18
    of the permeability 18
    of the porosity of the paste 18
    of SCM 343
    of self desiccation 480, 481

of the spacing factor 485, 512, 514
of the type of binder 18
of the type of the cement 482
of the water/binder ratio 18, 511
of usual concrete 478
Fresh concrete temperature
  effect
    on temperature increase 64
    on thermal gradients 64
  maximum 285, 290
Future trends 556

Gabbro 200
Galvanized rebars 489, 490
Glacial gravel 201
Glass fibre-reinforced rebars 491, 492
GOODMAN diagrams 451, 452
Grain size distribution of aggregates 45, 96, 97
  optimization 547, 548
Granite
  effect
    on abrasion resistance 474, 475
    on compressive strength 200, 202
Gravel
  cleanliness 9, 201
  fluvioglacial 58, 201
  pea gravel 58
  semi crushed 66
  smoothness 9, 201, 202
Griffith theory 87, 88
Grinding aids 113, 549
Gypsum 9, 112, 136
  effect on cement/superplasticizer interaction 131

Heat release during hydration 64, 69, 313, 316, 349, 365, 366
Heating HPC 69, 276
Heavyweight HPC 20, 526, 527
  compressive strength 526
  elastic modulus 527
  shrinkage 527
  unit mass range 526
Hemihydrate 101, 112, 131, 136
  effect on cement/superplasticizer interaction 131
HIBERNIA offshore platform 70, 307
High early strength 198
High initial strength *see also* early strength 38
High performance concrete (HPC)
  advantages
    economical 7
    environmental 8
    technical 7
  competition with steel 39, 52
  definition 1, 2, 4, 7, 8
  historical perspective 7
  marketing 39
  principle 8

Subject index

profitability 39
promotion 39
rationale 7, 35
savings in construction cost 42
structures
 final cost 49
 horizontal type 38
sustainable development 8, 70
use in pavements 37
High range water reducers *see also*
 Superplasticizers 8, 137
High rise building
 swaying 30, 36, 37, 56, 57
 use of HPC 23, 30, 31, 32, 33
 volume of HPC 39
High strength
 concrete
  definition 1, 4, 23
  historical perspective 7, 22
  market 560
Historical perspective 7, 23
Horizontal mixer 269
Horizontal-type structure 38
Hot weather concreting 51, 62, 291, 361
 use of a retarder 62, 299, 346, 368
Hydrated cement paste 85, 87, 90, 91, 92, 94
 effect
  on compressive strength 8
  on tensile strength 8
 improving the strength 88, 89
 influence of W/C ratio 92
Hydration 115, 116, 117, 118, 119, 120, 130, 131, 132
 degree 92
 in presence of superplaticizer 130, 131, 132, 133
 reaction 313, 314
  heat release 316
  of pure phases 314
  strength gain 315
Hydraulic structures 477
Hydrogarnet 117
Hysteresis curve
 influence of nature of the aggregate 445
 in stress strain curves 442
Hysteresis loop
 of the stress strain curve 16

Ice
 abrasion 76, 472, 477
 belt 70
 cooling 334
 shield 76
ÎLE DE RÉ Bridge 53
Impact crusher
 effect on the shape of the particles 200
Impurities
 effect on compressive strength 89
Industrial slabs

use of roller-compacted concrete 534, 535, 539
Inhomogeneities
 effect on compressive strength 89
 microstructural inhomogeneities 94
 reducing inhomogeneities 93
Initial temperature
 influence on compressive strength 426
Inner product 89, 90, 91, 92, 93
Instantaneous deformation 55
Internal bleeding
 cracking due to freezing and thawing 18
 effect on transition zone 463, 465
 influence
  of overvibration 463
  of vibration 94
Interstitial phase
 content 105, 106, 107
 crystallinity 110
 effect
  on initial stiffening 164
  on rheological behavior 85
  on rheology 165
 role in the formation of ettringite 116
 reactivity 9
Isothermal curing 314
Insulated
 blankets 361, 369
 forms 69, 369

JACQUES-CARTIER Bridge 275, 278, 291, 367
JOIGNY Bridge 59
Jumping forms 306, 346

Laumonite 202
Leaching 496
LE CHATELIER contraction *see also*
 Volumetric contraction 547
 partial elimination 547
Life cycle 561
Lightweight aggregate 72, 307, 516
 absorption 524, 525
 dry lightweight aggregate 517
 effect
  on compressive strength 20, 516, 517
  on elastic modulus 20
  on slump loss 525
 influence of the state 517, 525
 presoaked lightweight aggregate 517, 525
Lightweight HPC 516
 bond strength 520
 cement dosage 518
 compressive strength 519, 520
 direct tensile strength 520
 elastic modulus 520
 fatigue resistance 521
 influence of the nature of the fine aggregate 518

modulus of rupture   520
post-peak behaviour   521
quality control   512
shrinkage and creep   520, 521
splitting strength   520
strength range   516
thermal characteristics   521, 522
unit mass range   516
uses of   522
Lightweight structure   46
Lignosulfonate   22, 41, 92, 113, 122, 123, 549
  based superplasticizer   173
  effect on flocculation   124
  entrainment of air   22, 25, 122
  retardation   22, 25, 122
  simultaneous use with a superplasticizer   26
  in superplasticizer   170, 190
  use in combination with naphthalene and melamine   173
Limestone   200, 202
  filler   2, 559
    effect
      on the cost of HPC   10
      on the rheology   10
Liquid nitrogen cooling   51, 367
Liquid phase *see also* Interstitial phase   105, 106
Loading sequence   51
Lollipop test   492
Long term trends   561

Marsh cone method   175, 178, 179, 180
  water/binder to be used   181, 182
Mass loss during drying   320
Materials selection   163
Maturity
  meters   314
  method   55, 66
Maximum
  size
    of the aggregate   42, 58, 202, 203
    of the coarse aggregate   216, 219, 258
    selection   217, 219
  slump   271, 272
  temperature
    control of   366
    decrease of   199
    effect
      on cement hydration   349
      on thermal gradient   349
    of the fresh concrete   62
    in HPC elements   351, 363, 364
    influence
      of ambient temperature   364
      of cement content   349
      of cement dosage   357
      of the geometry   361, 362, 363
      of initial temperature   357, 361, 364, 366

  of the nature of the forms   363, 364
  in UC elements   350
MDF   93, 548, 550, 551, 553, 554
Mechanical bond
  influence of the coarse aggregate   58
Melamine-based superplasticizer   50, 69, 123, 134, 300
  advantage of using   171, 300
  redosage on the site   300
Metallic aggregates   473, 548
Metallic forms
  effect on compressive strength   427
Micro-climatic conditions   462
Micro-concrete   27
Microstructural difference
  between transition zone and bulk hydrated cement paste   94
Microstructure   5, 89, 90, 92, 93, 94, 469, 470
  influence
    of the heat release   356
    of supplementary cementitious materials   356
MINER's rule   451
Minislump method   175, 176, 177, 178
Mix proportionning   11
Mix design
  calculations   239, 240, 241
  limitations   255, 256
  method   215, 216, 233, 234, 257, 258, 259, 260, 261, 262
  sample calculation   241, 242, 243, 244, 245
  sheet   237, 238, 245
Mixer
  BANBURY   551
  central   269
  dry-batch   11, 62, 269, 281
  high-speed   72
  horizontal   269
  tilt   269
Mixing sequence   72, 269, 279, 280, 281, 282, 283
  effect on construction cost   280, 281
  influence of water content of aggregate   269
  length of mixing time   270
  optimization   11
Mixing water
  estimation   217, 220, 258
Mix proportioning *see* Mix design method
Mixture proportioning *see* Mix design method
Mock up   293
Modulus of elasticity *see* Elastic modulus
Modulus of rupture   16, 374
  of fibre-reinforced concrete   526
  as a function of f'_c 16, 432, 433
  of heavyweight concrete   526
  influence of coarse aggregate   16
  of lightweight HPC   520

Subject index 585

Moist curing *see also* Water curing 311, 312, 319, 323, 325, 326
    of different structural elements 329, 330, 331, 332, 333, 334
    effect on autogenous shrinkage 12
    necessity of 13
    timing 13
Moisture
    adjustment 219, 262
    content 223
Monitoring 42, 59, 62, 64
Monosulfoaluminate 119, 120, 314
MONTÉE ST-REMI Bridge 62, 275, 367
Mould
    cardboard 392
    disposable 392
    effect on compressive strength 391
    plastic 392
Mushrooming 306

Nanostructure 563
Naphthalene superplasticizer 26, 62, 72, 80, 123, 171, 172
    advantage of use 171, 172
    calcium salt 172
    electrostatic repulsion 134
    manufacturing 126
    neutralization 128
    polymerization 127
    sodium salt 172
National codes 461
Neoprene pads 384
Night delivery 57, 63, 266, 267, 299
Nitrogen adsorption
    specific surface area measurement
        of cement 143
        of silica fume 143
Nitrogen cooling
    effect on the temperature of fresh concrete 51, 275

Offshore platforms 36, 42, 43, 70
    use of lightweight aggregate 522, 523
On-site prefabrication 38, 49, 55, 78
Optimal packing 547
Optimum
    composition 61
    dosage of a superplasticizer 189
    influence
        of rheological requirements 189
        of strength requirements 189
Outer product 89, 90, 92
Ovalization of the mould
    effect on compressive strength 392

Packing 547
Pack-set 113
Passivation of steel 487
Pavement applications 37, 476
Pea gravel 58
P.E.I. Bridge *see* Confederation Bridge

Permeability
    air permeability 413
    influence of the W/B ratio 412
    measurement 412
    rapid chloride ion permeability 15, 413
Permeable forms 466
Penetration of chloride ions 470
Periclase 114
Personalized training 59
Phase composition of portland cement 102, 110
Pilot test *see also* Field test 291, 300
Placeability
    influence of entrained air 20, 93
Placing 267, 301, 302, 303
    influence
        of air entrainment 93, 304, 306
        of superplasticizer on placing cost 561
    rate of placing 267
    roller compacted concrete 309
    with a bucket 69
Plastic
    moulds 392
    shrinkage 63, 312, 329, 330
Plywood forms
    effect on early strength 427
POISSON's ratio 433, 435, 440, 442
    measurement 442
Polyacrylates 126
    electrostatic repulsion 135
    steric repulsion 134
Polycondensate of naphthalene sulfonate *see* Naphthalene superplasticizer
Polymorphism
    of C_3A 107
Polynaphthalene *see* Naphthalene superplasticizer 26
Polyvinyl alcohol 551
Ponding 324
PONT DE NORMANDIE Bridge 66
Pore shape
    effect on compressive strength 89
Porosity 471
    effect
        on compressive strength 88, 89, 94
        on strength 545
        on tensile strength 88
    how to decrease it 91
    influence of the W/B ratio 468, 469
Portland cement
    acceptance tests 113, 114
    chemical composition 102, 113
    control of the quality 169
    effect
        on rheology 164
        on strength 164, 165
    grinding 112, 113
    hydration 115, 116, 121
    influence
        of ASTM type 167

of fineness 165
initial stiffening 164
manufacturing 111
phase composition 102, 165
selection 163, 164, 167
type of cement 106
Portlandite 120, 469
PORTNEUF Bridge 282, 286
Post-peak behaviour
  of confined HPC 531, 532, 533
  of fibre-reinforced HPC 529
  of lightweight HPC 521
Post tensioning ducts 56
Pozzolanic reaction 139, 140
  influence of the temperature
    increase 356
  in lightweight concrete 518
Precasting on the job site 38, 49, 53, 76, 78
Preconstruction meeting 12, 287, 288
Prequalification test 12, 52, 53, 285, 288, 289
Prestressing 49, 55
Production
  of HPC 11, 265
    in a dry batch plant 11
    extra cost 265, 266
    rate of production 267
  rate 61
Pumping 301, 308
  air entrained HPC 11, 12, 57, 61, 69, 70, 301, 302
  effect on spacing factor 301
  influence of ambient temperature 12

Quality
  assurance manual 59
  control
    of a blended cement 194
    charts 295
    at the concrete plant 292
    evaluation 295
    of fly ash 196
    of the fresh concrete 293
    of HPC 512
    of HPC production 280
    at the jobsite 293, 294
    of materials used 267
    of a particular cement 182, 183, 268
    on Portneuf Bridge 282
    programme 45, 50, 52, 55, 73
    of SCM 268
    of silica fume 193
    of the slag 197
    of superplasticizers 173, 174, 182, 183, 268
  financial implications 39
  insurance 285
  of the moulds
    influence
      of ovality 295
      of perpendicularity 295
Quenching
  of portland cement clinker 105, 106
  of slag 149, 150, 151

Rapid chloride-ion permeability 15, 413
  influence of the W/B ratio 413, 414
  scale 414, 415
Reactive powder concrete 21, 549, 551, 555
  confined 552
  how it works 552
  how to make it 551
  mechanical characteristics 552
  scale effect 552
Redosage of superplasticizer 59, 300
Research needed 562, 563
Resistance
  to chemical attacks 496
  to fire 500
    improving the 501, 502
    influence of the W/B ratio 501
    spalling 501
    use of plastic fibres 502
Retardation 338, 344, 345, 346
  causes 339
  due to
    lignosulfonate 22, 122
    superplasticizer 138, 299
  how to decrease it 346
Retarder 27, 62, 188, 190, 278, 299
  economical advantage 280, 281
  effect on maximum temperature 367, 368
  use of 299
Retempering 271, 300
Reticulated structures 552
Rheological behavior
  differences in the 183, 184, 185
  influence
    of alkali sulfates 184
    of amount of calcium sulfate 184, 185
    of cement type 164, 186, 187
    of efficiency of mixing 184
    of fineness of the cement 184, 186
    of fly ash 10
    of initial temperature of the water 184
    of interstitial phase 85
    of limestone filler 10
    of the phase composition of the cement 184
    of a retarder 187, 188
    of silica fume 10
    of the sand 199
    of the shape of the coarse aggregate 200
    of slag 10
    of the water/binder ratio 184
Rheology of the cement paste

Subject index 587

improving the 342
influence
    of the chemical reactivity 342
    of entrained air 342, 343
    of the grain size distribution 342
    of the shape of cement particles 342
    of the sulfates 342
Rheometer 14, 271
Rigidity of the testing machine
    effect on compressive strength 381
Rock testing methods 373
    air permeability 413
    ends preparation 383
    stress strain curve 399
    water permeability 412
Roller compacted HPC 20, 21, 309, 534, 535, 536, 537, 538, 539, 540
    freeze-thaw durability 534
    industrial slabs 534, 535, 539
    manufacturing 534, 536
    placing 536, 537, 538
    unit mass control 538

Sampling of HPC 25, 295
    influence of the type of mold 25
RPC *see also* Reactive powder concrete 549
SAL treatment 168
Sample preparation 295
Sand
    content 219
    effect
        on segregation 199
        on workability 199
    fineness 42, 58, 199
    grading 200
    influence on the unit mass
        of heavyweight concrete 526
        of lightweight concrete 527
    manufactured sand 45, 200
    natural sand 200
    selection of 190
Sand box 66, 383, 384
Saturated surface dry state 221, 222
Saturation point 180, 181
    effect on superplasticizer dosage 189
Saturation ratio
    of sulfonated sites 127
Savings 61, 64
    in construction cost 42
    in foundation cost 47
    initial 66
Scaling resistance 18, 485
    influence
        of non air entrained concrete 18
        of the skin of concrete 485
        of spacing factor 18
SCOTIA PLAZA 50, 274, 281, 301, 346
Sea water 497
Segregation 26, 56, 272, 338
    causes 339

improving resistance to
    segregation 272
influence
    of colloïdal admixtures 272
    of silica 272
    of superplasticizer 26
    of a superplasticizer overdosage 272
Selection
    of the aggregate 96, 199
    of ASTM type 167
    of the cement 41, 163, 164, 167
    of cementitious system 190, 191
    of fine aggregate 199
    of the fineness 165
    of fly ash 195
    of maximum size 202
    of silica fume 192, 193
    of slag 197, 198
    of the superplasticizer 164, 170
        acceptance tests 170
        calcium naphthalene 172
        lignosulfonate-based 173
        melamine 171
        naphthalene 171, 172
    of the water reducer 41
Self desiccation 312, 315, 320, 220
    development 321
    effect
        on freezing and thawing resistance 480, 481
        on shrinkage measurement 402, 403
    in high performance concrete 13
    influence of the compactness 321
    initial
        compressive strength increase 404
        mass increase 404
    role of menisci 320
    in usual concrete 12
Self-healing 468
Sequence of addition 59, 279
Service life 76, 463, 473
Set retardation *see* Retardation
Setting time 115, 119
Shape of the specimen
    effect on the compressive strength 390
Shearing failure 295
Short-term trends 558
Shrinkage
    autogenous shrinkage 445, 446
    comparison with specimen 448, 449
    development
        in HPC 402, 445
        in UC 402
    different types 312, 329
    drying shrinkage 446, 447, 448
    initial mass increase 404
    in large beams 331
    in large columns 329
    long term results 446, 447, 448
    measurement 400
        difference between UC and HPC 402

influence
    of autogenous shrinkage 401, 402, 403
    of high W/B ratio concrete 402
    of low W/B ratio concrete 402
    of self desiccation 402, 403
  initial length 401
  new procedure 405, 406
  present procedure 401
  reference length 401
in small beams 332
in thick slabs on grade 333
in thin slabs 332
self desiccation 404
total shrinkage 311
Silica fume 2, 51, 140, 141, 142, 143, 144, 191
  blended cement 64, 69, 72, 80
  chemical composition 143, 144
  commercial availability 144
  effect
    on Blaine fineness 194
    on bleeding 146
    on blended cement 193
    on compressive strength 147 206, 208, 209, 210
    on the cost of HPC 10, 191, 206, 209, 211
    on densified silica fume 193
    on dosage 194
    on lime crystallisation 145, 146
    on microstructure 147
    on packing 547
    on quality control 193
    on segregation 272
    on selection 192, 193
    on slurry 193
    on superplasticizer dosage 206
    on transition zone 146
  filler effect 145
  grout 56
  influence of the type of alloy produced 143
  in microconcrete 27
  size of particle 143, 145
  specific
    gravity 143
    surface area 143
  use in lightweight concrete 518
  use of 28, 29, 51, 93, 96, 97, 286
  variability 192
Silicate phases
  in clinker 119
  effect on strength properties 85, 165
Size effect 424
Size of the specimens
  effect on compressive strength 374, 375, 376, 386, 387, 392, 393, 394
Skin of the concrete
  effect on scaling resistance 485
Slag 2, 51, 140, 147, 148

chemical composition 148, 149
cooling 148, 150
dosage 197
effect
  on cold weather concreting 198
  on the cost of HPC 10
  on early strength 197
  on the rheology 10
hydration 356
quenching 149, 150, 151
selection 197, 198
specific gravity 148
use of in HPC 50
Slipforming 42, 4, 71, 273, 307, 308
  rate 72
Slump 44
  adjustment 216
  control of the 59
  factors influencing the 340, 341
  influence
    of cement/superplasticizer compatibility 341
    of entrained air 342
  loss 24, 123, 138, 338, 343
    influence
      of the absorption of aggregates 223
      of the cement 24, 27
    use of SCM 343
  measurement 13
  precision 341
  retention 124, 133
  selection 217, 219
  test 271
SO_3 content of Portland cement 112
Social benefits 36
Soft particles
  effect on compressive strength 9
Soluble sulfates 167
Solubility
  alkalis 9
  of the different forms of calcium sulfate 9
Spacing factor 64, 286, 287, 290, 478
  critical value
    for HPC 480
    for UC 479
  dosage of the air-entraining admixture 514
  effect
    on freeze-thaw durability 18, 301, 303, 307, 512
    on scaling resistance 18
  efficiency of the air-entraining admixture 514
  influence
    of the number of cycles 483, 484
    of the type of superplasticizer 514
    of W/B ratios on the critical value 480, 483, 484
  modification after pumping 301, 307
Span length 49, 53, 62, 66, 76

Subject index

Special HPCs 19
Specifications 278, 286, 288, 289, 290
  for curing 335
Specific gravity
  of the aggregate 217
  of portland cement 225
  SSD state 225
  of superplasticizer 227
Specified compressive strength 52, 277, 280
Spherical head diameter
  effect on compressive strength 377, 378, 379
  measurement 380
Splash zone 71
  use of air entrained HPC 25
Splitting tensile strength 16, 62, 66, 431
  as a function of f'_c 16, 432
  of lightweight HPC 520
Stainless steel rebar 489
Standard deviation 45, 53, 63, 66, 74, 269, 296
Standardized unit cost of a bridge deck 65
Starting mix design 215
Steel
  fibres
    effect on reactive powder concrete 21
  passivation
    role of lime 140
Steric repulsion
  polyacrylates 134
Stiffening of the paste 118
Strength
  of aggregates
    effect on compressive strength 94, 95
    of lightweight concrete 518
  influence of porosity 545
  properties
    influence of silicates 85
    regression 429, 430
    retrogression 16
Stress-strain curve 15, 397, 398, 399, 400, 442, 443
  of confined HPC 531
  equation 397, 398, 444
  hysteresis loop 16
  influence
    of aggregate 442, 445
    of coarse aggregate 15, 16
  measurement 442
Stress transfer 16, 424
Structure stiffness 29
Sulfonation 126
  α position 126, 127
  β position 126, 127
  position of the sulfonate group 126
Sulfur 103
Superplasticizers 4, 5, 8, 25, 122
  adsorption 133
  commercial superplasticizer 170, 173
  delayed addition 190
  different types 170
  dosage 62, 226, 227, 236, 262
  effect on early strength 426, 427
  efficiency 170
  influence
    of a cold temperature 272
    of the dosage
      on air entrainment 26
      on the cost 559
      of inhomogeneity 93
      of SCM 97
    of the length of the chains 128
    of an overdosage on segregation 272
    of retardation 26
  effect
    on the cost of HPC 11
    on flocculation 124, 547
  lignosulfonate combinations 170
  mass of water 220
  materials 164
  optimum dosage 189, 207
  reduction of the dosage due to the use of SCM 191
  selection 164
  solids content 227, 230
  specific gravity 227
  use of a retarder 190
  viscosity 130
Superplasticizer addition
  delayed 270
  field 270, 300
  initial 270
Superplasticizer content *see* Superplasticizer dosage
Supplementary cementitious materials (SCM) 98, 139
  advantage of using 266
  content 225, 226
  hydration 356
  influence
    on the calculation of W/B ratio 2, 3, 8
    on chloride ion mobility 487
    on the cost of HPC 10
    on freeze-thaw durability 18
    on rheology 97
    on the temperature increase 14
  segregation of 466
  use of 8
Sustainable development 8, 40
SVANEN 76, 79
Syenite 200
SYLANS and GLACIÈRES viaducts 46

Technical advantage of using HPC 7
Temperature increase 62, 349, 358
  comparison between a UC and a HPC 350
  consequence of 351
  difference with UC 353

effect
  on compressive strength 352, 357, 396, 397
  on different factors 317
  on mechanical properties 14
  on microstructure 352, 354, 355
elastic modulus 357
factors affecting the 350
how to control
  crushed ice 62, 275, 367
  liquid nitrogen 51, 275, 367
  use of
    hot water 369
    a low heat of hydration cement 369
    retarder 367
    SCM 368
influence
  of ambient temperature 64, 360, 361, 364
  of cement dosage 64, 357
  of fresh concrete temperature 64
  of geometry 64, 361, 362, 363
  of initial temperature 357, 364
  of nature of the forms 64, 363
  of SCM 14
inhomogeneity 353
on outer and inner product formation 352, 355
tensile strength 357
Temperature of the fresh concrete
  average temperature 64, 274, 275
  control of 64, 272
  effect
    on hydration reaction 273
    on maximum temperature 272
  increase of 273
  influence
    of entrained air 272
    of heating mixing water 273, 274
Temperature rise *see* Temperature increase
Tensile strength
  influence
    of the hydrated cement paste 8
    of porosity 88
Tensioactive agent
  in lignosulfonate 125
Ternary compositions 195
Ternary phase diagram 102, 103
Testing HPC 15, 373, 391
  compressive strength 374, 375, 395, 396
  difference between usual concrete testing 374
  frequency of sampling 294
  fresh concrete 294
  influence
    of curing 15
    of specimen size 15
    of testing machine rigidity 15
  permeability measurement 374
  procedures 294

  review 12
Testing machine capacity 374, 375, 376
Thermal characteristics
  of lightweight concrete 521, 522
Thermal gradients 69, 330, 349, 370
  how to control 69, 369, 370
  influence 353
    of ambient temperature 64
    of the forms 330, 363, 369, 370
    of fresh concrete temperature 64
    of the rate of cooling 353
    on thermal stresses 354
  smallest 64
Thermal insulation of HPC 69, 353, 369
Thermal shrinkage 94, 313, 321, 322, 330
  prediction 330
Tilt mixer 269
Time to initiate cracking (corrosion) 492, 493
  influence
    of initial cement 495
    of the thickness of the covercrete 493
    of the W/B ratio 493
Tolerant concrete 45
Total shrinkage 311, 321, 322
  difference of origin 328
  influence of curing 328
  minimization 324
  role of aggregate 322
Transgranular fracture 202
Transition zone 85, 87, 88, 93, 94, 95, 203, 424, 464
  effect on elastic modulus 17
  influence of compressive strength 17
Trial batch 215, 219, 259
  adjustment 262
  from trial batch to $1\,m^3$
    composition 246, 247, 248, 249, 250, 251
Trial mix *see* Trial batch
Trial mixture *see* Trial batch
Transporting HPC *see* Delivery of HPC
Triangulated truss 46, 47, 49
Trisulfoaluminate *see* Ettringite
Trowel finish 304
TWENTY MILE CREEK Bridge 290
TWO UNION SQUARE 56

Ultra high strength cement based materials 92, 545
Unhydrated cement particles 98, 356, 468
Unit cost 38, 39, 40
  influence of the cost of admixture 45
Unit mass
  control of 340
  influence of the air content 340
  of heavyweight concrete 526
  of lightweight concrete 516, 517, 519, 522
  measurement for lightweight HPC 522, 523, 524

Subject index

modified normal weight 72
of roller-compacted concrete 538
Use of HPC
  bridge decks 33
  cost savings 23
  for durability 1
  in high rise building 23, 29, 50
  pavements 37, 476
  piers 33

VAN DER WALLS forces 90, 94
Vebe time 539
Verticality of the mould axis
  effect on compressive strength 391
Vibrating screed 304
Vibration 303
  effect on concrete strength 303, 304
Volume contraction 91
  effect on autogenous shrinkage 12
Volume of HPC
  in CONFEDERATION Bridge 80
  in an offshore platform 44, 70
  in the pylons of a cable stay bridge 66
Volume of solids in a superplasticizer 230
Volumetric
  changes 317
  contraction 313, 314, 315, 317
    influence of curing 319

Wall effect 94, 465
  effect on shrinkage 322
Washing water 268
Water/binder law 97
Water/binder ratio 3, 4, 5
  effect
    on compressive strength 425
    on creep measurement 409
    on freeze-thaw durability 511
    on porosity 468, 469
  influence
    of SCM on the calculation of 3
    of strength of hydrated cement paste 17
    of water/binder ratio 17
Water/binder selection 215, 235, 258
Water/cement ratio
  definition 1, 2, 3, 4
  law 27
Water/cement selection 215, 218, 220, 235
  effect on compressive strength 256
Water/cementitious material ratio
  definition 2, 3
Water content
  definition 221, 223
  of the sand 268
  selection 235, 261
  of a superplasticizer 228
Water curing 312, 319, 323, 325, 326
  different means 311
  duration 322
  effect
    on compressive strength 429
    on life cycle 13
    on volumetric changes 320
  how to 324, 325
  implementing 334, 335
Water reducer 26, 122, 342
  dosage 231
  efficiency 24
  influence
    of the dosage on compressive strength 25
    on flocculation 124
  lignosulfonate 22, 26
  nature 122
WATER TOWER PLACE 40
Weak aggregate 85, 86
Weakest link 95, 201, 256, 424, 546
Wear resistance *see* Abrasion resistance
WEIBULL theory 87
WÖHLER diagrams 450, 451
Workability 14
  control of 271
  influence
    of the absorption of lightweight aggregate 517
    of entrained air 20, 515, 516
    of the grain size distribution of cementitious particles 97
    of the shape of aggregates particles 97

YOUNG modulus *see* Elastic modulus